Metallized Plastics 5&6
Fundamental and Applied Aspects

Of related interest:

Polymer Surfaces and Interfaces: Characterization, Modification and Application
Editors: K.L. Mittal and K.-W. Lee

Polymer Surface Modification: Relevance to Adhesion
Editor: K.L. Mittal

Fundamentals of Adhesion and Interfaces
Editors: D.S. Rimai, L.P. DeMejo and K.L. Mittal

Adhesion Measurement of Films and Coatings
Editor: K.L. Mittal

Plasma Surface Modification of Polymers: Relevance to Adhesion
Editors: M. Strobel, C.S. Lyons and K.L. Mittal

Contact Angle, Wettability and Adhesion
Festschrift in Honor of Professor Robert J. Good
Editor: K.L. Mittal

Silanes and Other Coupling Agents
Festschrift in Honor of Dr. Edwin P. Plueddemann
Editor: K.L. Mittal

Acid-Base Interactions: Relevance to Adhesion Science and Technology
Festschrift in Honor of Dr. Frederick M. Fowkes
Editors: K.L. Mittal and H.R. Anderson, Jr.

Journal of ADHESION SCIENCE and TECHNOLOGY
Editors: K.L. Mittal and W.J. van Ooij

METALLIZED PLASTICS 5&6:
FUNDAMENTAL AND APPLIED ASPECTS

Editor: K.L. Mittal

///VSP///
Utrecht, The Netherlands, 1998

VSP BV
P.O. Box 346
3700 AH Zeist
The Netherlands

First published in 1998

ISBN 90-6764-290-8

Printed in The Netherlands by Ridderprint bv, Ridderkerk.

CONTENTS

Preface

This volume chronicles the proceedings of the 5th and 6th symposia on **Metallized Plastics: Fundamental and Applied Aspects**, held under the aegis of the Dielectric Science and Technology division of the Electrochemical Society in Los Angeles (May 5-19, 1996) and Paris (August 31-September 5, 1997) respectively. These events reflected a continuation of the series of symposia initiated in 1988 to update on a regular basis the R&D activity in the arena of metallized plastics which over the years has evinced tremendous interest from both fundamental and applied points of view. The previous symposia in this series were held as follows: the premier one in Chicago, October 10-12, 1988, the second in Montreal, Canada, May 7-10, 1990, the third in Phoenix, Arizona, October 13-18, 1991, and the fourth in Honolulu, Hawaii, May 17-21, 1993. The proceedings of these symposia have been well documented.

Both these symposia elicited an excellent response which is a testimonial to the brisk activity and interest apropos of metallized plastics. The technical programs clearly reflected an interest and need to ameliorate the existing methods and to devise new and more efficient ways to metallize plastics. Also, novel techniques to characterize metal-polymer interfaces or to monitor properties of metallized plastics were discussed. The programs for both symposia comprised both invited and contributed papers which were both overviews and original research in nature.

This volume contains a total of 29 papers (15 from the Los Angeles Symposium and 14 from the Paris Symposium) addressing many aspects and new developments in the field of metallized plastics. Of course, many more than 29 papers were presented at these symposia; but the remaining papers are not included in this volume for a variety of reasons. It should be recorded that all manuscripts were properly reviewed and suitably revised before inclusion in this volume. Also it should be underscored that the authors of the Los Angeles Symposium were asked to update the information. So the information presented in this book should be fresh and up-to-date.

This volume is divided into three parts: Part 1 "Metallization Techniques and Properties of Metal Deposits"; Part 2 "Spectroscopic Investigation of Interfacial Interactions"; and Part 3 "Surface Modification and Adhesion Aspects". The topics covered include: various metallization techniques for a variety of plastic substrates and simplification of electroless method by using plasma or UV laser pretreatment; various properties of metal deposits; investigation of metal-polymer interfaces using a variety of spectroscopic techniques; interaction of metals with self-assembled monolayers; study of early stages of metal-polymer interface formation; surface modification of plastics by a host of techniques including plasma, excimer laser, ion beams and characterization of modified plastics surfaces; surface modification of polymers used in the low Earth Orbit space environment; adhesion aspects of metallized plastics including a quantitative adhesion test for metal coated polymer fibers and nondestructive techniques for monitoring metallized plastics adhesion.

This and the preceding volumes (1-4) offer a fountain of information and represent current commentary on the R&D activity taking place in the technologically highly important field of metallized plastics. Yours truly sincerely hopes that anyone interested in the fundamental or applied aspect of metallized plastics will find this and earlier volumes a valuable source of current information and a fountainhead of new ideas.

Acknowledgements
This is the most enjoyable part of writing a Preface, as this gives an opportunity to thank those who helped in many different ways. First, it is my sincere pleasure to mention that the Los Angeles Symposium was jointly organized by Dr. Robert Opila (Bell Labs., Lucent Technologies) and myself; and the Paris Symposium was the culmination of the efforts of Dr. F. Arefi-Khonsari and Dr. J. Amouroux (Université Paris VI) and yours truly, and my thanks are extended to these three individuals. Thanks are also due to the appropriate officials of the Dielectric Science and Technology Division of the Electrochemical Society for sponsoring these events. My sincere thanks go to the reviewers for their time and efforts in offering many valuable comments which definitely improved the quality of this publication. My appreciation goes to my wife, Usha, for making corrections in certain manuscripts. Of course, the most important individuals responsible for a book are the authors and I deeply appreciate their co-operation, interest and contributions in this endeavour.

K.L. Mittal

References:

1. K.L. Mittal and J.R. Susko (eds), *Metallized Plastics 1: Fundamental and Applied Aspects*, Plenum Press, New York (1989).
2. K.L. Mittal (ed.), *Metallized Plastics 2: Fundamental and Applied Aspects*, Plenum Press, New York (1991).
3. K.L. Mittal (ed.), *Metallized Plastics 3: Fundamental and Applied Aspects*, Plenum Press, New York (1992).
4. K.L. Mittal (ed.), *Metallized Plastics: Fundamentals and Applications*, Marcel Dekker, New York (1998).

Part 1

METALLIZATION TECHNIQUES AND PROPERTIES OF METAL DEPOSITS

Metallized Plastics 5&6: Fundamental and Applied Aspects, pp. 3-23.
K.L. Mittal (Ed.)

ELECTROLESS METALLIZATION OF POLYMERS: SIMPLIFICATION OF THE PROCESS BY USING PLASMA OR UV-LASER PRETREATMENT

M. Romand, M. Charbonnier, M. Alami and J. Baborowski

Laboratoire de Sciences et Ingénierie des Surfaces
Université Claude Bernard - LYON 1,
69622 Villeurbanne Cedex, France

A new method for promoting chemical adhesion of electroless metals to polymer substrates is described. Firstly, the process consists of grafting nitrogenated functional moieties onto polymer surfaces using cold-plasma or UV-laser techniques allowing sample treatment in nitrogen-containing atmospheres (N_2 or NH_3). Then, this step is followed by a surface activation process during which chemisorption of palladium (catalyst for the subsequent plating reaction) is accomplished using only a dilute acidic $PdCl_2$ solution. This method, which is based on the strong chemical affinity of palladium towards nitrogen, avoids the use of unfriendly chemicals. X-ray photoelectron spectroscopy was employed to investigate ex-situ chemical changes on the polymer surfaces (poly(styrene) and poly(tetrafluoroethylene) samples in the present work) during processing. Adhesion performance of the metal-polymer interfaces was studied by subjecting the metallized specimens to a fragmentation test.

INTRODUCTION

Metallized plastics are used today in a large variety of technological applications. This explains why metallization of polymers has attracted considerable attention and research over the last few years[1-5]. Obviously, understanding and improving metal/polymer adhesion is one of the primary concerns for obtaining stability and long term reliability of the relevant manufactured products.

First, distinction should be made between the methods commonly used to metallize plastics. One of these involves metal atom deposition (e.g. via thermal evaporation or sputtering process) on the surfaces of polymers under high or medium vacuum. Different metals such as Al, Mg, Ti, Cr, Cu, Ni, Ag, Au ... have been so deposited on various polymer substrates (poly(ethylene) (PE), poly(propylene) (PP), poly(styrene) (PS), poly(ethylene terephthalate) (PET), polyimide (PI) when not specified or Kapton® when specified, poly(ether imide) (Ultem®), poly(ether ether ketone) (PEEK), poly(ethylene oxide) (PEO), poly(vinyl acetate) (PVAc), poly(methyl methacrylate) (PMMA) ...), and much detailed chemical information (particularly in the case of metal films deposited in ultra-high vacuum) about initial growth mechanisms, interfacial reactions and influence of substrate pretreatments has been obtained using, among others, in-situ x-ray photoelectron spectroscopy (XPS). In addition to the use of the conventional physical vapor deposition (PVD) techniques mentioned above, metallization of polymer substrates can also be carried out by electroless deposition. This wet phase process involves mainly Ni or Cu deposits on catalytically active surfaces. A crucial aspect of this technology is not the method of metallization itself, but the way in which the surfaces of polymer materials can be rendered catalytically active via suitable surface pretreatments. Obviously, the problem of adhesion between the electrolessly deposited metal and the polymer surface is also strongly dependent on the type of polymer involved as well as on the nature of the process which is used to render the polymer surface catalytically active for the electroless plating (see section "State-of-the-Art"). In this context, characterization by XPS of the electrolessly deposited metal/polymer interfaces and, therefore, understanding of adhesion mechanisms appear much more complicated than in the case of evaporated or sputtered metal/polymer interface given that a wet deposition process is not compatible with direct (i.e. in-situ) UHV investigations.

As already noted, and whatever may be the application sought, electroless processes require extensive pretreatments of polymer surfaces before deposition can be carried out. The objective of the present work is to describe the potential of a new process for activating polymer surfaces and obtaining strong metal-polymer adhesion. This process involves a specific plasma or UV laser-induced chemical modification of the polymer surfaces and a subsequent palladium chemisorption. The studies carried out in the present work were mainly conducted on poly(styrene) (PS) and poly(tetrafluoroethylene) (PTFE) substrates whose surfaces were investigated by XPS after each stage of treatment. PS and PTFE were selected for the present investigation because of the absence of oxygen in their chemical structure (see reason for this in section "Results and Discussion"). Besides, PTFE was chosen because this prototype of the family of fluorinated polymers is well known for its high chemical inertness and thermal stability which leads to the fact that adhesion of any material to it requires very drastic surface pretreatments. In addition, some data concerning poly(carbonate) (PC) samples were also obtained by using time-of-flight secondary ion mass spectrometry (TOF-SIMS) in an attempt to confirm some XPS results. The data obtained as a whole have led us to develop a simplified method for palladium chemisorption and electroless metallization and to better understand adhesion mechanisms at the corresponding metal-polymer interface.

STATE-OF-THE-ART

Electroless deposition is an autocatalytic process which consists of a redox reaction in aqueous solution between the ions of the metal to be deposited and a strong reducer. However, in the case of insulating samples, such a final operation has to be preceded by an activation step leading to the formation of catalytic sites (e.g. via Pd^{2+} chemisorption) on the surface to be metallized.

Figure 1 (route (a)) describes the conventional chemical way which, historically speaking, was the first to be followed for metallizing polymer materials. This route starts generally by using a chromic or sulfochromic etching to oxidize the surface and increase its roughness. Subsequently, a two-step process is performed via a surface treatment, successively in an acidic $SnCl_2$ solution (sensitization step), then in an acidic $PdCl_2$ solution (activation step).

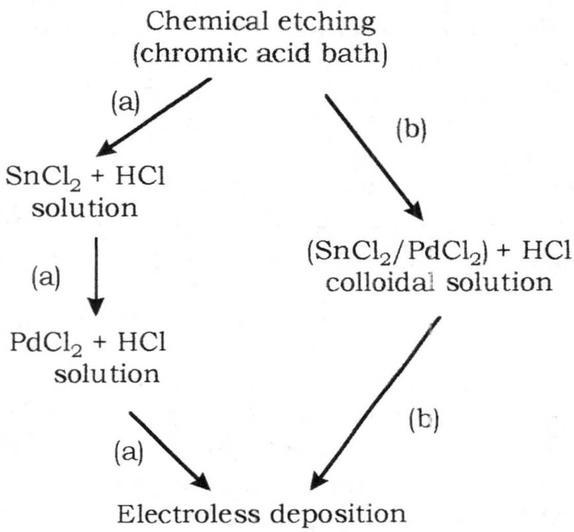

Figure 1. Schematic diagram of electroless plating processes using chemical etching pretreatment and a two-step (route a) or a one-step (route b) operation leading to surface activation (Pd chemisorption).

This two-step process is represented by some authors[6-9] as follows:

$$\text{Substrate} + \text{Sn(II)}_{aq} \rightarrow \text{Substrate - Sn(II)} \quad (1)$$

$$\text{Substrate - Sn(II)} + \text{Pd(II)}_{aq} \rightarrow \text{Substrate - Pd(0)} + \text{Sn(IV)}_{aq} \quad (2)$$

A simplification of this two-step operation was further developed by using a one-step process (Fig. 1, route (b)) which involves substrate treatment in an acidic colloidal $SnCl_2/PdCl_2$ solution which directly leads to surface activation[10-13]. The resulting reaction in this one-step process can be represented as follows:

$$\text{Substrate} + [\text{Sn(II)} / \text{Pd(II)}]_{aq} \rightarrow [\text{Substrate - SnPd(0)} / \text{Sn(II)}] + \text{Sn(IV)}_{aq} \quad (3)$$

In this latter case, the so-pretreated substrates were also submitted to an acceleration step by immersion in an HCl solution. In light of this one-step operation, it is often proposed[10,11] that Sn(II) ions be used to reduce Pd(II) ions to form colloidal particles which consist of a Sn / Pd alloy core surrounded by a protective layer of Sn(II) ions. The acceleration step is then carried out to remove this protective layer which makes the colloidal particles available to catalyze the subsequent electroless process. Despite the great complexity of the operating mode of mixed $SnCl_2/PdCl_2$ solutions and the difficulties associated with their ageing in terms of metallization efficiency, this simplified approach for obtaining surface activation is commonly used and many applications in various areas are still currently being developed[12].

Whatever the selected process (one or two-step routes) may be, reactions (2) and (3) lead us to think that palladium is chemisorbed as Pd(0) which is required to catalyse the further electroless deposition. However, other authors[13-15] do not agree with this interpretation and consider that the ultimate reduction to Pd(0) becomes efficient only after a certain initiation time in the plating bath which contains a strong reducer. This initiation time which is reported to vary over a large range (some seconds to some hundreds of seconds) is mentioned by most authors. Previous works carried out in our laboratory clearly support the fact that Pd^{2+} reduction to the atomic state is a necessary step and that the reaction initiates in the solution, level with the interfacial zone between the solution and the activated substrate[16]. Under these conditions, the corresponding reducing step can be represented by the reaction:

$$\text{Substrate - Pd(II)} + 2 \text{ electrons} \rightarrow \text{Substrate - Pd(0)} \quad (4)$$

In the particular case of nickel coatings, electrons necessary for the reduction are generally provided by different reducing agents such as sodium hypophosphite ($NaH_2PO_2.H_2O$), sodium boron hydride ($NaBH_4$), hydrazine (N_2H_4), or dimethylamine borane[17]. Electroless deposits obtained in these conditions correspond to "Ni-P", "Ni-B" alloys and "pure" Ni, respectively. In the same way, electroless copper deposition can be carried out from ionic solutions containing formaldehyde, hypophosphite, borohydride, hydrazine, dimethylamine borane ... as reducing agents[18].

Whatever the selected route (a) or (b), the initial stage generally consists, as already said, of an etching of the polymer surface with a strong oxidant such as chromic acid or sulfochromic acid. For example, our previous works[19] involving various polymer materials were performed by immersing the corresponding samples in a sulfochromic bath ($K_2Cr_2O_7$: 79 g dm^{-3} ; H_2SO_4: 868 ml dm^{-3}) at 75°C for 1 min. Nevertheless, it should be noted that some researchers particularly concerned with environmental problems over chromium have attempted to restrict the use of chromic-containing processes. So, Karas et al.[20] described an adhesion promotion process for poly(ether imide) (Ultem®) involving a chromium-free mild etch (treatment with potassium hydroxide after a sulfuric acid step) followed by application of a chemical adhesion promoter (via immersion into an aqueous thiourea solution). In this example, authors used a proprietary tin/palladium catalyst system and an electroless copper film was deposited.

In addition to the conventional (a) and (b) routes represented in figure 1, other processes have also been investigated and/or proposed as candidates for obtaining catalytically active polymer surfaces. For example, Viehbeck et al.[21] describe a seeding process for activating the surface of polyimide (Kapton®) and other electroactive polymers. This process consists of reducing electrochemically the outer region of such materials when these materials are brought into contact with an electrolyte solution containing a strong organic reducing agent. In this way, the electroactive surface is used to provoke electron transfer to metal ions in solution causing metals (e.g. Pd, Pt ...) to be deposited at the surface. The deposition of such metals renders the polymer surface active towards further metal deposition from conventional electroless plating baths. In addition, Baum et al.[22,23] describe a selective process based on the photoreduction of potassium iron (III) oxalate in the presence of palladium (II) chloride which results in the formation of an active palladium catalyst. This process is reported to operate well on a variety of dielectric materials including polyimide (PI) films when the iron-palladium treated dielectric films are irradiated with a 500 W Hg-Xe deep UV lamp.

Another example recently reported by Rye et al.[24-26] pertains to the adhesion of electrolessly deposited copper on patterned PTFE. This technology was achieved via an area-selective irradiation (electrons or x-rays) of the PTFE surface to produce crosslinking and a subsequent selective etching of the non-irradiated areas with a sodium naphthalenide solution. Cu deposition takes place in etched areas by using a $SnCl_2/PdCl_2$ one-step process followed by a Cu (II) / formaldhehyde plating bath.

Several other techniques have also been reported for improving surface wettability and bondability of metal films electrolessly-deposited on polymer substrates. This was particularly stimulated in the field of microelectronics whose evolution over the last few years has encouraged new specific approaches and the replacement of purely chemical surface pretreatments. Laser-stimulated chemistry has thus become an important tool for surface chemistry and thin film processing. So, a palladium deposition process was carried out for the first time in 1989[27] via a localized decomposition of a Pd-acetylacetonate spin-on film deposited from a chloroform solution on polyimide (PI). In this experiment, a continuous wave (CW) argon ion laser (351 nm) was used. The decomposition of the thin solid organometallic precursor film can be considered here as a photothermal process. Further, other experiments again with a CW argon laser (515, 488 or 458 nm)[28] and with UV-excimer lasers (ArF* at 193 nm, KrF* at 248 nm, XeCl* at 308 nm, XeF* at 351 nm)[29,30] were conducted on various polymer substrates (PI, PTFE, poly(phenylquinoxaline) (PPQ) ...) using as precursors mainly Pd-acetylacetonate ($Pd(C_5O_2H_7)_2$) and Pd-acetate ($Pd(CH_3COO)_2$). Depending on the laser wavelength and laser fluence, the corresponding photodecomposition processes are shown to be photothermal, photolytic or both. Clearly, this technique which combines a photo-induced process with electroless plating offers great potential for producing Pd prenucleation patterns (via the use of a screen-mask to define the laser irradiation area or via a direct writing of microscale features by scanning of the laser beam focused on the specimen surface) and fast deposition, in particular, of copper lines (high activity of Pd prenucleation sites in commercial Cu plating solutions).

Prenucleation of polymer surfaces with Pd from thin metal (Pd)-organic films has also been carried out using newly developed incoherent excimer UV and vacuum UV (VUV) sources[31-40]. Operation of such devices is based on excimer formation from a dielectric barrier discharge (silent discharge) in an atmosphere of a rare-gas or a rare-gas/halogen-gas mixture. Typically, emitted radiations (e.g. Ar_2* at 126 nm, Kr_2* at 146 nm, Xe_2* at 172 nm, ArF* at 193 nm, KrCl* at 222 nm, etc.) provide an alternative to the use of excimer lasers as excimer sources allow to extend the photon emission, in particular, towards the deep UV range. In addition, it is clearly shown that decomposition processes (e.g. of organometallic precursors) are only of a photolytic nature which allows low temperature processing on thermally unstable substrates. It is also worthwhile pointing out that excimer sources can be very useful in large area processing. Under these conditions, VUV light-induced decomposition of spin-on Pd-acetate films on various substrates (polyamide (PA), PMMA, PET ...) was accomplished through metal contact masks and patterned Pd films were obtained. The latter were subsequently reinforced by electroless copper and nickel plating processes[41-43].

As mentioned before, processes using appropriate lasers or incoherent excimer sources can lead to the direct deposition of metals (e.g. Pd) from thin solid metal-organic films. An alternative activation procedure could be the direct UV or VUV light-induced metal deposition from the gas phase. As an example, surface films obtained by adsorption from a suitable gas phase precursor (such as Cu-acetylacetonate) can be photodecomposed with the aid of excimer lasers[44,45]. However, this laser-induced CVD (LCVD) technique is reported to have several drawbacks (e.g., complicated CVD equipment, limited growth rates) and in light of this, the route involving decomposition of thin solid metal-organic films remains largely preferable for activating a polymer surface for subsequent electroless plating[39].

Also, an interesting work carried out in gas phase by excimer laser processing deserves to be highlighted. This work due to Niino and Yabe[46] concerns fluoropolymers (PTFE and fluorinated ethylene propylene copolymer (FEP)) and deals with their surface chemical modification in hydrazine gas (N_2H_4) photolyzed with ArF* (193 nm) excimer laser irradiation. On the basis of hydrophilic behaviour of the so-treated surfaces, selective-area electroless plating of nickel metal was performed by using the conventional two-step process (see Fig. 1, route (a)). Although the mechanism of metal deposition was not explained, the authors suggested that the hydrophilic behaviour of chemically modified surfaces was due to the presence of amino groups. As will be shown further (see section "Results and Discussion") this fact is of prime importance. Indeed, the present work will highlight the specific affinity of palladium (the activator of the metallization reaction) towards nitrogenated species grafted on polymer surfaces.

In light of the key role played by UV and VUV irradiation in the processing techniques described above, one is surprised at the fact that plasma methods have not been more developed for pretreating polymer surfaces for further electroless metallization. As far as we know, besides the work carried out in our laboratory[16, 19,47,48], only a few investigations dealing with this field can be found in the literature. As an example, Ehrbar and Ganguillet[49] report a method for preparation of plastic materials (PE, acrylonitrile-butadiene-styrene copolymer (ABS), etc.) involving successively glow discharge (corona) pretreatment, immersion of the sample into a reducing solution, then into the Sn(II)/Pd(II) solution (see Fig. 1, route (a)) or into a colloidal Pd solution (see Fig. 1, route (b)) and finally into a bath of chemical copper. In the same vein, Shirk and Ceresa[50] describe a process for preparing a "sensitized or activated" polyimide (PI) by contacting the surface with a flame followed by the application of an electroless metal catalyst. In addition, these authors note that $PdCl_2$ is the preferred catalyst as the flame-sensitized film does not require the supplemental $SnCl_2$ sensitizer. Furthermore, Buchwalter et al.[51] show that polymers (PI, PTFE, other fluorinated polymers, etc.) can be functionalized by means of a plasma (in H_2, H_2O, O_2, O_3, N_2O, NH_3, H_2S, N_2H_4 ... gaseous atmospheres) and that a metal-organic complex should be formed when the functionalized surface is immersed in an appropriate aqueous metal-salt solution (e.g. $PdCl_2$ activation solution). Copper deposition is then accomplished by using a conventional Cu plating bath. It should be noted that whatever the plasma gas used, these authors do attribute process efficiency to the formation of oxygen-containing functionalities. We will see in section

"Results and Discussion" the reasons why adhesion mechanisms between Pd and polymer surfaces activated with a $PdCl_2$ solution do not involve oxygenated species.

In summary, the literature data cited above point out that activity of electroless catalysts is largely dependent on surface chemistry of pretreated polymer substrates. Accordingly, Mance et al.[52] describe how the surface chemistry modification of plastic substrates (PS, ABS ...) determines the efficiency of electroless catalysts and, therefore, allows metallization. This effect was demonstrated for polymer surfaces modified by photooxidation (254 nm UV light from a low pressure Hg lamp, plasma oxidation, or ozonation) by using subsequently a conventional colloidal $SnCl_2/PdCl_2$ solution (Fig. 1 / route (b)).

EXPERIMENTAL

All plasma and laser treatments and subsequent chemical metallizations were performed on PS and PTFE commercial samples supplied by Goodfellow Cambridge Limited, Cambridge, England. Planar sheets 2 mm in thickness were cut into 10 x 50 mm coupons and cleaned ultrasonically in ethanol before further processing.

The conventional sensitization/activation treatment using the two-step route and leading to palladium chemisorption was carried out by immersing the substrate for 3 min in a solution containing 0.1 g dm^{-3} $SnCl_2$ and 0.1 ml dm^{-3} (37 %) HCl, rinsing it in deionized water, immersing it for 3 min in a solution containing 0.1 g dm^{-3} $PdCl_2$ and 3.5 ml dm^{-3} (37 %) HCl, and rinsing it before carrying out the metallization itself. In addition, the one-step process was also carried out. It consisted of immersing the substrate for 3 min in the mixed solution whose composition (12 g dm^{-3} $SnCl_2$, 0.25 g dm^{-3} $PdCl_2$, 60 ml dm^{-3} HCl) was that used by Osaka et al.[7], rinsing it in deionized water and immersing it in an acceleration solution (0.07 M HCl) for 3 min. On the basis of the literature data and taking into account the low stability of the solutions containing $SnCl_2$, the mixed solutions used here were freshly prepared (they were discarded 48 h after their preparation) and kept in the absence of light, under nitrogen atmosphere. Similarly, acidic $SnCl_2$ solutions used in the two-step route were prepared daily.

When the simplified process developed in our laboratory was used, samples were immersed in a $PdCl_2$ (0.1 g dm^{-3}) acidic solution for 3 min and rinsed in deionized water.

Plasma treatments of PS and PTFE samples were performed in a parallel plate RF reactor (diameter: 30 cm ; interelectrode distance: 8 cm) operating in the Reactive Ion Etching (RIE) mode. The samples to be treated were positioned on the RF powered electrode (17 cm in diameter) which was capacitively coupled, through an impedance matching network, to a 13.56 MHz RF power supply. The experimental conditions were fixed as follows : gas (O_2, N_2, NH_3) flow = 100 sccm, pressure = 100 mTorr, power density = 0.5 W cm^{-2} and variable treatment duration. The pressure limit in the reactor was 1 mTorr. O_2, NH_3 and N_2 gases used were 99.998, 99.999 and 99.995 % pure, respectively.

Laser treatments of a poly(carbonate) (PC) sample were performed previously to the present work with a Quantel Nd:YAG instrument (5th harmonic at λ = 213 nm ; pulse duration = 16 ps ; energy density = 0.10 to 0.80 mJ cm^{-2} ; beam diameter at the polymer surface = 7 mm). In the present work, the background pressure of the reaction chamber was around 0.5 Torr and the working pressure was 7.5 Torr. The distance between the silica window and the sample surface was varied from 0.5 to 2 mm. This value corresponds to the distance travelled by the laser beam through the reactive gas. Similar experiments are in progress on PTFE samples by using an ArF excimer laser (EMG 103 MSC model) from Lambda Physik (λ = 193 nm ; pulse duration = 20 ns ; energy density up to 200 mJ cm^{-2} ; laser beam not focused). In this work, the working pressure was 0.1 Torr, the energy density about 10 mJ cm^{-2} and the distance between the silica window and the sample surface about 20 cm.

In each case, the metallization step was carried out at 85°C using the following solution ($NiSO_4$, $6H_2O$: 36 g dm^{-3}, NaH_2PO_2, H_2O: 10 g dm^{-3}, lactic acid: 29 ml dm^{-3}, thiodiglycolic acid: < 0.1 g dm^{-3}) whose pH was adjusted to 5. Under these conditions, Ni-P films were electrolessly deposited onto polymer surfaces.

Immediately after each treatment, the samples were analysed by XPS in using the Riber SIA 200 spectrometer equipped with a non-monochromatic twin Al/Mg excitation source and a MAC 2 analyzer. Analyses of PS-based substrates were carried out using Mg Kα radiation while those of PTFE-based substrates done using Al Kα radiation.Spectra were recorded at a take-off angle of 25° with respect to the sample surface. Survey scans were recorded at an overall resolution of 2.3 eV and core level spectra at 0.9 eV. All XPS peaks were referenced to the C 1s signal, at a binding energy of 285 eV representative of the C-C and C-H environments of the polymer for PS and at 292.5 eV representative of the CF_2 environment for PTFE. Note that the use of a non-monochromatic source involves the presence of small satellite peaks due to Al or Mg K$\alpha_{3.4}$ excitation radiation. Such satellites, which appear approximately at 10 and 12 eV (for Al), and at 9 and 10 eV (for Mg) on the low-binding energy side of the main photoelectron peaks, obviously need not be considered in the present study. The peak identification was performed according to Beamson and Briggs'data[53]. The curve fitting was carried out after background subtraction by the Shirley method, using a data-system software which allows to fix, for each component, peak energy and intensity, full-width-at-half-maximum and Gaussian-Lorentzian percentage. Samples were exposed to laboratory atmosphere for less than 2 min between plasma, or laser pretreatment and introduction in the XPS chamber.

In addition, TOF-SIMS experiments were carried out on PC substrates using a Perkin-Elmer PHI 7200 spectrometer equipped with 8 keV Cs^+ ion source and reflector-type TOF analyser. The incident ion beam was pulsed at 10 kHz with a pulse duration of about 1 ns. Analyses were carried out by rastering the ion beam (100 × 100 mm) over the sample surface. Positive and negative ion spectra were obtained under static conditions (total working ion dose of about 1×10^{12} ions cm^{-2}). Surface charge neutralization by a pulsed electron flood was not required in the present work because of the absence of surface charge built-up due to the reduced primary dose used and to the rastering of the incident ion beam. At a mass of 28 D, mass resolution $m/\Delta m$ was about 10000.

RESULTS AND DISCUSSION

Figure 2 represents the procedures involving plasma (routes (a), (b) and (c)) and UV-laser (route (d)) processes which were explored both in previous studies and in the present work and which have led to the development of a new and simplified process (routes (c) and (d)). Route (a) using successively Sn(II) and Pd(II) solutions and leading to surface activation is a two-step operation similar to route (a) shown in Figure 1. Accordingly, route (b) using a colloidal Sn(II)/Pd(II) solution is a one-step operation similar to route (b) as represented in Figure 1.

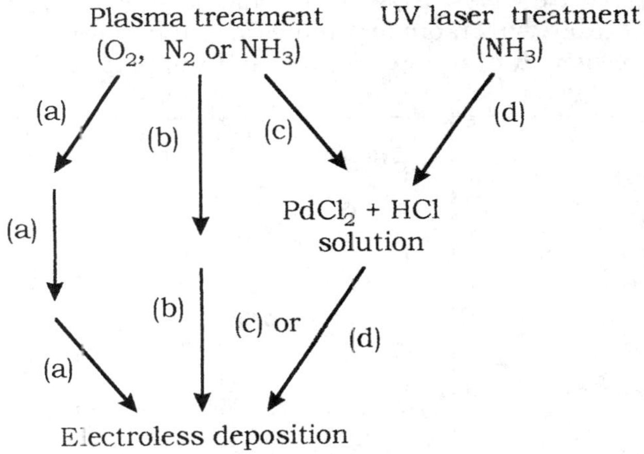

Figure 2. Schematic diagram of electroless plating processes using plasma or UV-laser pretreatment and a two-step (route (a)) or a one-step (routes (b), (c) or (d)) operation leading to surface activation (Pd chemisorption).

As shown in Figure 2, polymer surfaces were first pretreated by exposure to O_2, N_2 or NH_3 plasmas. It is well known that O_2 and oxygen-containing plasmas are commonly used as important industrial processes for modifying polymer surface characteristics such as wettability and adhesion without altering the bulk properties of the polymer. Depending on the operation parameters a variety of oxygenated functional groups, including C-OH, C-O-C, C=O, O-C=O and CO_3, have been reported to be surface-grafted[54-58]. Accordingly, nitrogen-containing plasmas give rise to primary amino groups (C-N) but the presence of C=N (imine) and C≡N (nitrile) species has also been reported[58-61].

Figure 3 shows XPS survey spectra of ethanol degreased PS and PTFE samples (spectra PS (1) and PTFE (1), respectively) and the same after exposure to an O_2 plasma (spectra PS (2) and PTFE (2), respectively). As expected, the degreased samples reveal only the presence of the C1s peak (at a binding energy of 285 eV) in the first case and those of C1s (292.5 eV), F1s (689.7 eV) and F2s (~ 35 eV) peaks in the second case, all arising from the CF_2 units of PTFE. Additional peaks are assigned to "satellites" arising from the use of a non-monochromatic source.

The net result of the interaction between PS and the reactive oxygen plasma species is the surface oxidation of the polymer. This leads to the presence of an intense O1s peak. As shown in a previous work[47], surface oxidation proceeds through opening of the aromatic ring (strong reduction of the π - π^* C1s satellite) and grafting of C-OH, C-O-C and O-C=O functional groups. It should be noted here that oxygen uptake can be attributed to reactions of carbon-centred radicals, which are formed on the polymer surface and sub-surface during the plasma treatment, with oxygenated species created in the plasma as well as with O_2 when the treated samples are further exposed to the atmosphere. On the other hand, a quite different behaviour is observed in the case of PTFE substrates. Indeed, the O_2 plasma-treated samples show only a small residual

O1s peak, indicating that PTFE is resistant to oxygen attack. This result confirms the data previously published concerning both PTFE and other fluoropolymers[62-65]. A reason put forward[66] is that surface radicals generated by the plasma process react preferably with radicals from adjacent chains to form crosslinking rather than with oxygen-containing species from the plasma. However, as no

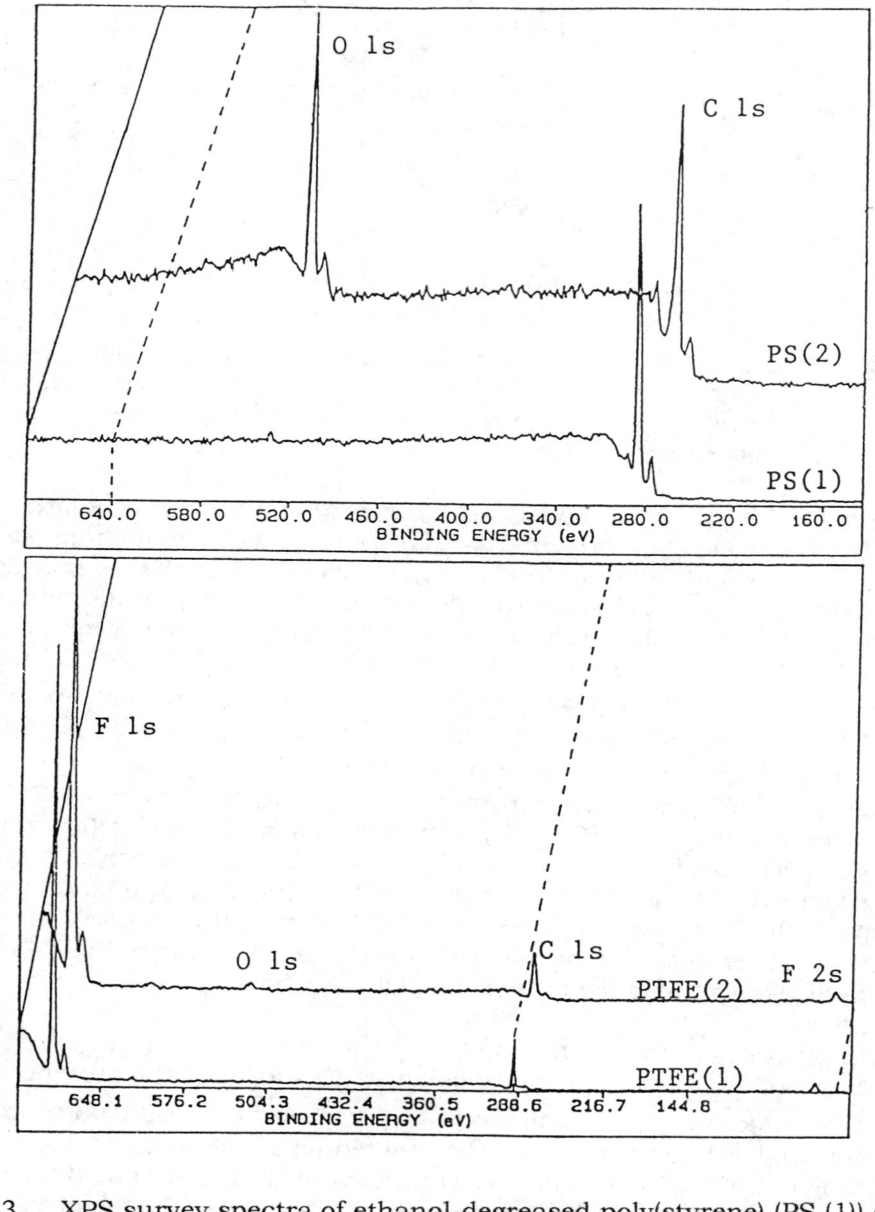

Figure 3. XPS survey spectra of ethanol-degreased poly(styrene) (PS (1)) and poly(tetrafluoroethylene) (PTFE (1)) samples. The same spectra for O_2-plasma-treated samples (PS (2) and PTFE (2), respectively) are also shown (treatment time = 1 min).

defluorination is observed, another reason which appears more plausible is that the main UV radiation (130.5 nm) emitted by atomic oxygen in the O_2 plasma is not strongly absorbed by PTFE and, therefore, does not cause breaking of surface bonds[67].

Figure 4 shows XPS spectra of ethanol degreased PS and PTFE samples after exposure to N_2 (spectra PS (3) and PTFE (3), respectively) and NH_3 (spectra PS (4) and PTFE (4), respectively) plasmas. As for PS samples, PS (3) and PS (4) spectra reveal that both nitrogenated and oxygenated functionalities are surface-grafted. The presence of oxygenated species is not at all surprising given that a number of previous studies have already detected high surface concentrations of oxygen when polymer surfaces were plasma-treated with gases not containing oxygen[60,68-70]. This can be explained by the presence of residual O_2 and H_2O in the reactor during the plasma treatment and, as already stated, by the fact that free radicals remaining on the polymer surface after the plasma treatment are still able to react with atmospheric oxygen when samples are transferred outside the reactor. Under these conditions, the exact nature of the nitrogenated functional groups which were incorporated on the polymer surface is not easy to determine via examination of the high-energy side features of the C1s spectrum. Nevertheless, amine (C-N), amide (N-C=O) and possibly imine (C=N) functionalities should, among others, be identified[48]. Similar results concerning N_2-plasma-treated PS samples have already been reported[58,59]. In addition, the N1s peak of N_2 or NH_3-plasma-treated PS samples located at 399.0 eV is shown to be characteristic of N-C bonds[51].

Plasma-treated fluorinated polymers have been studied to a much lesser extent than plasma-treated hydrocarbon polymers. As for the effects of N_2 or NH_3-plasma treatments on fluorinated polymer surfaces, also much less has been reported in the literature[71-76] in comparison with the case of hydrocarbon polymers. Of particular interest to us is the work of Gengenbach et al.[75] relative to NH_3-plasma-treated PTFE and fluorinated ethylene propylene copolymer (FEP). Indeed, during NH_3-plasma treatments PTFE and FEP surfaces are reported as undergoing considerable defluorination and nitrogen uptake as amino groups. After the plasma treatment carbon-oxygen functional groups are formed as a result of complex air-oxidation phenomenon occurring when samples are exposed to the atmosphere. The aged surfaces are then reported to become oxygen-enriched. PTFE (3) and (4) spectra in Figure 4 confirm partly Gengenbach et al.' work. In both cases, the plasma process incorporates amino groups on the fluoropolymer backbone. Oxygenated functionalities are also observed probably as a result, as mentioned earlier, of the reaction of atmospheric oxygen with the free radicals created on the PTFE surface during the plasma process. However, significant differences between the N / O ratios as determined by XPS can be seen. Clearly, for a similar treatment time, the NH_3-plasma treatment incorporates much more nitrogen (7 %) than the N_2-plasma treatment (3 %). On the other hand, oxygen uptake (about 3 %) is quite similar in both cases. To summarize, surface functionalization of PTFE with nitrogenated species is easier with NH_3-plasma treatment than with N_2-plasma treatment. These results highlight the predominant role of dissociation phenomenon in the gas-phase during the plasma process and the subsequent influence of the interactions of plasma species (ions, electrons, free radicals, UV and VUV radiations) with polymer surfaces. A useful example to illustrate such effects is relative to nitridation of poly(ethylene) (PE) samples, which were exposed to MW and MW-RF

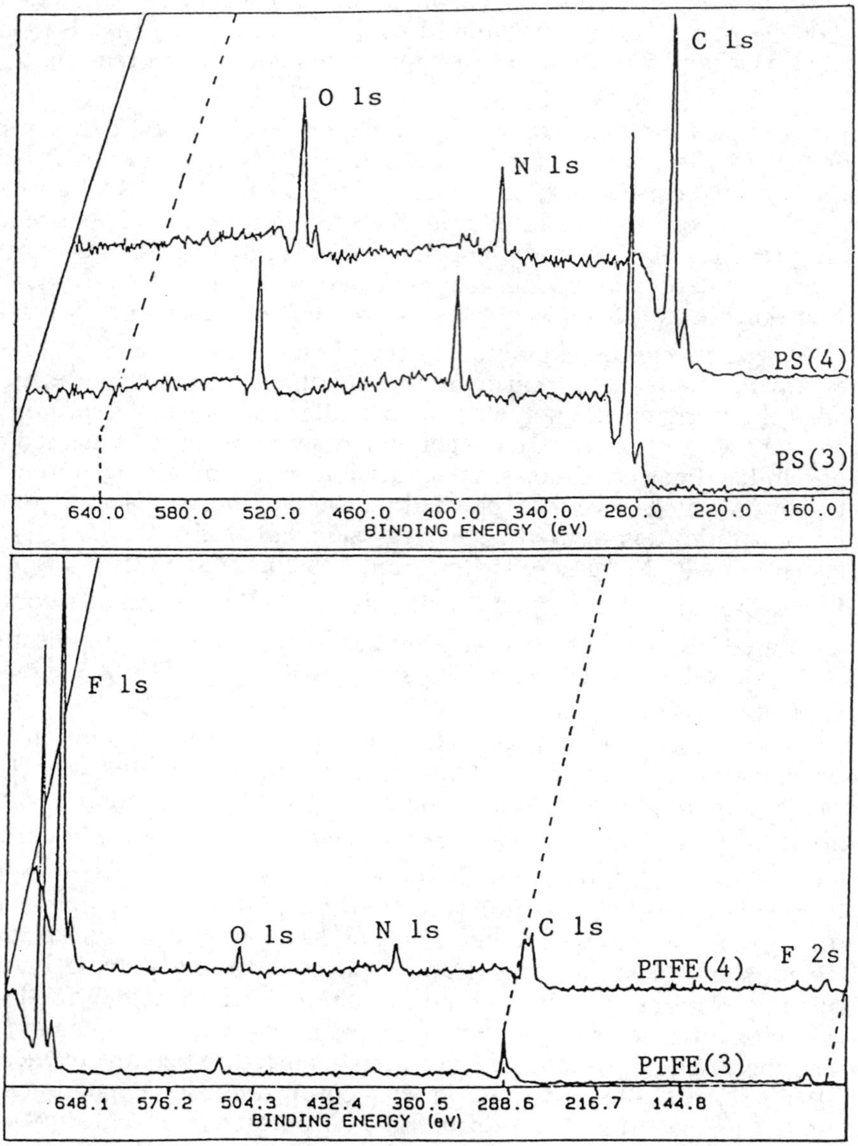

Figure 4. XPS survey spectra of ethanol-degreased PS and PTFE samples after N_2 (PS (3) and PTFE (3)) and NH_3 (PS (4) and PTFE (4)) plasmas exposure (treatment time = 1 min).

plasmas in N_2 or NH_3 environments. In these experiments[77,78], Klemberg-Sapieha et al. using various optical filters have shown that in the case of NH_3 plasma a very pronounced nitrogen uptake in the form of amine (C-N) groups is attributed to the influence of an intense VUV radiation (at 121.5 nm) emitted by atomic hydrogen.

In addition, XPS results shown in Figure 4 suggest some chemical modifications concerning carbon environment due, in particular, to the NH_3 plasma treatment of PTFE. Without discussing the details[79], Figure 5 represents the C1s spectrum of a NH_3-plasma-treated PTFE sample. This spectrum is evidently more complex than that of a virgin PTFE which contains only one component arising from the CF_2 units at 292.5 eV. As can be seen, additional contributions appear at low binding energies, which can be associated partly with defluorination process (e.g. distinguishable signals at about 291 and 288 eV) and partly with nitridation (intense signal at about 286.1 eV). The peak at 286.1 eV is assignable both to amine (C-N) and alcoholic (C-OH) functionalities. The latter functionality would result from the reaction of free radicals created on the polymer surface with the residual oxygenated species present in the reactor. Returning now to Figure 4, it can also be observed that exposure of PTFE to a N_2 plasma does not lead to an intense low binding energy 'composite" band containing, in particular, a substantial contribution of C-N functionalities. Clearly, these results indicate that incorporation of nitrogen on the PTFE surface with NH_3 plasma is worthy of further detailed investigation.

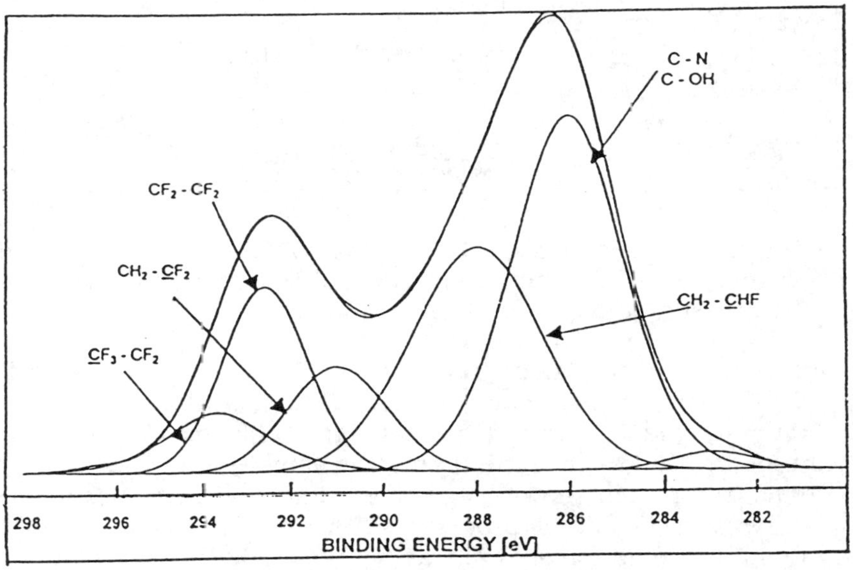

Figure 5. C1s spectrum of a PTFE sample after exposure to NH_3 plasma (treatment time = 5 min).

Let us consider now the effects of the sensitization / activation process (two-step operation / route (a)) on O_2, N_2 or NH_3-plasma-treated PS and PTFE substrates. The results obtained by using the one-step process (route (b)) are not given here because of the difficulties in interpreting the chemisorption mechanisms of the colloidal particles. Nevertheless, it should be noted that smaller Sn/Pd ratios are obtained when N_2 or NH_3 plasmas are used instead of O_2 plasmas. Figure 6 shows the XPS survey spectra of plasma-treated PS samples after the sensitization / activation process (route (a)).

As can be seen, the O_2-plasma-treated sample (spectrum PS (2) / a) strongly chemisorbs Sn, and to a lower extent, Pd. On the other hand, the N_2 or NH_3-plasma-treated samples (spectra PS (3) / a and PS (4) / a, respectively) do not adsorb any Sn but they adsorb significant concentrations of Pd. Clearly, these results highlight the selective affinity of tin towards oxygen species and, more important for the purpose of the present work, that of palladium towards nitrogenated species.

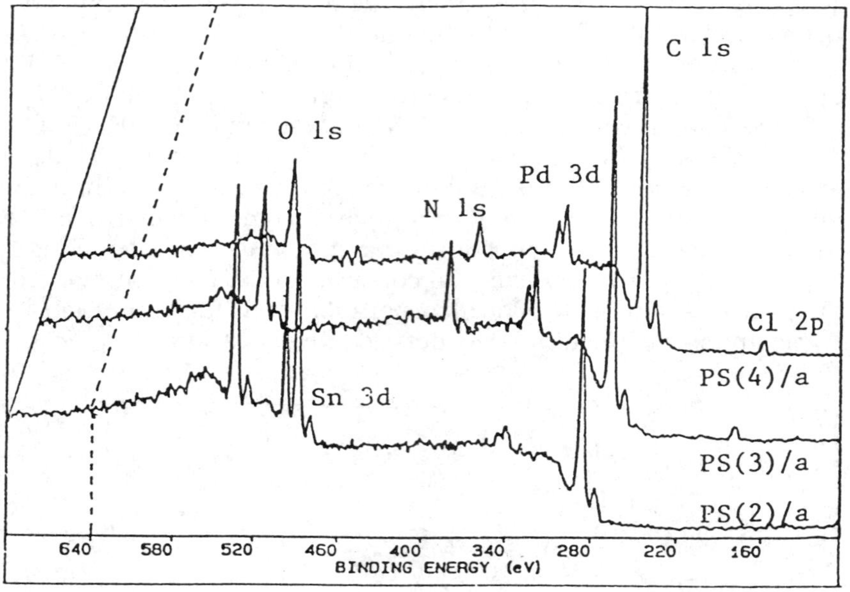

Figure 6. XPS survey spectra of O_2 (PS (2) / a), N_2 (PS (3) / a) or NH_3 (PS (4) / a) plasma-treated PS samples after sensitization / activation by using the two-step process (route (a), Figure 2).

To confirm these observations, the O_2, N_2 and NH_3-plasma-treated PS samples were then immersed in a simple acidic $PdCl_2$ solution (route (c)). As can be seen in Figure 7, samples previously treated in O_2 plasma do not adsorb any Pd (spectrum PS (2) / c). On the other hand, the samples on which nitrogenated species were plasma-grafted show a particular ability for chemisorbing Pd. Obviously, and irrespective of the process used (routes (a) or (c)), samples only degreased in ethanol do not present any activity towards the sensitization / activation (routes (a) or (b)) and the direct activation (route (c)) processes. Briefly, Pd chemisorption is very dependent on the nature of functional groups present or created on the polymer surface.

A similar experimental approach was then taken with PTFE substrates. O_2-plasma-treated specimens were not considered in the data provided hereafter because such substrates do not chemically react with oxygen. Figures 8 and 9 represent XPS survey spectra of PTFE samples treated by nitrogenated-plasma, then sensitizated / activated by using the two-step process ($SnCl_2$, then $PdCl_2$ solutions / route (a)) and after activation using the new way (simple $PdCl_2$ solution / route (c)), respectively. As can be seen in Figure 8, the N_2-plasma-treated sample (spectrum PTFE (3) / a) strongly chemisorbs Sn and, to a lower extent, Pd. This is in accordance with the presence of a relatively high oxygen amount on the PTFE substrates after exposure to N_2 plasma and with the fact that nitrogen is incorporated to a much lesser extent (see spectrum PTFE (3) in Figure 4). On the other hand, spectrum PTFE (4) / a in Figure 8 relative to the NH_3-plasma-treated sample does show, as one might expect, a lower uptake of Sn and a higher concentration of Pd. Again, the presence of nitrogenated groups chemisorbed (see spectrum PTFE (4) in Figure 4) on PTFE substrates strongly support the formation of chemical bonds between nitrogen and palladium

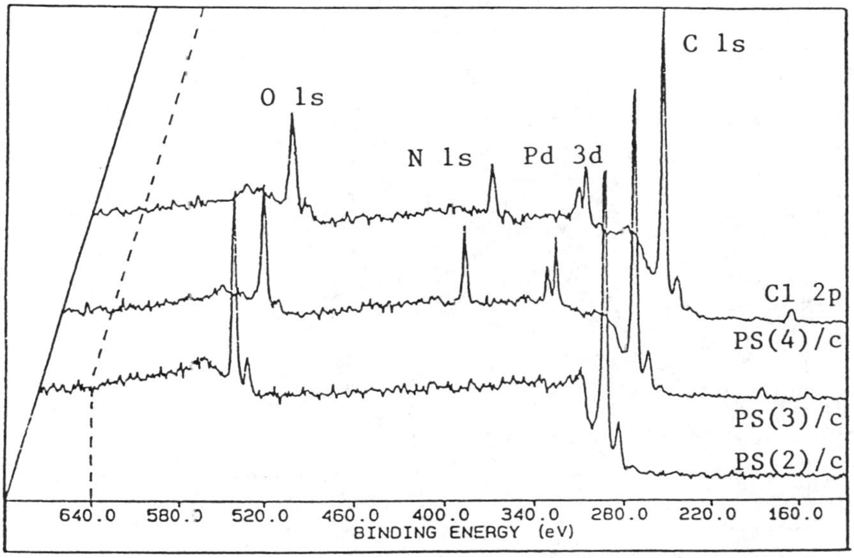

Figure 7. XPS survey spectra of O_2 (PS (2) / c), N_2 (PS (3) / c) or NH_3 (PS (4) / c), plasma treated PS samples after activation in a simple dilute acidic $PdCl_2$ solution (route (c), Figure 2).

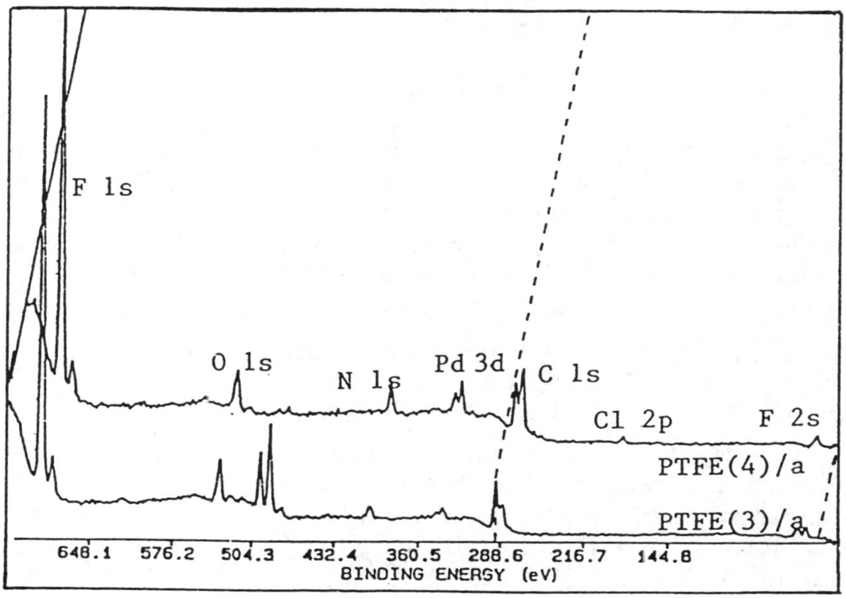

Figure 8. XPS survey spectra of N_2 (PTFE (3) / a) and NH_3 (PTFE (4) / a) plasma-treated PTFE samples after sensitization/activation by using the two-step process (route (a), Figure 2).

species. To confirm the suggested mechanism, N_2 and NH_3-plasma-treated samples were also immersed in a simple acidic $PdCl_2$ solution (route (c)). As can be seen in Figure 9 (spectrum PTFE (4)/c), Pd is chemisorbed on the PTFE surface, when the latter has been nitrogen-enriched via a NH_3-plasma treatment (see spectrum PTFE (4) in Figure 4).

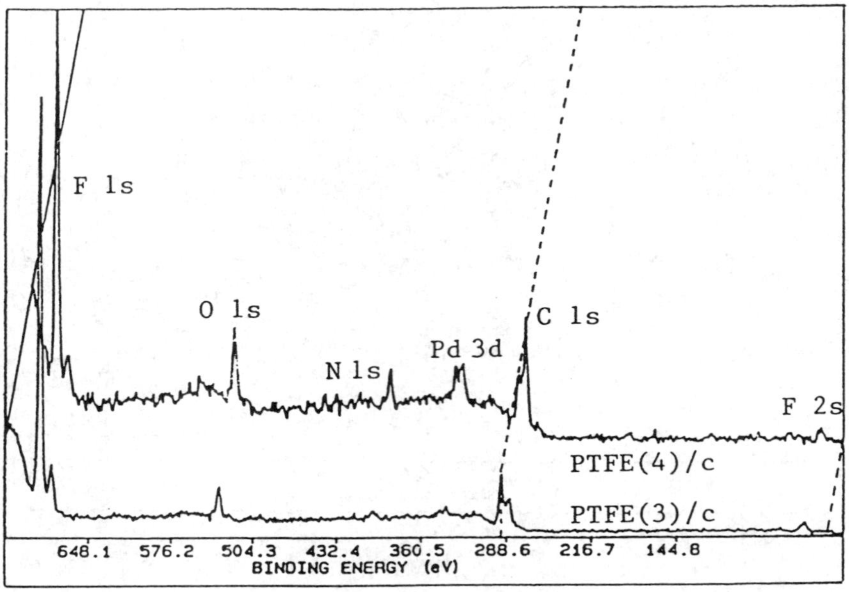

Figure 9. XPS survey spectra of N_2 (PTFE (3) / c) or NH_3 (PTFE (4) / c) plasma-treated PTFE samples after activation in a simple dilute acidic $PdCl_2$ solution (route (c), Figure 2).

As mentioned above in the section "State-of-the-Art", photochemically-induced surface modification of various polymers has recently stimulated much interest and many applications are in progress today in various industrial areas (microelectronics, optics, biomedical technology, etc.). In this field, UV and VUV laser radiations offer a large potential because of their capabilities to photodissociate many molecules in the gaseous state as well as to break chemical bonds at the surface of polymer materials, and such mechanisms lead to the possibility of grafting new functionalities[80]. As an example, Okoshi and Murahara[81], Murahara and Toyoda[82] using an ArF excimer laser have reported how to defluorinate a PTFE surface and substitute F atoms by CH_3 groups from irradiation in a gaseous $B(CH_3)_3$ atmosphere. In a similar way, we have recently shown[83] that nitrogen atoms could be grafted on poly(carbonate) (PC) surfaces by laser irradiation through an ammonia atmosphere. In our work, a Nd:YAG laser (5th harmonic, $\lambda = 213$ nm ; pulse duration = 16 ps) was used. Electroless plating was then carried out on the nitrogen-grafted surfaces by using the simplified route (d) described in Figure 2. It should be noted that this process consists of an extension of route (c) which involves a plasma pretreatment.

In the present work, an ArF excimer laser ($\lambda = 193$ nm) was used to functionalize the surface of PTFE substrates from an ammonia atmosphere. In this case, nitridation (or amination) is thought to proceed through the direct photodissociation of NH_3 molecules and the formation of amino radicals in the gas phase. It should be borne in mind that the energy of ArF excimer laser photons (6.4 eV) is higher than that of the N-H bond (4.03 eV) in ammonia molecules. In comparison, direct photodissociation is not possible in the case of a nitrogen atmosphere because the energy of the N≡N bond is 9.82 eV. Figure 10 represents XPS survey spectra of a degreased PTFE sample prior to and after laser irradiation under NH_3 atmosphere (spectrum PTFE (5)). After the activation process via immersion of the treated substrate in a simple acidic $PdCl_2$ solution (route (d)), metallization is achieved as for NH_3 or N_2 plasma-treated PTFE. As can be seen here again, nitrogenated species are grafted on the PTFE surface during the laser treatment and subsequently Pd is chemisorbed owing to its strong affinity towards nitrogen.

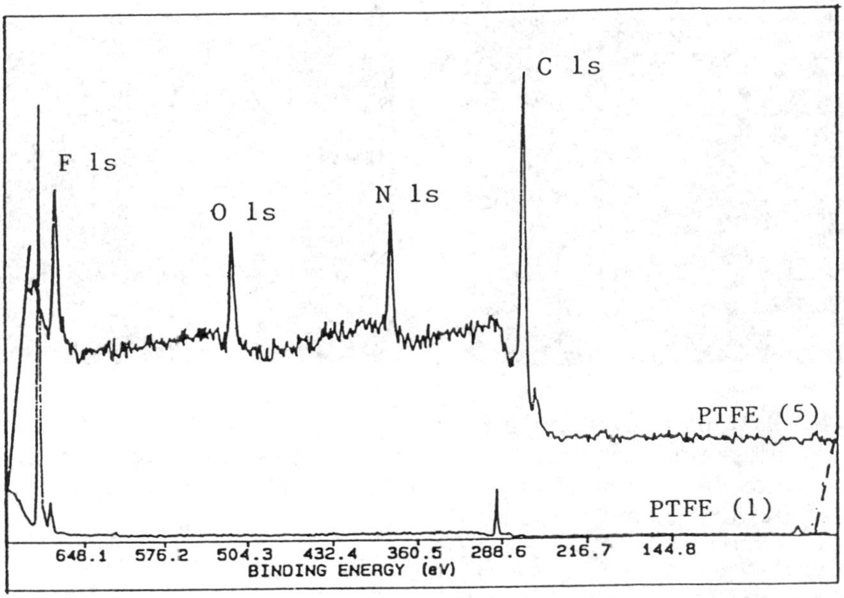

Figure 10. XPS survey spectra of a degreased PTFE sample before (PTFE (1)) and after (PTFE (5)) an ArF excimer laser treatment in a NH_3 atmosphere (treatment time: 5 min).

Irrespective of the method used (routes (a), (b), (c) or (d)) for surface activation and subsequent electroless deposition, the adhesion of the metallic layer on the polymer substrate is one of the primary concerns for achieving stability and long term reliability of the corresponding "composite" system. In the present case, it is clear that adhesion properties of the electroless plating are firstly determined by the bond strength of the catalytic nuclei to the treated-polymer surface. As for any adherate-adherend combination, several techniques are, in principle, available for measuring the practical adhesion[84-87] of the specific adhering system considered in this work. Qualitatively, the adhesion of electroless coatings deposited after plasma or UV-laser pretreatment and activation process can be assessed as good, as no samples using the conventional Scotch tape test failed in our investigations. In addition, the adhesion of electroless deposits was investigated using a fragmentation test. Specimens subjected to this test were polymer substrates cut into a dog-bone shape and coated by the electroless film. In this test, which is particularly well-adapted to the case of thin and slightly deformable films deposited on ductile substrates, the coated material is strained in uniaxial tension and the development of the crack pattern is observed as a function of the applied strain[88]. As an example, Figure 11 represents the fragmentation patterns obtained in the case of a NH_3-plasma-treated PTFE sample after deposition of an electroless Ni-P coating about 500 nm in thickness. (A) and (B) SEM micrographs are relative to samples prepared by using the conventional two-step process (route (a)) and the simplified one-step process (route (c)), respectively. In both cases, a distribution of parallel fragments is initiated upon the fragmentation operation. As can be seen, the average critical length of fragments is smaller for specimens prepared by using the simplified one-step process, which means a better adhesion of the coating to the polymer substrate.

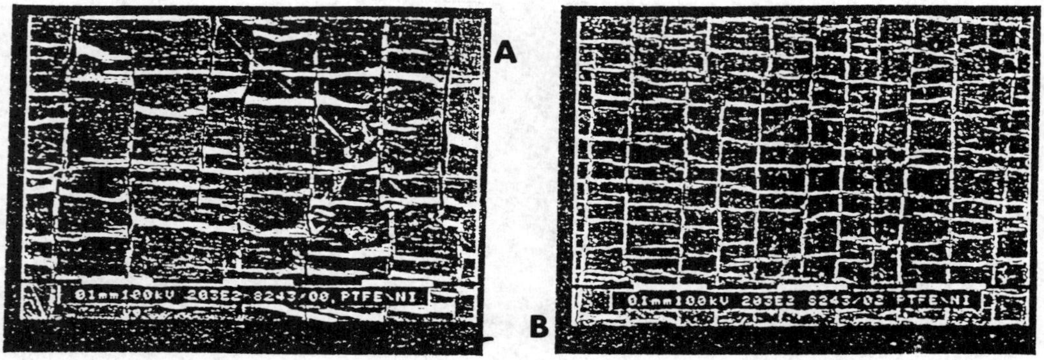

Figure 11. SEM micrographs of a fragmentation pattern (Ni-P / PTFE system) after application of a 12 % strain. The Ni-P coating is obtained (A) after the two-step process (route (a)), (B) after the one-step process (route (c), Figure 2).

CONCLUSIONS

Taking advantage of the considerable affinity of palladium for nitrogen, a new method for metallizing polymer surfaces through an electroless process has been developed. This method is based on the direct surface activation of polymer materials using (i) a gas-phase pretreatment through plasma excitation, or UV-laser irradiation in nitrogenated atmospheres and (ii) immersion of the so-treated samples in an activation bath constituted only of a dilute acidic $PdCl_2$ solution. In addition, it is noteworthy that the advantage of this simple process goes beyond obtaining strong adhesion at the metal-polymer interface since it makes possible, (i) the elimination of strong corrosive solutions (chromic and sulfochromic acid baths) conventionally used to pretreat polymer surfaces, which avoids costs associated with environmental constraints, and (ii) the non-use of acidic tin chloride solutions in which chemical reactions may occur (e.g., oxidation, hydrolysis, complex formation).

Based on the above considerations, it is, of course, imperative to pursue the present work with a view to optimizing the nitridation (amination) of polymer surfaces, in particular, by using plasma, or UV-laser process in the NH_3 environment. Indeed, in both treatments, it should be reminded that many parameters such as the nature of excitation sources, the geometry of reactors and the experimental conditions selected by the operator, crucially influence the gas phase and, therefore, the surface chemistry of the treated samples.

REFERENCES

1. K. L. Mittal and J. R. Susko, Editors, "Metallized Plastics 1: Fundamental and Applied Aspects," Plenum Press, New York, 1989.
2. E. Sacher, J. J. Pireaux and S. P. Kowalczyk, Editors, "Metallization of Polymer," ACS Symposium Series No 440, American Chemical Society, Washington DC, 1990.
3. K. L. Mittal, Editor, "Metallized Plastics 2: Fundamental and Applied Aspects," Plenum Press, New York, 1991.
4. K. L. Mittal, Editor, "Metallized Plastics 3: Fundamental and Applied Aspects," Plenum Press, New York, 1992.
5. K. L. Mittal, Editor, "Metallized Plastics: Fundamentals and Applications," Marcel Dekker, New York, 1998.
6. R. L. Cohen and K. W. West, J. Electrochem. Soc., 120, 502 (1973).
7. T. Osaka, H. Takematsu and K. Nihei, J. Electrochem. Soc., 127, 1021 (1980).
8. J. Horkans, J. Kim, C. Mc Grath and L. T. Romankiw, J. Electrochem. Soc., 134, 300 (1987).
9. M. C. Burrell, G. A. Smith and J. J. Chera, Surf. Interface Anal., 11, 160 (1988).
10. R. L. Cohen, J. F. D'Amico and K. W. West, J. Electrochem. Soc., 118, 2042 (1971).
11. B. K. W. Baylis, A. Busuttil, N. Hedgelock and M. Schlesinger, J. Electrochem. Soc., 123, 348 (1976).
12. G. O. Mallory and J. B. Hajdu, Editors, "Electroless Plating: Fundamentals and Applications," American Electroplaters and Surface Finishers Society, Orlando, 1990.
13. C. H. de Minjer and P. F. J. V. D. Boom, J. Electrochem. Soc., 120, 1644 (1973).
14. R. L. Meek, J. Electrochem. Soc., 122, 1478 (1975).
15. M. C. Burrell, B. R. Karas, D. F. Foust, W. V. Dumas, E. J. Lamby, W. T. Grubb and J. J. Chera, in "Metallized Plastics 1: Fundamental and Applied Aspects," K. L. Mittal and J. R. Susko, Editors, p. 223, Plenum Press, New York, 1989.
16. M. Charbonnier, M. Alami and M. Romand, J. Appl. Electrochem., 28, 449 (1988).
17. D. Singh, R. Balasubramaniam and R. K. Duke, Corrosion, 51, 581 (1995).
18. Y. Shacham-Diamond, V. Dubin and M. Angyal, Thin Solid Films, 262, 93 (1995).
19. M. Alami, M. Charbonnier and M. Romand, J. Adhesion, 57, 77 (1996).
20. B. R. Karas, D. F. Foust, W. L. Dumas and E. J. Lamby, J. Adhesion Sci. Technol., 6, 815 (1992).
21. A. Viehbeck, C. A. Kovac, S. L. Buchwalter, M. J. Goldberg and S. L. Tisdale, in "Metallization of Polymers," E. Sacher, J. J. Pireaux and S. P. Kowalczyk, Editors, p. 394, ACS Symposium Series No 440, American Chemical Society, Washington, DC, 1990.
22. T. H. Baum, D. C. Miller and T. R. O' Toole, Chem. Mater., 3, 714 (1991).
23. T. H. Baum, D. C. Miller and T. R. O' Toole, in "Metallized Plastics 3: Fundamental and Applied Aspects," K. L. Mittal, Editor, p. 9, Plenum Press, New York, 1992.
24. R. R. Rye and A. J. Ricco, J. Electrochem. Soc., 140, 1763 (1993).
25. R. R. Rye, G. W. Arnold and A. J. Ricco, J. Electrochem. Soc., 140, 3233 (1993).

26. R. R. Rye, A. J. Howard and A. J. Ricco, Thin Solid Films, 262, 73 (1995).
27. Y. S. Liu and H. S. Cole, Chemtronics, 4, 209 (1989).
28. H. . Cole, Y. S. Liu, J. W. Rose and R. Guida, Appl. Phys. Lett., 53, 2111 (1988).
29. T. J. Hirsch, R. F. Miracky and C. Lin, Appl. Phys. Lett., 46, 1357 (1990).
30. G. Shaveev, W. Marine, H. Dallaporta, L. Bellard and A. Cros, Thin Solid Films, 241, 52 (1994).
31. B. Eliasson, M. Hirth and U. Kogelschatz, J. Phys. D: Appl. Phys., 20, 1421 (1987).
32. B. Eliasson and U. Kogelschatz, Appl. Phys. B, 46, 299 (1988).
33. H. Esrom and G. Wahl, Chemtronics, 4, 216 (1989).
34. A. Bauer, J. Ganz, K. Hesse and E. Kölher, Appl. Surf. Sci., 46, 113 (1990).
35. H. Esrom and U. Kogelschatz, Appl. Surf. Sci., 46, 158 (1990).
36. U. Kogelschatz, Pure Appl. Chem., 62, 1667 (1990).
37. B. Gellert and U. Kogelschatz, Appl. Phys. B, 52, 14 (1991).
38. U. Kogelschatz, Appl. Surf. Sci., 54, 410 (1992).
39. H. Esrom and U. Kogelschatz, Thin Solid Films, 218, 231 (1992).
40. P. Bergonzo, P. Patel, I.W. Boyd and U. Kogelschatz, Appl. Surf. Sci., 54, 424 (1992).
41. H. Esrom, J. Demny and U. Kogelschatz, Chemtronics, 4, 202 (1989).
42. I. W. Boyd and J. Y. Zhang, Nucl. Instrum. Methods B, 121, 349 (1997).
43. J. Y. Zhang, H. Esrom and I. W. Boyd, Appl. Surf. Sci., 96-98, 399 (1996).
44. G. Wahl and H. Esrom, Appl. Surf. Sci., 46, 96 (1990).
45. J. Y. Zhang and H. Esrom, Appl. Surf. Sci., 54, 465 (1991).
46. H. Niino and A. Yabe, Appl. Phys. Lett., 63, 3527 (1993).
47. M. Charbonnier, M. Alami and M. Romand, J. Electrochem. Soc., 143, 472 (1996).
48. M. Alami, M. Charbonnier and M. Romand, Plasmas and Polymers, 1, 113 (1996).
49. J. P. Ehrbar and C. Ganguillet, U.S. Patent 4,165,394 (1979).
50. A. Shirk and M. Ceresa, U.S. Patent 3,954,570 (1976).
51. L. P. Buchwalter, S. L. Buchwalter, C. R. Davis, R. D. Goldblatt, J. E. Heidenreich, S. L. Nunes, J. M. Park, R. R. Thomas, D. Tortorella and L. M. Ferreiro, European Patent, EP 0 478 975 A2 (1991).
52. A. M. Mance, R. A. Waldo, S. W. Gaarenstroom and R. R. Witherspoon, in "Metallized Plastics 3: Fundamental and Applied Aspects," K.L. Mittal, Editor, p. 29, Plenum Press, New York, 1992.
53. G. Beamson and D. Briggs, "High Resolution XPS of Organics Polymers," The Scienta ESCA 300 Database, John Wiley and Sons, Chichester, 1992.
54. S. Wu, "Polymer Interface and Adhesion", Marcel Dekker, New York, 1982.
55. D. Briggs, C. R. Kendall, A. R. Blythe and A. B. Wootton, Polymer, 24, 47 (1983).
56. L. J. Gerenser, J. F. Elman, M. G. Mason and J. M. Pochan, Polymer, 26, 1162 (1985).
57. J. M. Pochan, L. J. Gerenser and J .F. Elman, Polymer, 27, 1058 (1986).
58. L. J. Gerenser, J. Adhesion Sci. Technol., 7, 1019 (1993).
59. R. Foerch, N. S. Mc Intyre, R. N. S. Sodhi and D. H. Hunter, J. Appl. Polym. Sci., 40, 1903 (1990).
60. J. Lub, F. C. B .M. van Vroonhoven, E. Bruninx and A. Benninghoven, Polymer, 30, 40 (1989).
61. F. M. Petrat, D. Wolany, B. C. Schwede, L. Wiedmann and A. Benninghoven, Surf. Interface Sci., 21, 274 (1994).

62. M. A. Golub, T. Wydeven and R. D. Cormia, Polymer, 30, 1576 (1989).
63. M. A. Golub, T. Wydeven and R. D. Cormia, Langmuir, 7, 1026 (1991).
64. M. K. Shi, A. Selmani, L. Martinu, E. Sacher, M. R. Wertheimer and A. Yelon, in "Polymer Surface Modification: Relevance to Adhesion," K. L. Mittal, Editor, p. 73, VSP, Utrecht, 1996.
65. H. J. Griesser, D. Youxian, A. E. Hughes, T. R. Gengenbach and A. W. H. Mau, Langmuir, 7, 2484 (1991).
66. D. Youxian, H. J. Griesser, A. W. H. Mau, R. Schmidt and J. Lieselang, Polymer, 32, 1126 (1991).
67. E. M. Liston, in "The Interfacial Interactions in Polymeric Composites," G. Akovali, Editor, NATO ASI Series, Series E: Applied Sciences - Vol. 230, Kluwer Academic Publishers, Dordrecht, 1993.
68. H. Yasuda, H. C. March, S. Brandt and C. N. Reilley, J. Polym. Sci., Polym. Chem. Ed., 15, 991 (1977).
69. D. S. Everhart and C. N. Reilley, Anal. Chem., 53, 665 (1981).
70. N. Inagaki, S. Tasaka and H. Kawai, J. Adhesion Sci. Technol., 3, 637 (1989).
71. J. R. Hollahan, B. B. Stafford, R. D. Falls and S. T. Payne, J. Appl. Polym. Sci., 13, 807 (1969).
72. G. C. S. Collins, A. C. Lowe and A. Nicholas, Eur. Polym. J., 9, 173 (1973).
73. H. Yasuda, J. Macromol. Sci. Chem., A 10, 383 (1976).
74. M. Kawakami, H. Koya and S. Gondo, Biotechnol. Bioeng., 32, 369 (1988).
75. T. R. Gengenbach, X. Xie, R. C. Chatelier and H. J. Griesser, in "Plasma Surface Modification of Polymers," M. Strobel, C. S. Lyons and K. L. Mittal, Editors, p. 123, VSP, Utrecht, 1994.
76. S. L. Kaplan, E. S. Lopata and J. Smith, Surf. Interface Anal., 20, 331 (1993).
77. J. E. Klemberg-Sapieha, L. Martinu, O. M. Küttel and M. R. Wertheimer, in "Metallized Plastics 2: Fundamental and Applied Aspects," K. L. Mittal, Editor, p. 315, Plenum Press, New York, 1991.
78. J. E. Klemberg-Sapieha, O. M. Küttel, L. Martinu and M. R. Wertheimer, J. Vac. Sci. Technol., A 9, 2975 (1991).
79. M. Charbonnier, M. Romand and M. Alami, in preparation.
80. M. Chtaïb, E. M. Roberfroid, Y. Novis, J. J. Pireaux, R. Caudano, P. Lutgen and G. Feyder, J. Vac. Sci. Technol., A 7, 3233-3237 (1989).
81. M. Okoshi and M. Murahara, J. Mater. Res., 7, 1912 (1992).
82. M. Murahara and K. Toyoda, J. Adhesion Sci. Technol., 9, 1601 (1995).
83. M. Charbonnier, M. Alami, M. Romand, J. P. Girardeau-Montaut and M. Afif, Appl. Surf. Sci., 109/110, 206 (1997).
84. K. L. Mittal, Electrocomponent Sci. Technol., 3, 21 (1976).
85. K. L. Mittal, in "Adhesion Measurement of Thin Films, Thick Films, and Bulk Coatings," K. L. Mittal, Editor, p. 5, ASTM STP 640, American Society for Testing and Materials, Philadelphia, 1978.
86. K. L. Mittal, J. Adhesion Sci. Technol., 1, 247 (1987).
87. K. L. Mittal, in "Adhesion Measurement of Films and Coatings," K. L. Mittal, Editor, pp. 1-13, VSP, Utrecht, The Netherlands, 1995.
88. Y. Leterrier, Y. Wyser, J.-A. E. Månson and J. Hilborn, J. Adhesion, 44, 213 (1994).
89. Y. Leterrier and J.-A. E. Månson, J. Mater. Sci. Lett., 16, 120 (1997).
90. Y. Leterrier, L. Boogh, J. Andersons and J.-A. E. Månson, J. Polym. Sci. Part B, Polym. Phys., 35, 1449 (1997).
91. Y. Leterrier, J. Andersons, Y. Pitton and J.-A. E. Månson, J. Polym. Sci. Part B, Polym. Phys., 35, 1463 (1997).

Metallized Plastics 5&6: Fundamental and Applied Aspects, pp. 25-39.
K.L. Mittal (Ed.)

PULSED LASER DEPOSITION OF TiN ON PMMA AND POLYIMIDE: A NOVEL ROUTE FOR POLYMER METALLIZATION

R.D. Vispute, J. Narayan, and K. Jagannadham
Department of Materials Science and Engineering
North Carolina State University, Raleigh, NC 27695.

We have developed a novel method for metallization of polymeric materials by employing a pulsed laser deposition technique. Highly metallic, adherent and hard TiN films were deposited using a KrF excimer laser ablation of high purity, stoichiometric and hot pressed TiN target at a base pressure of 1×10^{-7} Torr, and the ablated species were allowed to condense on the polymeric substrate materials. The processing parameters such as the laser fluence, pulse repetition rate, substrate-to-target distance, and substrate temperature were optimized for the fabrication of high quality metallic and adherent TiN films on PMMA and polyimide. The films were characterized by Auger electron spectroscopy, Raman spectroscopy, scanning and transmission electron microscopies, and four-point-probe electrical resistivity. Auger spectroscopy revealed that the films were close to stoichiometric TiN. The TiN films deposited on polyimide and PMMA were smooth, golden in color, and were found to be polycrystalline with an average grain size of ~12 nm. The adhesion of TiN films was measured by the direct pull-off method. The films were highly adherent to the polymer substrates as compared to Cu and Au films deposited by sputtering. Four-point- probe electrical resistivity measurements showed characteristic metallic behavior with a low value of resistivity, ~30-75 μΩ-cm, at room temperature. The specific resistivity was found to be a strong function of deposition temperature and reached the lowest value of ~ 15 μΩ-cm at room temperature for epitaxial TiN films on Si (100) deposited at 600°C by pulsed laser deposition. We also present results on the laser etching of TiN films on polymers and on the development of insulator/metal/polymer multilayer structures (AlN/TiN/polymer, DLC/TiN/polymer) for multichip modules in electronics packaging applications.

INTRODUCTION

Polymeric materials such as polyimides and poly(methyl methacrylate) (PMMA) are increasingly being used in advanced microelectronics since they have many desirable properties such as light weight, moldability, ability to form intricate shapes, corrosion resistance, and versatile electronic properties[1,2]. In this context, metallization of polymers is an important concern in electronic applications because of demands for increased packaging density and greater reliability[3]. One of the major challenges in building a reliable metal/polymer structure is the adhesion between the metal and the polymer since the long term stability of the devices depends critically on the adhesion experienced at the metal-polymer interface and hence this has become an important technological problem[4]. The adhesion between a metal and a polymer is strongly dependent on the specific type of polymers and metals involved, as well as on the deposition process under which the interface between the two is formed. The most commonly used elements for metallization of polymers are Cu, Al, Ti, Ni, and Cr, and the methods of deposition have involved thermal evaporation and sputtering processes[3-4]. Here we propose titanium nitride metal layer on polymers since TiN films have unique and desirable properties such as low electrical resistivity, extreme hardness, and excellent mechanical and thermal properties[5]. TiN crystallizes in a B1 (NaCl-type) structure, and has complex interatomic bonding comprising of covalent, metallic, and ionic contributions. The low electrical resistivity is attributed to the contribution of nine valence electrons from the N 2p and Ti 3d orbitals[6]. The present interest in TiN thin films also stems from their desirable thermodynamical and electrical properties which open up potential applications in the microelectronics industry. For example, recent studies have shown that TiN is the most promising diffusion barrier material and is being evaluated for use in both Si and III-V compound device technologies in multilevel metallization schemes involving Al and Cu as the second layer[7]. The TiN films on polymers seem to be technologically viable for metallization for protective coating applications. The major impediment in the applications of these films on polymeric materials is the high growth temperatures employed in most commonly used chemical vapor deposition (CVD) techniques to activate the desired chemical reactions on the substrate to form a film[8], and hence the process cannot be employed for the polymers which have low softening temperatures, for example, the softening temperatures of PMMA and polyimide are 80°C and 300°C, respectively. The reactive sputtering technique provides an alternative to this problem, and is utilized for TiN deposition at low temperatures[9,10]. However, since polymers are soft and have a low damage threshold for ion bombardment, this technique may lead to the degradation of surfaces and adversely affect the adhesion of the deposited films.

In this paper, we report our detailed work on the processing and characterization of conducting and adherent TiN films deposited by pulsed laser deposition (PLD) on polyimide and PMMA. We show that the PLD technique is a low temperature thin film processing method which can be well adopted for metallization of polymers at room temperature, in-situ processing, and multilayer fabrication of advanced electronic materials on polymers.

EXPERIMENTAL

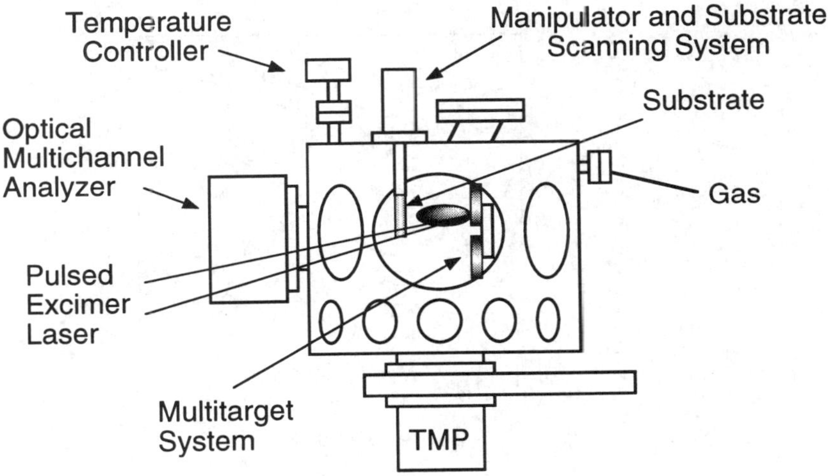

Figure 1. Schematic of the PLD technique for the fabrication of TiN thin films on polymers.

The TiN films were deposited inside a stainless steel vacuum system evacuated by a turbomolecular pump to a base pressure of 1×10^{-7} Torr. A schematic diagram of the PLD chamber is shown in Fig. 1. Radiation from a KrF excimer laser (λ = 248 nm and τ = 25 ns) was used to ablate the stoichiometric and hot pressed TiN target with an energy density ranging from 2 to 10 J/cm^2. The target was irradiated by a laser beam through a quartz window at an angle of 45° to the target surface. The laser induced vaporized material was then allowed to condense onto the polymer substrates kept at a distance of about 5 cm from the target surface. A commercially available PMMA and polyimide (supplied by DuPont) were cleaned sequentially in an ultrasonic bath of soap solution, ethyl alcohol, and isopropyl alcohol before deposition. The films were grown at various laser energy densities, substrates temperatures, and pulse repetition rates. The laser beam was also scanned over target during deposition over an area of 1 x 1cm^2 to achieve uniform films over 2.5 x 2.5 cm^2 area. The multilayer structures such as AlN/TiN/polymers and diamond-like-carbon (DLC)/TiN/polymers were fabricated in-situ by the PLD method using a rotating target system. Laser deposited TiN films were characterized by Auger and Raman spectroscopies, plan-view transmission electron and scanning electron microscopies, electrical resistivity, and x-ray diffraction (XRD) technique. Auger electron spectroscopy was carried out using a JEOL JAMP 30 microprobe. The primary electron beam current was 0.3 mA at 10 keV with a beam diameter of less than 1 mm. The samples were sputter etched during analysis using a rastered 3 keV Ar^+ ion beam. Raman spectroscopy was carried out using green light (514.5 nm line) from Ar^+ ion laser. The incident light was made grazing along the plane of the TiN film and the scattered light was collected in the direction normal to the film/substrate interface. The scattered light was dispersed in a one meter 1704 spex monochromator with a resolution of 0.4 cm^{-1} and the counts were taken for one minute at each wavelength with a photomultiplier in photon counting mode. The film thickness and uniformity were also verified using α-step 200-Tencor Instrument. The adhesion measurements were carried out using the pull-off technique employing a tensile tester machine (ATS Inc).

RESULTS AND DISCUSSION

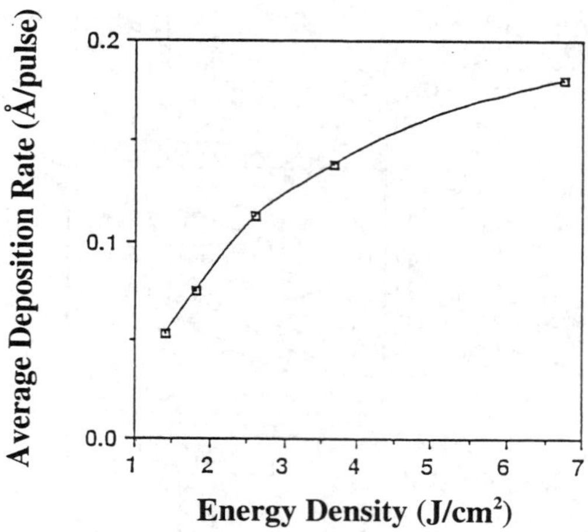

Figure 2. Deposition rate of TiN film on PMMA substrate versus laser energy density.

Pulsed laser ablation and deposition of TiN as a function of laser fluence was studied. The laser energy density significantly affects the deposition rate, and the particulate emission from the target. Figure 2 shows the average deposition rate of TiN on polymer substrates at room temperature and a base pressure of $1x10^{-7}$ Torr as a function of laser energy density. The curve indicates that the deposition rate increases with the laser fluence. However, at higher laser fluence, the rate of ablation is affected by laser absorption in the plasma plume. For all laser fluences, the TiN films were golden in color, which is a primary characteristic of TiN films. It should be noted that the ablation and the deposition were carried out at a base pressure of 1×10^{7} Torr, without a nitrogen partial pressure and at room temperature. In all experiments, the TiN stoichiometric composition of the target was maintained due to highly nonequilibrium laser ablation process[11,12]. The highly reactive nature of Ti metal species in the plasma plume is an additional factor for maintaining the TiN stoichiometry. The qualitative information on the composition of these films was obtained using Auger electron spectroscopy. Figure 3 shows typical derivative Auger electron spectrum of laser deposited TiN film on polyimide at room temperature. The two major peaks exist in the AES spectrum of TiN: one at 381 eV and the other at 418 eV. The chemical analysis of TiN using Auger electron spectroscopy is complicated because the principal Auger electron emission from nitrogen ($KL_{23}L_{23}$) occurs at an energy (381 eV) that overlaps with the Ti ($L_3M_{23}M_{23}$) transition at 383 eV. To circumvent this problem, the nitrogen content was characterized by the positive excursion of the combined N and Ti peaks at 382 eV and the titanium content by the negative excursion of the Ti peak at 420 eV, which essentially gives a quantitative analysis of TiN[13]. The analysis of the nitrogen stoichiometry in the film is facilitated by the following procedure. First, the ratios of the peak heights (Ti+N)/Ti, Ti/O, and (Ti+N)/O are considered to be almost independent of Auger electron signal detection. Second, the nitrogen content in the deposited film is considered high enough so that a

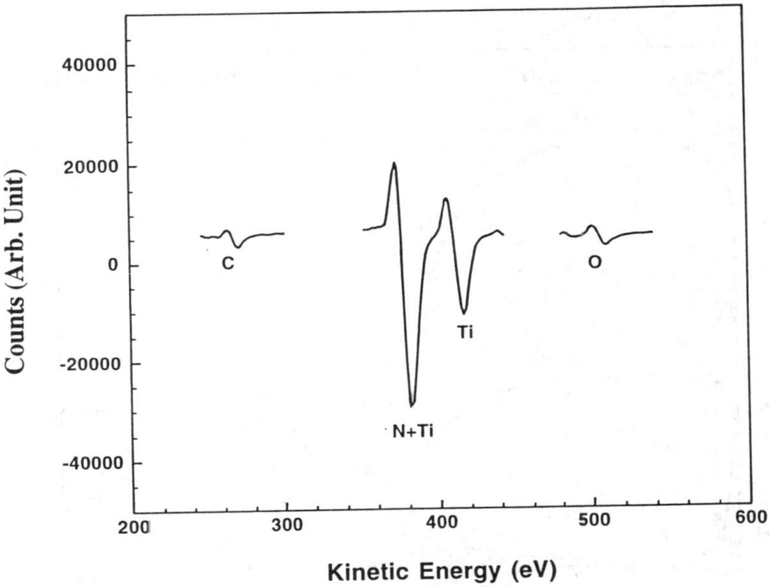

Figure 3. Auger electron spectrum of TiN film deposited at room temperature on polyimide.

slight change in the (Ti+N) peak intensity is linear with that of Ti. The Auger analysis was performed at the surface as well as in the bulk after sputtering TiN for 3 minutes to eliminate the surface oxide. The oxygen peaks which appear after sputtering are a result of gettering of oxygen present in the spectrometer by Ti. The (Ti+N)/Ti ratio remained almost constant with an average value of 1.926 over the range of temperature used in the present work. The observed carbon and oxygen lines are expected due to the atmospheric exposure of the samples. The Auger peak intensities for the Ti+N (381 eV), Ti (420 eV), O (503 eV) and C (270 eV) in the depth profiling experiment are shown in Fig. 4. This result clearly shows that the TiN films of uniform composition are formed on polymers by laser deposition from stoichiometric TiN target at room temperature and base pressure of 10^{-7} Torr.

The TiN films were also characterized by Raman spectroscopy. The scattering in the acoustical range is primarily determined by the vibrations of the Ti ions and that in the optical range by the vibrations of the N ions. Thus, the ratio of the accustical to the optical scattering intensities provides a measure of the Ti:N ratio or stoichiometry. The spectrum of TiN polycrystalline material is characterized by acoustic peaks[14] below 370 cm^{-1} and optical peaks above 500 cm^{-1}. The first order defect induced acoustic and optical phonon peaks were recorded for the TiN films deposited on PMMA at room temperature, as shown in Fig. 5. The Raman spectrum shows a main peak in the range 200-350 cm^{-1} which is associated with Ti atom vibrations in the absence of nitrogen. Another peak in the range of 450-550 cm^{-1} is associated with the titanium vacancies in the presence of nitrogen ions. These features are similar to that of the TiN target and the films obtained on (100) Si substrates. Transmission electron microscopy (TEM) studies were carried out to investigate the microstructure of the TiN films. The TEM

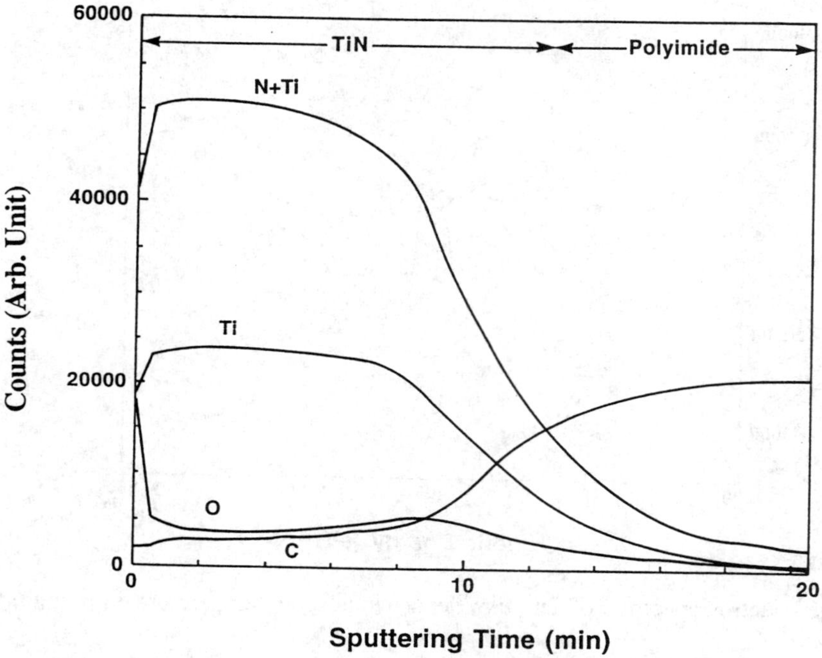

Figure 4. AES depth profiles of TiN film deposited at room temperature by PLD on polyimide.

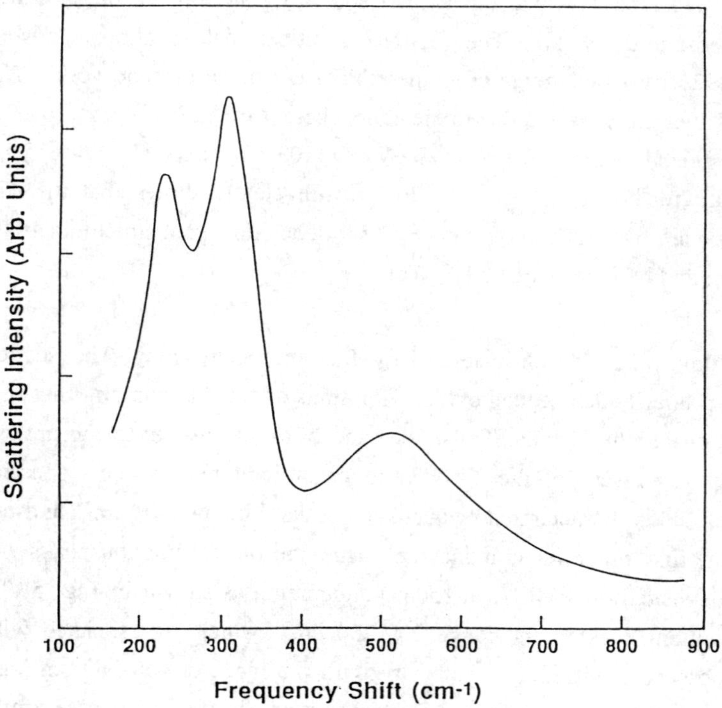

Figure 5. Raman spectrum of laser deposited TiN film at room temperature on PMMA substrate.

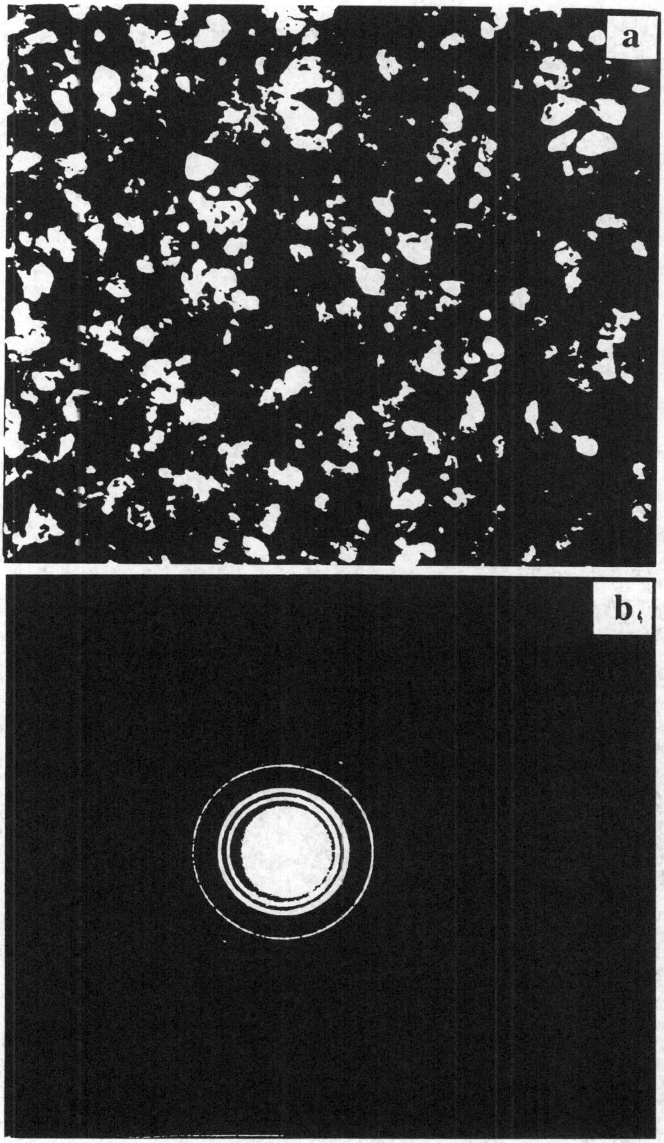

Figure 6. (a) Bright field planar TEM image of TiN film deposited at room temperature on PMMA. (b) Corresponding SAED pattern showing polycrystallineTiN film.

micrograph in Fig. 6 (a) shows that the TiN films are polycrystalline with an average grain size of 12 nm. The corresponding selected area electron diffraction (SAED) pattern is shown in Fig. 6 (b). The SAED pattern clearly shows diffraction rings corresponding to (111), (200), (311), (400), and (331) planes of TiN. From the electron diffraction pattern, the lattice constant of TiN was found to be 0.421 nm which is very close to that of stoichiometric bulk TiN.

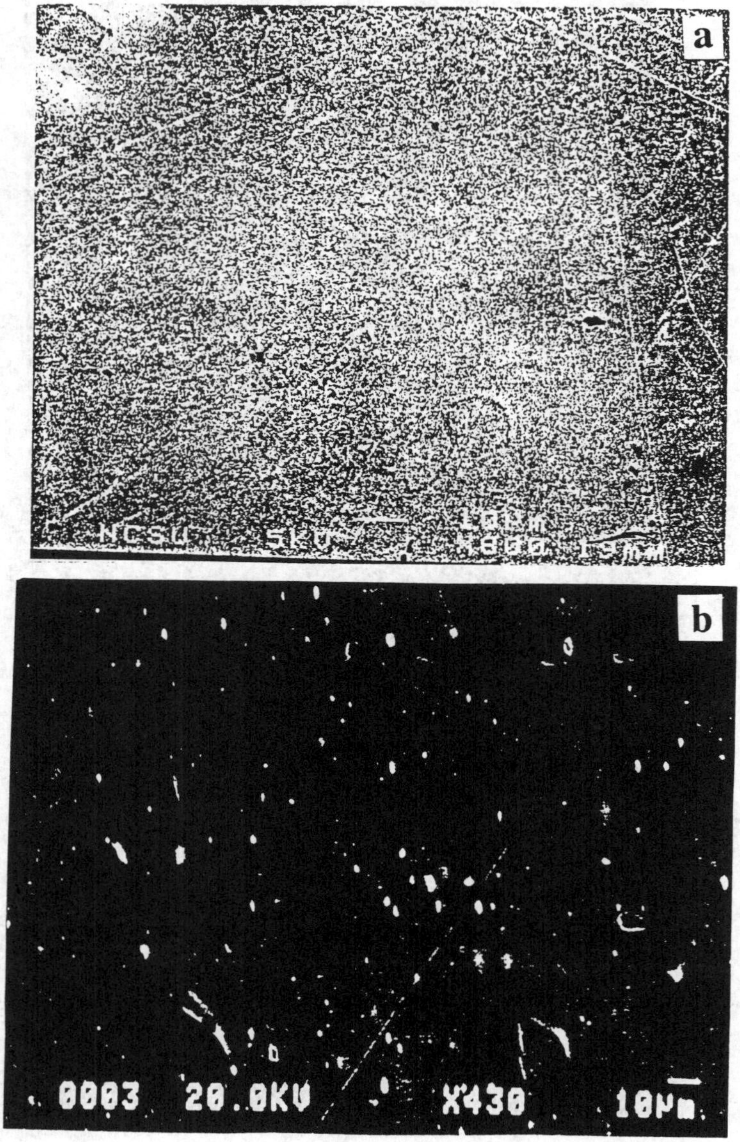

Figure 7. SEM micrographs of the TiN films on (a) polyimide and (b) PMMA deposited at room temperature by PLD.

The TiN films deposited on PMMA and polyimide substrates were examined by optical and scanning electron microscopies to study the surface smoothness of the laser deposited films. The SEM micrographs of the 50-60nm thick TiN films deposited on polyimide and PMMA at room temperature, laser fluence of 4-5 J/cm^2 and base pressure of $1x10^{-7}$ Torr, are shown in Fig. 7(a) and (b), respectively. The films were found to be quite smooth and uniform. The

Figure 8. SEM micrographs of thicker (>100 nm) TiN film on PMMA showing peeling-off of the film.

high absorption of KrF laser by the TiN target is responsible for the uniform films with relatively low density of particles. The average particle density in all depositions at 4-5 J/cm^2 was found to be less than $5x10^3$ $particles/cm^2$. No significant effect of the pulse repetition rate on the surface morphology was observed, which indicates that the films can be grown at higher rates using the PLD technique. We have noted that as the TiN film thickness increased above a critical value, the volume strain accumulated which resulted in the peeling of the film. In the case of TiN films on PMMA, peeling of the film was observed above 80-85 nm. The films peeled off, as shown in Fig. 8, when the thickness was more than 80-85 nm. While in case of films deposited on polyimide, the interfacial bonding is much stronger than for TiN/PMMA resulting in smooth surface morphology without any peeling or cracking of the TiN films. These results are interesting since the adhesion of the metallized coatings on polymers is an important aspect of polymer metallization. It should be noted that the adhesion of a film on a substrate primarily depends on the method of thin film deposition (physical state of the depositing species), chemical properties of the participating materials (interatomic potentials and the nature of chemical bonding of the film with the substrate), physical characteristics of the substrate surface and its roughness, and substrate temperature. In our experiments, the strong adhesion of the TiN films with the polymer is expected due to the deposition of energetic (typically 20-50 eV) and excited Ti-N radicals. These radicals can form strong bonds at the interface by releasing their energies on the polymer surfaces at room temperature. To measure the adhesion of the laser deposited TiN films on polymers, the direct pull-off method[4,15] was employed. A pulling device was

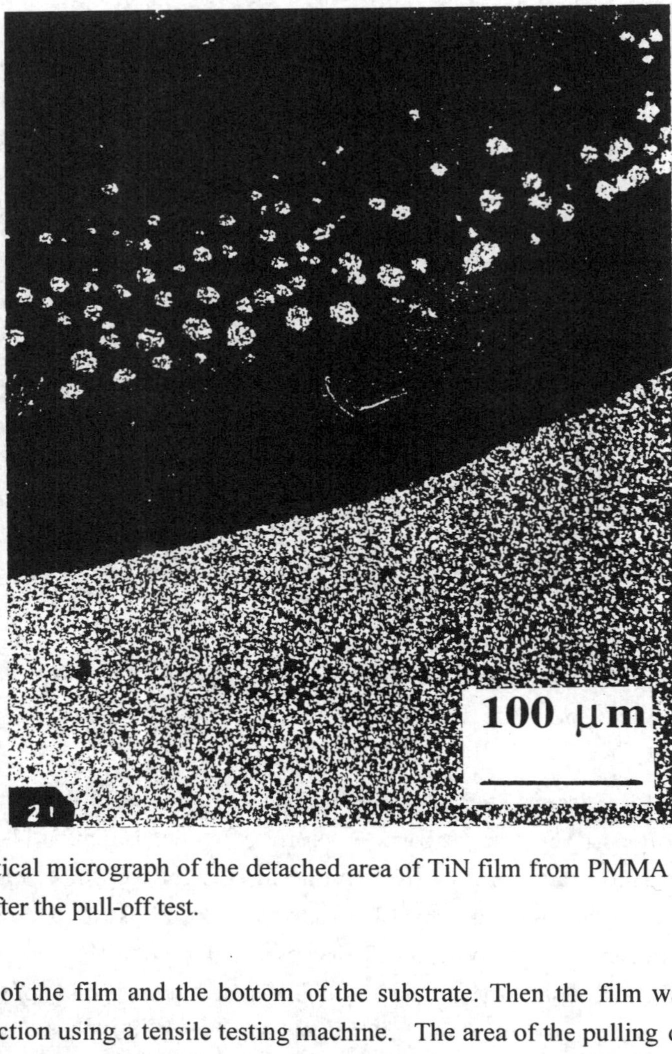

Figure 9. Optical micrograph of the detached area of TiN film from PMMA substrate after the pull-off test.

glued onto surface of the film and the bottom of the substrate. Then the film was pulled in the perpendicular direction using a tensile testing machine. The area of the pulling device was 4.31 cm^2. A force of increasing magnitude was applied to the surface until the film was removed and the maximum force applied was measured. The area of the film detached from the substrate was measured by optical microscopy or SEM. The optical micrograph of the detached area of TiN film from the PMMA substrate is shown in Fig. 9. The pull-off forces for the TiN on PMMA and polyimide were found to be 1.85×10^7 and 5.61×10^7 Pa, respectively. These results clearly indicate the strong interfacial bonding of TiN with polymer substrates as compared to surface modified polymers and metals deposited by thermal evaporation and sputtering methods[16,19]. We believe that the formation of energetic and excited Ti species during TiN laser ablation and the formation of a TiC bond at the interface with the polymer are likely to be responsible for the good adhesion of the TiN films. In fact, the formation of Ti-C bonds at the interface of Ti with polymers is reported for thermally evaporated Ti metal[18-19]. A low formation energy and the highly reactive nature of Ti are responsible for Ti-C bonding at the Ti-polymer interface.

Now we focus on the electrical properties of the laser deposited TiN films on polymers. The electrical resistivity was measured using a four-point technique in a temperature controlled cryostat capable of reaching a temperature as low as 10K. The contacts on TiN films were made with gold wire by introducing silver dots. The resistivity of all the films decreased linearly with a decrease in temperature and reached an asymptotic value below 50K. The magnitude of the intercept at the ordinate was found to decrease with increase in deposition temperature. The room temperature resistivity of these films was in the range of 30-75 μΩ-cm as compared to 15 μΩ-cm obtained for the films deposited at 600°C on Si(100)[20]. The smaller grain size of TiN films on polymers is considered to be responsible for the higher electrical resistivy of the TiN films since the grain size significantly affects the electron scattering mechanism in TiN films[6].

We have also investigated a dry etching method which involves the use of laser ablation for an etching of TiN films and to form metallic lines on polymer substrates. This approach has been shown to permit etching and patterning of hard and brittle materials without causing significant damage to the adjacent material. The laser etching can potentially lead to improvement in spatial resolution of the resulting etched features due to its monochromatic and coherent characteristics. There are several advantages of laser etching over other processing techniques and these are: 1) etching rates are generally high, 2) etching selectivity an be very high, 3) direct etching or patterning is possible, 4) laser etching is a clean process and therefore no contamination occurs, and 5) it is possible to automate it and hence the process can be rendered compatible with modern technology. In the present case, a KrF excimer laser was used for the etching of TiN films on polymers. The schematic of the laser etching process for a single microbridge fabrication is shown in Fig. 10 (a). The beam was focused through a quartz lens to a 1.5 x 0.7 cm^2 area on the film surface and was made incident perpendicular to the sample surface. A wide range of incident energy fluence was explored for etching experiments. Incident laser fluence was varied from 0.1 to 1 J/cm^2. The experiment was performed in air and the substrate was held at room temperature. In Fig. 10 (b) is shown a 100 μm wide microbridge of TiN on PMMA obtained by laser etching at 0.4 J/cm^2 laser energy density and using a simple stainless steel shadow mask.

We also have investigated PLD processing of protective and dielectric layers of aluminum nitride (AlN) and diamond-like carbon on TiN/polymer substrates. The AlN has hexagonal wurtzite structure with lattice constant of a = 0.311 nm and b = 0.498 nm, a band gap of 6.2 eV, thermal conductivity of 320 W/mK, resistivity of 10^{13} Ω-cm, and dielectric strength of 10MV/cm[21]. These properties along with its high chemical stability make AlN a suitable candidate for applications in microelectronics, optoelectronic, high temperature devices and for electronic packaging. Since it is hard, stable, and insulating, it can also be used for protective coating applications on polymers. It will be very interesting to use the overlayer of insulating AlN on the TiN metallized polymer for isolation. In this context, we have made an attempt to integrate nitride based insulator and metallic multilayer structures on polymers. For these applications PLD technique can be exploited for the integration of materials and the development of the multilayer structures on the polymers via in-situ processing at room temperature.

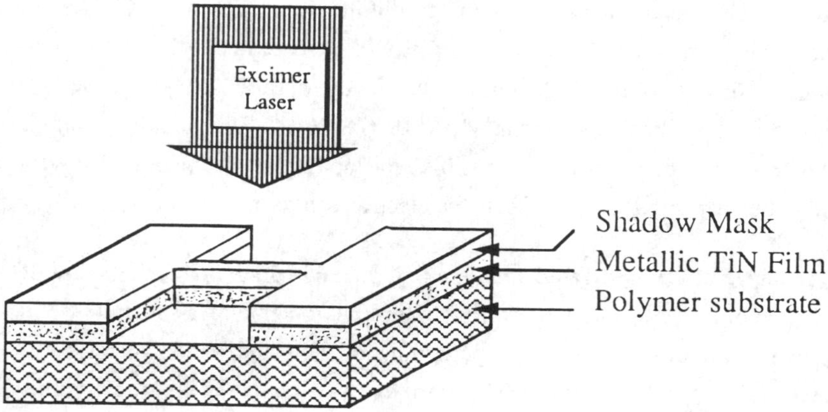

(a)

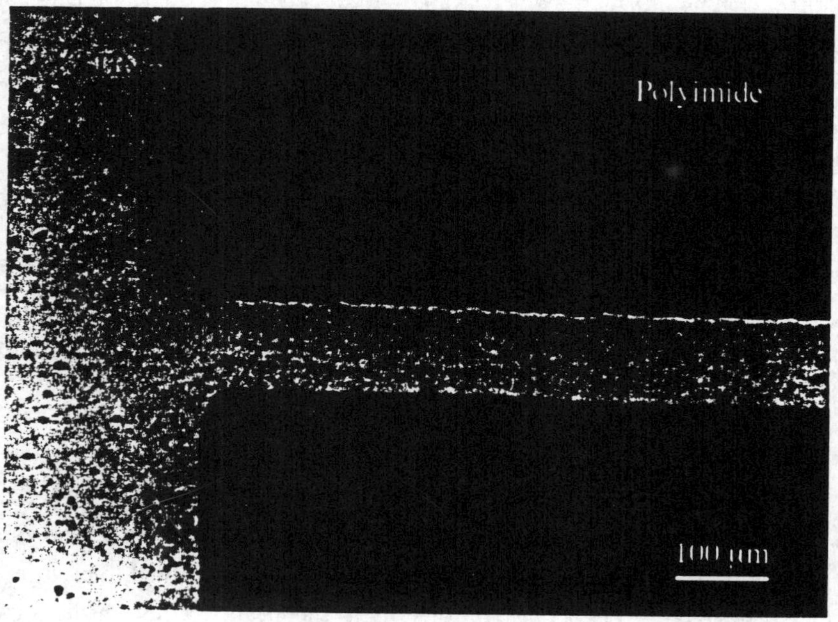

(b)

Figure 10. (a) Schematic of etching of TiN film using pulsed excimer laser. (b) TiN metal line (~100 μm width) on polyimide obtained by laser etching technique.

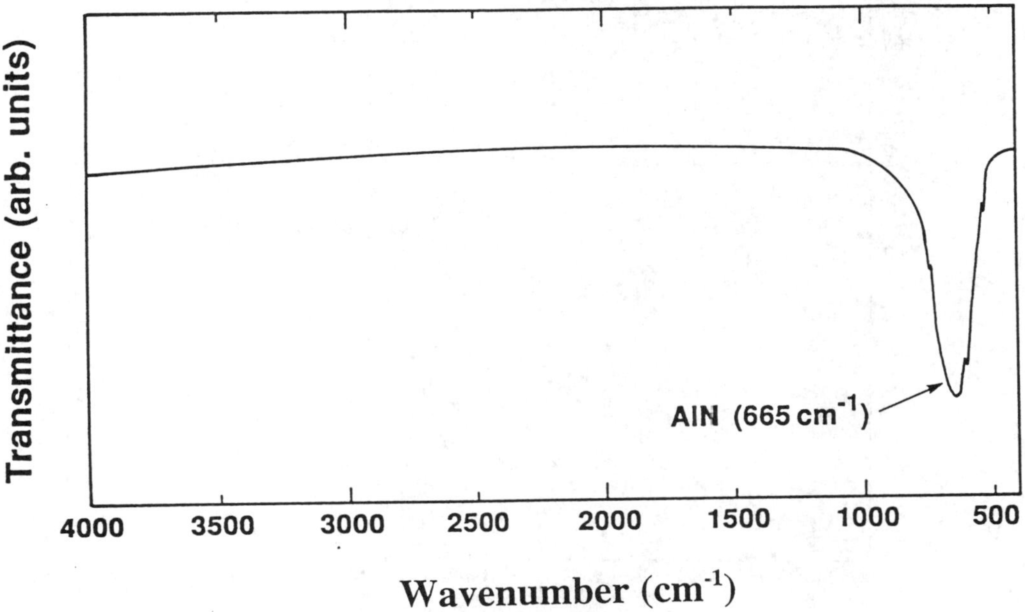

Figure 11. FTIR spectrum of pulsed laser deposited AlN film at room temperature on polyimide. (The data were obtained by subtracting the spectrum from the polyimide substrate).

Figure 11 shows the FTIR transmittance spectrum of the laser deposited AlN film on polyimide at room temperature. The data were obtained by the subtracting the spectrum of the polyimide substrate. The strong absorption peak at 665 cm^{-1} is due to the transverse optical phonon modes of AlN[22]. Diamondlike carbon (DLC) films have many unique and desirable properties for wear resistance coatings because of their superior hardness and mechanical properties. In general, the deposition of DLC requires the depositing species to be energetic. PLD is well studied for DLC[23]. In our experiments, graphite was ablated at 15-20 J/cm^2 to achieve a high fraction of sp^3 bonding in DLC films. The films deposited at 15 J/cm^2 were uniform, light-brown, and translucent. Raman spectroscopy was also used for the characterization of DLC films. Fig. 12 shows the Raman spectrum for the DLC film deposited on PMMA. The presence of a broad peak around 1550 cm^{-1} is a characteristic of a DLC phase. The more commonly observed Raman peaks for graphite or glassy carbon at 1365 and 1590 cm^{-1} are absent indicating a high fraction of sp^3 bonding in the DLC film. We have deposited DLC films on TiN/PMMA substrates by in-situ laser deposition. A 50 nm thick TiN layer was deposited followed by 50 nm DLC film. The films were found to be smooth and adherent to the polymers. The fabrication of high quality insulator/metal/polymer multilayer structures such as AlN/TiN/ polyimide and DLC/TiN/polyimide are useful for multichip module technology.

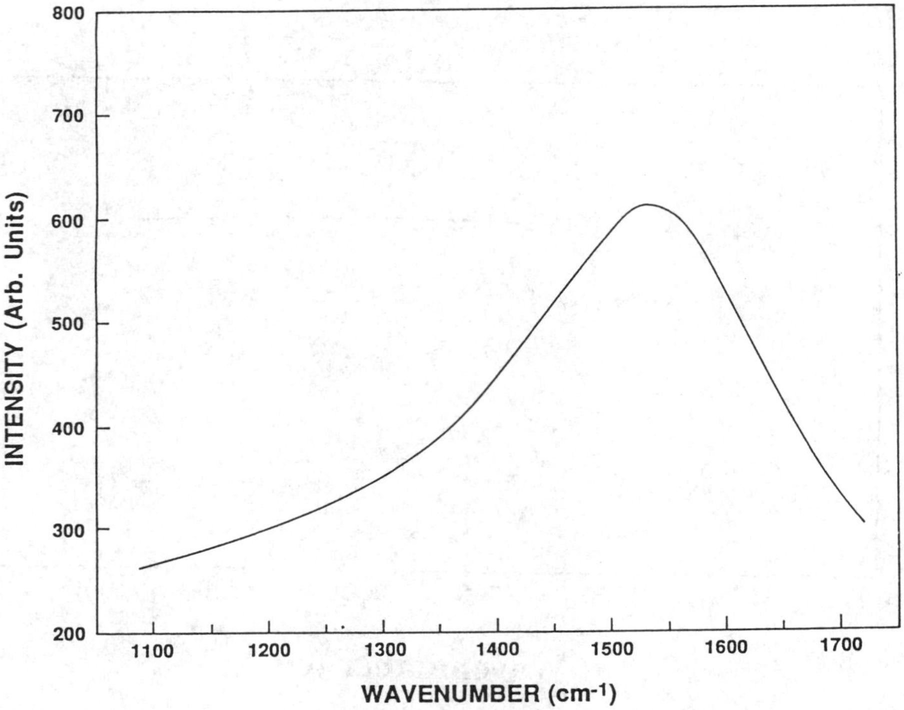

Figure 12. Raman spectrum of the pulsed laser deposited diamond-like-carbon film on PMMA at room temperature.

CONCLUSION

In summary, we have used a pulsed laser deposition technique for depositing TiN layer at room temperature. The processing parameters have been optimized for the fabrication of conducting TiN metal coatings on polymers. The electrical resistivity of these films was found to be in the range of 30-75 μΩ-cm at room temperature. The metallic TiN films were strongly adherent to the polymer substrates as compared to metals deposited by thermal evaporation and sputtering techniques. The TiN films were found to be smooth and polycrystalline with a fine grain size of 12 nm. The AlN/TiN/polymer and DLC/TiN/polymer structures were fabricated in-situ at room temperature. Thus, hard and adherent layers of AlN and DLC films deposited on TiN/polymer system provide protective coatings as well as isolation layers on conducting TiN for advanced technological applications of polymers. The use of the dry processing technique of laser etching was described for the formation of TiN metallic lines on the polymers.

ACKNOWLEDGMENT

This research was supported by the National Science Foundation.

REFERENCES

1. Mater. Res. Soc. Symp. Proc., **304** (1993).
2. M.K. Ghosh and K.L. Mittal, Editors, “Polyimides: Fundamentals and Applications”, Marcel Dekker, New York, 1996.
3. E. Sacher, J.J. Pireaux, and S.P. Kowalezyk, editors, "Metallization of Polymers", ACS Symp. Ser. No. 440, American Chemical Society, Washington D.C. 1990.
4. K.L. Mittal, Electrocomponent Sci. Technol. **3**, 21 (1976) .
5. N. Biunno, J. Narayan, S.K. Hofmeister, A.R. Shrivasta, and R. K. Singh, Appl. Phys. Lett. **54**, 1519 (1989); Ibid **55**, 405 (1989).
6. L. Porte, L. Roux, and J. Hanus, Phys. Rev. B. **28**, 3214 (1983) .
7. R.D. Vispute, R. Chowdhury, P. Tiwari, and J. Narayan, Appl. Phys. Lett. **65**, 2565 (1994).
8. R.I. Hegde, R. W. Fiordalice, E. O. Travis, and J. Tobin, J. Vac. Sci. Technol. B **11**, 1287 (1993).
9. W.Tsai, M. Delfinc, J.A. Fair, and D. Hodul, J. Appl. Phys. **73**, 4462 (1993).
10. N. Savvides and B. Window, J. Appl. Phys. **64**, 225 (1988).
11. Mater. Res. Soc. Bull., XVII **2**, (1992).
12. D. B. Chrisey and G. K. Hubler, Editors, “Pulsed Laser Deposition of Thin Films”, Wiley-Interscience, New York, 1994.
13. J. E. Sundgren, B. O. Johansson, and S. E. Karlsson, Surface Sci., **128,** 265 (1993).
14. W. Spengler, R. Kaiser, A. N. Christensen, and G. Muller-Vogt, Phys. Rev. B, **3,** 16 (1978).
15. J. Valli, J. Vac. Sci. Technol. A **4**, 3007 (1986).
16. E. M. Liston, J. Adhesion, **30**, 199 (1989).
17. B.J. Love, and P.F. Packman, J. Adhesion, **40**, 139 (1993).
18. N.J. Chou and C. H. Tang, J. Vac. Sci. Technol. A **2**, 751 (1984).
19. F.S. Ohuchi and S. C. Freilich, J. Vac. Sci. Technol. A 4, 1039 (1986).
20. J. Narayan, P. Tiwari, X. Chen, J. Singh, R. Chowdhury, and T. Zheleva, Appl. Phys. Lett. **61**, 1290 (1990) ; US Patent # 5406123 (April 11, 1995).
21. S. Strite and H. Morkoc, J. Vac. Sci. Technol. B **10**, 1237 (1992) .
22. L.E. McNeil, M. Grimsditch, and R. H. French, J. Am. Ceram. Soc. **76**, 1132 (1993).
23. R.D. Vispute, J. Narayan, and K. Jagannadham, J. Electronic Mater. **25**, 151 (1996).

Metallized Plastics 5&6: Fundamental and Applied Aspects, pp. 41-50.
K.L. Mittal (Ed.)

METALLIZATION OF POLYMERS USING PLASMA ENHANCED CHEMICAL VAPOR DEPOSITED TITANIUM NITRIDE AS INTERLAYER

A. Weber, A. Dietz, R. Pöckelmann and C.-P. Klages

Fraunhofer Institute for Surface Engineering and Thin Films
Bienroder Weg 54 E, D-38108 Braunschweig, Germany

A low temperature process for titanium nitride (TiN) deposition by means of an electron cyclotron resonance (ECR) plasma CVD process was applied to poly(tetrafluoroethylene) (PTFE), polyimide (PI 2611), benzocyclobutene (BCB, CYCLOTENE 5021), and poly(butylene terephthalate) (PBT, Crastin S 600) . Tetrakis(dimethylamido)titanium (TDMAT) or titanium(IV)isopropoxide (TIP) introduced into the downstream region of a nitrogen ECR plasma were used as precursors for TiN deposition at 100°C. The thin TiN film (thickness 15-30 nm) acts as a conductive layer for the electroplating process or as adhesion promoter for sputtered titanium films. Prior to the deposition of the TiN interlayer, the samples, positioned on a biased substrate table, were exposed to an argon ECR plasma in order to enhance the adhesion of the TiN interlayer by Ar ion bombardment. This metallization procedure avoids the use of toxic and polluting etching agents for polymer activation as well as electroless metal deposition. The maximum adhesion of the electroplated copper on PTFE and PBT was established to be 13 N/mm^2 and 21 N/mm^2, respectively. For sputtered Ti films on PI and BCB maximum adhesion was 13 N/mm^2 and 23 N/mm^2, respectively. As shown by atomic force microscopy (AFM), TiN grains were formed on the polymer surface. Film composition was investigated by secondary ion mass spectrometry (SIMS).

INTRODUCTION

Polymers such as PTFE, polyimide (PI) and benzocyclobutene(BCB) are of great interest for the microelectronics industry because of their desirable characteristics: low dielectric constant, low dissipation factor, high thermal stability, and high chemical resistance. Poly(butylene terephthalate) (PBT) is a good candidate for three-dimensional moulded interconnection devices (3D-MID) where the conventional printed circuit board is replaced by a patterned metal layer directly on the injection moulded plastics.

The low surface energy of fluoropolymers and their chemical inertness leads to tremendous difficulties regarding the metallization of such polymers[1]. Thus, a surface modification of fluoropolymers is required to improve the adhesion[2]. Hazardous and polluting etching agents such as sodium naphthalenide or sodium in liquid ammonia are necessary to activate the poly(tetrafluoroethylene) PTFE surface before metallization by electroplating[3]. These treatments lead to the removal of fluorine leaving unsaturated sites like C=C or C≡C which can be oxidized to provide a higher surface energy for the subsequent metallization step.

Besides these conventional chemical surface treatments some work has been done regarding interactions of non-equilibrium plasmas, ion beams, and radiation interactions with polymer surfaces[4]. Enhanced copper-PTFE adhesion was achieved by presputtering treatment of PTFE prior to the sputtering of copper[5]. The presputtering with Ar^+ ions leads to an increase of the surface roughness. The same effect was also observed with x-ray damaged PTFE[4] and by bombarding the PTFE surface with Xe atoms in the keV range[6]. The latter method leads to the formation of PTFE filaments on the surface.

Haag and Suhr[7] reported improved adhesion of electroplated copper onto plasma etched PTFE. Prior to the electroplating, a thin metallic layer was deposited by employing a plasma enhanced chemical vapor deposition (PECVD) process. Organometallic compounds of Pd, Pt, Au and Cu were used in this process.

Previously we reported the electroplating of PTFE by employing tetrakis(dimethylamido)-titanium (TDMAT) as precursor in an ECR plasma process to deposit TiN as interlayer[8]. Here we report for the first time the use of titanium(IV)isopropoxide (TIP)[9] for TiN deposition as a conductive layer for subsequent electroplating process or as an adhesion layer for sputtered titanium suitable for various polymers.

EXPERIMENTAL DETAILS

Depositions were carried out using an ASTeX HPM/M ECR source mounted on a reaction vessel equipped with a substrate holder which is connected to a radio frequency (13.56 MHz) generator. A detailed description of the experimental set up was published previously[10]. The following polymer substrates were used: PTFE, thickness 1 mm, purchased from Goodfellow; spin-on Teflon AF 1600

(DuPont), thickness 5 μm on silicon; spin-on polyimide (PI 2611 DuPont), thickness 3 μm on silicon; spin-on benzocyclobutene (BCB, CYCLOTENE 5021 Dow), thickness 1 μm on silicon; poly(butylene terephthalate) (PBT, Crastin S 600 DuPont), thickness 2 mm. The substrates were placed four inches below the ECR source flange.

Prior to the deposition of TiN, the substrates were treated with an ECR argon plasma using 15 sccm Ar and a MW power of 400 W. The samples were biased to -100 V for this treatment by connecting the substrate table to a radio frequency (rf) generator operating at 13.56 MHz.

For TiN deposition, a flow of 15 sccm [standard cm^3/min] N_2, regulated by a mass flow controller, was introduced through the ECR cavity while TDMAT $\{Ti[N(CH_3)_2]_4\}$ or TIP $\{Ti[O\text{-}CH(CH_3)_2]_4\}$ vapor was introduced through a gas ring into the downstream region between the plasma cavity and the substrate table. The liquids TDMAT (99%, purchased from Solvay Germany) and TIP (98 %, purchased from Aldrich) were kept in a double glass wall container to allow precise temperature control. The temperature was thermostatically maintained for TDMAT and TIP at 25°C and 35°C, respectively, to produce a flow of 2.5 sccm. The microwave (MW) power was set to 400 W and the pressure was 0.1 Pa. No external heating of the substrate table was applied. After a process time of 20 minutes the substrate temperature rose to about 100°C. Growth rates were typically 1.5 nm/min.

After the deposition of the thin TiN film, 10 μm copper was deposited by electroplating on the PTFE and PBT substrates using a commercial acidic copper bath (Glanzkupferbad RUBIN WK, purchased from Kiesow, Detmold) at a current density of 5 A/dm (deposition rate 1 μm/min). Titanium was sputtered on the pretreated PI and BCB substrates by means of a rf sputter process (rf power 100 W, voltage 1 kV, argon flow 13 sccm, pressure 1 Pa).

The adhesion was checked by the Scotch tape test and a z-axis pull-out test[11]. In the latter test a progressive load was applied to a piston glued to the metal surface until the sample failed. The joint was prepared with an epoxy resin at 100°C.

Atomic Force Microscopy (AFM) investigations were performed with a commercial AFM device under ambient conditions using silicon nitride pyramidal tips with a tip radius of 50 nm, tracking forces of about 150 nN, and a scan frequency of 2 Hz. For the determination of root mean square (RMS) roughness values, a subtraction of a 3rd order polynom, fitted to the experimental curves, from linear scans over 3 μm distances was applied in order to account for substrate bending and long-wavelength roughness. Secondary Ion Mass Spectrometry (SIMS) investigations were performed with a CAMECA ims 5f spectrometer.

RESULTS

Figure 1 shows the influence of an argon plasma pretreatment on the surface roughness of the polymers used. The AFM images of the surface morphologies of the polymers are displayed in Figure 2. Figure 3 represents the relation between the surface roughness and the adhesion strength of electroplated copper on PTFE using TDMAT as precursor to deposit the TiN interlayer (deposition time 20 minutes). The use of a commercially available copper bath RUBIN WK allowed us to electroplate copper on PTFE and PBT without a predeposition of electroless copper as decribed previously[8].

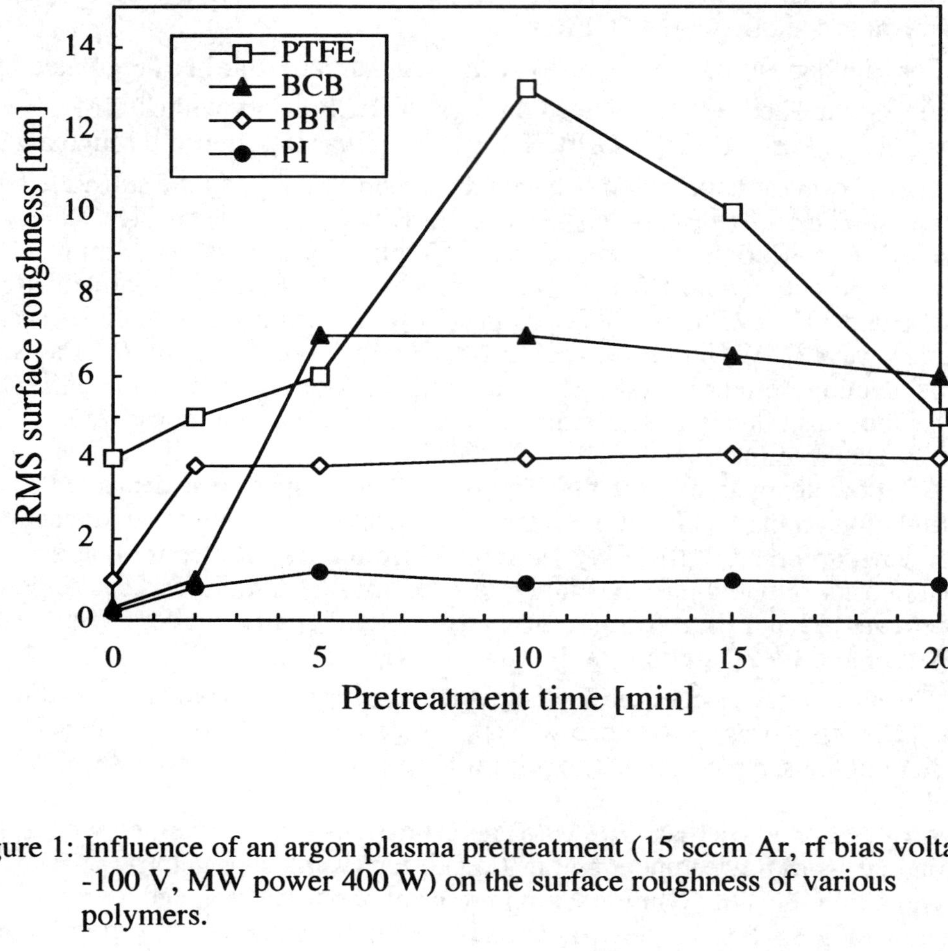

Figure 1: Influence of an argon plasma pretreatment (15 sccm Ar, rf bias voltage -100 V, MW power 400 W) on the surface roughness of various polymers.

The maximum adhesion strength of copper electroplated onto PBT was measured to be 21 N/mm^2. This result was achieved after an argon plasma pretreatment time of 5 minutes followed by the TiN deposition employing TIP as the precursor (deposition time 20 minutes). The measured adhesion strength of sputtered titanium film (thickness 2 μm) on untreated and treated PI and BCB is shown in Figure 4. The argon plasma pretreatment and the TiN interlayer deposited using TIP leads to a substantial increase of the adhesion strength up to 23 N/mm^2 in case of BCB.

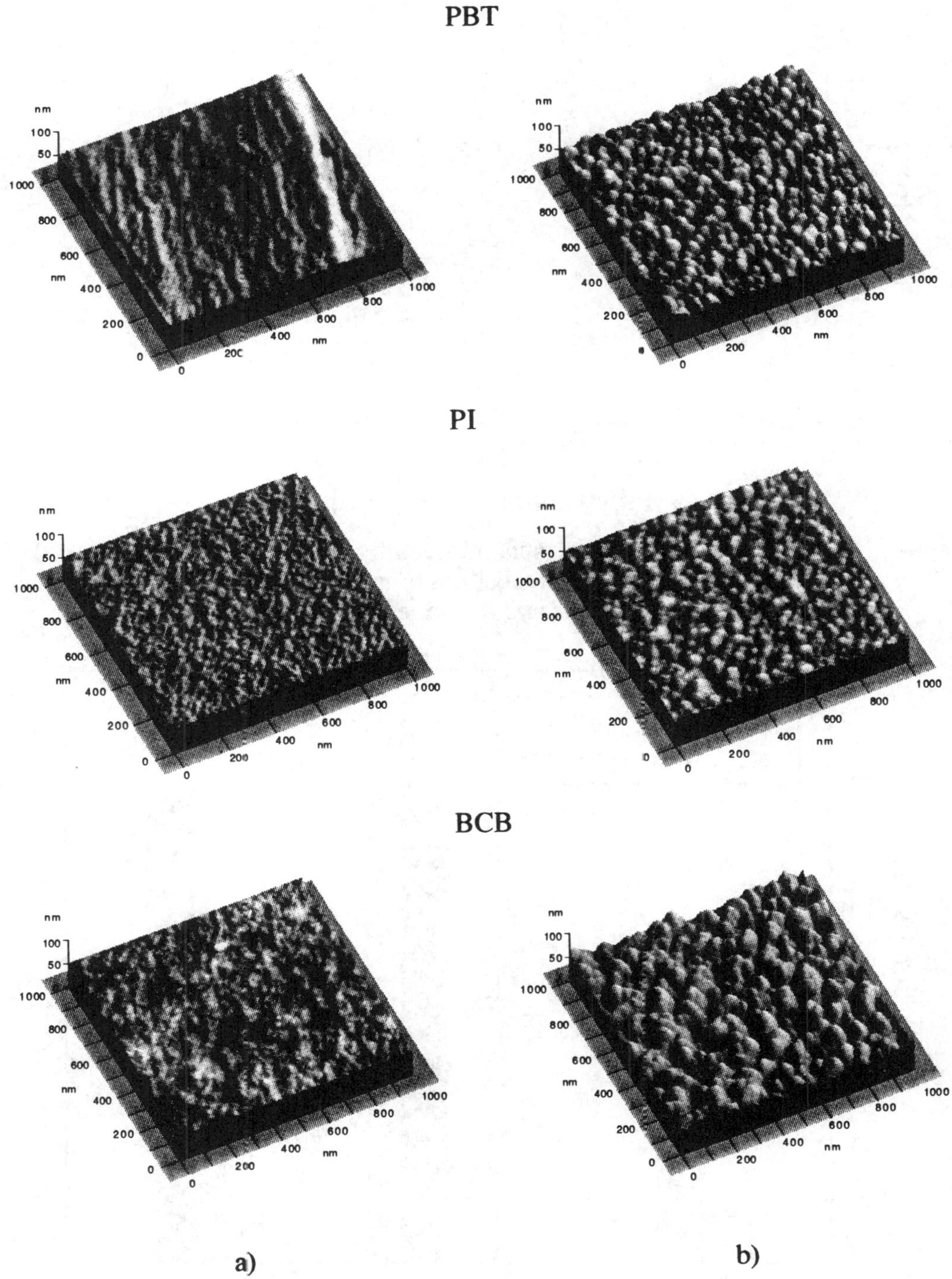

Figure 2: AFM images (1x1 μm^2) of (a) untreated polymer surfaces; (b) polymer surfaces pretreated with Ar plasma (5 min, for conditions see Fig. 1).

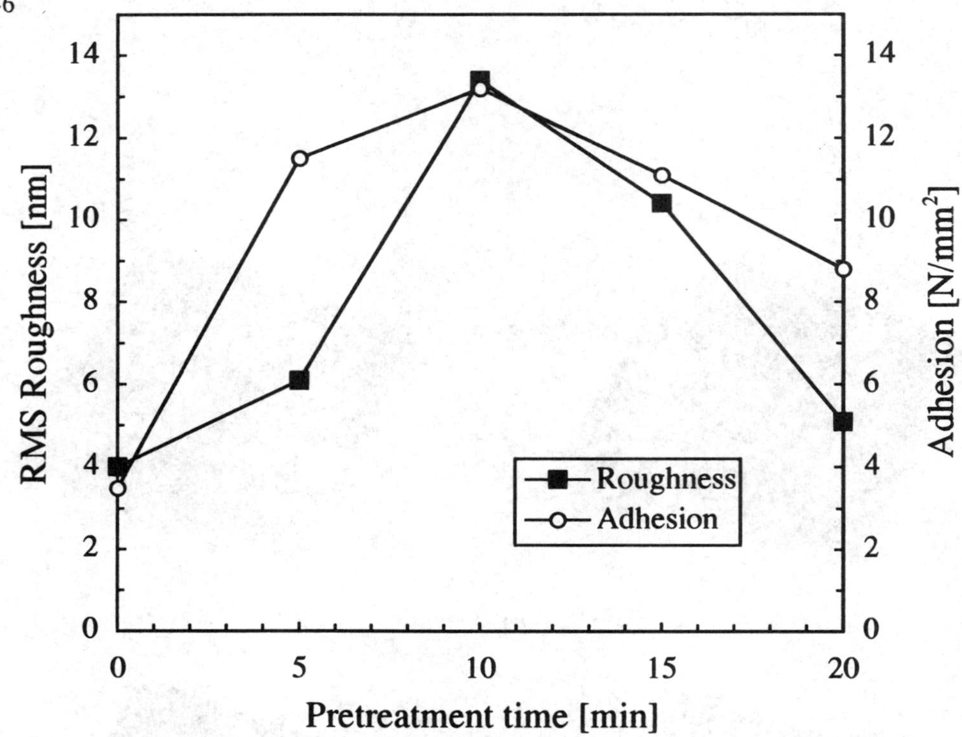

Figure 3: Influence of an argon plasma pretreatment (15 sccm Ar, rf bias -100 V MW power 400 W, TiN deposition time 20 min using TDMAT) on the surface roughness and the adhesion strength of electroplated copper on PTFE.

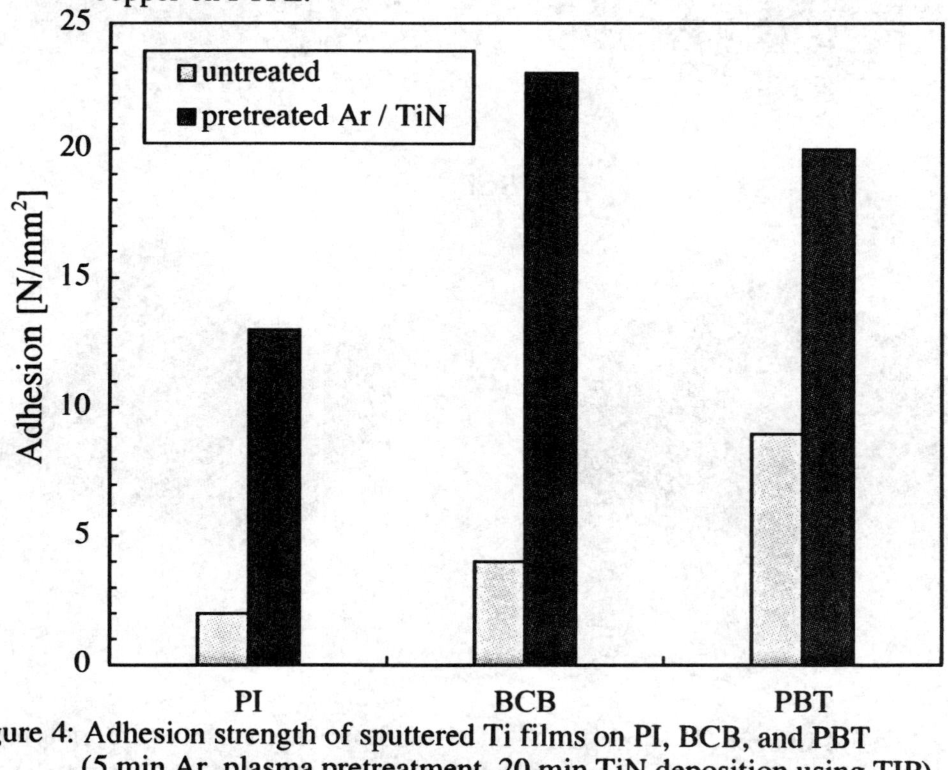

Figure 4: Adhesion strength of sputtered Ti films on PI, BCB, and PBT (5 min Ar plasma pretreatment, 20 min TiN deposition using TIP).

The deposition rate of TiN using TDMAT or TIP as precursors with the parameters described above was 1.5 nm/min. The measured sheet resistances of TiN layers on SiO_2 and various polymers after a deposition time of 20 minutes are shown in Figure 5.

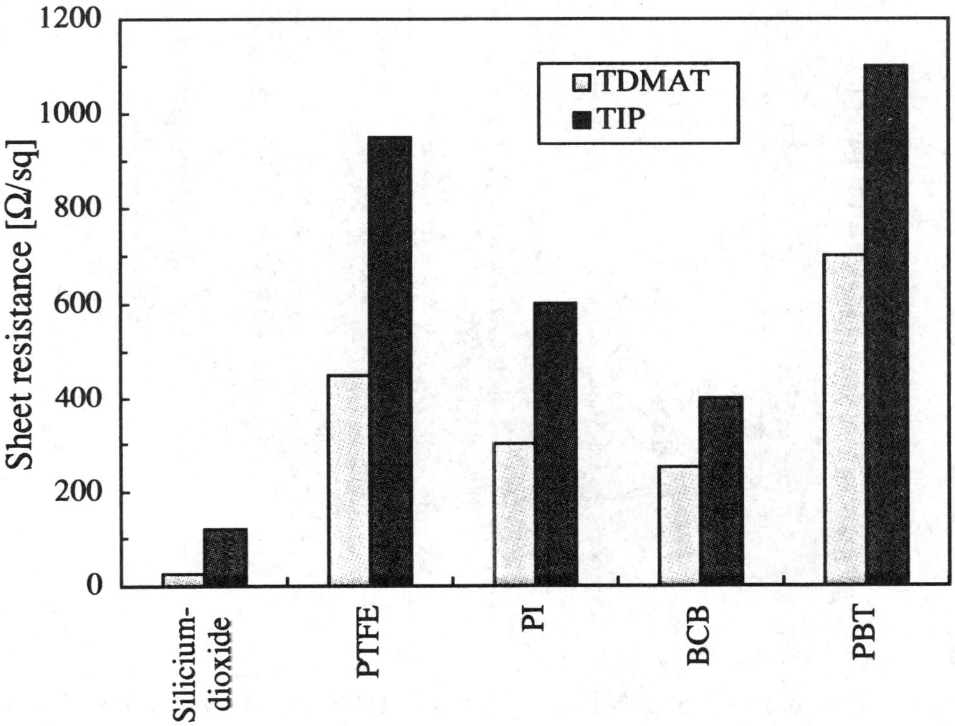

Figure 5: Sheet resistance of TiN interlayer on SiO_2 and various polymers using TDMAT or TIP as precursors after deposition time of 20 min. (conditions as described in text).

SIMS investigations were performed using the smooth spin-on polymers as substrates. The compositions of the TiN interlayers as revealed by SIMS on SiO_2 and various spin-on polymers are plotted in Figure 6. The AFM images in Figure 7 show the surfaces of SiO_2 and PTFE after TiN deposition using TDMAT as precursor.

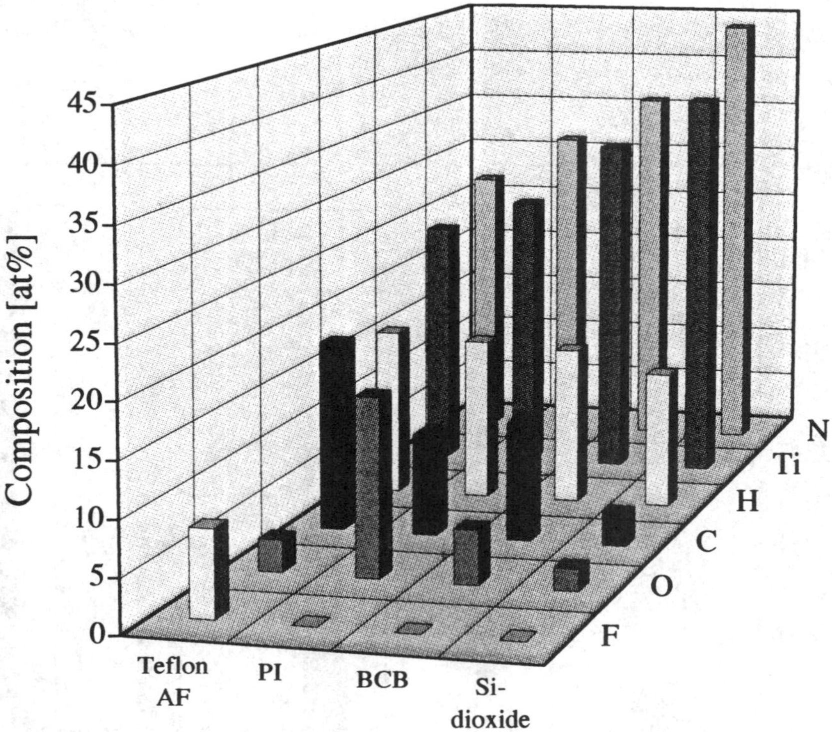

Figure 6: Composition of TiN interlayers (using TIP as precursor) on SiO_2 and various polymers as revealed by SIMS.

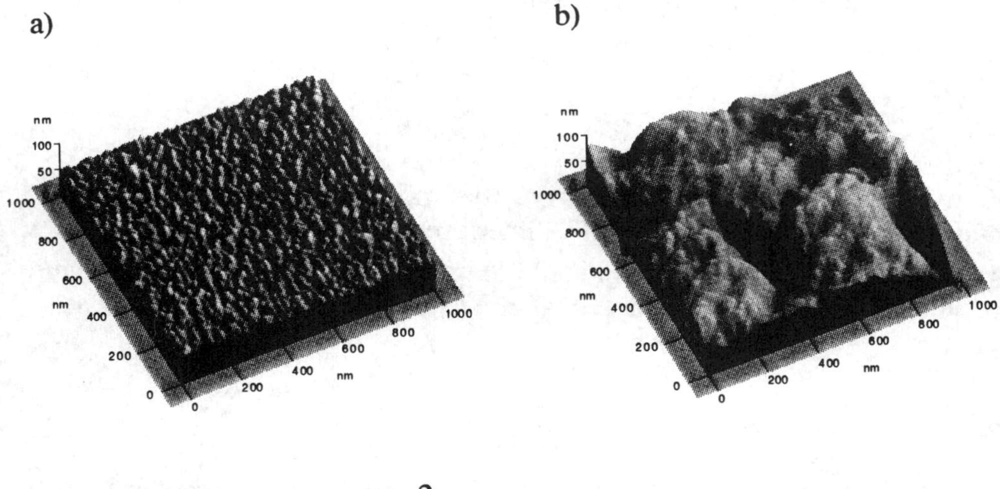

Figure 7: AFM images (1x1 μm^2) of TiN (thickness 30 nm, using TDMAT as precursor) deposited on (a) SiO_2 and (b) PTFE.

DISCUSSION

As indicated in Figure 1 the maximum surface roughness on PTFE was achieved after 10 minutes whereas for PI, BCB and PBT a pretreatment time of 5 minutes was sufficient to increase the surface roughness. The increase in surface roughness is due to the bombarment of the polymer surface with Ar^+ ions [5]. At a pretreatment time above 10 min the surface roughness decreases in case of PTFE or remains constant for the other polymers. This effect is presumably caused by the increase of the surface temperature of the polymer samples due to heating in the downstream plasma. The increasing surface temperature probably leads to the formation of volatile carbon species. These species act as monomers for the redeposition of a thin plasma polymer layers which inhibits the progress of the surface roughness. The argon plasma pretreatment leads to an increase of the adhesion strength.

Figure 3 shows clearly the relationship between surface roughness and metal adhesion on PTFE. Samples without argon plasma pretreatment had already failed the Scotch tape test. The failure of the metal / TiN / fluoropolymer joint always occurred at the TiN / PTFE interface. An influence of the interlayer thickness on the adhesion strength could not be found. The measured adhesion force of the electroplated copper of 13 N/mm^2 is in the range found for PTFE treated with Na/NH_3 [12]. Similar effects could be found after the sputtering of Ti films on untreated and treated PI and BCB as shown in Fig. 4.

As indicated in Fig. 5 the quality of the deposited TiN depends on the precursor and the substrate material. The oxygen containing TIP molecule forms TiN with a higher resistivity as compared to the oxygen-free TDMAT precursor. However, the achieved sheet resistance on the polymers using TIP as precursor is sufficient for the subsequent electroplating process. TIP is in contrast to TDMAT less air sensitive, less toxic, more stable during storage and heating and does not pollute the exhaust gases with amines. Thus, TIP allows an easier and safer precursor handling. Furthermore, TIP is much cheaper than TDMAT and thus an interesting alternative for TiN deposition in plasma activated processes.

Due to the low deposition rate of TiN, a simultaneous plasma etching process takes place. The formation of volatile carbon species depends on the thermal stability and the chemical structure of the polymer leading to carbon, oxygen and fluorine incorporations during the growth of TiN as shown in Fig. 6. This explains the high sheet resistances and the grey color of the "TiN" on the polymers compared to the gold colored TiN on SiO_2 in Fig. 5. The highest oxygen content was found in the case of PI. This can be explained by the preferred cleavage of the carbonyl group in PI (O content 10 at%). Teflon AF contains also about 10 at% oxygen but the cyclic C-O-C ether group is more stable under the applied plasma conditions. BCB consists only of 2 at% oxygen located in Si-O-Si bonds which explains the low oxygen content of TiN deposited on this polymer. The different growth conditions for TiN on polymers as compared to SiO_2 affect also the crystallite formation. The TiN layer on SiO_2 (Figure 7a) exhibits distinct well-shaped small crystallites whereas the polycrystalline morphology is much less pronounced on PTFE (Figure 7b).

SUMMARY

Thin TiN films deposited on PTFE, PI, BCB and PBT in a low temperature plasma enhanced process can be used as interlayers for the electroplating of copper or metal deposition by sputtering. To achieve a good adhesion of metal layers a pretreatment with argon ions at a substrate potential of -100 V is required. The ion bombardment increases the surface roughness leading to an enhanced adhesion of the interlayer and of the subsequent deposited metal layer. The thin conductive interlayer, with a thickness of 15-30 nm, is suitable for a subsequent electroplating process. Thus, no palladium activation and electroless copper deposition as used in today´s metallization of polymers is required.

In case of PTFE and PBT a maximum adhesion strength of 13 N/mm^2 and 21 N/mm^2, respectively, was achieved for electroplated copper. The adhesion of sputtered Ti films on PI and BCB was substantially improved to 13 N/mm^2 and 23 N/mm^2 compared to untreated polymers. The described combination of the plasma based pretreatment steps followed by an electroplating process is a promising approach to avoid polluting and hazardous chemicals used in today´s metallization of polymers.

ACKNOWLEDGMENT

The authors are indepted to Dr. K. Schiffmann and Dr. P. Willich for AFM and SIMS investigations, respectively, I. Hippel, T. Höing, and N. Nissen for technical assistance, and Dr. Spindler (Siemens AG) for polymer coated silicon wafers. This work was supported by the BMBF under grant no. 13N6517/1

REFERENCES

1. P. K. Wu, G.-R. Yang, X. F. Ma, and T.-M. Lu, Appl. Phys. Lett. **65**, 508 (1994)
2. E. Sacher, Progr. Surface Sci. **47**, 273 (1994)
3. L. M. Siperko and R. R. Thomas, J. Adhesion Sci. Technol. **3**, 157 (1989)
4. D. R. Wheeler and S. V. Pepper, J. Vac. Sci. Technol. **20**, 442 (1982)
5. Chin-An Chang, Appl. Phys. Lett. **51**, 1236 (1987)
6. R. Michael and D. Stulik, J. Vac. Sci. Technol. **A 4**, 1861 (1986)
7. C. Haag and H. Suhr, Appl. Phys. **A 47**, 199 (1988)
8. A. Weber, A. Dietz, R. Pöckelmann, and C.-P. Klages, Appl. Phys. Lett. **67**, 2311 (1995)
9. A. Weber, R. Pöckelmann, and C.-P. Klages, Appl. Phys. Lett. **67**, 2934 (1995)
10. A. Weber, R. Nikulski, C.-P. Klages, M. E. Gross, W. L. Brown, E. Dons, and R. M. Charatan, J. Electrochem. Soc. **141**, 849 (1994)
11. F. Garbassi, M. Morra, E. Occhiello, "Polymer Surfaces", p. 341 (John Wiley, New York, 1994)
12. H. Schonhorn and F. W. Ryan, J. Adhesion **1**, 43 (1969)

Metallized Plastics 5&6: Fundamental and Applied Aspects, pp. 51-59.
K.L. Mittal (Ed.)

DEPOSITION OF TRIBOLOGICAL HARD COATINGS ON PLASTICS BY REACTIVE GAS FLOW SPUTTERING

T. Kälber and T. Jung

Fraunhofer-Institute for Thin Films and Surface Engineering
Bienroder Weg 54 E
D-38108 Braunschweig, Germany

A new sputter deposition process is described which enables the coating of polymers with metallic and compound wear resistant thin films at a low vacuum level (10–100 Pa) and at rates higher than standard magnetron sputtering. Since no high vacuum equipment is needed, this process represents an inexpensive alternative to traditional coating techniques. We discuss the substrate heating mechanisms in the gas flow sputtering process and demonstrate how substrate heating can be effectively reduced. Coatings of titanium and chromium as well as their oxides, nitrides and carbides have been deposited successfully on various plastics such as PA (polyamide) and PEEK (polyetherether-ketone). Most promising for many tribological applications are metal-containing hydrogenated carbon films (Me-C:H). Thin film adhesion was assessed with a z-axis pull test and on both PA and PEEK samples very good adhesion was achieved (~20 N/mm^2).

INTRODUCTION

The metallic coating of polymers for decorative, electronic and other applications has received considerable attention over the past few years[1-4]. However, the use of compound thin films on plastics in order to improve their tribological properties has not been studied as extensively. Plastics often have advantages as compared to metals when used as materials for construction parts, among them are low cost, light weight and the relative easy fabrication of complex geometries.

However, their application is often limited due to the surface wear of the polymer, and in the case of fiber reinforced plastics also due to the wear of the tribological counterpart. Such limitations can be overcome by the plasma assisted deposition of hard coatings [5,6].

The best quality wear resistant thin films are usually obtained by sputtering processes, followed by arc, evaporation and chemical vapor deposition techniques. In general, however, these techniques exhibit a relatively low productivity and high investment costs. Our aim was, therefore, to develop a deposition process which produced hard and wear resistant coatings of physical vapor deposition quality but with lower cost. This is done by increasing the deposition rate and thus the productivity, and by reducing the vacuum requirement and thus the equipment costs.

THE REACTIVE GAS FLOW SPUTTERING PROCESS

In the reactive gas flow sputtering (RGFS) process, the metal is sputtered from the targets in a hollow cathode d.c. glow discharge and subsequently transported to the substrate by the flow of the inert working gas argon (Fig. 1)[7,8]. For an intense and stable hollow cathode glow discharge a working pressure of 10 to 100 Pa has proved as an optimum.

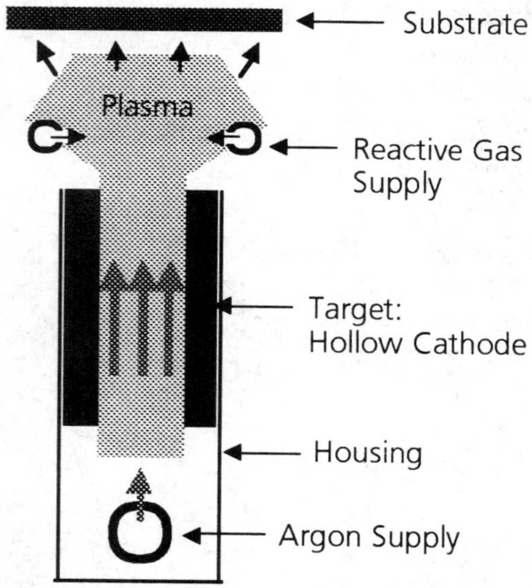

Figure 1. Schematic set-up of the reactive gas flow sputtering process.

The deposition rate is related to the cathode geometry and increases, in a wide interval, linearly with the gas flow velocity and the discharge power. To realize a high throughput of gas, a combination of a roots pump and a one-stage rotary vane pump is used. Due to the increase of the charge carrier density and, consequently, of the discharge power by the hollow cathode effect, inexpensive d.c. excitation can be used without any support by magnetic fields. Figure 2 shows a linear gas flow sputtering source of 10 cm length in operation. We have studied linear sources of up to 47 cm active target length and it seems that there is no limit on linear upscaling.

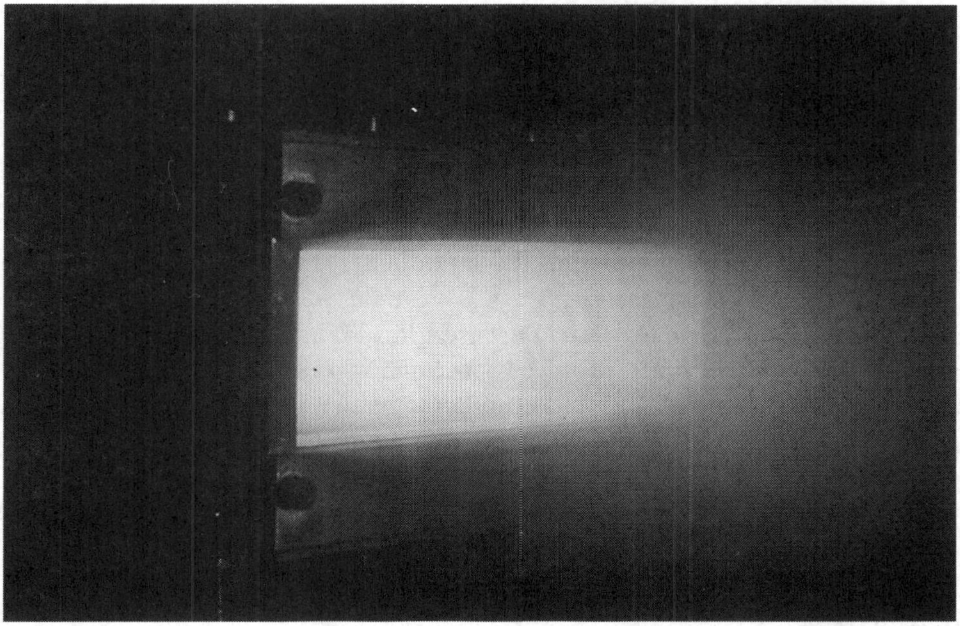

Figure 2. Linear gas flow sputtering source in operation.

A reactive gas, if required, is injected between the sputtering source and the substrate (Fig. 1). Due to the argon counter-flow the penetration of the reactive gas into the source is strongly suppressed. In this way, its reaction with the target surface is reliably suppressed, which is difficult to avoid in standard magnetron sputtering. Target surface contamination would result in a considerable decrease of the deposition rate and, depending on the target material and the reactive gas, also in discharge instabilities (arcing).

ANALYTICAL TECHNIQUES

Thin film adhesion was measured with a z-axis pull test, according to the European Standard EN 582. In this test a pull stud of 7 mm^2 contact area is glued onto the coated substrate. The stud is then pulled off perpendicularly from the substrate with a slowly and steadily increasing force. The force is constantly

Monitored and the critical value at which interfacial failure between the substrate and film occurs is used to represent the thin film adhesion in terms of critical force per contact area. Generally very good adhesion was achieved, even without prior plasma activation of the polymers. Typical values were around 20 N/mm².

The abrasive wear of the hard coatings was examined by the CALOTEST® (CSEM, Neuchâtel, Switzerland). In this test a rotating steel ball of 30 mm diameter grinds with 300 rpm and a load of 0.5 N into the surface of the coating. A suspension of 1 μm Al_2O_3 particles in glycerol is used as the abrasive. The abrasive wear rate is expressed in terms of worn volume (volume of the cap of the sphere ground into the coating) divided by the load on the ball and the total sliding distance (i.e. in m^3/Nm).

A ball-on-disc test was used to measure the coefficient of friction. For standard testing a 100Cr6 steel ball of 4.6 mm diameter slides in circular motion over the coated substrate at a typical load of 1 N. No lubricant or abrasive was used in these tests. The ball-on-disc test was also used to obtain further insight into the wear behavior of the hard coatings. It gives information mainly about adhesional wear, typical to many applications, and thus complements the results on abrasive wear of the CALOTEST®.

The atomic composition of the coatings was examined by Electron Probe Micro Analysis (EPMA).

HARD COATINGS ON PLASTICS

We have deposited successfully a whole range of hard coatings with the RGFS process on various plastic substrates. Among the substrates used were PA (polyamide), PEEK (polyetheretherketone), EP (epoxy resin), and PF (phenol formaldehyde resin). Coatings were metallic, oxide, nitride, and carbide films of titanium and chromium. Most of these films show good wear resistance, but also a rather high coefficient of friction and high counterbody wear.

The best choice for many tribological applications is, therefore, metal-containing amorphous carbon films (Me-C:H)[9,10]. They not only exhibit low surface wear, but also low counterbody wear and a low coefficient of friction (generally below 0.2). Coatings of Ti-C:H were deposited successfully with the reactive gas flow sputtering process on the above polymers, using a 100 to 200 nm Ti-interlayer to improve adhesion.

Figure 3 shows the dependence of abrasive and adhesional wear on the metal content of Ti-C:H films. Between 10 and 40 at.% Ti the wear rate of the Ti-C:H films decreases with increasing Ti-content. Films with the best overall properties have been obtained in the range between 20 and 40 at.% titanium, depending on the choice of the reactive gas and other process parameters. The values shown in Figure 3 refer to films which have been deposited without bias and with ethine as reactive gas.

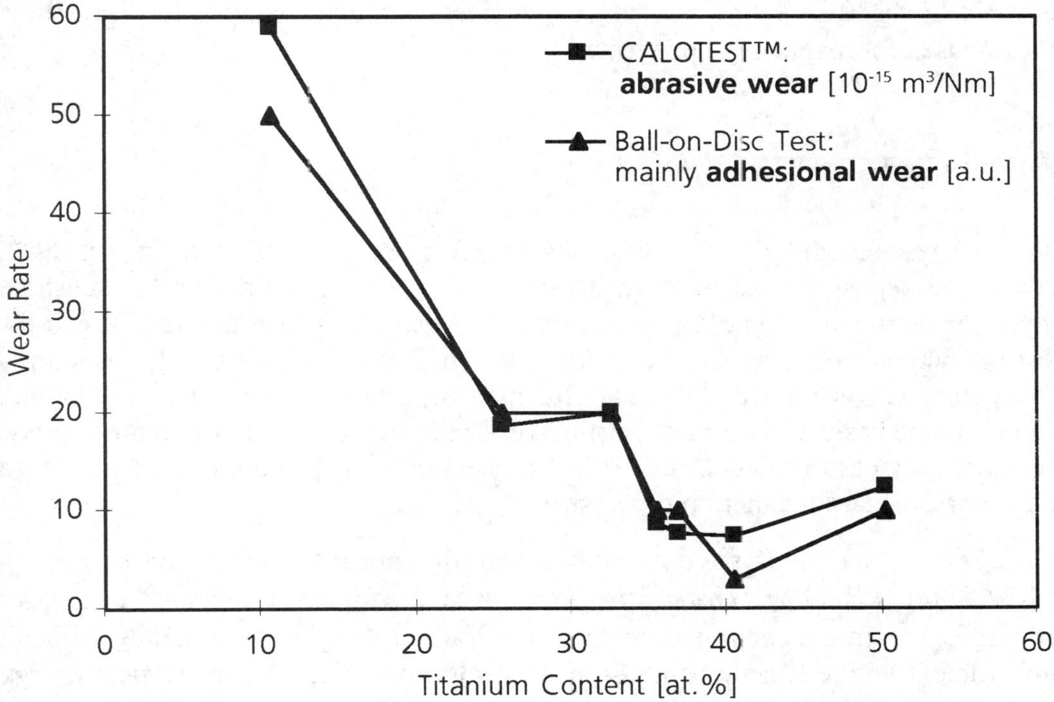

Figure 3. Abrasive and adhesional wear of Ti-C:H films deposited without bias on polyamide with the RGFS process. For details about measurement setup and parameters refer to the Analytical Techniques section.

Ti-C:H films deposited with a d.c. bias on the Ti-interlayer show a similar dependence of wear rate on Ti-content as illustrated in Figure 3. The use of substrate bias, however, leads to a substantial increase in substrate heating and thus reduces the maximum possible deposition rate. In addition, substrate bias would be difficult to apply when coating bulk material. Our aim was, therefore, to optimize film quality without employing a substrate bias. The fact that we could obtain rather good quality film even without substrate bias can be mainly attributed to the rather high plasma density of the hollow cathode discharge and the resulting intense ion flux to the growing film. Values of the plasma density as derived from Langmuir measurements were in the range of 10^{18} m^{-3} to 10^{20} m^{-3}, depending on discharge parameters. Electron temperatures were around 0.2 to 1 eV which correspond to ion energies of approximately 0.8 to 4 eV of the ions impinging on a floating substrate.

The wear rates of Ti-C:H films were further reduced by using methane instead of ethine as the reactive gas[11,12]. The improvement of film quality when using methane is also reflected in a more compact film morphology, as observed by SEM. This could be explained by a tendency of ethine to polymerize already in the gas phase after plasma excitation: this phenomenon would be more pronounced at the elevated pressures of the RGFS process, in comparison with the relatively low pressures typically used for magnetron sputtering.

SUBSTRATE HEATING

To optimize the productivity of the RGFS process, we tried to find means to increase the deposition rate without increasing the heat flux and thus the substrate temperature at the same time. Or, to put it another way, to reduce the heat flux to the substrate without reducing the deposition rate. The latter is the case, e.g., if simply discharge power is reduced. For a well-aimed reduction of the substrate heating, insight into the basic heating mechanisms is essential. We carried out Langmuir probe and heat flux measurements in the RGFS process and compared the results to those obtained in standard magnetron processes.

The evaluation of the data showed that the substrate heating mechanisms in the RGFS process differ substantially from those predominant in typical magnetron discharges[12,13]. In magnetron processes substrate heating occurs mainly through bombardment with energetic neutrals and fast electrons (for data on particle energy and thermalization see, e.g., references 14 and 15). Energetic neutrals are neutralized ions reflected at the target surface. Fast electrons are secondary electrons emitted from the target surface and accelerated across the sheath, which are able to escape from areas where the magnetic field is perpendicular to the target surface. Heat input from the plasma plays only a minor role in magnetron discharges, except in cases where very high bias voltages are used.

In the RGFS process fast electrons are trapped inside the hollow cathode where they oscillate between the targets and lose their kinetic energy by inelastic collisions. Similarly, there are no energetic neutrals striking the substrate, as they become thermalized within a few centimeters . Due to the high working pressure (10 to 100 Pa) and the very effective hollow cathode excitation, plasma densities typically found in the RGFS process are around two orders of magnitude higher than those in standard magnetron sputtering. That is why in the RGFS process the plasma plays an important role in heating the substrate. A major contribution to the plasma heating comes from the ion-electron recombination resulting from the constant flux of charged particles to the floating substrate. In the case of bias deposition the contribution of the kinetic energy of ions striking the substrate surface can exceed considerably the recombination energy and cause important additional substrate heating. Therefore, we have concentrated on optimizing film quality without applying a substrate bias, as outlined above.

The second major contribution to the total heat flux in the RGFS process is convectional substrate heating. This might sound surprising at first, since in the conventional high vacuum glow discharges neither gas heating nor convectional heat transport plays an important role[12,13]. At the elevated pressures of the RGFS process, however, a substantial heating of the working gas occurs in the hollow cathode. The working gas is heated, in part, by the thermalization of energetic neutrals and sputtered atoms. Yet the biggest contribution comes from inelastic collisions of accelerated ions in the target sheath[12,13].

IMPROVEMENT OF PROCESS PRODUCTIVITY

In the case of static, continuous deposition, the ratio of deposition rate to heat flux in the RGFS process can be improved by two means:

1. The deposition rate increases with decreasing process pressure, while substrate heating decreases at the same time. Therefore, process pressure should be chosen as low as possible, just high enough to realize the desired discharge power.
2. The deposition rate increases, in a wide interval, linearly with the gas flow velocity inside the hollow cathode, while substrate heating is hardly affected. Since the gas flow velocity is proportional to the pumping capacity (not to the gas flow itself), this fact should be considered when a new coating machine is designed.

Changes of the deposition arrangement, e.g. the use of straight or L-shaped canals between the sputtering source and the substrate have not been shown to result in a higher ratio of deposition rate to heat flux. The reason is the strong loss of sputtered material due to the coating of the inner canal walls, while the gas temperature decreases only slowly. The decrease of the sputtered material and the plasma density was similar in these experiments.

The changes in the discharge power result in almost proportional variation in both heat flux and deposition rate. The discharge power is adjusted in the way that the substrate temperature does not exceed its critical value during the deposition process. The critical heat flux, however, depends, among other things, on the substrate geometry (especially the substrate thickness; Fig. 4). In the case of a thermally isolated substrate, substrate cooling occurs almost exclusively through radiation. If the surface of such a substrate is exposed to a constant heat flux, it heats up until it reaches an equilibrium temperature, as given by the Stefan-Boltzmann equation. The maximum heat flux which a very thick substrate can bear is only about half of that of a very thin substrate. This is because the total radiating surface of a very thin substrate (front and back) is about twice as large as its heat collecting surface (mainly front).

Due to the heat transport into the bulk of thick substrates, however, they reach their equilibrium temperature much later than their thin counterparts.

A further increase in productivity of the plastic coating process, besides the measures outlined above, can be achieved by substrate rotation and/or discontinuous deposition. This allows to increase the heat flux (and thus the deposition rate) to the substrate or the substrate part being coated, while other substrates or concealed substrate parts cool down through radiation.

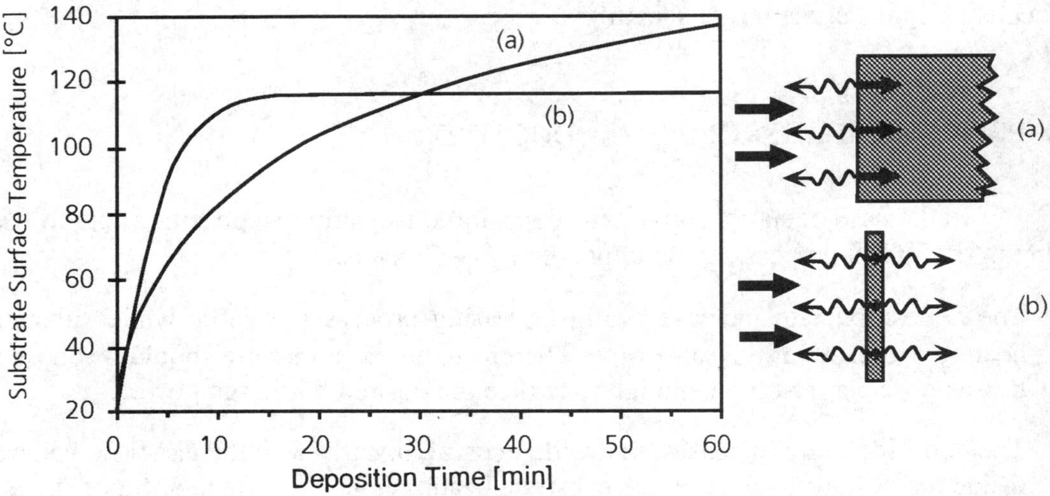

Figure 4. Heating of a PA66 substrate with a constant heat flux of 150 mW/cm^2 (calculated). (a): thick substrate (semi-infinite solid: thermal relaxation time >> deposition time, i.e. substrate thickness d >> 23 mm for a deposition time of 40 min); (b): thin substrate ($d \approx 4$ mm). Thick arrows indicate heat flux, undulating arrows stand for radiation.

CONCLUSION

The RGFS process has been shown to be a highly suitable means to coat plastics with metallic or wear resistant hard coatings. As deposition rates are higher and equipment costs lower than with, e.g., standard magnetron sputtering, the RGFS process represents an inexpensive alternative to conventional sputtering techniques.

ACKNOWLEDGMENTS

The financial support of the BMBF under grant No. 13N6466 is gratefully acknowledged.

REFERENCES

1. K.L. Mittal and J.R. Susko (Eds.), *"Metallized Plastics 1: Fundamental and Applied Aspects,"* Plenum Press, New York, 1989.
2. K.L. Mittal (Ed.), *"Metallized Plastics 2: Fundamental and Applied Aspects,"* Plenum Press, New York, 1991.
3. K.L. Mittal (Ed.), *"Metallized Plastics 3: Fundamental and Applied Aspects,"* Plenum Press, New York, 1992.
4. K.L. Mittal (Ed.), *"Metallized Plastics: Fundamental and Applied Aspects,"* Marcel Dekker, 1998.
5. O. Knotek and F. Löffler, in *"Metallized Plastics 2: Fundamental and Applied Aspects,"* K.L. Mittal (Ed.), p. 141, Plenum Press, New York, 1991.
6. C.T. Wan, K.A. Taylor, D.L. Chambers, and G.T. Susi, in *"Metallized Plastics 2: Fundamental and Applied Aspects,"* K.L. Mittal (Ed.), p. 81, Plenum Press, New York, 1991.
7. T. Jung and A. Westphal, *Mater. Sci. Eng. A*, **140**, 528 (1991).
8. T. Jung and A. Westphal, *Surface Coating Technol.*, **59**, 171 (1993).
9. H. Dimigen, H. Hübsch, and R. Memming, *Appl. Phys. Lett.*, **50**, 1056 (1987).
10. H. Dimigen and C.-P. Klages, *Surface Coating Technol.*, **49**, 543 (1991).
11. T. Kälber and T. Jung, *Surface Coating Technol.*, in press.
12. T. Kälber, *"Hohlkathoden-Gasflußsputtern zur Verschleißschutzbeschichtung von Kunststoffen,"* Dissertation TU Braunschweig, Fraunhofer IRB Verlag, Stuttgart, to be published.
13. T. Kälber and T. Jung, in *"Metallized Plastics: Fundamental and Applied Aspects,"* Marcel Dekker, New York, 1998.
14. H. Windischmann, *Crit. Rev. Solid State Mater. Sci.*, **17** (6), 547 (1992).
15. R.E. Somekh, *J. Vac. Sci. Technol. A*, **2**, 1285 (1984).

Metallized Plastics 5&6: Fundamental and Applied Aspects, pp. 61-73.
K.L. Mittal (Ed.)

SUBSTRATE HEATING DURING SPUTTER COATING OF POLYMERS

T. Kälber and T. Jung

Fraunhofer-Institute for Thin Films and Surface Engineering
Bienroder Weg 54 E
D-38108 Braunschweig, Germany

Based on data from Langmuir probe and optical emission spectroscopy (OES) measurements the contributions of different substrate heating mechanisms to the total heat flux are discussed for two sputtering processes: gas flow sputtering (GFS) and magnetron sputtering. The variation of these contributions with deposition parameters shows ways to reduce the heat flux to the substrate. In the GFS process the ratio of deposition rate to heat flux could be greatly increased by a proper choice of process parameters, notably process pressure. The main contribution to substrate heating in a typical GFS process results from the recombination of electron-ion pairs and convection heating by the hot process gas. In unbalanced magnetron sputtering of heavy target atoms most of the heat input into the substrate surface comes from reflected, energetic neutrals and high energy electrons.

INTRODUCTION

Polymers are employed today in a continuously growing number of applications. The outstanding characteristics, which make them an attractive choice of material, include their low cost, low density, and excellent mouldability. However, their surface properties often require modification in order to meet the demands of specific applications. Among the properties often sought to be improved are their wear and friction, electrical conductivity, optical appearance, barrier properties, chemical resistance, weathering resistance, resistance against ionizing radiation, biomedical properties, wettability and adhesion.

Some of these properties can be achieved by lacquering, electrochemical coating or simply surface modification. However, many of these properties require the deposition of thin solid films using a deposition process which involves energetic particle bombardment of the growing film. This is the case in sputtering processes.

A main drawback of sputtering processes – especially for the coating of inexpensive polymeric parts – is their relatively high cost. In order to reduce deposition costs and make sputtering processes more attractive for widespread industrial use in coating polymers, it is essential to increase process productivity. The productivity is mainly determined by the deposition rate which, in turn, is often severely limited when coating heat sensitive polymers: In many cases process power and thus deposition rate have to be reduced drastically in order to prevent overheating of the polymer substrate.

A useful parameter to compare and optimize deposition processes with respect to their productivity is the ratio of deposition rate to heat flux. A knowledge of the different substrate heating mechanisms is crucial in order to increase this ratio. The information about substrate heating mechanisms allows a proper choice of deposition technique and process parameters. It also helps in the modification of existing techniques or the development of new deposition processes. It is important to distinguish between the particle bombardment which is essential for thin film formation, and the bombardment which only heats the substrate. An optimum arrangement would reduce the latter to a minimum and allow the former only as much as necessary to obtain the desired thin film properties.

The substrate heating mechanisms and their variation with process parameters will be discussed for two different sputtering processes. The main focus of this work has been on the gas flow sputtering (GFS) process. In the case of reactive sputtering it represents an interesting high rate alternative to traditional sputtering techniques, such as magnetron sputtering [1-3]. For comparison, substrate heating in a magnetron sputtering process will be discussed. Due to the better reliability of Langmuir probe measurements, both sputtering techniques have been investigated in a metal deposition process (no reactive gas).

PROBE MEASUREMENTS

The heat flux to the substrate was measured through the heating of a thermally isolated copper block which was part of the substrate holder. The plasma density n_i, electron temperature T_e, and plasma and floating potentials were determined using cylindrical Langmuir probes. The temperature of the argon sputtering gas was estimated by measuring the molecular rotational temperature in a pure nitrogen discharge under otherwise identical conditions. The molecular rotational temperature was determined by comparing unresolved N_2 and N_2^+ molecular bands in optical emission spectra with simulated rotational band spectra. For further details concerning these measurements refer to ref. 4.

GAS FLOW SPUTTERING

The GFS source consists of a hollow cathode dc glow discharge, working at pressures of several tens Pascal. A voltage of 400-800 V dc is applied to the two targets facing each other (Figure 1). The oscillation of the fast primary electrons between the targets leads to a very high plasma density and therefore to an efficient sputtering of the target material. The inert working gas (argon) flowing through the hollow cathode both transports the sputtered material towards the substrate and prevents the reactive or residual gas from penetrating into the source. The hollow cathode used for this work measured 60 mm in depth, 250 mm in length, and had a target-to-target distance of 30 mm. The substrate-to-source distance was kept at 50 mm.

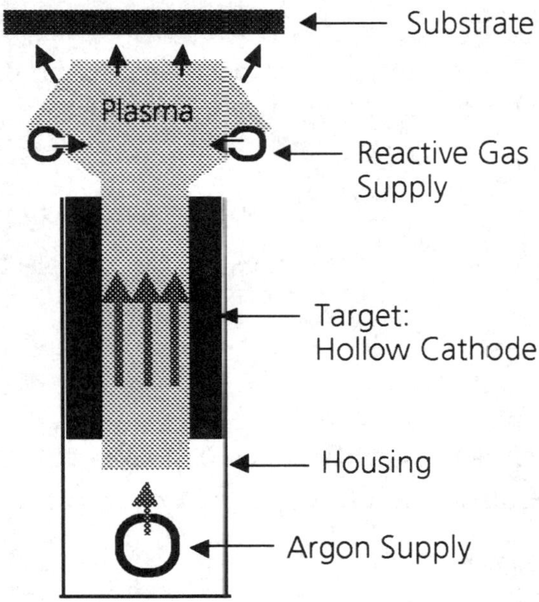

Figure 1. Set-up of the reactive gas flow sputtering process (schematic).

In the following, the different substrate heating mechanisms will be discussed in their order of importance for the gas flow sputtering process.

Heat of Recombination

With plasma densities of several 10^{18} m^{-3} the heat flux J_{rec} delivered to the substrate by the recombination of electron-ion pairs is rather important. It is assumed that about 90% of the ionisation energy $E_i = 15.6$ eV is deposited as heat onto the substrate, which leads to

$$J_{rec} \approx 0.9 \cdot \frac{j_i}{e} \cdot E_i \tag{1}$$

The incident ion current density j_i scales with the plasma density and the square root of the electron temperature; e denotes the elementary charge.

Convection Heating

There are several mechanisms responsible for heating the gas inside the hollow cathode: the dissipation of sputtered atom and energetic neutral kinetic energy and collisions of ions in the target sheath. Energy transfer through elastic collisions with thermal electrons plays only a negligible role. Gas rotational temperatures, as determined by optical emission spectroscopy (OES) in a pure nitrogen discharge, rise with increasing discharge power and are of the order of 700 K for a moderate power of 1.4 kW.

The resulting heat flux J_{conv} to the substrate is proportional to the difference between the gas temperature T_g and the substrate temperature T_s, i.e.

$$J_{conv} = \alpha \cdot (T_g - T_s) \tag{2}$$

The heat transfer coefficient α varies with the process pressure p and the gas flow velocity u as [5]:

$$\alpha \sim \sqrt{p \cdot u} \tag{3}$$

For $p = 30$ Pa and $u = 27$ m/s, α can be estimated [5] as 0.85 Wm^{-2}K^{-1}. At a gas temperature of 700 K this would result in a heat flux J_{conv} of 35 mW/cm^2.

Radiant Heating

The heat flux values presented in this paper have been measured while sputtering from indirectly cooled titanium targets. The targets were screwed tightly onto the water cooled backing plate with a thin carbon foil in between to improve heat transfer. At 4 kW discharge power the target surface reached a temperature of approximately 900 K, giving rise to a substantial substrate heating via radiation. The heat flux resulting from radiant heating was measured with a modified heat flux probe which was covered with a thin pane of quartz glass. Taking into account the spectral transmission of the quartz glass, these measurements yielded a radiant heating of 200–350 mW/cm^2 at 4 kW discharge power. Heat transfer calculations gave corresponding results. The substrate heating through radiation from the plasma is negligible compared with the heat fluxes from thermal radiation [6].

High Energy Electrons

Most of the secondary electrons produced at the cathode by ion impact and accelerated across the target sheath oscillate between the targets of the hollow cathode and contribute to the ionisation of the process gas. However, some of these high energy electrons can be reflected towards the substrate at outer target corner. They lose some of their energy via inelastic collisions on their way to the substrate. This energy loss has been calculated based on ionisation cross section data from [7,8] and an energy loss of 26 eV per ionisation process [9]. Its variation with process pressure is illustrated in Figure 2.

Ion Kinetic Energy

Due to the high plasma density of the gas flow sputtering process, a high film quality can often be achieved without applying an external substrate bias [10]. In a deposition process with electrically floating substrate, the plasma ions are accelerated onto the substrate by the difference between plasma and floating potentials $U_p - U_f$. The resulting heat flux is

$$J_i = j_i \cdot (U_p - U_f) \sim n_i \cdot T_e^{3/2} . \qquad (4)$$

At a typical ion current density j_i of 5 mA/cm^2 and $U_p - U_f \approx 1$ V eq. 4 yields a heat flux of 5 mW/cm^2. If a substrate bias is used, the term $(U_p - U_f)$ in eq. 4 has to be replaced by the bias voltage. This can lead to a significant increase in substrate heating. All the heat flux values presented in this paper refer to the sputter deposition of metal films without a substrate bias.

Heat of Condensation

For titanium the heat flux resulting from the heat of condensation is given by

$$J_{cond} = 11.6 \frac{\mathrm{W}}{\mathrm{m}^2} \cdot \frac{r}{\mu\mathrm{m/h}} \qquad (5)$$

where r is the deposition rate in µm/h. At 1.4 kW discharge power and a corresponding titanium deposition rate of 2 µm/h, the heat of condensation gives rise to a substrate heating of 2 mW/cm^2.

Kinetic Energy of Thermal Electrons

The heat flux J_e resulting from the kinetic energy of thermal electrons capable to overcome the potential difference $\Delta U \approx U_p - U_f$ of a floating substrate is given by

$$J_e = \frac{j_i}{e} \cdot \frac{\int_{e\Delta U}^{\infty} (E - e\Delta U) \cdot f(E)\ dE}{\int_{e\Delta U}^{\infty} f(E)\ dE}. \tag{6}$$

With $\Delta U \approx 1$ V, $j_i \approx 5$ mA/cm^2 and a Maxwellian electron energy distribution with $T_e \approx 0.25$ eV eq. 6 yields a heat flux J_e of 1.5 mW/cm^2.

Substrate Heating in a Typical GFS Process

In a typical low-power GFS process (1.4 kW), as employed for the coating of heat sensitive substrates, the heating mechanisms discussed above contribute to the total substrate heat flux (140 mW/cm^2) as follows:

- Heat of recombination: ~ 48 mW/cm^2
- Convection heating: ~ 43 mW/cm^2
- Radiant heating: ~ 26 mW/cm^2
  (only for hot, indirectly cooled targets)
- High energy electrons: ~ 15 mW/cm^2
- Ion kinetic energy: ~ 5 mW/cm^2
- Heat of condensation: ~ 2 mW/cm^2
- Kinetic energy of thermal electrons: ~ 1.5 mW/cm^2

Variation of Substrate Heating with Discharge Parameters

Figures 2 to 4 depict the variation of the different substrate heating mechanisms with discharge parameters. The graphs are based on the above calculations together with data from Langmuir probe measurements (yielding n_i, T_e, and j_i), OES (yielding T_g) and heat flux measurements. The calculated total heat flux (single solid line) is in excellent agreement with the measured heat flux (solid squares). It is of the same magnitude and shows the same variation with discharge parameters.

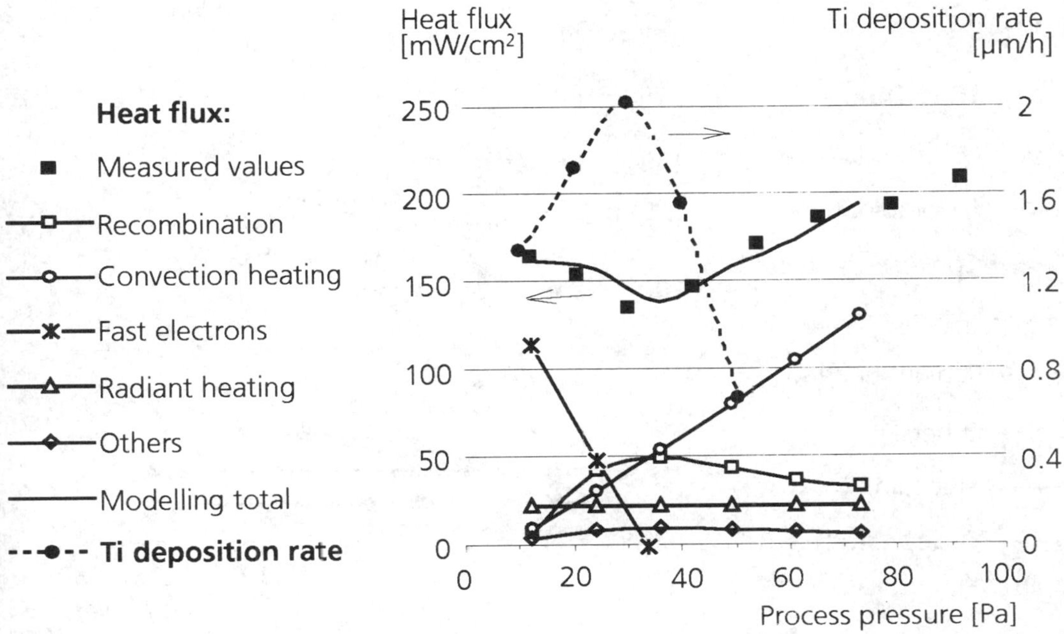

Figure 2. Variation of the heat flux to the substrate and deposition rate with process pressure during gas flow sputtering of titanium (1.4 kW discharge power; constant pumping speed; 18 m/s mean gas flow velocity).

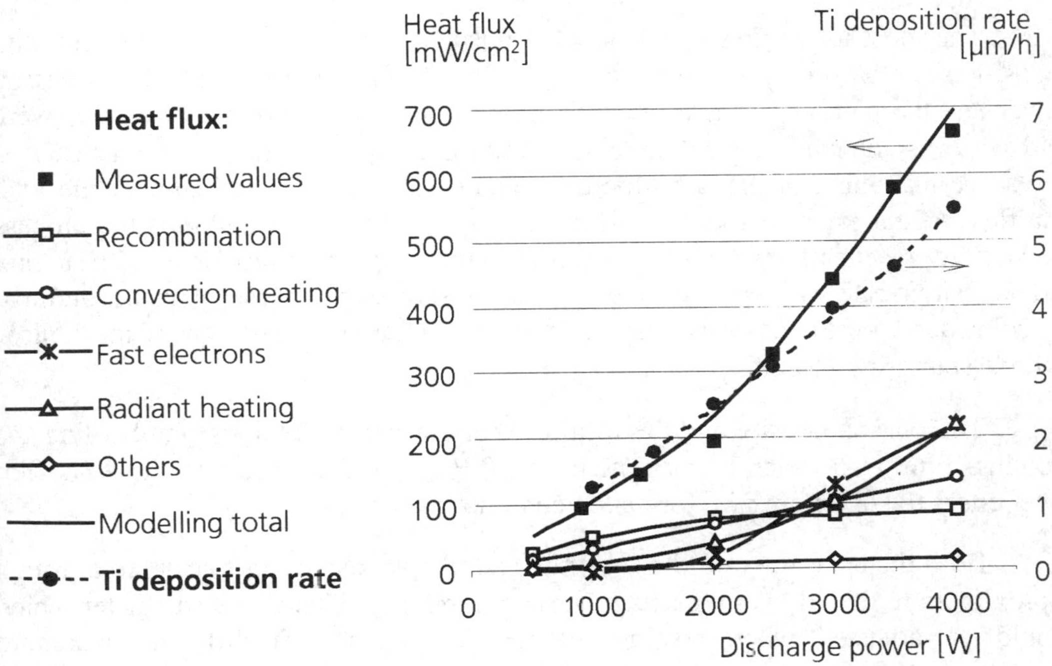

Figure 3. Variation of the heat flux to the substrate and deposition rate with discharge power during gas flow sputtering of titanium (30 Pa process pressure; full pumping speed; 27 m/s mean gas flow velocity inside the hollow cathode).

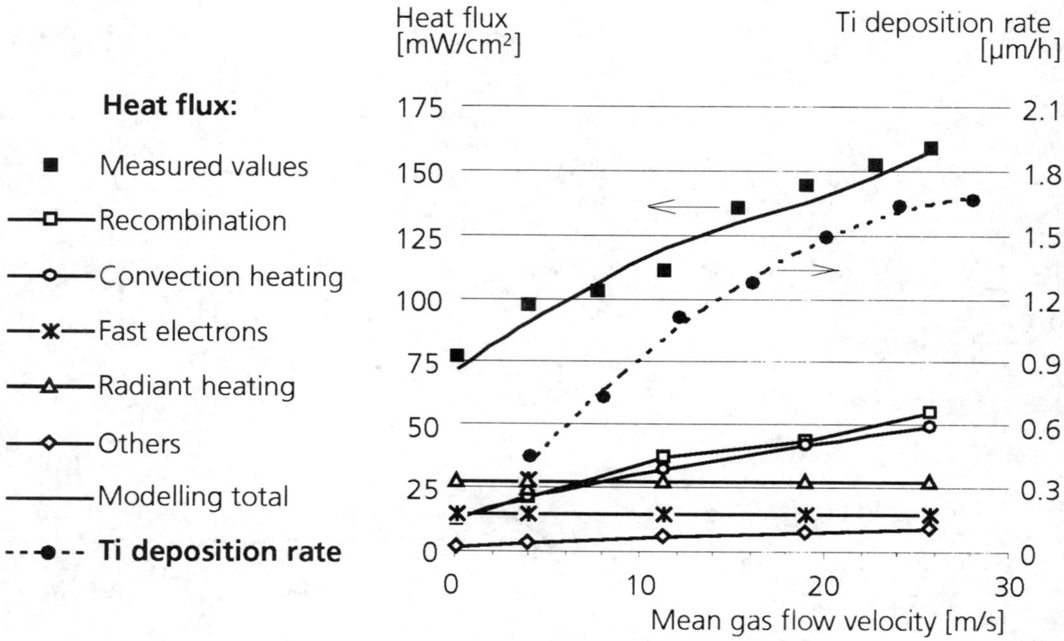

Figure 4. Variation of the heat flux to the substrate and deposition rate with the mean gas flow velocity inside the hollow cathode (gas flow sputtering of titanium; 1.4 kW discharge power; 30 Pa process pressure).

The most interesting parameter dependence is the variation of heat flux and deposition rate with process pressure (Figure 2). For these experiments the discharge power and the pumping speed (which is proportional to the gas flow *velocity*) were held at a constant level, while the pressure was varied by changing the argon flow. At 30 Pa the maximum of the deposition rate coincides with the minimum of the total heat flux. At this pressure the ratio of deposition rate to heat flux and thus the process productivity reaches a maximum. It should be noted that the variations of heat flux and deposition rate with process pressure are based on different mechanisms, and that the coincidence of the corresponding extremes depends on the design of the hollow cathode sputtering source.

The plasma density and the resulting current density of ions bombarding the growing film also reach a maximum at 30 Pa. At this pressure the thin film experiences the optimum film formation conditions.

Both the heat flux to the substrate and the deposition rate scale, to a rough approximation, linearly with discharge power (Figure 3). This is the parameter which should be adjusted when coating different polymers with different maximum allowable heat fluxes.

In the GFS process, the transport of sputtered material to the substrate is convection driven, and the deposition rate varies, in a wide interval, linearly with the gas flow velocity inside the source (Figure 4). Above a mean gas flow velocity of about 25–30 m/s this increase reaches a saturation. The heat flux, however, keeps increasing with the gas flow velocity. Therefore the ratio of deposition rate to heat flux goes through a maximum, which is located at a mean gas flow velocity of about 15–20 m/s. An increase of the pumping speed in order to attain higher gas flow velocities, and thus higher deposition rates, is therefore practical, but only for velocities below this maximum.

MAGNETRON SPUTTERING

The main difference between particle bombardment encountered with magnetron and gas flow sputtering is a higher flux of directed, energetic particles from the magnetron target to the substrate. This results from the different target arrangement and the reduced thermalization of energetic particles due to the lower process pressure. Process pressure, as well as plasma density, are about two orders of magnitude lower than in the GFS process. Hence the heat of recombination plays only a minor role for the substrate heating in magnetron processes. Gas temperature is also lower than with gas flow sputtering, and its contribution to the total substrate heating is negligible due to the low process pressure.

In the following only the heating mechanisms differing from those already mentioned in the GFS Section above will be discussed. The examples given refer to the sputtering of tungsten with an unbalanced magnetron of approximately the same size as the above hollow cathode.

Energetic Neutrals

A fraction R of the incident ion flux is neutralized and reflected at the target. After the reflection the particles still have a significant fraction γ of their incident energy. The particle reflection coefficient $R \sim (m_t/m_g)^2$ depends strongly on the target-to-gas mass ratio m_t/m_g and amounts to approximately 0.17 for tungsten [11]. The energy reflection coefficient $\gamma \sim (m_t/m_g)^{1/3}$ shows a weaker dependence on the target atom mass and comes to about 0.25 for tungsten [11]. The variation with the target-to-gas mass ratio of the heat flux generated by energetic neutrals is illustrated in Figure 5.

On their way to the substrate the energetic argon neutrals lose a fraction β_{ar} of their initial energy E_0 through collisions with gas atoms. According to a fit on data calculated by Somekh [12] the thermalization coefficient β_{ar} for energetic argon atoms in argon (for E > 1.7 eV) can be written as

$$\beta_{ar} \equiv \frac{E}{E_0} \approx \exp\left(-1.05 \cdot 10^{-2} \frac{p \cdot d}{\text{Pa} \cdot \text{mm}}\right) \tag{7}$$

With a process pressure p of 0.37 Pa and a substrate-to-target distance d of 220 mm eq. 7 yields $\beta_{ar} \approx 0.43$.

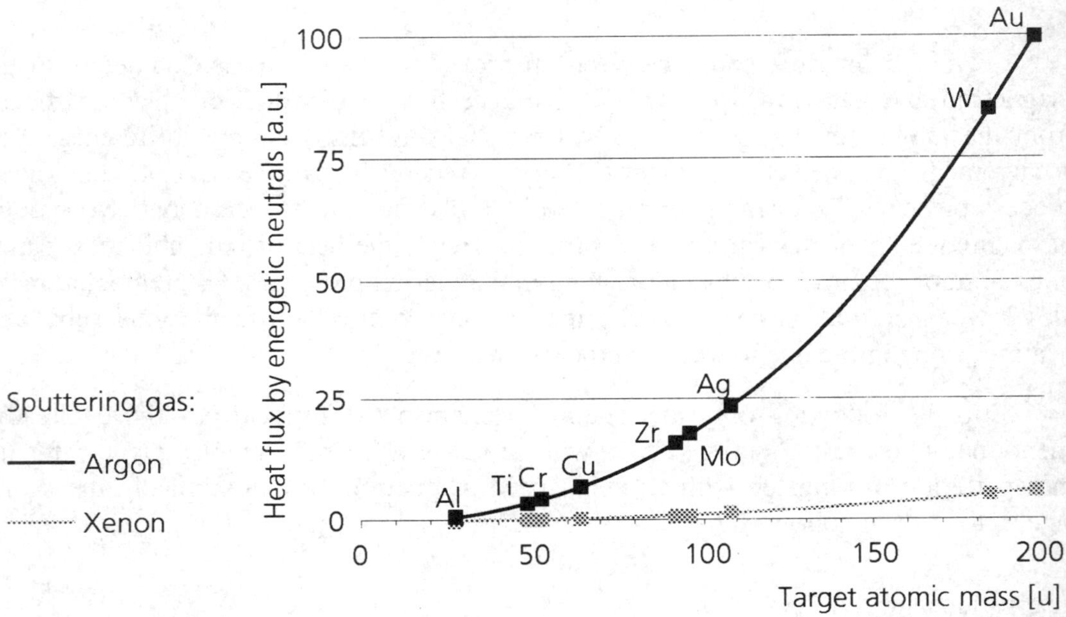

Figure 5. Heat flux to the substrate generated by energetic neutals and its variation with the target-to-gas mass ratio (magnetron sputtering of tungsten).

The energetic neutrals deposit a large fraction η of their incident kinetic energy into the substrate surface. In the present case, the energy transfer coefficient η amounts [11] to about 0.85.

All the processes discussed sum up to a linear variation of the heat flux J_{ar} caused by energetic neutral bombardment with the target power density P_t/A_t, i.e.

$$J_{ar} \approx \eta \cdot \beta_{ar} \cdot R \cdot \gamma \cdot \frac{P_t}{A_t} \tag{8}$$

In the present case, eq. 8 yields a heat flux of approximately 95 mW/cm^2, which amounts to about 70% of the total measured heat flux.

High Energy Electrons

Especially when unbalanced magnetrons are employed [13-15], a notable fraction δ of the high energy secondary electrons can escape from their magnetic field confinement in front of the targets and cause substantial substrate heating. The corresponding secondary electron emission coefficient for ion bombardment γ_i depends only weakly on the ion energy and amounts to approximately 0.1 for argon ions bombarding tungsten [16]. On their way to the substrate the electrons lose only little energy and the corresponding thermalization coefficient β_e is about 0.96–0.98.

The heat flux delivered to the substrate by fast electrons is given by

$$J_e \approx \delta \cdot \beta_e \cdot \gamma_i \cdot \frac{P_t}{A_t} . \tag{9}$$

The escape probability δ depends strongly on magnetic field design and constructional features as, for example, realized in Inset™ cathodes [17].

Sputtered Atoms

Sputtered atoms are emitted from the target with an average kinetic energy E_{me} which is proportional to the target atom mass m_t and which amounts to about 35 eV for tungsten atoms sputtered by 1.2 keV krypton ions [11]. The corresponding heat flux to the substrate is given by

$$J_{me} = r \cdot \frac{\rho}{m_t} \cdot \beta_{me} \cdot E_{me} , \tag{10}$$

where r denotes the deposition rate and ρ the mass density of the growing film. In the present case, with $\beta_{me} \approx 0.25$ [12] and $r \approx 1.7$ µm/h, the resulting heat flux J_{me} totals approximately 4.3 mW/cm^2.

Substrate Heating Encountered with Unbalanced Magnetron Sputtering of Tungsten

The substrate heating mechanisms given below correspond to unbalanced magnetron sputtering from directly cooled, 127 x 254 mm^2 tungsten targets at 2 kW discharge power, a process pressure of 0.37 Pa, and a substrate-to-source distance of 220 mm, without applying an external substrate bias. They are based on data obtained by Langmuir probe, heat flux and deposition rate measurements in conjunction with the above calculations. The total heat flux measured at these process parameters was 130 mW/cm^2. The individual substrate heating mechanisms contribute to the total heat flux in the following way:

- Energetic neutrals: $\leq 95\ mW/cm^2$
- High energy electrons: several 10 mW/cm^2
- Heat of recombination: $\sim 7\ mW/cm^2$
- Sputtered atom kinetic energy: $\sim 4.3\ mW/cm^2$
- Heat of condensation: $\sim 3.9\ mW/cm^2$
- Ion kinetic energy: $\sim 2.4\ mW/cm^2$
- Thermal electron kinetic energy: $\sim 0.8\ mW/cm^2$

In the present case of sputtering the heavy atom material ^{186}W by an unbalanced magnetron, the dominant substrate heating mechanisms are the bombardment by energetic neutrals and high energy electrons. However, in the case of a different target material and magnetron design this may change drastically.

The heat input from energetic neutrals, for instance, can be neglected when sputtering the light atom material ^{27}Al. If the flux of high energy electrons to the substrate is suppressed by constructional means (as in case of the Inset™ magnetron [17]), or by a different magnetic field design, the total substrate heating can be greatly reduced. However, such a procedure also reduces the plasma density at the substrate and hence the particle bombardment of the growing film. It is, therefore, not suitable for the deposition of thin films requiring intensive particle bombardment during growth.

CONCLUSION

The main substrate heating mechanisms in a typical gas flow sputtering process are the heat input from the recombination of electron-ion pairs and the convection heating by the hot process gas. The ratio of deposition rate to heat flux in gas flow sputtering can be greatly increased by a proper choice of process pressure. The optimum pressure for the hollow cathode investigated was found to be 30 Pa. At this pressure the ion current density – an important factor in thin film formation – also reaches its maximum.

In magnetron sputtering of heavy target atoms most of the heat flux to the substrate comes from the kinetic energy of energetic, reflected neutrals. This heat flux can be greatly reduced by using a heavier sputtering gas, like xenon. In the case of unbalanced magnetron sputtering there is also a substantial substrate heating by high energy electrons. The corresponding heat flux can also be reduced by a different magnetic field design or constructional measures. However, such procedures also reduce the ion bombardment of the growing film and are mostly unsuitable for the deposition of hard coatings.

The ratio of deposition rate to heat flux in the two sputtering processes is comparable for the deposition of metals. For reactive sputtering, however, the gas flow sputtering process exhibits a very much higher ratio of deposition rate to heat flux. For example, at a heat flux of 140 mW/cm^2 (maximum substrate temperature 120 °C), the deposition rate of titanium containing amorphous hydrocarbon films (Ti-C:H) was 5.3 µm/h in the GFS process and 1.7 µm/h in a magnetron process of comparable target size. Film properties, as abrasive wear, were comparable in both processes [10]. Due to its high ratio of deposition rate to heat flux, the reactive gas flow sputtering process is particularly suitable for the coating of heat sensitive materials as polymers.

ACKNOWLEDGMENTS

The financial support of the BMBF under grant No. 13N6466 is gratefully acknowledged.

:FERENCES

T. Jung and A. Westphal, *Mater. Sci. Eng. A*, **140**, 528 (1991)
T. Jung and A. Westphal, *Surface Coating Technol.*, **59**, 171 (1993)
T. Jung, T. Kälber, and V. v.d. Heide, *Surface Coating Technol.*, **86-87**, 218 (1996)
T. Kälber, Dissertation TU Braunschweig, Fraunhofer IRB Verlag, Stuttgart, to published
"VDI-Wärmeatlas," p. Gal, VDI-Verlag, Düsserldorf, 1991.
J.A. Thornton, *Thin Solid Films*, 54, 23 (1978).
B. Chapman, "Glow discharge Processes," p. 29, John Wiley & Sons, New York, 1980.
Y.P. Raizer, "Gas Discharge Physics,", p. 52, Springer-Verlag, Berlin, 1991.
I.G. Christophorou, "Atomic and Molecular Radiation Physics," p. 35, Wiley-Interscience, London, 1971.
T. Kälber and T. Jung, *Surface Coating Technol.*, in press.
H. Windischmann, *Crit. Rev. solid State Mater. Sci.*, 17 (6), 547 (1992).
R.E. Somekh, *J. Vac. Sci. Technol.* A, 2 (3), 1285 (1986).
B. Window and N. Savvides, *J. Vac. Sci. Technol.* A, 4, 196 (1986).
B. Window and N. Savvides, *J. Vac. Sci. Technol.* A, 4, 453 (1986).
B. Window, *Surface Coating Technol.*, 71, 93 (1995).
B. Chapman, "Glow discharge Processes," p. 84, John Wiley & Sons, New York, 1980.
W. Class and R. Hieronymi, *Solid State Technol.*, 25 (12), 55 (1982).

Metallized Plastics 5&6: Fundamental and Applied Aspects, pp. 75-84.
K.L. Mittal (Ed.)

HIGH RATE COATING OF PLASTIC FILMS AND PLASTIC SHEETS WITH CLEAR OXIDE LAYERS

S.Schiller, H.Morgner, N.Schiller and S.Straach,

Fraunhofer-Institut Elektronenstrahl- und Plasmatechnik
Zeppelinstraße 1
01324 Dresden, Germany

The high rate deposition of transparent layers onto plastic films and plastic sheets offers a broad application potential. Examples are optical layer stacks on plastics, thin clear Al_2O_3 barrier layers on plastic films for food packaging, and thick clear abrasion resistant layers on films for solar control stacks, solar mirrors, safety glass windows, furniture and other applications. Thick clear SiO_x layers are preferred on plastic sheets for automotive glazing, shop windows, and safety glazing.

Pulse Magnetron Sputtering (PMS) is used wherever high coating uniformity or special layer properties are required. The PMS technology enables a stable and long-term operation of the magnetrons for the deposition of highly insulating layers. Typical applications are solar control and low-emissivity coatings.

Barrier films for packaging applications can be produced by reactive evaporation of aluminum from boat evaporators. To improve the barrier properties, the Magnetron Activated Deposition (MAD) process has been developed. Two magnetrons are operated at a medium frequency supplying a very uniform plasma in the deposition chamber, which makes this technique best suited for coating of wide webs.

Thermoplastics have optical clarity and processing benefits, but lack abrasion resistance. Al_2O_3 as well as SiO_x layers perform very well as clear abrasion resistant layers. The typical layer thickness for these applications ranges from 1 to 4 µm.

For the deposition of dense layer structures at high rates the Hollow Cathode Activated Deposition process (HAD) has been developed. Typical deposition rates are 100-150nm/s for Al_2O_3 and 300-500nm/s for SiO_x

The deposited layers show a glassy, dense, amorphous structure and the microhardness (measured by nanoindentation) amounts to 6GPa for Al_2O_3 layers and 3GPa for SiO_x.

INTRODUCTION

In recent years, the coating of plastic films with layers made of chemical compounds, especially with oxides and nitrides, has increasingly gained in importance. Typical applications are solar control films and low-emissivity films for window panes, antiabrasion layers and solar reflecting films as well as barrier layers for packaging films [1].
All these applications have one thing in common, i.e. the necessity to coat plastic films of more than 1m width at the required quality and rather low cost. Low coating costs call for high web speed and for high transportation speeds in the case of deposition on plastic sheets.
Because of the requirements mentioned above, either the pulsed magnetron sputtering process or plasma activated high rate deposition processes are the methods of choice.
Magnetron sputtering is used wherever high coating uniformity or special layer properties are needed which cannot be obtained by means of high rate evaporation. In all other cases, high rate evaporation will be preferred because of the lower deposition cost.
Typical applications of the high rate evaporation concern the production of barrier layers for packaging films and the deposition of thick, transparent antiabrasion layers [2-4]. In either case, deposition rates between 50 and 1000nm/s are needed for economical reasons whereas a layer thickness uniformity of ±5% to ±10% is usually adequate.

PMS-PROCESS (PULSE-MAGNETRON-SPUTTERING-PROCESS)

For the coating of moving substrates, bipolar technology with a dual magnetron sputtering system is recommended (see Fig.1). Here the two magnetrons alternately act as anode and cathode. In this way a metallic anode is also available over long operation times despite the presence of the reactive gas. If the chosen frequency is high enough, charging on the insulating layers is limited which would otherwise result in troublesome arcing. Typically, pulse generators with a sine wave or a square wave shaped output with a medium frequency of 50-100kHz in bipolar mode are used as power supply.

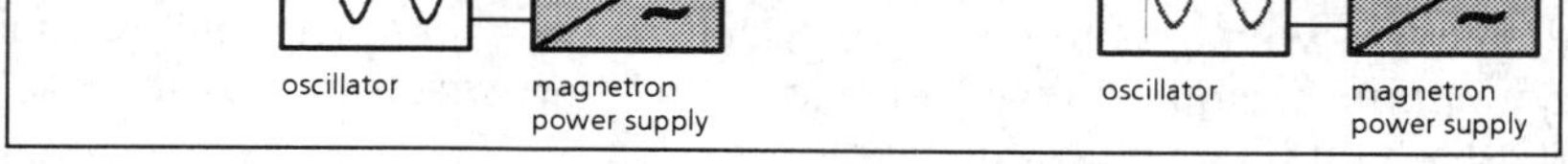

Fig.1. Basic arrangement and time behavior of bipolar pulsed magnetron sputtering (PMS) with a Dual Magnetron System (DMS).

The uniformity of the layer thickness of about 1.5% over the total width and the long operation time make this process a preferable method for the deposition of optical layers on plastic substrates.
The deposited highly insulating oxide or nitride layers are practically absorption free. Because of the excellent layer quality, the typical applications are optical layer stacks that perform as low emissivity coatings, solar protection coatings, or antireflection layers.
Table I shows the typical layer materials that are applied by the PMS-DMS process and the corresponding dynamic deposition rate. The dynamic deposition rate is defined for

moving substrates as the product of the layer thickness and the substrate transportation speed.

Table I. Layer Materials and Corresponding Dynamic Deposition Rate.

layer	dynamic deposition rate [nm · m / min]
TiO_2	45
SiO_2	120
Si_3N_4	60
Al_2O_3	80

MAD-PROCESS (MAGNETRON-ACTIVATED DEPOSITION-PROCESS)

Compared to sputtering, evaporation can give much higher deposition rates. Without additional plasma activation, acceptable layer properties can be obtained for some applications only. In most cases, plasma activation during the layer deposition is a must. This is true also for the deposition of thin transparent Al_2O_3 layers as barrier coatings on plastic films for food packaging purposes.
The high deposition rates of alumina in the range of some 100nm/s can be reached in an economical way by the reactive evaporation of Al with the aid of boat evaporators [5].
We have developed a process for the plasma activated high rate deposition by using two magnetron electrodes that, similar to the dual magnetron sputtering, are pulsed with a medium frequency. The corresponding arrangement is shown in Fig.2 and the practical arrangement in Fig.3. According to the length of the available magnetrons, a uniform plasma can be generated over a web width of more than one meter.

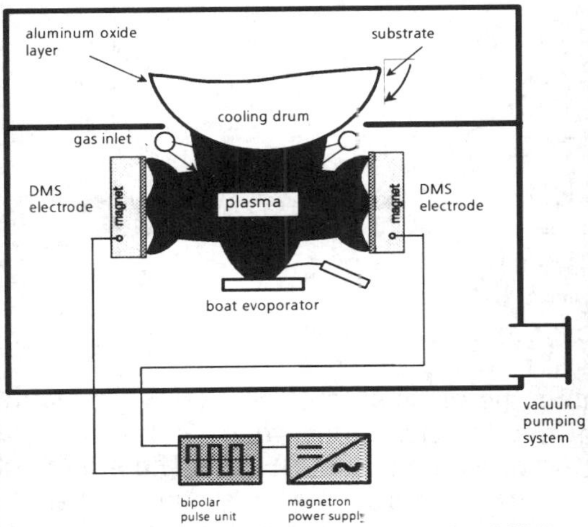

Fig.2. Principal arrangement of the MAD Process.

Fig.3. Practical arrangement of the MAD Process with cylindrical magnetron electrodes in an Al roll coater.

The plasma density in this process lies in the range of 10^9 to 10^{10} cm^{-3}.
The transparent barrier layers for flexible packaging films are 20 nm thick and the web speed is higher than 5m/s. The transparent packaging films are microwaveable and show low permeability for water vapor and oxygen. The typical numbers for the water vapor transmissivity, WVTR, are <1g/m²d (d:day) and for the oxygen transmissivity <3cm³/m²d for a coated 12µm PET film.

HAD-PROCESS (HOLLOW-CATHODE-ACTIVATED DEPOSITION-PROCESS)

The layer properties that can be achieved by using the MAD plasma activation are not sufficient especially for the deposition of 2-6µm thick abrasion resistant layers on plastic substrates at low temperatures. These demands can be fulfilled much better by the hollow cathode plasma activation.
The HAD process is based on the reactive evaporation of the oxide or the reactive evaporation of metals with high rates in combination with a hollow cathode plasma activation. Most developed up to now are the processes for the deposition of Al_2O_3 and SiO_X. Al_20_3 can be deposited in the best way by reactive evaporation of Al, whereas the best economical approach involves boat evaporation. The deposition rates range from 100-150nm/s. SiO_X layers are deposited by EB evaporation of SiO_2. The typical deposition rates range from 200-600nm/s.
The hollow cathodes generate plasma with a very high density. This can be explained by the electron energy distribution function in the hollow cathode plasma. Fig.4 shows the typical electron energy distribution which consists of the Maxwellian distribution of the isotropic electrons and the superimposed group of directed electrons [6,7] that is also called Low Voltage Electron Beam (LVEB).

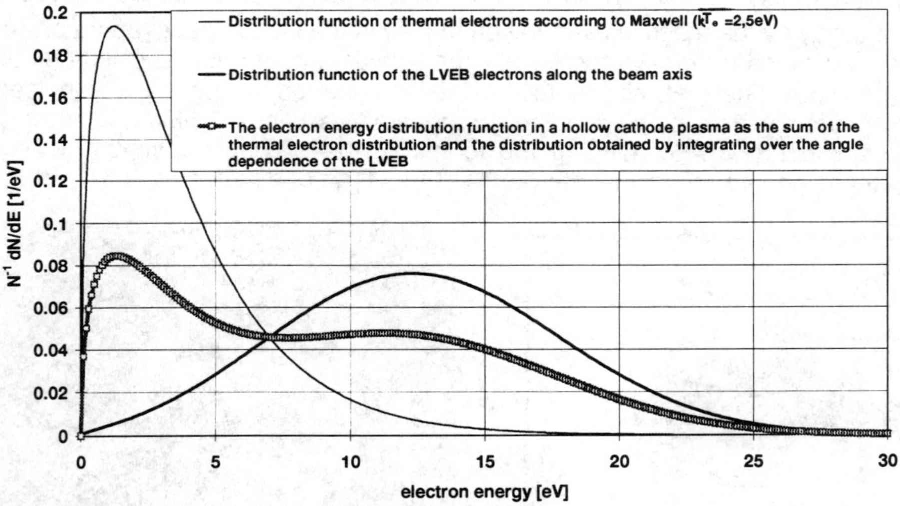

Fig.4. Typical Electron Energy Distribution Function in a Hollow Cathode Plasma.

The electron energy distribution of all electrons in the plasma contains a large portion of electrons that exceeds the ionisation energy of the particles in the vapor (6-16eV). So a very high plasma density in the order of 10^{12} cm^{-3} can be achieved (Fig.5). The energy of the tail in the electron energy distribution which charges the insulating substrate determines the self bias potential. The self bias potential is estimated to be about 15V and a transition of the electron distribution function to a Maxwellian distribution in the dense vapor is assumed.
The HAD process can be applied for the coating of plastic films and plastic sheets as well. The principal arrangement of the HAD process for the coating of films is shown in Fig.6. In every case the dense plasma is in a close contact with the substrates. Fig.7 shows the plasma activated reactive evaporation of aluminum as a practical example.

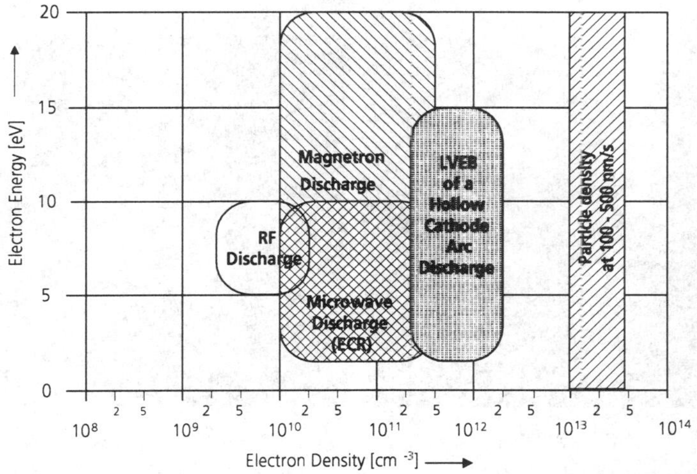

Fig.5. Range of Plasma Density and Electron Energy in Different Low Pressure Plasma Discharges (For comparison, the particle density near the substrate for deposition rates of 100 - 500 nm/s is given).

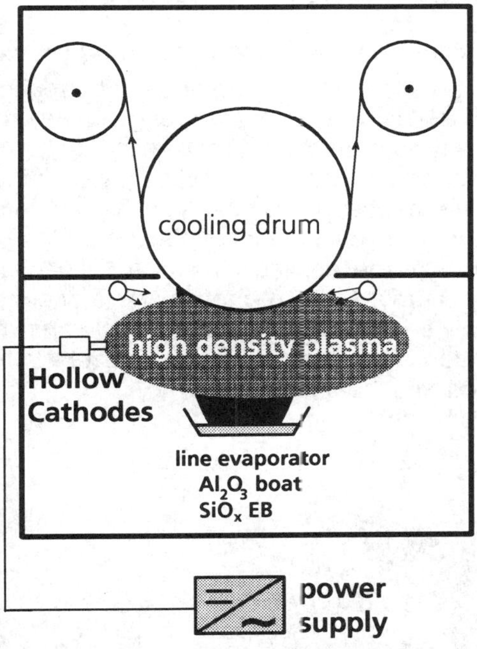

Fig.6. Principal Arrangement of the HAD Process.

Fig.7 Reactive Evaporation of Al and Plasma Activation by Means of Two Deflected Low Voltage Electron Beams for Web Coating (shutter closed).

Al_2O_3 Deposition Process

The deposition of Al_2O_3 can be performed by the reactive evaporation of Al_2O_3 as well as by reactive evaporation of Al (see Fig. 7). The Al evaporation process is more stable and reliable than the evaporation of the oxide. The latest variant is based on the evaporation of Al by boat evaporators.

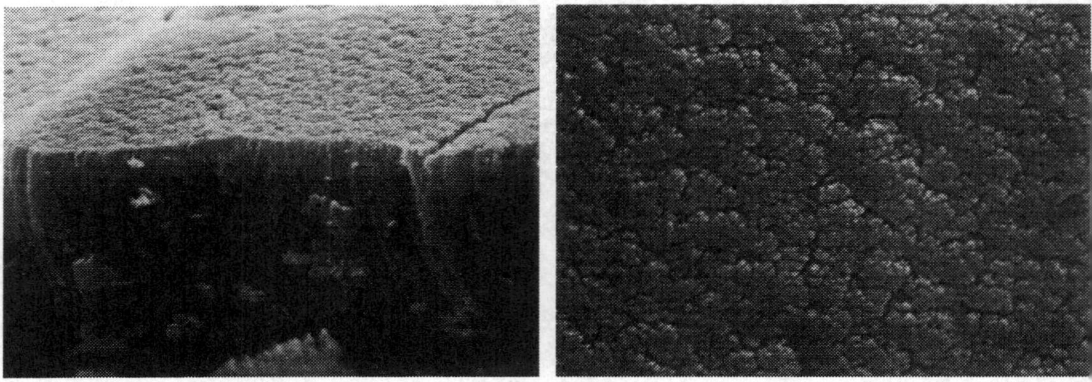

SEM - cross section (magnification 10, 000) | SEM - surface topography (magnification 20, 000)

Fig.8.Aluminum oxide deposited by **reactive evaporation of Aluminum without plasma** activation; deposition rate:180nm/s; estimated mean kinetic **particle energy:0.2-0.3eV**

The SEM micrographs in Figs.8 and 9 show how the structure of the deposited layers can be improved by the HAD plasma activation.
The structure is mostly influenced by the mean kinetic energy of the condensing particles [8,9]. The particle energy is determined by the evaporation temperature and will be reduced by thermalization by cold reactive gas. Therefore, the particle energy will range in the order of some tenths of an electron volt. This results in a typical structure as shown in Fig.8.
The layers are not dense, but columnar and show pronounced microcracks on the surface. These structures are typical for layers that are deposited without plasma activation irrespective of the evaporation of the oxide or the reactive evaporation of the metal.
The structure improves depending on the degree of the applied plasma activation and consequently the enhanced mean kinetic particle energy. A mean kinetic particle energy of 3.5eV improves the structure already essential. The layers are denser, still columnar but the microrcracks on the surface have disappeared. The abrasion resistance of such a coating is already significantly enhanced compared with the uncoated plastic.
The further increase in the mean kinetic particle energy up to 5eV results in dense amorphous layers with a smooth surface (see Fig.9).

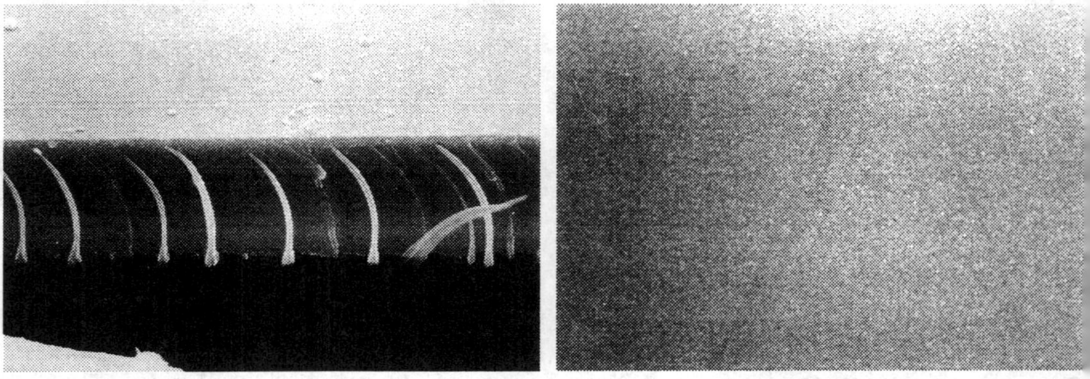

SEM - cross section (magnification 10, 000) | SEM - surface topography (magnification 20, 000)

Fig.9. Aluminum oxide deposited by **HAD plasma activated** reactive evaporation of Aluminum; deposition rate: 100nm/s; estimated mean kinetic **particle energy: ca. 5eV**

The amorphous Al_2O_3 layers show a high hardness (see Table II), a very good adhesion to the polyester film without any additional pretreatment and an excellent abrasion behavior (see Fig.10) for layers that are thicker than 2µm [10].

Table II. Properties of the Al_2O_3 Layers.

Structure:	dense amorphous structure
Refractive index:	≈ 1.6 - 1.65
Dynamic hardness:	≈ 6 GPa (measured by nanoindentation)
Adhesion:	peel force ≈ 6-14N/cm (on polyester film)
Abrasion resistance:	gloss loss ≈ 0-5% after 500 rev.

Furthermore, the excellent transparency of the coating can be seen in Fig.11. The absorption in the range of the visible light is nearly zero and the solar-weighted absorption amounts to 0.6% only.

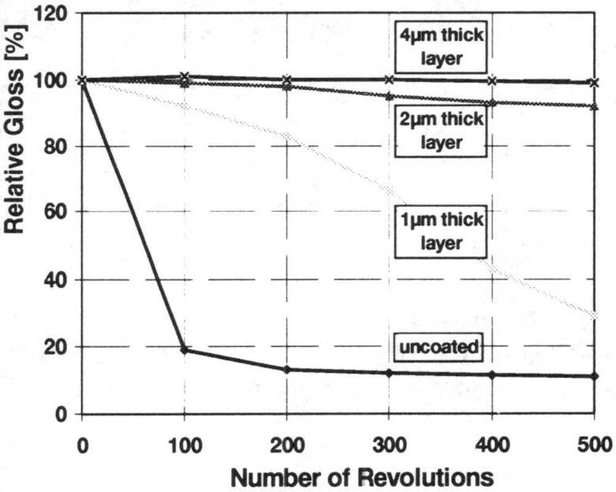

Fig.10. Taber Abrasion Test (CS10 wheels, 500g load, 500 rev.) Abrasion Resistance of Al_2O_3 Layers on Polyester Film (Gloss loss evaluation); Deposition Rate: ≈ 100nm/s.

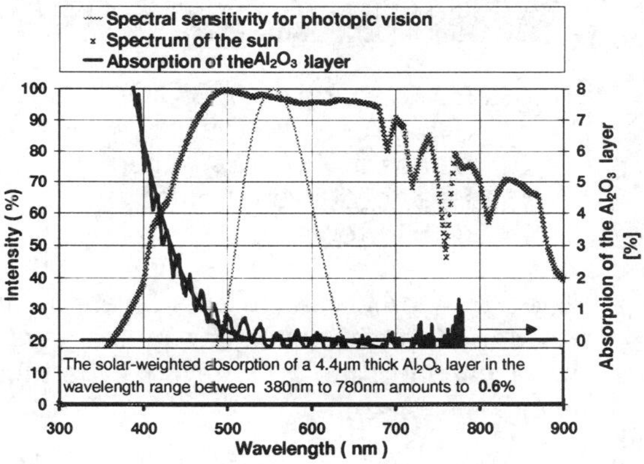

Fig.11. Absorption of a 4.4µm Thick Al_2O_3 Layer.

SiO_x Deposition Process

Fig.12 shows the EB-evaporation of SiO_2 from a rotating tube in combination with the HAD plasma activation.

The deposited SiO_x layers are glassy, dense and amorphous (see micrograph in Fig.13) similar to the Al_2O_3 layers. Despite the fact that the hardness is only half of the Al_2O_3 layers the abrasion resistance of the SiO_x layers (see Fig.14) is excellent, too. For 5-6µm thick layers an abrasion resistance similar to normal float glass can be achieved (2% haze after the Taber test). The abrasion resistance typically depends on the thickness of the coating (see Fig.15) so that 5-6µm layer thickness is necessary to obtain a float glass-like abrasion behavior for the coated polycarbonate sheets.

Fig.12. EB Evaporation of SiO_2 from a Rotating Silica Tube and Plasma Activation by Means of Two Low Voltage Electron Beams for Web Coating (HAD-Process).

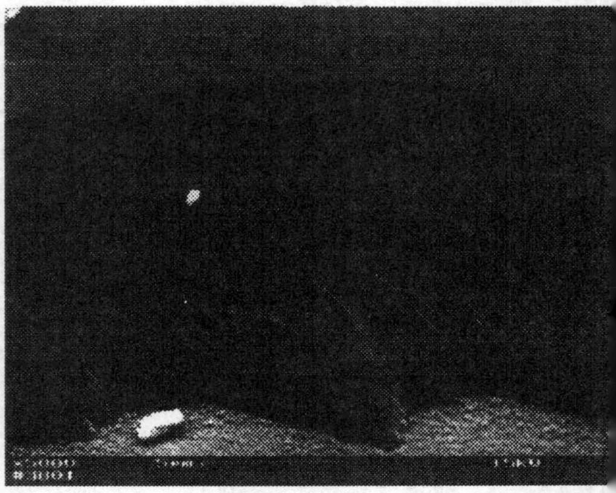

Fig.13. SEM - Cross Section of a 4 µm Thick SiO_X with Organic Modification. Deposition Rate: ≈ 300nm/s.

The typical properties of SiO_X layers are compiled in Table III.
An important improvement in the layer properties has been achieved by an organic modification of the SiO_X layers. The modification results in an enhanced elasticity of the layers.

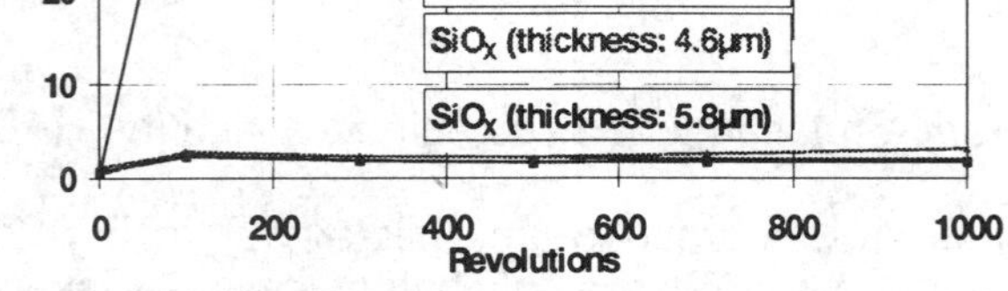

Fig.14. Abrasion Resistance of SiO_X Layers on Polycarbonate (Haze evaluation) Taber Abrasion Test (CS10F wheels, 500g load, 1000 rev.) Deposition Rate: ≈ 300nm/s.

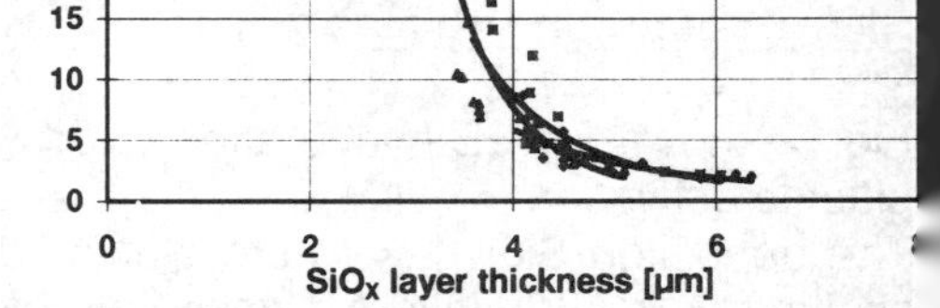

Fig.15. Abrasion Resistance of SiO_X Laye as a Function of the Layer Thickness.

Table III. Properties of the SiO_X Layers.	
Structure:	dense amorphous structure
Refractive index:	≈ 1.45
Dynamic hardness:	≈ 3 GPa (measured by nanoindentation) for pure inorganic layers ≈ 2.2-2.7 GPa for organically modified layers
Elasticity:	Strain to microcracking 1.8% for organically modified layers
Adhesion:	peel force ≈ 3N/cm (on polycarbonate)
Abrasion resistance: (Taber Abrasion Test)	Haze: 2%* after 1000rev. for organically modified layers (wheels: CS10F; load: 500g); layer thickness: 6μm * (the Taber test result for float glass is also 2% under the same conditions)

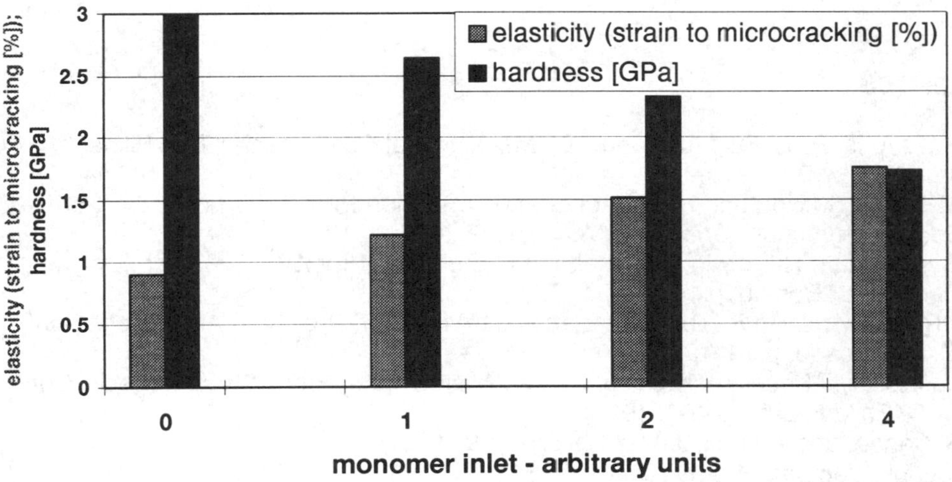

Fig.16. Enhanced Elasticity of Organically Modified SiO_X Layers.

The strain-to-microcracking ranges around 0.6-0.8% for pure oxide layers and can be enhanced up to nearly 2% for organically modified SiO_X layers, whereas a decrease in the hardness is observed (see Fig.16). But there is a process window of enhanced elasticity without deterioration of the abrasion resistance. The improvement of the elasticity is important for coatings on plastic substrates for outdoor applications because of the mismatch of the thermal expansion of the plastic and the oxide coating. A temperature variation range of 100 °C has to be taken into account which results in a thermal expansion of 0.7% for polycarbonate whereas the thermal expansion of silicon dioxide is almost zero.

SUMMARY

Coating by magnetron sputtering is an established technique for applications that call for a layer thickness uniformity of about 2%. To deposit highly insulating compound layers, it is necessary to employ pulsec magnetron sputtering (PMS) process. Substantial applications are the production of multilayer systems for solar control and low-e films.

For a layer thickness uniformity of about 10% the evaporation technique is recommended. The deposition rates cover a range of 100-1000nm/s. Therefore, they exceed those of magnetron sputtering by two orders of magnitude.

High deposition rates and the low condensation temperatures which are limited by the thermal stability of the plastic substrates are often linked to a poor layer quality. To obtain good layer properties also with high rate evaporation it is necessary to apply a plasma activation. Here the plasma density has to be adapted to the high deposition rate. In this paper we have presented and explained two plasma processes that match the high deposition rates.

In the HAD process the obtained plasma densities are in the order of 10^{12}per cm^3. This results in ion current densities on the substrate of about $30mA/cm^2$. This type of plasma activation allows the deposition of clear and glassy Al_2O_3 as well as SiO_X layers on plastic substrates. Both oxides perform excellently as abrasion resistant layers.

If lower requirements are imposed on the ion current densities, the MAD process will be the method of choice. This process excels in its ability to provide a very good uniformity of the plasma. The typical application field for this process is the deposition of Al_2O_3 barrier layers on plastic films.

REFERENCES

[1] H.Lievens, Proc. 22nd Intl. Conf. On Metallurgical Coatings and Thin Films, San Diego, 1995

[2] S.Schiller, M.Neumann, H.Morgner and N.Schiller, Proc 37th Ann. Tech. Conf. Society of Vacuum Coaters (SVC), Boston, p. 203, 1994

[3] G.Hoffmann, R.Ludwig, J.Meinel and G.Steiniger, Proc.37th Ann.Tech. Conf. SVC, Boston, p.155, 1994

[4] J.Ross,R.A.Hallman, D.J.Kester and J.D.Wisnosky, Proc.38th Ann. Tech. Conf. SVC, Chicago, p.13, 1995

[5] N.Schiller, J.Reschke, K.Goedicke and M.Neumann, Proc.39th Ann. Tech. Conf. SVC, Philadelphia, p.404, 1996

[6] S.Klagge, and A.Lunk, J.Appl.Phys.70, 99 (1991)

[7] A.Lunk, Vacuum, 41, 99 (1990)

[8] A.G.Dirks and H.J.Leamy, 47, 219 (1977)

[9] W.Olbrich and G.Kampschulte, Surface Coating Technol., 59, 274 (1993).

[10] M.Neumann, H.Morgner and S.Straach, Proc.39th Ann. Tech. Conf. SVC, Philadelphia, p.446, 1996

Metallized Plastics 5&6: Fundamental and Applied Aspects, pp. 85-96.
K.L. Mittal (Ed.)

PHYSICAL AND CHEMICAL SURFACE PRECONDITIONING AND METALLIZATION OF THERMOPLASTICS FOR ELECTRONIC COMPONENTS

H. Kupfer

Technical University of Chemnitz
Department of Physics
Reichenhainer Strasse 70
D - 09107 Chemnitz, Germany

and

R. Ostwald

Daimler-Benz AG
Research Center Ulm
Wilhelm-Runge-Strasse 11
D - 89081 Ulm, Germany

Demands of the electronic industry for IC packaging can mostly be fulfilled by injection moulding of tailored high temperature polymer materials. More stringent requirements in view of electrical conductivity as well as electromagnetic shielding, gas-tight barrier, corrosion protection, interconnection, and packaging processes have to be realized by the deposition of well-adhering metal layers. The complete coating of components of very complex shapes is the prerequisite for high quality packagings. It can be fulfilled by electroless plating or by magnetron sputtering. Well-adhering metal layers, which can withstand high thermomechanical stress, were deposited on mineral-filled polyethersulfone surfaces with defined microroughness. The microroughness could be produced not only by wet chemical etching but also by environmentally friendly plasma etching processes. An economically and ecologically attractive process for the fabrication of metallized high temperature polymer housings with high performance for electronic demands was found with a combination of r.f. plasma etching, magnetron sputtering of an adhesion improving layer, and reinforcing the latter by electroplating with copper, nickel, and tin.

INTRODUCTION

During the last few years new high temperature (HT)plastic materials have widely been used for many applications. In many cases, plastic components must be metallized to fulfill special requirements. For instance, injection-moulded polymeric housings for electronic components must be coated, e.g. by electroless plating[1], with metal layers for electromagnetic shielding[2-4], for a gas diffusion barrier[5], for corrosion protection, for interconnection and packaging technologies[6]. For this purpose, preconditioning and metallization processes are needed for the deposition of well-adhering metal layers on polymeric bodies of complex shapes. The metal-polymer interface has to withstand high thermomechanical stresses caused by the thermal expansion mismatch when high temperature packaging processes are applied. Therefore, a very high adhesion of the metal layer to the polymeric substrate is indispensable.

To provide metal films on conventional polymers with high peel strength, preconditioning methods such as chemical etching and metallization methods, e.g. electroless deposition and electroplating, are widely used in the industry[7]. For example, peel strength values of copper metal layers up to 1.5 N/mm on ABS polymer were obtained as a result of both chemical bonds and mechanical interlocking[8].

In comparison to wet chemical methods, plasma assisted processes for etching and metallization by physical vapour deposition (PVD)[9] have the advantage of low environmental pollution. This provides a compelling reason to investigate the reactive ion assisted r.f. etching process for tailored high temperature (HT) polymers such as mineral-filled polyethersulfone. The metallization of even complex shaped housings can be realized by a conventional magnetron sputtering process. The morphology providing strong adhesion is compared with the surface prepared by chemical etching methods.

EXPERIMENTAL

We have investigated the adhesion strength of Cu films on the surfaces of injection moulded mineral-filled polyethersulfone (PES, ULTRASON KR4110, BASF) substrates. The flat polymeric samples were cleaned in a 50/50(by volume) mixture of water/2-propanol, dried, and annealed at 150 °C for at least three hours.

The preparation processes are shown schematically in Figure 1. For the chemical preconditioning of the polymer surfaces, wet chemical etching methods on the basis of chromic acid were applied. After reduction and neutralization steps, the surfaces were rinsed in tap-water, in demineralized water, and dried at room temperature. The surface quality was monitored by optical microscopy, by Scanning Electron Microscopy (SEM) and by profilometry. The polymer surfaces were catalytically activated by tin chloride/palladium chloride process. A Cu film or a nickel phosphorous film (NiP) of < 1 μm thickness was electrolessly (chemically) plated from a common autocatalytically working copper sulfate-formaldehyde or nickel sulfate-hypophosphite bath. The quality, especially the conductivity and the thickness uniformity of the Cu and the NiP layers, was monitored by a four-point-probe measurement of the sheet resistance.

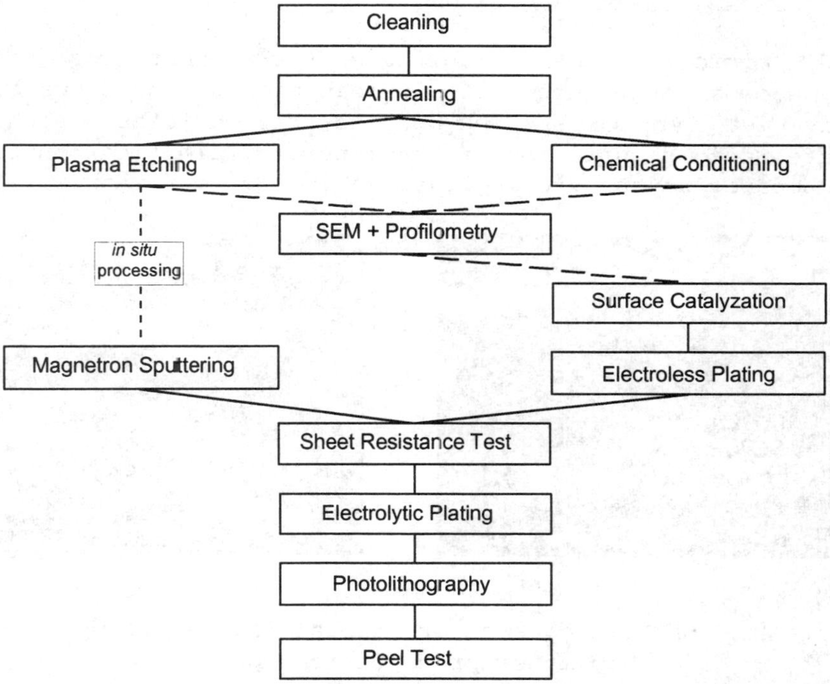

Figure 1. Process scheme for sample preparations and tests.

In contrast to the chemical procedure indicated, the substrates were exposed to a reactive low pressure plasma at various process pressures. For the plasma etching, the total pressure was varied in a range of 0.1 Pa to 20 Pa and the etching time between 10 min and 20 min. Different process gases were used (oxygen, nitrogen, argon and their mixtures). The surfaces of the etched substrates were investigated by Scanning Electron Microscopy (SEM), profilometry (roughness and etching rate measurements), and Atomic Force Microscopy (AFM).

A copper film (thickness = 3 μm) was deposited by magnetron sputtering *in situ*. It grows at a rate of 170 nm/min by applying a d.c. discharge power of 500 W. Argon was used as the process gas at p_{tot} = 0.4 Pa. The sheet resistance was monitored as stated above. For further investigations of the long-term adhesion and thermal ageing of the copper or nickel phosphorous layer-polymer systems, both the electrolessly plated and the magnetron sputtered copper films were electrolytically reinforced with Cu from a common copper sulfate/sulfuric acid bath up to 30 μm. The adhesion was determined by a peel test using 1 mm wide metal film stripes which were prepared by a photolithographic process.

RESULTS AND DISCUSSION

Chemical Preconditioning

SEM micrographs of the surfaces of PES substrates [Figure 2] clearly show a large number of microcavities produced due to the selective wet chemical etching process. The cavities are the result of a more rapid dissolution of the mineral filler (carbonate particles) compared to the etching rate of the polymer matrix.

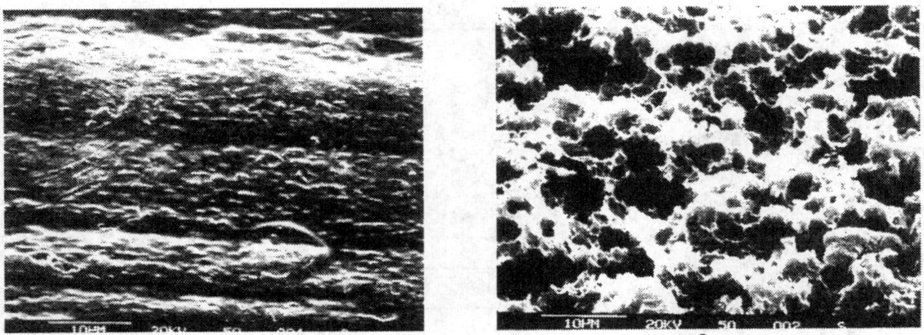

a b

Figure 2. SEM micrographs of mineral-filled PES untreated (a) and after preconditioning (b) with chromic acid, reduction treatment and rinsing.

The average microroughness of the substrate surface can be enhanced by a stronger chemical attack, especially by increased temperature, higher chromium(VI)-oxide concentration, and increased duration. A complete wetting of the hydrophilic polymer surfaces by the acidic solutions of tin chloride and palladium chloride, even in the depth of the smallest microcavities, is a prerequisite for the mechanical anchoring of the electrolessly deposited copper or nickel phosphorous. After the electrolytic reinforcement with Cu and after peel tests, many Cu particles could be found in the cavities on the PES surface. An increased microroughness yields higher peel strength [Figure 3]. Profilometry of the resulting surfaces is a very useful method for monitoring the uniformity of microroughness of the polymer surfaces. An average microroughness of the PES surface of 13 μm is responsible for a peel strength of normally 1.5 N/mm which can be enhanced with an adhesicn layer of 0.2 μm Cu to 2.5 N/mm by an annealing process for 48 hours at 150 °C (see figure 3).

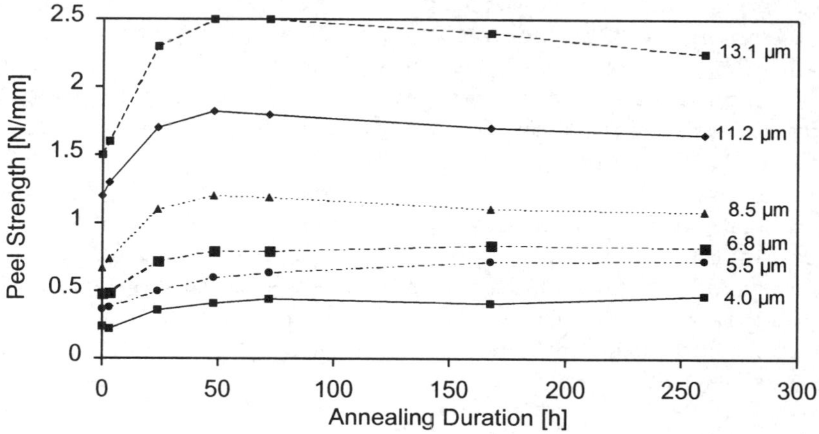

Figure 3. Effect of annealing time at 150°C on the peel strength of copper layers on wet chemically etched PES substrates having different microroughnesses.

In the case of a NiP adhesion layer of the same thickness, the peel-strength maximum after 48 hours at 150 °C is only 1.5 N/mm [Figure 4]. With continuation of thermal ageing, a very slight decrease of the peel strength takes place. Even after an extended ageing of 120 days, a peel strength of more than 1 N/mm still remains.

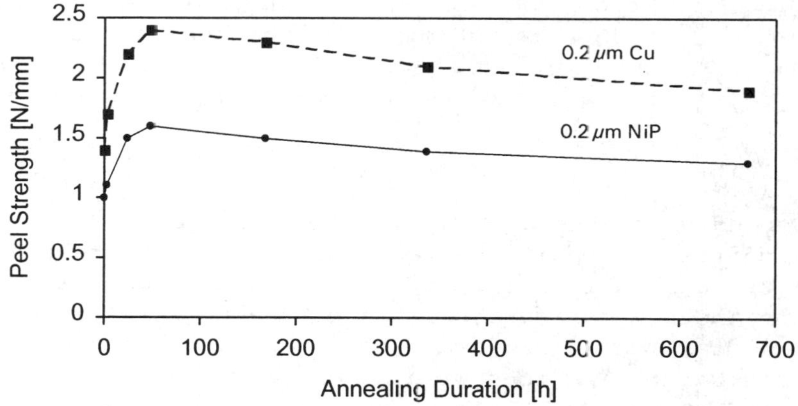

Figure 4. Influence of thermal ageing at 150°C on the peel strength of Cu layer electrodeposited on adhesion layers of electrolessly plated Cu or NiP on wet chemical treated PES substrates.

Reactive Ion Assisted Plasma Etching

The effect of ion sputtering. A low pressure r.f. plasma offers a different process than the chemical etching for the preconditioning of the polymer surfaces. Reactive neutral atomic particles (e.g. dissociated oxygen molecules), vacuum ultraviolet radiation (VUV radiation) and/or ions take part in a complex way.

An efficient use of the ions for physical sputtering requires a high acceleration voltage. This can be provided by arranging the substrates on the powered r.f. electrode and using a reactive r.f. discharge. Due to the high mobility of the electrons in a plasma, a resulting d.c. bias voltage in an r.f. discharge can be adjusted in the range of -200V to -1000V. This bias voltage accelerates the ions against the substrate and results in a sputtering effect. On the other hand, if an insulating substrate is positioned in the negative glow of the plasma discharge, the ions are accelerated only by the difference $U_{Plasma} - U_{floating}$ (-10V to -20V) [Figure 5].

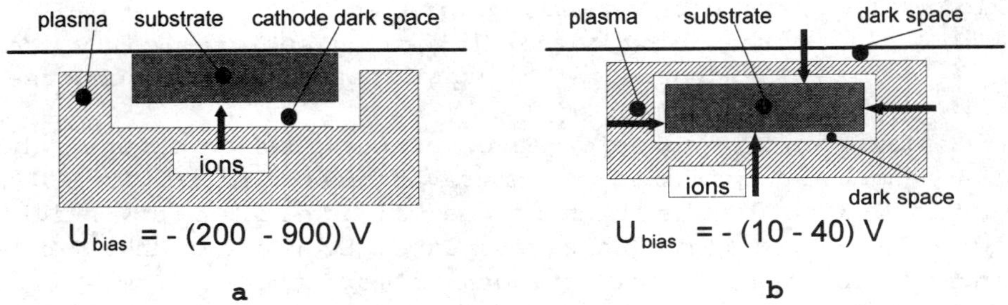

Figure 5. Ion bombardment of the substrates. (a) Substrates are arranged on the powered r.f. electrode - a high bias voltage accelerates the ions; (b) Substrates are arranged in the negative glow of the plasma - a low bias voltage is obtained.

Depending on the process gas and the position of the substrate, different conditions for the ions and the reactive particles of the plasma result in different etch rates [Figure 6]. By arranging the polymeric substrates in the negative glow of the plasma, a drastically reduced erosion coefficient was found in an oxygen discharge. The ions with an energy of about 10 eV to 20 eV cannot sputter the substrate. Likewise, a low polymer erosion was obtained by highly accelerated argon ions (E_i = 800 eV) of a non-reactive r.f. plasma (pure physical sputtering: point 1 in figure 6).

By combining the physical sputtering with the chemical and photochemical effects of a reactive r.f. oxygen plasma, a strongly increased etching of the substrates occurs which cannot be explained only by the sum of the physical sputtering and the chemical effects. Both increase the polymer erosion in a synergistic manner that can be related to cracking of polymer molecules by chemical reactions and especially by the VUV radiation from the plasma.

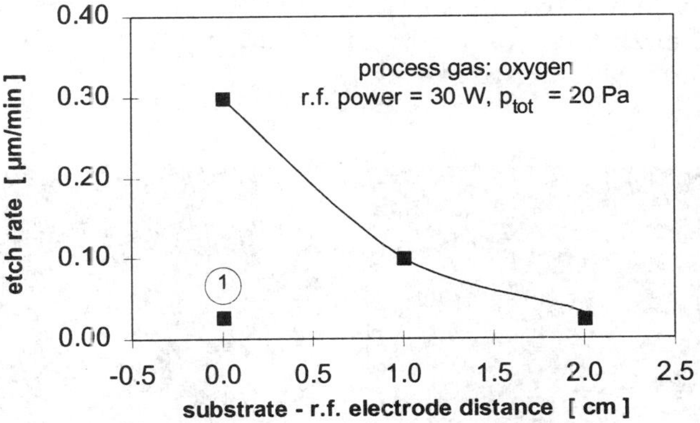

Figure 6. Etch rate of PES at different ion bombardment conditions in an oxygen r.f. discharge: The combination of ion bombardment with plasma chemical etching increases the etch rate drastically (substrate arranged at the powered r.f. electrode) in comparison to a pure plasma chemical etching (substrate arranged in the negative glow of the plasma at 2 cm distance from the r.f. electrode). Point 1: Etch rate by pure physical sputtering in argon process gas.

Preconditioning of PES by reactive ion assisted plasma etching. The chemically enhanced ion sputtering of the polymer develops specific morphologies on the substrate surface. In the case of the mineral-filled PES the dolomite particles act as hard caps and protect the underlying polymer against the sputtering due to the ions. This results in a typical columnar structure with some dependence on the total pressure. At a low total pressure of 0.1 Pa the ions are not scattered from the neutral gas molecules of the plasma discharge within the cathode dark zone. They hit the polymer surface in vertical direction without any influence by scattering and thus they develop sharp and straight columns at the surface [Figure 7a]. The columns are directed against the striking ions. This confirms the dominating role of the ions in this pressure range. In this case the adhesion strength of magnetron sputtered Cu film is less than 0.5 N/mm.

A completely different surface morphology can be obtained by increasing the pressure of the process gas [Figure 7b]. At 10 Pa to 20 Pa the scattering of the ions by the neutral gas particles in the cathode dark zone plays an important role. The scattered ions impinge upon the substrate surface from many directions and undercut the filler particles. In addition to this, the sputtering effect will be reduced by decreasing the sputtering coefficient due to ion scattering. On the other hand, the chemical reactions will be intensified by the increased particle density of atomic oxygen at this pressure. A cavern-like structure occurs which provides many adhesion sites by mechanical anchoring between the metal layer and the substrate.

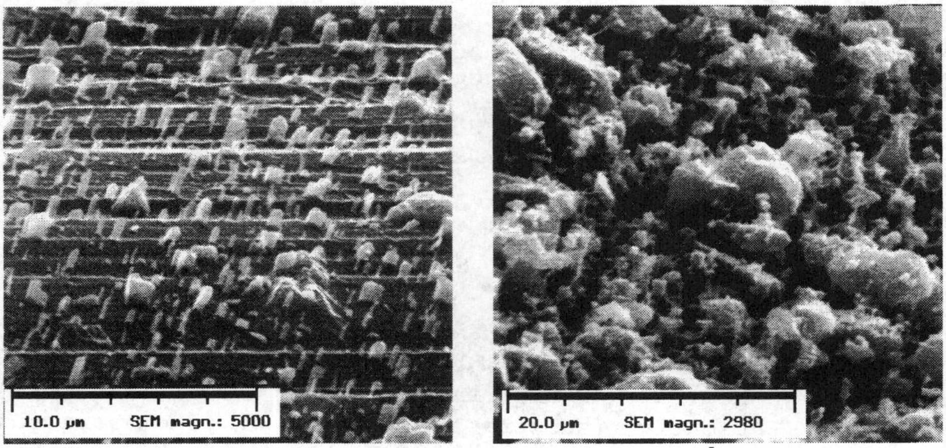

a b

Figure 7. SEM micrographs of mineral filled PES, plasma preconditioned at 0.1 Pa (a) and at 20 Pa (b) in an oxygen discharge.

After surface modification as described above, Cu films could be deposited by magnetron sputtering with an adhesion strength up to 1.5 N/mm. Unlike the chemical preconditioning of PES, after the plasma conditioning and magnetron deposition investigations of the peeled surface of the metal film show that polymer particles adhered at the back side of the metal stripes. This indicates a weak mechanical strength of the columns.

The addition of nitrogen to the oxygen process gas yields an intensified sputtering effect at high total pressures of 20 Pa. The cavern-like structure of the PES substrate surface described above is converted into a dense columnar structure. It is similar to the structure obtained in the pure oxygen case at about 1 Pa. The magnetron sputtered Cu films deposited on these surfaces exhibit a very high, about 2 N/mm, adhesion. [Figure 8]. As in the case of wet chemical etching also for the plasma etching the adhesion strength is maximum for highest microroughness. This confirms the dominant role of the mechanical anchoring.

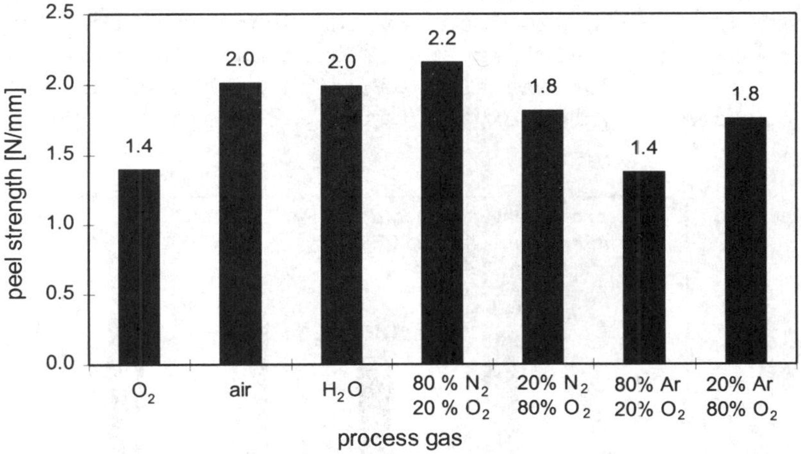

Figure 8. Adhesion of sputtered Cu films (thickness = 3μm) in dependence of the used process gas in the plasma preconditioning of PES polymer substrates.

substrate: mineral filled PES
preconditioning: p_{tot} = 20 Pa; r.f. power = 40 W; etching time = 20 min

Atomic force microscopic investigations show an increase in the effective surface area due to etching by a factor of 2.5 [Figure 9]. This indicates that the adhesion improvement by plasma conditioning is both due to the effect of the mechanical anchoring and the enlargement of the substrate surface area. The latter provides more sites for film adhesion by chemical bonds.

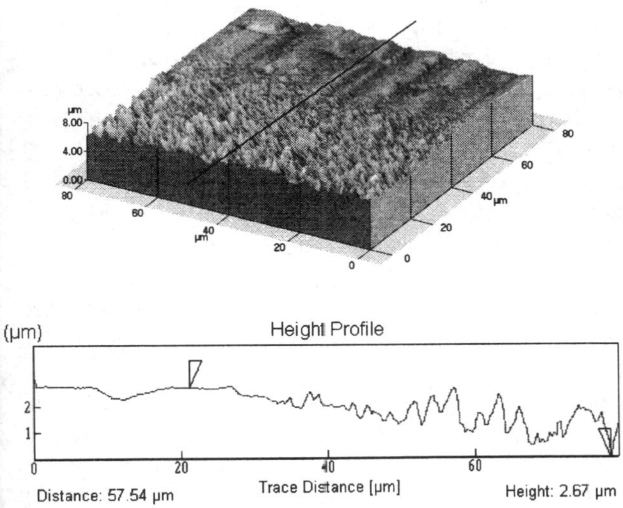

Enlargement of the substrate surface area by an increased microroughness using an oxygen r.f. plasma discharge:
A surface area of 244 μm^2 (unetched polymer) is enlarged up to 606 μm^2 (plasma-etched polymer).

Figure 9. Atomic Force Microscopy investigation of an r.f. plasma preconditioned PES surface.

This observation corresponds to the enhanced peel strength of about 1.7 N/mm which was found by the deposition of a CuO_x/Cu layer system on mineral-filled PES [Figure 10]. It agrees with the improved performance for the film adhesion on PES after plasma etching processes in which oxygen takes part (see figure 8).

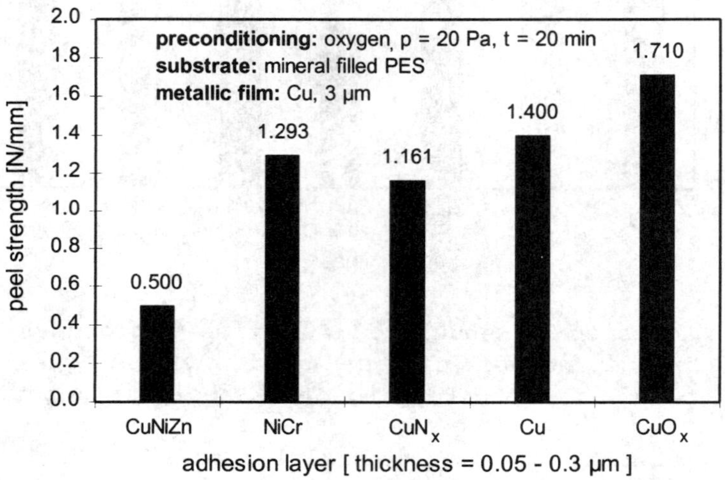

Figure 10. Adhesion improvement of sputtered Cu film on mineral filled PES with various adhesion promoting interlayers.

Metallization by Combined Processes

Metallic PVD films deposited on rough surfaces show a coarse-grained structure and a rough surface morphology. This increases the gas diffusion through the metallic layers. To protect sensitive electronic components reliably against water vapour adsorption, an electroplated thick layer is advantageous. For this reason, we favour an ecologically friendly plasma preconditioning process for the polymer, in combination with the deposition of a metallic adhesion layer by magnetron sputtering. Upon reinforcing this layer by electroplating of copper we can achieve high electromagnetic shielding and gas-tightness of the polymeric packages. If solderability is required, additional electroless, or electrolytic plating of a thin nickel barrier layer of 1 μm thickness would be important, first in preventing corrosion of the copper surface and second to provide a diffusion barrier[10] to the copper. Best results in solderability were obtained with electrolytically deposited top layer of tin with a thickness of about 20 μm.

Complex shaped plastic parts with small through-holes of at least 0.3 mm diameter and 3 mm length can be plasma etched and magnetron sputtered with copper. The Cu can be further autocatalytically reinforced by a direct electroless plating of Cu without any activation. A cross-sectional investigation of the holes demonstrates a uniform metallic layer.

CONCLUSIONS

The degree of microroughness of the polymer substrate is responsible for the anchoring mechanism which seems to be the most important contribution to metal adhesion. On the other hand, an enhanced peel strength can be obtained using an electrolessly plated Cu adhesion layer, compared to a NiP layer, even with the same degree of microroughness of the PES surface. This shows there exists a copper related bonding effect in addition to the "mechanical anchoring". Using a PVD process an adhesion promoting interlayer of CuO_x/Cu showed a definite enhancement of the peel strength due to chemical bonds between CuO_x and the PES surface. Therefore, it can be concluded that the affinity of Cu to the oxygen and/or the sulfur in the PES surface is responsible for chemical bonds resulting in enhanced adhesion of the Cu layer on the PES as compared to NiP layer. A combination of r.f. plasma etching, magnetron sputtering of an adhesion layer of CuO_x/Cu, and reinforcing the latter by electroless or electroplating is an economically attractive and environmentally friendly process for the fabrication of metallized HT-polymer housings with high performance for electronic requirements.

ACKNOWLEDGMENTS

We are thankful to Mrs. G. Voit and Mr. R. Schödlbauer from the Daimler-Benz Research Center Ulm, Germany, for sample preparation and adhesion measurements. The authors are also very grateful to Mr. M. Röder for SEM sample preparations and Mr. P. Schlott for many depositions.

Financial support by the Federal Minister for Research and Technology of Germany, under Contract No. 13N6555, is gratefully acknowledged.

REFERENCES

1. Y. Akonake and T. Osaka, Adv. Electrochem. Sci. Eng. 3, 55 (1994).
2. N.V. Mandich, Plating Surface Finish., 81, 60 (1994).
3. B. Viswanathan, Current Sci., 65, 537 (1993).
4. G.A. Krulik, Plating Surface Finish., 71, 56 (1984).
5. J. Springer and G. Schammler in "Metallized Plastics 2: Fundamental and Applied Aspects," K.L. Mittal, editor, p. 15, Plenum Press, New York, 1991.
6. M. Boheim, R. Dolp, U. Goebel, and W. Hager, Proc. 4th Intl. Conf. on Micro Systems Technologies 1994, 623 (1994), ICC Berlin, VDE Verlag GmbH Berlin and Offenbach, 1994.

7. N.V. Mandich and G.A. Krulik, Plating Surface Finish., 80, 68 (1993).
8. H. Meyer, Galvanotechnik, 86, 67 (1995).
9. D. Mann, J. Fessmann, G. Kampschulte, and K. Mertz, in "Metallized Plastics 3: Fundamental and Applied Aspects," K.L. Mittal, editor, p. 295, Plenum Press, New York, 1992.
10. M. Paunovic, P.J. Bailey, R.G. Schad, and D.A. Smith, J. Electrochem. Soc. 141, 1843 (1994).

Metallized Plastics 5&6: Fundamental and Applied Aspects, pp. 97-107.
K.L. Mittal (Ed.)

NEW GENERATION SPUTTERING AND PLASMA POLYMERIZATION PLANT FOR METALLIZATION OF PLASTICS

H. Grünwald, M. Jung, R. Kukla, R. Adam, S. Kunkel, and J. Krempel-Hesse

Leybold Systems GmbH, Wilhelm-Rohn-Strasse 25, D-63450 Hanau, Germany

Derived from the working principle of high-throughput single-disc CD coating machines, a new generation of coating plants for shaped parts such as car headlight reflectors with aluminum and a corrosion protection top layer was developed. The machines process small batches of parts at short repetition times passing them through a loadlock and active chambers for plasma pretreatment, sputtering, and finally plasma polymerization. The system offers new production opportunities as compared to the traditional large evaporation systems. In contrast to evaporation, in this machine alloys also can be deposited. The conformity of the layers to even very complex part geometries - even recesses - is excellent and the handling of parts can be automated easily.

INTRODUCTION

For the last four decades, reflectors of car headlights, car signal lights and a few lighting elements have been metallized by aluminum evaporation in large batch coaters[1,2]. In the mid-seventies, subsequent plasma polymerization for corrosion protection of the aluminum deposit was integrated into the procedure[3,4]. Nowadays, the process sequence usually also includes plasma pretreatment and sometimes a plasma polymerized base coat for better adhesion and corrosion resistance of the aluminum coating in the same chamber. Some manufacturers also add an evaporated SiO top coat in order to avoid fogging on car headlight reflectors. In contrast to this high degree of process integration, loading, evacuating and unloading of up to several hundred parts is still a very discontinuous (cycle time 30 to 45 minutes) and labour consuming process, interrupting materials flow in production. While the plastic reflector bodies come out from the moulds piece by piece or pair by pair each minute, lots of these parts are collected and stored near the aluminum batch coater before they are mounted manually onto the fixtures. These fixtures usually are tailored to specific parts, and if masking is also required then these become quite expensive. For mass production, hundreds of such fixtures, mostly of different shapes, have to be manufactured and stored.

So, for the sake of continuous materials flow and cutting costs for equipment and labour, a coating system having short cycle time for small batches (close to the moulding machine) is highly desirable. Loading and operation should be simple, and the complexity of substrate holders should be low.

TWO-CHAMBER SPUTTER COATER

Based on the experience of building hundreds of compact disc metallizers, Leybold first introduced the just-in-time coater "DynaMet 2H" (Fig. 1)

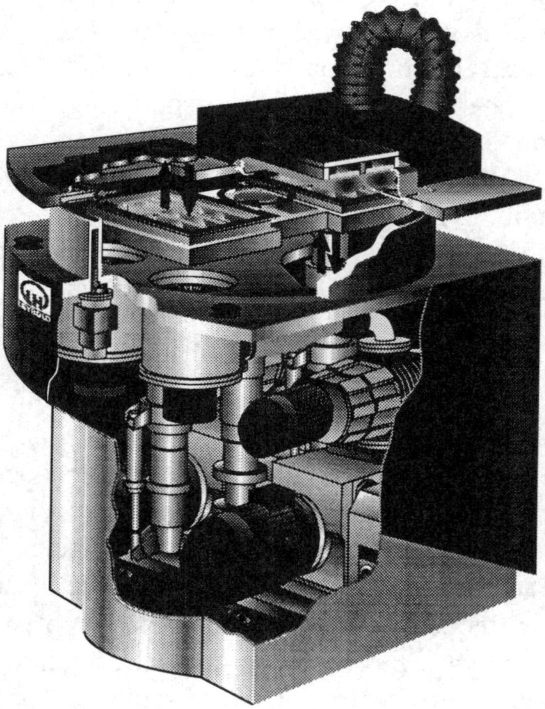

Fig. 1. Single substrate coater DynaMet 2H for deposition of metals and alloys.

that metallizes one or a small number of three-dimensional substrates within 40 seconds or less. A detailed description of this machine is given elsewhere[5,6]. The metal layer is deposited by sputtering using a special high rate magnetron sputter cathode, the Interpole Target-Hollow Magnetron (IPT-HM)[7]. The characteristics of sputtering offer new opportunities for production of reflectors or decorative parts.

In contrast to evaporation, it is possible to coat the parts underneath the metallizing source. Due to scattering of the Al by the Ar atoms during sputtering, even complex three-dimensional parts can be coated conformly without mechanical movement. As a consequence, the parts to be coated do not need to be held on rotating fixtures, they can be simply placed in some static holder such as a frame.

With this arrangement, 1.5 µm of copper can be sputtered onto plastic parts within 25 s, depending on shape, the limitation being the heat stability of the base material.

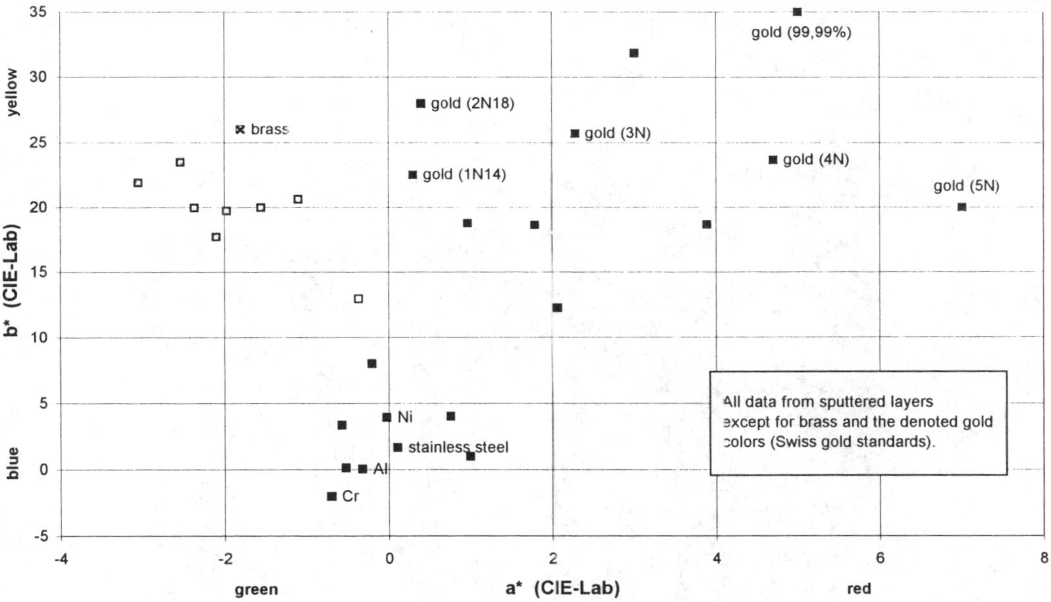

Fig. 2. Colors[8] of decorative coatings realized until now by sputtering of metals and alloys in the Dynamet 2 H: a*: redness (positive scale) - greenness (negative scale), b*: yellowness (positive) - blueness (negative scale). The level of lightness L* is between 85 and 95 for all samples.

The sputtered material is removed from the target layer by layer. Since no liquid phase is involved, alloys such as AlCu or brass for decorative coatings can be sputtered off and deposited without change of composition. Refractory materials are no problem so the DynaMet 2H can be used for depositing stainless steel, chromium or nickel. Fig. 2 displays the colors as expressed in the CIE Lab system[8] (measuring conditions: 2°, D65 including specular reflection) which have been realized so far in this coater by nonreactive sputtering of metal or metal alloy targets: a wide variety of colors can be produced, ranging from aluminum, chromium, stainless steel and nickel to brass-like and golden appearance. However, corrosion sensitive materials such as aluminum need some additional protection.

FOUR-CHAMBER COATING SYSTEM

Especially for high volume car lighting production, a four-chamber system was recently introduced into the market, offering a complete process sequence from degassing the substrates to the plasma polymerized top coat (see Figs. 3 and 4). In this system, the loadlock and the three process chambers are grouped in an upright position around a rotating substrate carrier drum.
The four substrate chambers inside the drum have a usable volume of 800 x 450 x 250 mm^3. The chamber walls are covered by metal shields which can easily be replaced for cleaning.

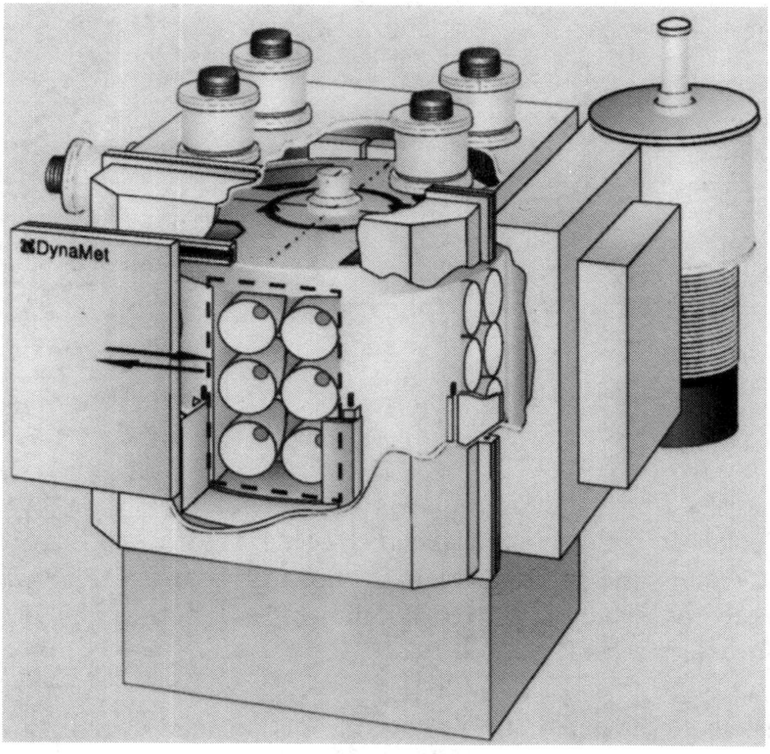

Fig. 3. Schematic view of the DynaMet 4V.

Each chamber is equipped with its own pump set. During processing of the substrates the chambers are separated from each other. The different processes in the four chambers all run parallel. After finishing the process steps, the chamber separation locks are opened and the substrate carrier drum rotates by 90° around a vertical axis. The cycle time of the machine is 36 s, including 5 s for substrate exchange.

Fig. 4. DynaMet 4V coater including roughing pumps, control cabinets and operation control panel. The loadlock is opened for exchange of substrates.

CHAMBER CONFIGURATIONS AND PROCESSES

Each set of substrates successively runs through the following steps:

Loadlock Chamber: The loadlock is closed and opened by tangential movement of the door. This results in completely free access for the loading and unloading operations. A powerful set of roughing pumps evacuates the loadlock in less than 24 seconds.

Preatreatment Chamber: In the basic configuration, the preatreatment chamber is just a desorption chamber, equipped with its own turbomolecular pump.If necessary, it can be equipped with a plasma source and the corresponding set of pumps that are needed for the higher throughput of gas.

Sputtering Chamber: In order to deposit the aluminum layer at high rate and good quality and high thickness uniformity all over the substrate volume of 800 x 450 x 250 mm^3, two Interpole Target-Hollow Magnetrons[7] are installed side by side, as shown in Fig. 5. Each cathode is powered by a 60 kW dc generator. Arcing, which might damage the surface of the aluminum target as well as the surface of the substrates, is effectively suppressed by an Arc Suppressor Interface (ASI)[5] for each magnetron. If an arc occurs, the ASI extinguishes the discharge within microseconds. This limits the energy content of the arc to less than 1 Ws, which is below the level where damage to target or substrates occurs. Targets with a thickness of 28 mm are used to reach a target lifetime of approximately 20,000 cycles.

Fig. 5. Sputtering chamber, opened for maintenance.

The magnetrons have a special target exchange mechanism that enables the operator to exchange the target-plus-backing plate unit within approx. 10 min per cathode. The sputter chamber is covered by metal shields that can be dismantled for cleaning. The basic sputter conditions are:

Base pressure	$< 5 \times 10^{-3}$ Pa
Ar pressure	$2 - 9 \times 10^{-1}$ Pa
Power per cathode	45 - 60 kW
Target thickness	28 mm
Target utilization	approx. 50 %
Target life	> 20,000 cycles

Cathode power and target lifetime depend on the shape and size (especially depth) of the substrates and on the required aluminum thickness. For deep substrates or substrates with planes normal to the target surface more power and longer deposition time are required compared to more flat substrates or substrates with planes parallel to the target surface. For a load of 5 headlamp reflectors of the Volkswagen "Golf" type per carrier, 60 kW per cathode are sufficient.

Top Coat Chamber: For plasma polymerisation the chamber is fitted with an electrode arrangement, driven by a 2.5 kW RF-generator. The system was optimized for very reliable plasma ignition and very homogeneous spatial plasma distribution. For high mass flows, the chamber is pumped by a 12,000 l/s oil diffusion pump. The monomer vapor (typically siloxanes) is supplied from the gas phase of a liquid container via a mass flow controller.

TEST PROCEDURES

Relative Aluminum thickness distribution in the front plane of the substrate chamber (Fig. 6) was determined by optical transmission of the Al layer (at 624 nm) deposited on a glass pane by a Macbeth TD 904 transmission measuring system. Thickness distribution on car headlight reflectors (Fig. 7 and 8) was measured by profilometry on glass pieces that were masked by an adhesive tape.

Adhesion was checked qualitatively by cross hatching the pretreated and aluminized (but not top coated) parts and subsequent manual 90° peel test using an adhesive tape.

Thickness distribution of the plasma polymerized top coat was determined by profilometer on silicon wafer pieces that were fixed on the reflectors by an adhesive tape.

Corrosion resistance was examined by three of the most demanding tests in the car industry:

- Keeping in a closed cabinet under 100 % relative humidity at 40 °C for up to 1000 hours.
- Immersion into sodium dodecyl sulfate solution (1 %) in 60 °C water for 24 to 72 hours.
- Immersion into 0.2 % aqueous sodium hydroxide solution at room temperature for 3 hours.

Surface energy was determined by applying test inks as used for corona treatment of plastic webs.

All these tests were conducted with the same type of VW "Golf" car headlight reflectors made from lacquered thermoset plastic material.

RESULTS

Aluminum Layer: The thickness distribution of aluminum measured in the front plane of the substrate chamber is shown in Fig. 6. Even without moving the substrate relative to the magnetrons during deposition, the Al thickness varies only between 70 % and 100 % (85 ± 15 %) over most of the area. At a small area in the center, the thickness reduces to somewhat less than 70 %.

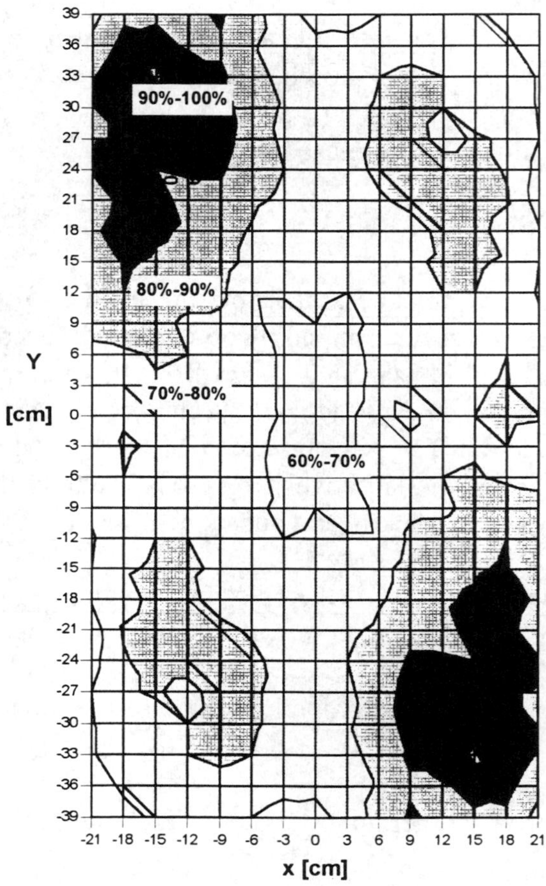

Fig. 6. Thickness distribution of aluminum deposit in the front plane of the substrate chamber.

The thickness distribution on real three-dimensional substrates was measured with five headlight reflectors of the VW "Golf" type, positioned as shown in Figs. 4 and 5. The Volkswagen reflectors are numbered from 1 to 5 from top to bottom. Fig. 8 shows that there is a variation by a factor of 2.3 between the front surfaces (level 1 in Fig. 7) and the parabolic areas which are nearest to the light source (level 4). In further process optimization this factor was reduced to 1.8. This is a better thickness ratio than the one achieved by evaporation. The thickness variation between equal positions at the reflectors (denoted 1.1 to 4.2) at different positions of the reflectors inside the chamber (denoted 1 to 5) is relatively small.

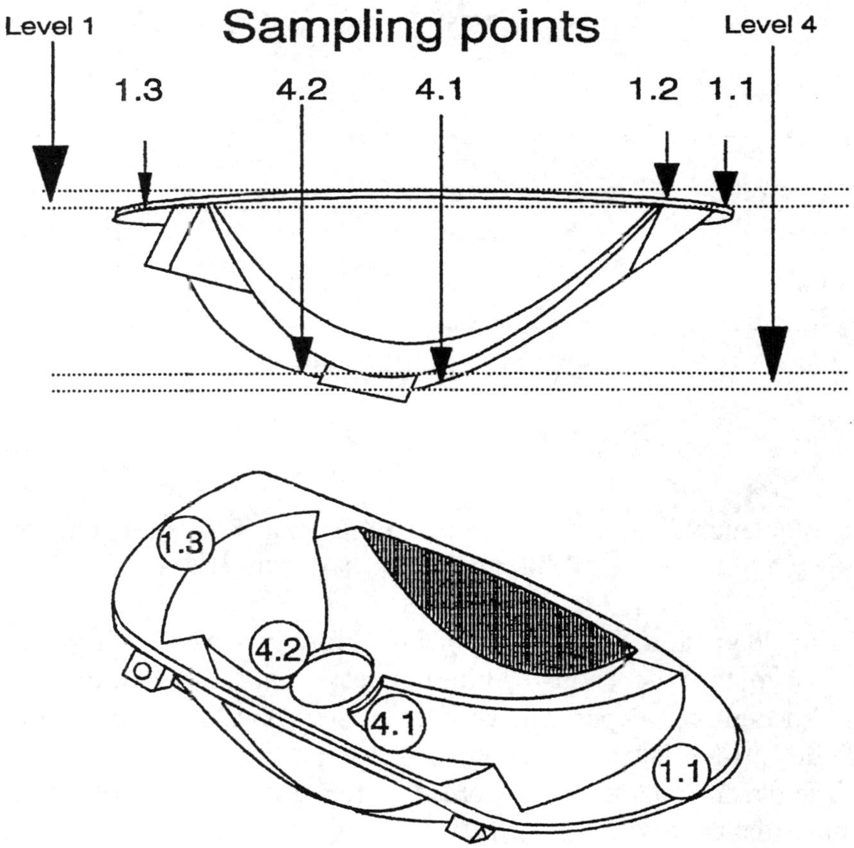

Fig. 7. Definition of sampling positions for measurement of aluminum and top coat thickness in a headlight reflector.

The adhesion of the Al-layer was found to be strongly dependent on the plastic base material and the pretreatment. So far, for each surface to be coated, lacquered or not, a suitable pretreatment could be identified.

Top Coat: The deposition rate of the siloxane plasma polymer was about 3 nm/s. The thickness distribution of the plasma polymer protective coating over the "Golf" headlight reflector (Fig. 7) was found to be as uniform as the Al layer. The ratio of the average film thickness on level 1 to the thickness on level 4 was found to be less than 2.1 after some optimization.

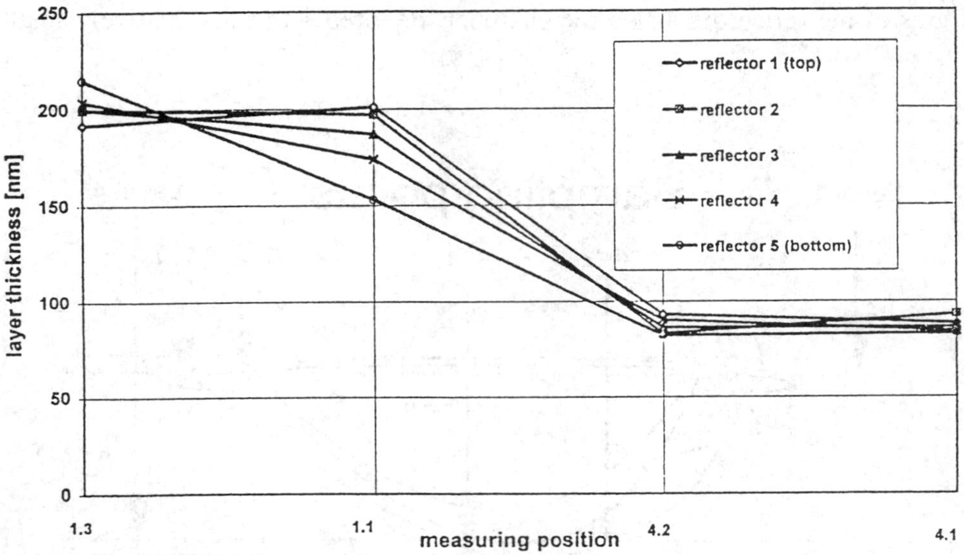

Fig. 8. Aluminum thickness distribution at 4 positions as denoted in Fig. 7 on each of 5 "Golf" reflectors (1-5) from the same batch.

The deposition time for efficient corrosion protective layers is in the range of 10 to 15 s, depending on the reflector shape. This means that there is enough time left for depositing another plasma polymer such as some more hydrophilic anti-fog layer if required.

The plasma polymerization process proved to be even more stable than in the common batch coaters. No difference in adhesion and corrosion test results could be found after 10, 500, 1000 and 2500 cycles. For the usual requirements of the car industry today (240 h humidity chamber, 24 h sodium dodecyl sulfate and 3 h NaOH bath) all tests were passed. A large process window was found for a large variety of differently shaped reflector parts made from lacquered thermoset materials or non-lacquered thermoplastics such as ABS, polycarbonate (PC) and ABS/PC blends. By some parameter optimization, considerably longer exposure times before failure were achieved, such as > 1000 h humidity chamber (test stopped) for the "Golf" headlight reflector.

The surface energy of the corrosion protective top coat was measured as approximately 32 mJ/m², depending on process conditions. It could be shown that this is sufficient for direct glueing of the lens to the reflector. However, surface energy may not be high enough for some other adhesives. Also, some outgassing from the reflector and other components could condense on the corrosion protective layer and form droplets, such that reflectively of the headlight would fade. In order to avoid this, some overlayers were developed with higher surface energy (between 44 and 70 mJ/m²) in order to spread the condensing materials evenly on the reflector surface. These anti-fog layers deposited from a variety of hydrocarbons within 12 s did not influence optical properties of the reflectors. Colors were neither induced by absorption nor reflectance decreased.

SUMMARY AND CONCLUSION

We have presented a successful realization of a just-in-time coater concept for car headlight reflectors. Using sputtering instead of evaporation and with a newly developed plasma polymerization source for the top coat, a cycle time of 36 s was achieved for a substrate volume of 800 x 450 x 250 mm^3. This equals a throughput of e.g. 5 "Golf" headlamp reflectors per 36 s. Both deposition processes run reliably and show a uniform thickness distribution, even on complexly shaped substrates. All relevant corrosion tests in car industry are passed within a comfortably large process parameter window. In production now for more than a year, the plant proved to operate very reliably. The sputter chamber and plasma polymerization chamber can all be operated well beyond 2500 cycles (three shifts) before cleaning. So the uptime is expected to be high. The coater can be integrated into automatically working production lines. Production cost can be cut significantly by automating (un)loading and by smaller stocks of parts and substrate holders.

REFERENCES

1. F.J. Horn, and H.B. Hebble, Jr. in "Vapor Deposition," C.F. Powell, J.H. Oxley, and J.M. Blocker, Jr. (eds); 2nd Ed., Chapter 17. Wiley, New York (1967)
2. H. Winterhalter in: "Veredeln von Kunststoff-Oberflächen," K. Stoeckhert (ed.), Chapter 4, Hanser, Munich (1974), (in German)
3. H. Stein, E. Zehender, K. Kerner, and B. Blaich, DE Pat. 26 25 448 (1976)
4. G. Dittmer, *Ind. Res. Devel.*, 169-183 (Sept. 1978).
5. D. Hofmann, G. Ickes, U. Patz, and R. Kukla, Proc. Society of Vacuum Coaters, 37th Ann., Techn. Conf., 250 (1994)
6. R. Kukla, D. Hofmann, G. Ickes and U. Patz, *Surface Coatings Technol.*, **74-75**, 648 (1995)
7. R. Kukla, T. Krug, R. Ludwig, and K. Wilmes, *Vacuum*, **41**, 1968 (1990)
8. R.S. Hunter and R.W. Harold in "The Measurement of Appearance," 2nd Ed. Chapter 8, Wiley, New York (1987).

Metallized Plastics 5&6: Fundamental and Applied Aspects, pp. 109-117.
K.L. Mittal (Ed.)

ELECTRODEPOSITION OF COPPER ON PVC USING SOLUBLE POLYANILINE

C.C.M. Fornari Jr., C. Heilmann and C.A. Ferreira

Laboratório de Materiais Poliméricos, Escola de Engenharia, Universidade Federal do Rio Grande do Sul, Av. Osvaldo Aranha, 99 sala 701, 90035-190, Porto Alegre, Brazil.

The conducting polymer polyaniline (PAni) in its emeraldine conductive state doped with camphorsulfonic acid (CSA) was applied on the poly(vinyl choride) (PVC) substrate in the form of a solution in N-methylpyrrolidinone (NMP). The formation of the PAni conductive film occurred by solvent evaporation. The copper deposition was carried out on PAni film with an electrolytic cell containing the $CuSO_4$ / NaCl solution and the Cu deposits with the highest adhesion were obtained at 45°C and at a current density of 15 $mA.cm^{-2}$.

INTRODUCTION

The chemical metallization[1] is a technique that allows covering of plastics with a metallic layer, using normally a process involving two steps. The first one is an electroless metallization, where metallic ions (Cu ions, for example) are reduced by a reaction involving Sn atoms catalysed by Pt, Pd or

Au and allowing the formation of a thin metallic layer of Cu that increases the conductivity of the polymer surface. In the second step, the thickness of the Cu layer is increased by electroplating, and thus one can obtain very good quality metallic films. The metallized plastics obtained by this process present several advantages when compared with the metallic equivalents such as lighter weight and lower cost of fabrication. Among the several existing plastics only ABS (Acrylonitrile-Butadiene-Styrene terpolymer) can be metallized by using the chemical metallization. In this work we present the metallization process on cheaper polymer, poly(vinyl choride) (PVC) using a conductive polymer polyaniline (PAni) dissolved in N-methylpyrrolidinone (NMP).

PAni, the structure of which is presented in figure 1, is an intrinsically conductive polymer and thus offers many possibilities of technological applications[2].

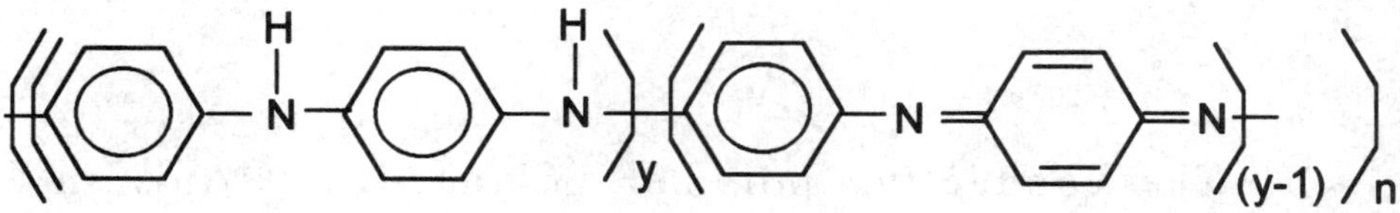

Figure 1. Structural formula of polyaniline (PAni).

In its totally reduced leucoemeraldine form (y=1) or totally oxidized pernigraniline form (y=0), PAni is easily soluble in NMP, but presents the inconvenience of being an electrical insulator similar to classical polymers. When doped at the intermediate oxidation states ($0.4 < y < 0.6$) the emeraldine form of PAni becomes conductive, but is insoluble[3].

According to Epstein and MacDiarmid[4], the PAni in its conductive state and in contact with common solvents like chloroform is in the compact coil structure, hindering the interaction of the chains with the solvent. To increase its solubility, its treatment with a specific doping agent: camphorsulfonic acid (CSA)[4,5] has been proposed. This treatment increases the conductivity of PAni up to 100 times than that of the films doped with HCl through the complete protonation of the imine nitrogen atoms, as well as it induces its solubility in solvents like m-cresol due to the effect of the CSA counter-ion[4], by changing the PAni molecular conformation from the compact coil structure to the expanded coil structure.

The PAni films produced from m-cresol solution have conductive properties. However, Epstein and MacDiarmid[4] assumed that in chloroform, where PAni is in a compact coil form, its conductivity is in the order of 10^{-1} $S.cm^{-1}$ and it has low crystallinity. The treatment of this compact form with CSA makes the conformation of PAni more extended when in contact with m-cresol. So these authors claim there are two factors that affect the conductivity and solubility of PAni: the primary and the secondary doping. The primary doping is related to the conformation of the polymer chains influenced by the action of the doping acid HCl, CSA, TSA (p-toluene sulfonic acid), DBSA (dodecylbenzenesulfonic acid)[6,7] and the secondary doping is related to the solvent-polymer interaction through hydrogen bonds[8].

Once soluble PAni is obtained in the conductive form, it can be applied as a paint on insulating materials, forming conductive films thereon. One possible application of these films is in the metallization of conventional polymers.

We have proposed a technique to provide conductivity to the surfaces of classical polymers like HDPE (high density polyethylene) by immersing the polymer sample in the reaction bath during the chemical synthesis of PAni[9]. This procedure produces a film of PAni on the surface of the polymeric material, and after that one can obtain a copper film by electrochemical procedure. However, this process presents the inconvenience of producing an excessive amount of insoluble conductive polymer. The PAni and the copper films deposited on PAni/HDPE presented excellent adhesion to the substrate.

The use of soluble PAni offers some advantages because it facilitates the control of the film thickness, the application on pieces of different sizes and shapes, and the possibility of total use of PAni.

EXPERIMENTAL

Aniline (Aldrich) was bi-distilled under N_2 before polymerization. All the others chemicals were of pro Analysis grade and were used without further purification.

The PVC was cut in plates with dimensions of 2 x 2.5 cm. The plates were cleaned with a detergent solution, dried and placed quickly over a liquefied petroleum gas flame, in order to produce a hydrophilic surface. After that the PAni/NMP solution was brushed onto the PVC plates which were placed in an oven at 30°C for complete solvent evaporation.

The copper deposition was carried out in a solution containing 250g/l of $CuSO_4$ and 157.7g/l of NaCl in distilled water[10] with slightly acidic pH, i.e. ~ 6.0. A copper bar was used as the counter-electrode and all potentials were referred to a standard saturated calomel electrode (SCE).

The morphologies of copper deposits were characterized by scanning electron microscopy using a Philips model XL20 microscope.

The adhesion tests of the PAni and the copper films to the PVC substrate were carried out in accordance with the DIN 53151 method. The method consists in scratching six parallel lines separed by 1 mm in the specimen and then another set of six parallel lines scratched at right angles to the previous ones. The second step is to apply an adhesive tape to the specimen surface and pull it off quickly. The adhesion is evaluated by the number of squares pulled off by the tape (100% adhesion if no squares are pulled off).

RESULTS

Obtaining Conductive Soluble PAni

The chemical polymerization of aniline was carried out from a 0.05M aniline solution using 0.1M ammonium persulfate $(NH_4)_2S_2O_8$ as the oxidizing agent in 1M HCl solution. The reaction was carried out in an ice bath for a period of 2 hours[11], under magnetic stirring. The de-doping of the polymer was done in 0.1M NH_4OH for 24 hours under magnetic stirring. The aniline polymerization process was accompanied through the open circuit potential technique[12] and the results are shown in figure 2. The signal presents the same shape as described previously[13] showing a maximum at 0.77 V/SCE.

De-doped PAni was then re-doped with CSA in an adequate proportion to protonate 50% of the nitrogen atoms of the chains. This protonation occurred through the physical mixture of PAni and the acid in an agate mortar under inert atmosphere, and CSA doped PAni was then dissolved in NMP in a concentration of at least 1%.

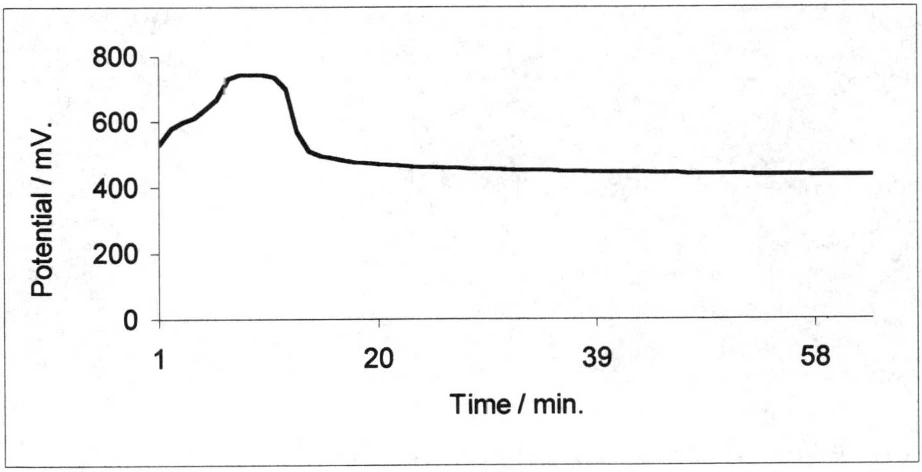

Figure 2. Open circuit potential (mV/SCE) profile for PAni polymerization using $(NH_4)_2S_2O_8$ as oxidizing agent.

Deposition of Cu on PAni/PVC.

The PAni solution was brushed onto the flame treated PVC samples, in order to render the PVC surface conductive. After the solvent evaporation, the tape test showed 100% adhesion of the PAni films on the PVC surface.

The plates of PVC treated by flame, the PAni film deposited on PVC surface and the electrodeposited copper film on the PAni/PVC are presented in figure 3.

The copper deposition on the PAni/PVC composite was carried out with an electrolytic cell containing the $CuSO_4$/NaCl solution. Several current densities were applied as shown in figure 4. The adhesion of the copper film to the PAni/PVC composite substrate depends on the current density and the temperature of the electrochemical bath. The total time of deposition was 20 minutes with a lineal speed of propagation of the growing copper film equal to 1.2 $mm.min^{-1}$. The best adhesion results were obtained at a current density of 15 $mA.cm^{-2}$ and at 45^0C.

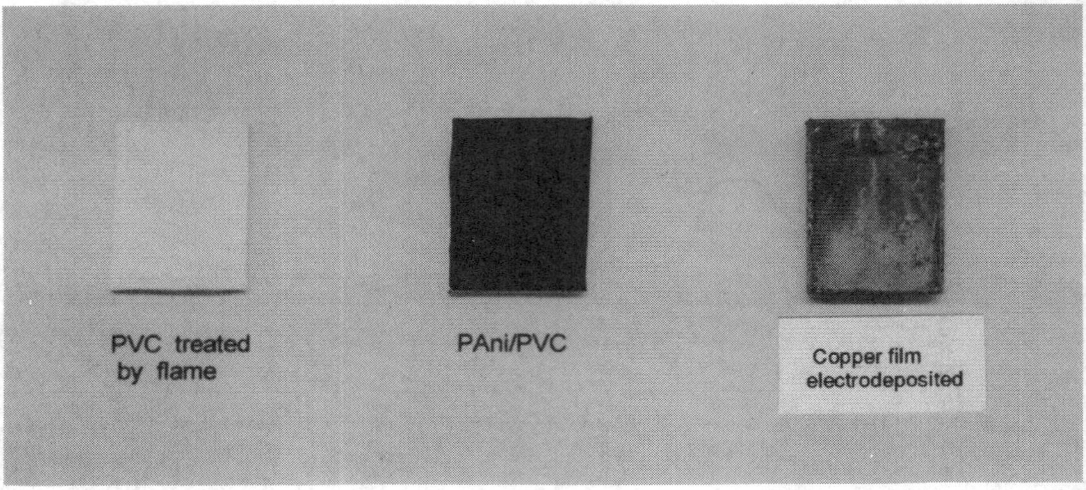

Figure 3 – Samples of flame-treated PVC, PAni on PVC and electrodeposited copper film on PAni/PVC.

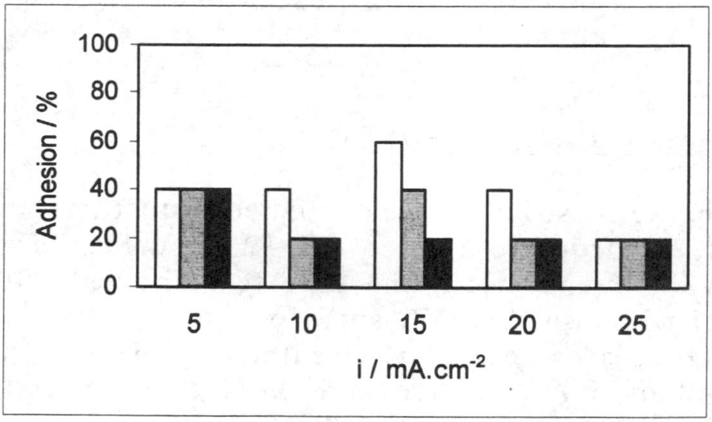

Figure 4 – Adhesion of copper films to the PAni/PVC as a function of the current density (i) and the temperature of the electrochemical bath. The white bar, 45^0C; the gray bar, 30^0C; and the black bar, 25^0C.

The deposition of the copper film begins on the wire that makes the electrical contact with the PAni/PVC plate and it grows on the whole surface developing a thickness gradient (~50μm at the top and ~30 μm at the bottom of the PAni/PVC sample), as shown in figure 5. The wire should be partially immersed in the electrolytic solution during the copper electroplating, based on previous works[10,14].

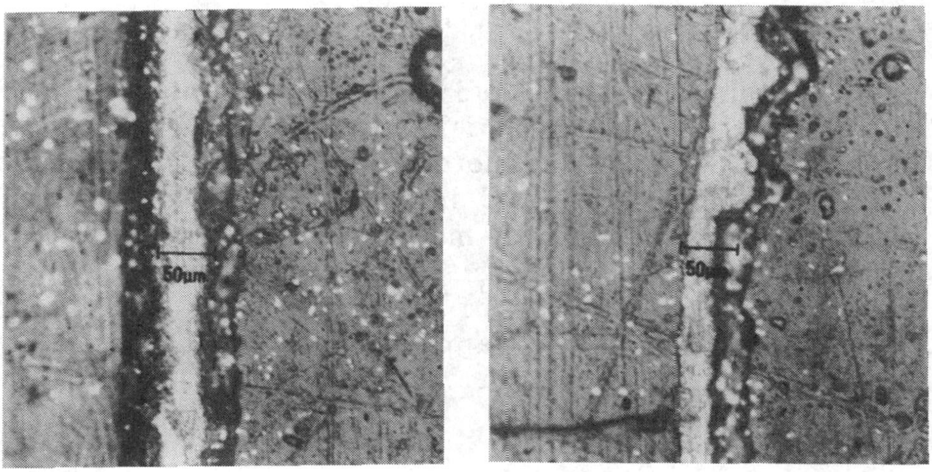

Top Bottom

Figure 5 – Cross-sectional view of a Cu/PAni/PVC sample. Pictures were taken near the wire contact (top) and near the bottom of the cell (bottom).

Figure 6 – SEM micrograph of copper film electrodeposited on the PAni/PVC composite.

The electroplated copper film shows excellent brightness and uniformity. Figure 6 shows the SEM image of the copper film obtained.

CONCLUSIONS

We have shown in this study that it is possible to metallize PVC making use of a solution of PAni. After the preparation of a conductive PAni solution the metallization process requires three steps: the treatment of the PVC surface, the application of PAni, and the electroplating of the metallic film. The copper films deposited on PAni/PVC composite show a very good appearance and uniformity. The adhesion of the copper film is related both to the bath conditions and the applied current. The best adhesion results were obtained at 45^0C at a current density of 15 $mA.cm^{-2}$.

ACKNOWLEDGMENTS

The authors are grateful to FAPERGS, CAPES and CNPq for financial support.

REFERENCES

1. W. Goldie, "Metallic Coatings of Plastics", vol.1, pp 15-73, Electrochemical Publications Limited, Middlesex, England, 1968.
2. L. H. C Matoso, Química Nova, **19**, 388 (1996).
3. Y. Wei, X. Tang and Y. Sum, J. Polym. Sci., **27**, 2385 (1989).
4. J. A. Epstein and A. G. MacDiarmid, Synth. Met., **65**, 103 (1994).
5. Y. Cao, P. Smith and A.J. Heeger, Synth. Met., **48**, 91 (1992).
6. M. K. Jang, K. Seok and I. J. Chung, Synth. Met., **69**, 139 (1995).
7. K. G. Neoh and M. Y. Pum, E. T. Kang and K. L. Tan, Synth. Met., **73**, 209 (1995).
8. J. A Heeger, P. Smith, A. Andreatta, Y. Cao, E. Österholm, O T. Ikkala, J. Väkipartta, E. Virtanen, H. Järvinen, T. Taka and P. Passiniemia, Synth. Met., **69**, 97 (1995).
9. C. Yang , P. Smith and Y. Cao, Synth. Met., **69**, 191 (1995).
10. Galvanotécnica Prática, Edited by Polígono Ltda., São Paulo, vol. 2, 1973.

11. C. A. Ferreira, D. M. Lenz and M. E. R. Schultz, J. Polym. Eng., **16**, 295 (1996).
12. N. Gospodinova, P. Mokreva and L. Terlemezyan, Polymer, **14**, 3102 (1994).
13. J. A. Epstein, A. G. MacDiarmid and L. H. C. Mattoso, Synth. Met., **68**, 1 (1994).
14. J. M. Park, R. J. White, M. Angelopoulos and W. S. Huang, Mol. Cryst. Liq. Cryst., **189**, 227 (1990).

Metallized Plastics 5&6: Fundamental and Applied Aspects, pp. 119-124.
K.L. Mittal (Ed.)

STUDY OF MECHANISM OF DIRECT COPPER PLATING ONTO NONCONDUCTORS VIA Pd/Sn ACTIVATION AND SULFUR PROMOTION

C. H. Yang, Y. Y. Wang and C. C. Wan

Department of Chemical Engineering
National Tsing-Hua University
Hsin-chu, Taiwan, 30043, R.O.C.

C. J. Chen

Gold Circuit Electronics Ltd.
113, Shi Yuan Rd, Chung Li IND. PARK.
Tao-yuan, Taiwan, 32036, R.O.C.

Direct copper electrodeposition can be carried out on a nonconductor if treated with Pd/Sn activator and sodium sulfide solution. A model based on the bridging ligand concept is proposed to explain why this process is possible. Several new ligand compounds were also found effective and could probably serve as substitute for sodium sulfide.

INTRODUCTION

Electroless copper deposition is a key technology for fabricating of printed circuits. However, there is always concern regarding its bath control and the toxicity of the chemicals involved. So several processes have been developed as possible substitutes. They are usually referred to as "direct metallization" or "direct plating" since the nonconductive substrate can be copper electroplated directly after certain activation step without the need for a conductive electroless copper film. These processes can be classified into three groups. The first kind is based on the application of a conductive polymer film, whereby subsequent electrodeposition can be effected. The second kind uses colloidal dispersion of carbon instead. The underlying principle is similar since the carbon colloid can also form a conductive film on the nonconductor surface. Since the conductivity of the surface can be increased from another source, electroless copper can consequently be eliminated.

The third type of process uses a Pd/Sn catalyst coupled with a sulfide solution dip. The best known process is the Crimson Process developed by Shipley Co.[1-5]. The first part of the process resembles the first part in electroless copper.

However, after undergoing activation in a Pd/Sn catalyst solution and acceleration, the substrate is dipped in a sodium sulfide bath. The substrate becomes readily platable after this immersion. It seems reasonable to visualize the plating mechanism of this process based on the same principle as the previous two kinds.

Researchers[6--8] have envisioned that the palladium sulfide formed on the surface constitutes a conductive layer and, consequently, copper electroplating can be done later. However, we found the surface conductivity to be very low when we actually measured it after the sulfide treatment (resistivity over 200 MΩ). In fact, the surface conductivity after this step is about two orders of magnitude lower than a surface treated with a conductive polymer film. But when copper plating is applied to it, the rate of lateral copper growth is at least ten times as fast as it is on conductive polymer films. Obviously, another totally different mechanism is needed to explain this kind of direct metallization process. Weng and Landau[9] did propose a model for a surface covered with discrete palladium catalyst particles. However their model, also based on the concept of conductivity, seems unable to offer an explanation why the lateral growth rate can be dramatically enhanced if the palladium particles are treated with a sodium sulfide solution. The growth rate in their study was very slow, i.e. about 0.02 cm/min, which is slower by at least an order of magnitude than the process in industrial practice.

PROPOSED MODEL

A sulfide-treated nonconductor substrate was electrically connected to a wire as shown in Fig. 1. Once current was applied to the working electrode, copper grew laterally from the tip of the connection wire in the form of a disk. Why does this sulfide-treated surface, although still remaining practically nonconductive, can become readily platable can be explained in terms of the following two reactions model,

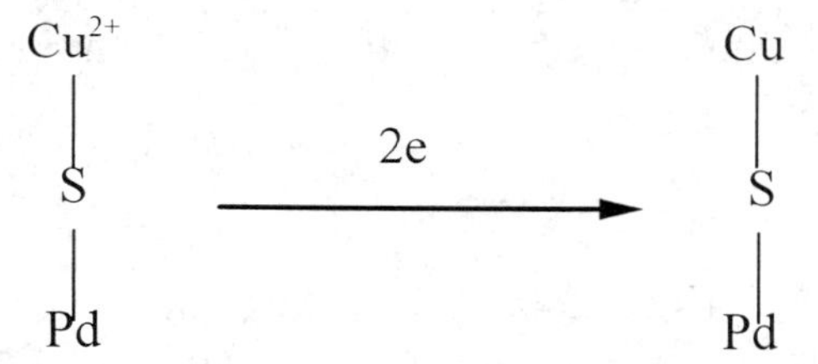

(1)

$$Cu^{2+} \xrightarrow{2e} Cu \qquad (2)$$

Palladium sulfide is adsorbed on the nonconductor surface. Its surface conductivity remains very poor. However, the sulfide ion is a bi-functional ligand in this case. Hence it has the ability to attract cupric ions from the bulk solution and force them to be attached to the surface, rendering the subsequent electroreduction possible. Furthermore, the sulfide ion being a ligand can remove hydration water molecules from the cupric ion to complete the reduction reaction[10]. These properties facilitate

Reaction(1) to occur in the vicinity of a current connection tip.

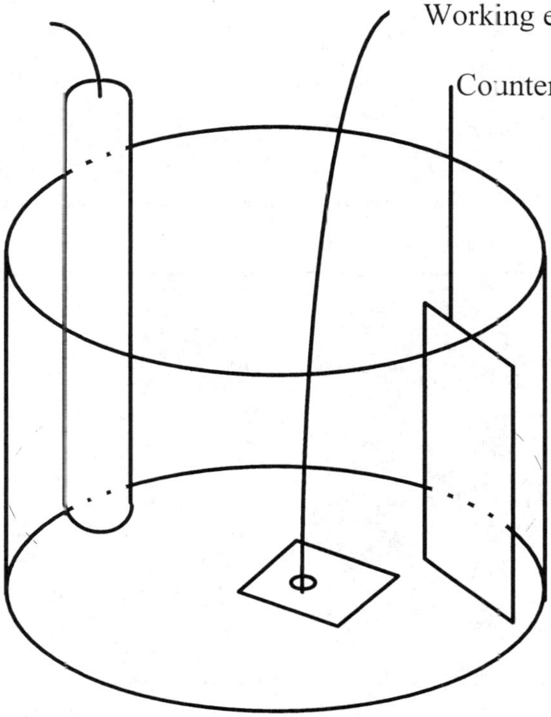

Fig 1. Direct Plating Cell.

Once Reaction (1) starts, the newly formed metallic copper makes the surface conductivity at that very spot increase quickly. So electrons can move from the connection wire to the spot easily and Reaction (2) can also proceed vertically. We can envision the spot with much improved conductivity due to newly formed copper deposit as an extension of the connection tip. Without Reaction (1), Reaction (2) can only proceed very slowly as Weng and Landau[9] described.

The combination of Reaction (1) and Reaction (2) makes the copper deposit extend laterally very fast and the thickness of the deposit also increases simultaneously. This is why this system becomes industrially feasible.

A schematic diagram is shown in Fig. 2 to illustrate how copper deposit grows in three dimensions via this bridging ligand property.

DISCUSSION

Detailed electrochemical studies have been made to support the proposed model[10]. If the model is correct, the two reactions should show a mixed potential characteristic. This means we should be able to observe a plating potential less polarized than the electrode potential corresponding to Reaction (2). Furthermore, if the current were interrupted, the two reactions would then form a local-cell type couple. In other words, Reaction (1) would remain cathodic while Reaction (2) would go anodic and proceed in the reverse direction. This means that we should

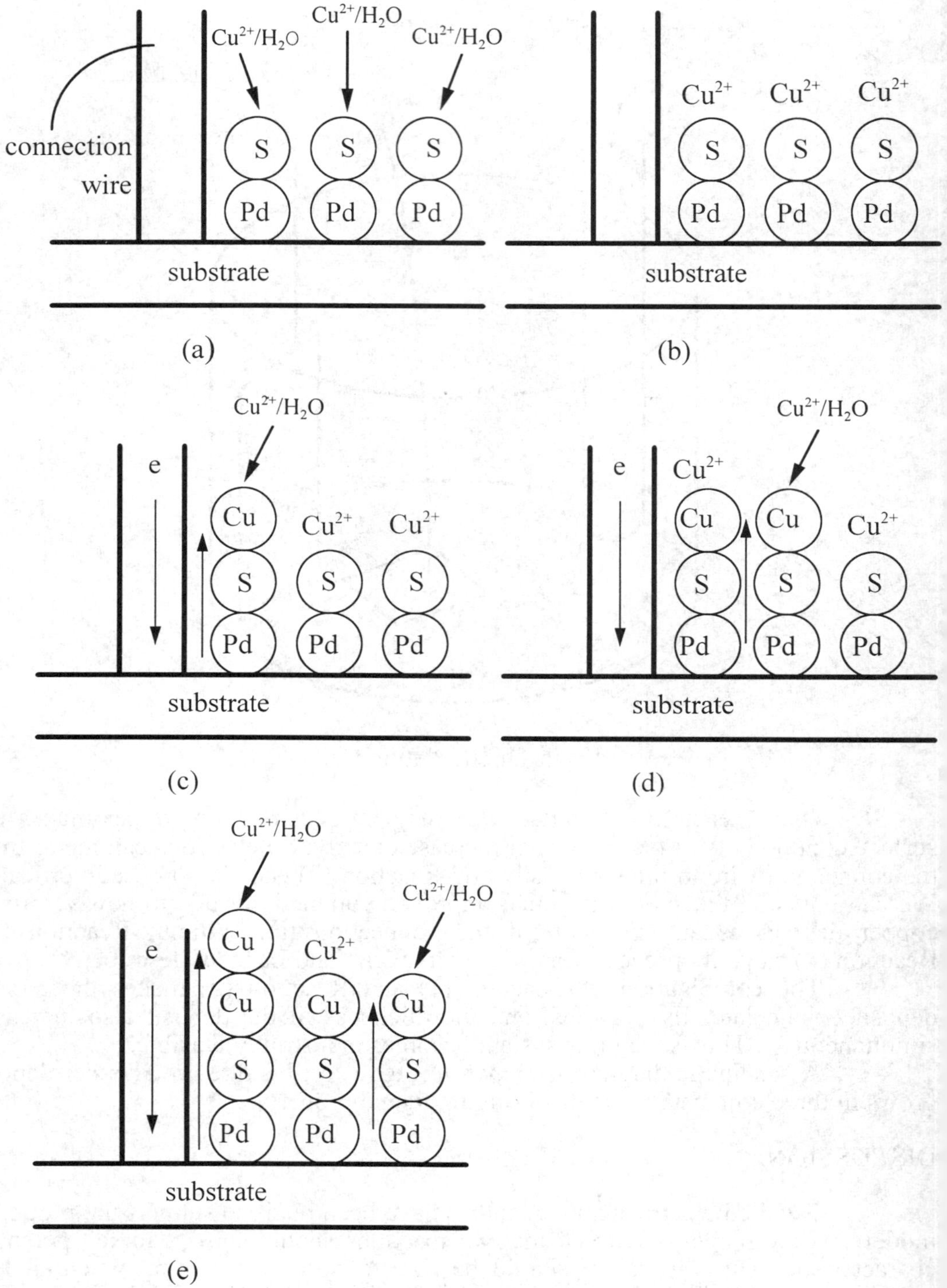

Fig. 2. Proposed two-reaction model. (a)&(b) -- Cu^{2+} ions attach onto S from the plating solution. (c)&(d) -- adsorbed Cu^{2+} ions are reduced by charge transfer process (Reaction (1)). (e) -- Reaction (1) and Reaction (2) occur spontaneously.

observe a continuous lateral copper growth even after the current is switched off and, more interestingly, we should simultaneously observe a mild dissolution of copper in the area already deposited. All the above statements have been confirmed experimentally[10].

Another logical deduction from this ligand concept would be that there should be other compounds exhibiting similar function if they also possess this bi-functional ligand property. In fact, we found that compounds like NaSCN, $(NH_2)_2CS$, NaCN and KI could serve similar function as Na_2S. currently employed[10]. The results are shown in Fig. 3. It is clear that on applying a current to the substrate, all of the above mentioned ligands can exert depolarizing ability. Among the ligands studied, sodium sulfide is still the most effective based on deposition rate. However, these other ligands may be of industrial interest from other point of view. For instance, some ligands do not contain sulfur which tends to form undesired strong bond with copper and needs additional etching to remove it.

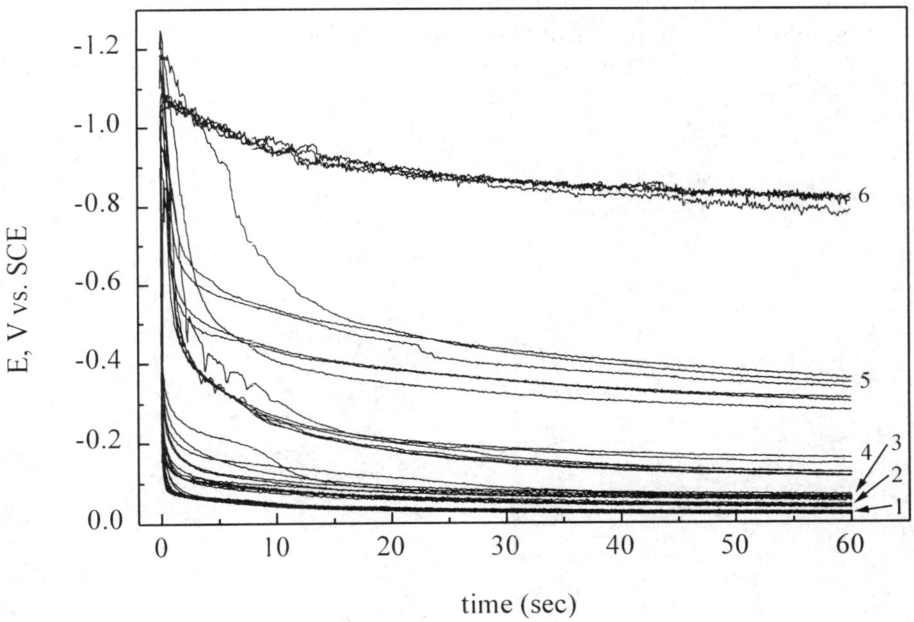

Fig. 3. Potential decay curves for constant current direct plating after different promotion processes: 1. Na_2S, 2. $(NH_2)_2CS$, 3. NaSCN, 4. NaCN, 5. KI and 6. blank without any promotion process.

CONCLUSION

A model based on the bridging ligand concept was proposed to explain why direct plating via Pd/Sn activation and sodium sulfide dip can readily proceed.

Ligands other than Na_2S were also found effective for direct plating although their potential for industrial application needs to be investigated further.

REFERENCES

1. M. Gulla, D. F. Foust and G. K. Philipose, U. S. patent 4,810,333 (1989).
2. J. J. Bladon, U. S. patent 4,895,739 (1990).
3. J. J. Bladon, U. S. patent 4,919,768 (1990).
4. J. J. Bladon, N. Robinson and M. Rousseau, U. S. patent 4,952,286 (1990).
5. J. J. Bladon, U. S. patent 5,007,990 (1991).
6. G. Allardyce, S. Hirst, J. J. Bladon and J. Rasch, *Trans Inst. Metal Finish.*, **72** (1), 7 (1994).
7. H. Meyer, R. J. Nichols, D. Schroer and L. Stamp, *Electrochim. Acta,* **39**, 1325 (1994).
8. K. H. Dietz, *Plating Surf. Finish.*, **82 (7)**, 60 (July. 1995).
9. D. Weng and U. Landau, *J. Electrochem. Soc.*, **142**, 2598 (1995).
10. C. H. Yang, Y. Y. Wang, C. C. Wan and C. J. Chen, *J. Electrochem. Soc.*, **143**, 3521 (1996).

Metallized Plastics 5&6: Fundamental and Applied Aspects, pp. 125-130.
K.L. Mittal (Ed.)

A DUAL-REACTION MECHANISM FOR DIRECT COPPER PLATING ON NONCONDUCTORS VIA Pd ACTIVATION AND SULFUR PROMOTION

C. H. Yang, Y. Y. Wang, and C. C. Wan

Department of Chemical Engineering
National Tsing-Hua University
Hsinchu 300, Taiwan

C. J. Chen

Gold Circuit Electronics Co.
113 Shi Yuan Rd., Chung Li Industrial Park
Taoyuan 320, Taiwan

A dual-reaction mechanism for direct copper plating using Pd/Sn catalyst and Na_2S solution as promoter was previously proposed and substantiated by several quantitative experiments. In this study, polarization results were employed to evaluate the kinetic parameters of these two reactions proceeding in parallel. The reaction involving bridging ligand was found to be more favorable in low copper concentration systems.

INTRODUCTION

Autocatalytic electroless copper plating has been widely used to metallize a nonconductor surface. One of its most successful applications is the plated through-hole process (PTH) for the printed circuit board industry. However, environmental concern has pressured the industry to develop alternative process which needs no toxic compounds like formaldehyde. These processes can in general be categorized into three groups.[1] The first alternative is based on colloidal carbon. The second group uses a conductive polymer film to replace the electroless copper film. The third group uses a metal-colloid catalyst followed by a promotion step. Since the basic function of electroless copper is to apply an electrically conductive film on the nonconductor surface so that subsequent electroplating can be made possible, it is natural to rationalize that the three alternative processes provide different conductive films.[1,2] In fact, direct metallization (DM) has been defined as an operation whereby the nonconductive surface of PTH/vias is rendered conductive so that electroplating can be directly applied.[2] This is applicable to the first two processes. However, this explanation is not totally acceptable for the third group. Hupe and Iwan[3] used poly(vinyl pyrrolidone) as promoter and Pd/Sn as catalyst. In their patent, they confessed that why the surface became directly platable was still unknown. Presumably, they have measured the surface conductivity and found its high resistance simply cannot support the concept commonly held by many workers. Shipley company has developed a series of processes based on Na_2S solution as promoter and Pd/Sn catalyst.[4-7] The Pd catalyst is converted to PdS by contacting the Na_2S solution. The process is industrially named the Crimson Process and the promoter is named "Enhancer" by the supplier. The surface after activation is practically unplatable and exhibits high surface resistance. However, the surface becomes readily platable after promotion although the surface still shows high resistance. The reason for the dramatic increase of platability needs a logical answer. Yang et al.[8] first proposed a dual-reaction model to explain this apparent paradox. The surface after treatment with the Pd/Sn catalyst and Na_2S promoter can proceed with the following two reactions in parallel.

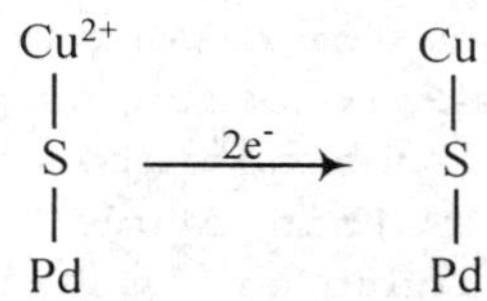

(1)

$$Cu^{2+} \xrightarrow{2e^-} Cu \qquad (2)$$

According to this model, the adsorbed PdS on the surface provides little additional conductivity, hence the surface still exhibits high resistance. However the sulfide ion is a bi-function ligand which can attract cupric ions from the solution. In addition, the sulfide ion being a bridging ligand can pass current and induce the occurrence of Reaction (1). Since PdS is uniformly adsorbed on the surface, Reaction (1) starts from the spot in contact with the external current and proceeds laterally along the surface very rapidly. Once a region is covered with copper by Reaction (1), its surface conductivity increases dramatically and Reaction (2) begins in that region and copper can be deposited vertically to increase the deposit thickness.

This model has been substantiated by many qualitative experiments.[8] However it seemed natural to seek quantitative kinetic information related to Reaction (1) and Reaction (2). So the object of this study was to design a method to differentiate the response of these two reactions to applied potential.

EXPERIMENTS

The direct plating procedure practiced in this study is as follows.

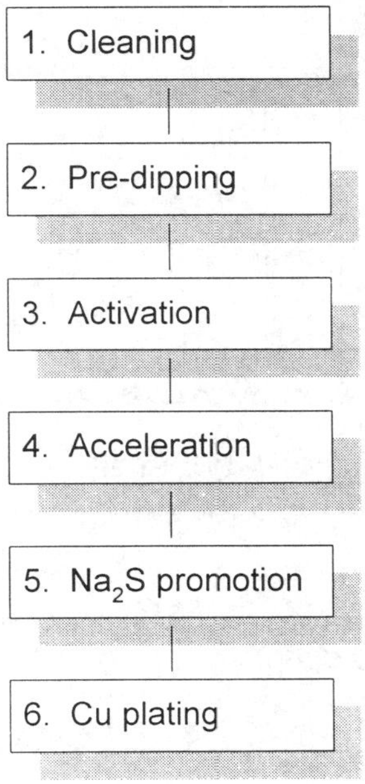

So the sequence basically follows Crimson process. The composition of the activation solution was palladium chloride 10 g/L, stannous chloride 500 g/L, and hydrochloric acid 300 mL/L. The concentration of the palladium was much higher than in the conventional electroless copper process so that the commonly required conditioning step could be skipped. This is because conditioning step would introduce organic compounds onto the board, which may interfere in the subsequent surface analysis. After dipping and rinsing with water, the test board, which was an FR-4 inner layer board with no copper coating, went through the acceleration step by dipping in 4 g/L sodium hydroxide solution for five minutes. The board was then immersed in a promoter solution (Na_2S, 0.5 g/L) which sped up the rate of direct plating to an acceptable level.

The copper plating procedure was the same as reported previously.[8] However for kinetic studies, we used four different concentrations of copper sulfate: 0.8 M, 0.3 M, 0.03 M, and 0.006 M.

The reaction model is conceptually shown in Figure 1.

As the tip of the connection wire was attached to the surface, Reaction (1) would start from the tip to form a lateral growth of copper deposit (i_r), which would quickly increase the local electrical conductivity and induce a vertical growth of copper (i_z). Without an effective PdS film to catalyze the surface, only Reaction (2) can occur at the tip. In this case, i_r is practically zero and copper deposit would not spread along the surface. This is why Na_2S dipping is so critical for this system.

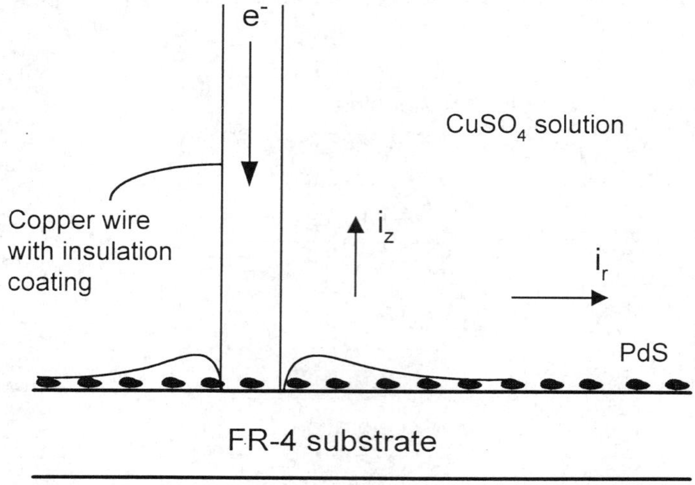

Figure 1. Conceptual reaction model of direct copper plating on palladium sulfide.

RESULTS AND DISCUSSION

According to the proposed model, the direct plating current (i_t) can be envisioned as a sum of two parts.

$$i_t = i_r + i_z \tag{3}$$

If we control the system under low polarization condition, the current should be linearly related to the cathodic overpotential as follows,

$$i = -i_0 \frac{nF}{RT} \eta \tag{4}$$

where i_0 is the exchange current density, η is the overpotential, and n is the number of charges transferred. On a board without the PdS layer, the current obtained should be equal to i_z since i_r is zero in this case. So by subtracting this i_z from the i_t measured from a catalyzed board, we could obtain i_r in response to η. Since i_z and i_t were found to follow equation (4), the relation between i_r and η is naturally also linear.

A representative linear relation between these partial currents and potential is shown in Figure 2. We could calculate the exchange current density of these two reactions from the slopes of these straight lines. The results are shown in Table I. The $(i_z)_0$ values are consistent with the data of Mattson and Bockris,[9] which demonstrates the fundamental validity of our model and method. A closer look at Table I shows that $(i_r)_0$ is less affected by the concentration change of $CuSO_4$ than $(i_z)_0$. For instance, between 0.3 M and 0.8 M, $(i_r)_0$ and $(i_z)_0$ are comparable in magnitude. However, $(i_r)_0$ is four times as large as $(i_z)_0$ if the concentration of $CuSO_4$ is reduced to 0.006 M. This difference is understandable. In the case of low concentration, copper ions near the surface were already attached to sulfur ligands, which facilitated the quick lateral growth. Hence a major portion of the current went to i_r. In the case of high copper concentration, the concentration polarization due to copper ions in the solution is less pronounced and more current could be diverted to Reaction (2). Hence in a system in which surface coverage is top priority, a low copper concentration is recommended.

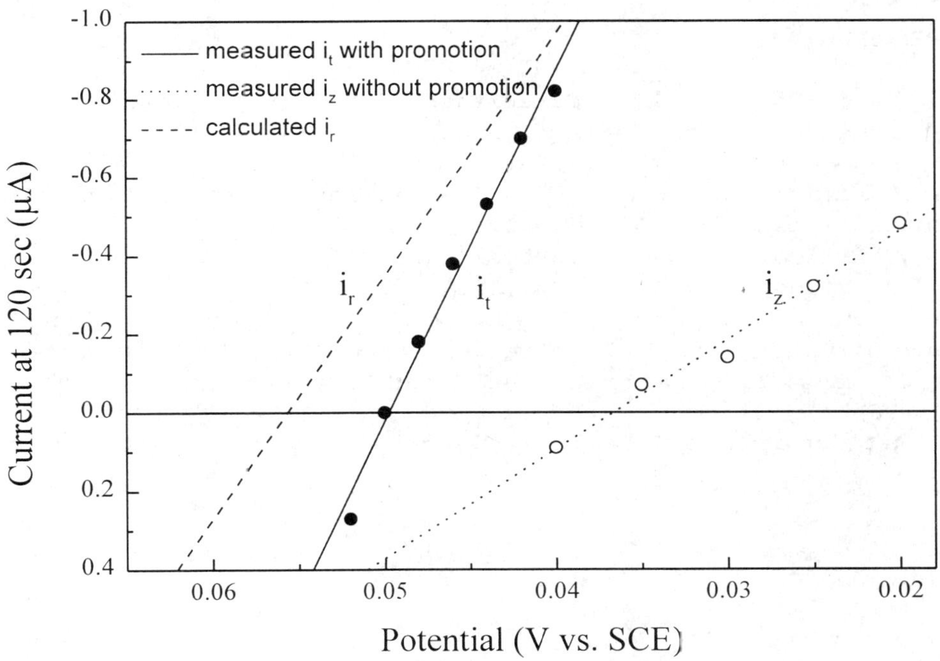

Figure 2. Experimental and calculated linear relations between the partial currents and the potential in 0.03 M copper sulfate electrolyte.

Table I. The Exchange Current Densities $(i_r)_0$ and $(i_z)_0$ in Different Concentrations of Copper Sulfate Solution.

$[CuSO_4]$, (M)	$(i_r)_o$, (mA/cm^2)	$(i_z)_o$, (mA/cm^2)	$(i_r)_o/(i_z)_o$
0.006	0.48	0.12	4.0
0.03	0.77	0.34	2.3
0.3	1.02	1.17	0.9
0.8	1.27	1.17	1.1

CONCLUSION

A mechanism of direct copper plating via Pd/Sn activation and Na_2S promotion was proposed based on a dual-reaction model and bridging ligand concept. This mechanism was further quantitatively verified by evaluation of the exchange current density of the two copper reduction reactions.

REFERENCES

1. K. Nargi-Toth, Printed Circuits Fab., 34 (Sept. 1992).
2. J. Langan, Plating Surface Finish., 46 (Jan. 1997).
3. J. Hupe and H. Iwan, U.S.Pat.5,145,572 (1992).
4. G. Allardyce, S. Hirst, J. J. Bladon, and J. Rasch, Trans. Inst. Metal Finish., 70, 135 (1992).
5. J. J. Bladon, U.S.Pat.4,895,739 (1990).
6. J. J. Bladon, U.S.Pat.5,007,990 (1991).
7. J. J. Bladon, A. Lamola, F. W Lytle, W. Sonnenberg, J. N. Robison, and G. Philipose, J. Electrochem. Soc., 143, 1206 (1996).
8. C. H. Yang, Y. Y. Wang, C. C. Wan, and C. J. Chen, ibid., 143, 3521 (1996).
9. E. Mattson and J. O'M. Bockris, Trans. Faraday Soc., 55, 1586 (1959).

Metallized Plastics 5&6: Fundamental and Applied Aspects, pp. 131-136.
K.L. Mittal (Ed.)

ECONOMICAL SELECTIVE METALLIZATION OF INSULATING SURFACES

Jim Y. Lee and C.Q. Cui[1]

Department of Chemical Engineering
National University of Singapore
10 Kent Ridge Crescent
Singapore 119260

An alternative process for depositing metals on selected areas of an insulating surface is suggested here. The selected areas are defined by non-photolithographic imaging techniques using carbon based printing materials. Metallization ensues immediately after treatment in an alcoholic KOH solution followed by chemical sensitization of the surface by silver ions. Electroplating of the same or another metal can then be used to build up the desired thickness and finishing. This process has good line resolutions limited only by printing, is a more generalized and simplified procedure, and uses inexpensive and non-proprietary chemicals. It is most adept at rapid prototyping and the fabrication of flexible printed circuits.

INTRODUCTION

Metallization of insulators is an effective way to render insulators conductive without incurring the weight and cost of the bulk metal. It is also occasionally used to impart metallic sheen and lustre to an otherwise unappealing look of the plastics. Depending on applications, the metal layer can vary in thickness from 0.5 μm to 40 μm. Metallization can be carried out by physical means such as vapour deposition, direct application of a conductive ink to the surface, or by chemical methods which

---

[1] Present address: Institute of Microelectronics, Singapore

sensitize the surface for chemical deposition of metal in subsequent steps [1-4]. The cost associated with the physical methods of deposition is relatively high and as such the methods are limited to applications that do not call for heavy build-up of thickness. In contrast, the chemical methods are more adaptable to engineering applications that require sufficiently thick deposits of metal only in selected areas of the surface, e.g. metallization of circuit patterns on a printed-circuit board (PCB). Such process involves many steps which invariably begin with the photolithographic generation of selected areas for metal deposition on the insulating surface, use of palladium-tin colloidal systems to sensitize the selected areas, and the activation of the palladium-tin colloids by an activating agent (e.g. a mild acid) to form an adherent layer of palladium metal that catalyses the ensuing process of electroless metal (e.g. copper) deposition . The processing cost depends on the metallization area, and could be prohibitive if the metal coverage on the surface is extensive. In this paper, we present an alternative process which can achieve the same end results while using a much simplified procedure and relatively inexpensive starting materials. The process is particularly attractive to prototyping and the low cost production of flexible circuits.

EXPERIMENTAL

The formation of metallization patterns on the insulating surface makes use of printing methods instead of photolithography. The graphic patterns to be transferred to the insulating surface can be produced in a number of ways. The original artwork of the pattern can be enlarged or reduced from the master film according to the size of the final image. The final image is printed on the insulating surface with a carbon based ink or paint following established printing techniques inclusive of both impact and nonimpact methods. Subsequent metallization is accomplished by electroless deposition where the rate depends on the chemistry and quality of the ink and/or paint, with the conductive form of carbon producing the best results. The printed film should undergo specific surface treatments as described in the following examples to accelerate the deposition process if a less satisfactory ink is used. However, the use of palladium-tin colloids is totally unnecessary. The carbon in the print suffices as a catalyst for the electroless deposition of copper, nickel and silver and hence only areas imprinted with the carbon ink will be rendered conductive by selective metal deposition. After a sufficient thickness is obtained, electrolytic deposition of a similar or dissimilar metal can be used to build up the thickness at a lower cost. Subsequent finishing procedures for the metal electroplate, if necessary, then follow the standard practice in electroplating. A concise description of some example applications for the production of metallized graphic patterns on poly(ethylene terephthalate) (PET) based films, and on less flexible substrates, is given below. Examples 1 -3 make use of common xerography as the vehicle for image transfer, while other examples explore even more prosaic forms of printing (brushing, painting and sketching). The great latitude in the choice of substrates and media for image transfer is demonstrative of the generality of the method.

Example 1. The artwork of an electrical circuit was transferred to a PET based transparent film by means of common xerography. The film after xerography was treated in a 10 wt% alcoholic KOH solution for approximately 5 minutes followed by a thorough deionised water rinse. Thereafter the film was sensitized in an ammonical silver solution containing 10g/*l* of $AgNO_3$ for 30 seconds to a minute. The sensitized surface was then rinsed and immersed with the printed face up in a commercial stabilized electroless copper bath (e.g. MacDermid Electroless Copper 22). Depending on the quality of carbon in the toner, the printed area was covered by copper deposit both visible to the naked eye and conductive in 30 minutes. Limited coverage of the non-printed areas by copper was only noticeable after a longer plating time (e.g. 60 minutes). Such copper was, however, poorly adherent and could easily be removed in a simple rinsing operation. Only the copper deposited catalytically on the carbon in the printed area remained adherent and suitable for subsequent build-up of thickness by electrolytic methods. The procedure is well suited for rapid prototyping work. For multilayered circuits, the patterns on the individual layers can be transferred to different films which are then processed independently and interconnected by external conductor wires. Registration requirement can, therefore, be kept to a minimum.

Example 2. The procedure of Example 1 was followed with the exception that a 1mM $PdCl_2$ solution was used for sensitization and electroless plating was carried out by means of a commercial electroless nickel bath (MacDermid ELNIC 110) operating at 85°C. The printed area was covered with bright and conductive electroless nickel in 10 minutes.

Example 3. The same procedure as of Example 2 was followed with the exception that a 10 g/*l* $SnCl_2$ solution acidified with 40 m*l*/*l* of concentrated HCl was used for sensitization, and electroless silver was deposited from a solution of the following composition:

$AgNO_3$	3.5 g/*l*
Ammonia solution	in sufficient amount to dissolve precipitated Ag_2O
NaOH	2.5 g/*l*
Glucose	45 g/*l*
Tartaric acid	4 g/*l*
Alcohol	100 m*l*

The entire surface of the film was covered with electroless silver in 20 minutes but only the silver on the printed area remained adherent after a thorough washing.

Example 4. A graphic pattern was hand drawn with a 2B pencil on an unclad epoxy-glass laminate used in the manufacture of printed circuit boards. The laminate was treated in the alcoholic KOH solution, followed by the ammonical silver solution, and with good rinses in deionised water in between, before finally immersed into the commercial electroless copper bath of Example 1. Conductive copper with good adhesion was deposited only on the drawn pattern after 30 minutes.

Example 5. A graphic pattern was painted on a poly (methyl methacrylate) substrate using a household enamel paint loaded with graphite powder. The same procedure as of Example 4 was followed. Conductive electroless copper was deposited on the painted area in 20 minutes.

Example 6. One side of a smooth Teflon® substrate was brushed with graphite powder and the procedure for cleaning, sensitization and electroless deposition followed that of Example 4. Electroless copper deposited only on the side with graphite in 10 minutes. The deposit was conductive and adherent to the substrate.

Other examples can be found in reference 5.

After electroless deposition and as and when it is necessary, electroplating of a similar, a dissimilar, or multiple metals can then proceed according to existing techniques including all the necessary pre- and post- treatments of the surface for a good finished product. A clear lacquer compatible with the substrate material can also be applied to the final product for protection against oxidation in air.

DISCUSSION

The process here makes use of the catalytic properties of carbon for electroless metal deposition. However, there are relatively few reports of such carbon properties in the open literature [6], let alone any discussion about the reaction mechanism. The Olin Blackhole process [7] for through-hole plating, and a number of patents from Sharp on lithium secondary batteries [8-9] are applications clearly demonstrative of the catalytic properties of carbon. A review of the patent literature does reveal several processes which bear certain similarities to ours. A technique for creating a flexible circuit by xerography was mentioned in which a modified photocopier charged with proprietary toners was used to imprint circuit patterns that were catalytic towards electroless copper plating [10]. An early IBM patent [11] used a substrate made from pyrolyzable polymer materials but ensuing metallization operations depended primarily on the conductivity of pyrolysed polymers rather than their catalytic properties, if any. These commercial processes of selective metallization often have to resort to proprietary chemicals and somewhat complex methodologies. Their successful implementation depends, to a large extent, on system-specific chemistries and hence is lacking in generality compared to the approach described here.

The carbon in ordinary photocopier toner was adequate to catalyse the deposition of electroless copper, nickel and silver. Electroless deposition of gold from a gold borohydride bath [12] was not as successful. The rate of deposition depends on the conductivity of carbon and the activity of the electroless bath, and the time for deposition varies from 10 minutes for a graphite image in a fast commercial electroless copper bath, to over 12 hours for an untreated photocopied image in a generic electroless copper bath. Pretreatment in alcoholic KOH accelerates the deposition. Selective leaching might have occurred during the treatment, leaving carbon in a more porous and purer form, which promotes electroless deposition by virtue of an increase in carbon conductivity and a better penetration of the reactive components of the bath. Adsorbed silver ions, particularly the silver amine complexes in an ammonical silver solution, are found to enhance the catalytic effects further. Without the presence of carbon, these silver species are inconsequential in the electroless deposition of the metals.

Qualitative peel tests were conducted to test the adhesion of the electrodeposit on the substrate. Detachment from the substrate occurred invariably at the carbon-substrate interface and metal was not separated from carbon in any circumstance. Prolonged treatment in alcoholic KOH solution deteriorated the carbon-substrate adhesion, and was accountable for most of the observed failures. A moderate treatment time is recommended so that the increase in deposit rate is not accompanied by weakened adhesion to the substrate. For a very thin layer of metal deposited from an electroless bath, the metallized pattern should be connected to the power supply before it is immersed into the plating bath. Alternatively, an alkaline bath may be used.

CONCLUSIONS

Electroless copper, nickel and silver were successfully deposited on selected areas of an insulating surface which were imprinted with a carbon based printing material. The rate of deposition increases with carbon quality and the activity of the electroless bath. Pretreatment in alcoholic KOH followed by the adsorption of ionic silver species also aids in electroless deposition, but excessive exposure in the former may lead to a deleterious weakening of the carbon-substrate bonding. The electrolessly deposited metal can be electroplated in subsequent steps to build up the thickness to desired specifications.

REFERENCES

1. K.L. Mittal and J.R. Susko, Editors, “Metallized Plastics 1: Fundamental and Applied Aspects,” Plenum Press, New York, 1989.
2. K..L. Mittal, Editor, “Metallized Plastics 2: Fundamental and Applied Aspects,” Plenum Press, New York, 1991.
3. K.L. Mittal, Editor, “Metallized Plastics 3: Fundamental and Applied Aspects,” Plenum Press, New York, 1992.
4. K.L. Mittal, Editor, “Metallized Plastics: Fundamentals and Applications,” Marcel Dekker, New York, 1998.
5. J.Y. Lee, C.Q. Cui and A.Q. Zhang, UK Patent GB 2,287,472B, (1997).

6. P.A. Thrower, Editor, "Chemistry and Physics of Carbon," Marcel Dekker, New York, 1965.
7. K.L. Minten, A.J. Battisti, and G. Pismennaya, US Patent 4,631,117, (1986).
8. T. Mitate, N. Nishimura, A. Umemoto, H. Okamoto, K. Yamada and T. Yoneda, US Patent 5,432,029, (1995).
9. K. Yamada, H. Tanaka, T. Mitate and M. Yoshikawa, US Patent 5,595,838, (1997).
10. J. Ejelstad and K. Gilleo in "Handbook of Flexible Circuits," K. Gilleo, Editor, Chapter 3, Van Nostrand Reinhold, New York, 1992.
11. M.F. Levy and T.F. Saunders, UK Patent 1,194,112, (1968).
12. Y. Okinaka, in "Electroless Plating: Fundamentals and Applications," G.O. Mallory and J.B. Hajdu, Editors, Chapter 15, American Electroplaters and Surface Finishers Society, 1992.

Metallized Plastics 5&6: Fundamental and Applied Aspects, pp. 137-144.
K.L. Mittal (Ed.)

PLASTICS METALLIZATION USING A DYNAMIC CHEMICAL PLATING PROCESS

A.FARES KARAM and G.STREMSDOERFER

Ecole Centrale de Lyon.
Laboratoire Ingénierie et Fonctionnalisation des Surfaces.
36 Avenue Guy de Collongue, P.O.Box 163.
69131 Ecully Cedex, France.

This paper describes a new process of metallizing plastics that combines the advantages of the static wet plating techniques and the simplicity of the spray gun processes. The principle of this technique is based on the controlled sequential spray of a reducing and an oxidizing agents. Compact and adherent Ni-B metallic deposits were obtained on polymers such as PVC and PET. It appeared that the wettability of the surfaces of these polymers and the number of the electrons exchanged on the surface during the redox reaction had a significant influence on the growth of the metallic film. The different experimental parameters were investigated and are discussed as well as a clear explanation of the method's principle is given. Some examples are given to illustrate some of its possibilities and future prospects are summarized.

INTRODUCTION

Electroless plating is one of the most used methods to metallize plastics and nonconductors in general. This plating technique is a unique approach to obtain perfectly homogeneous and uniform metal deposit on a catalytic surface, even if it has complex and intricate shape . Although the mechanism is purely catalytic, activation step for nonconductors is a crucial stage before plating[1]. Also, it features several economical advantages like the absence of an external current source. In spite of all these advantages, electroless plating is still suffering from several practical problems such as [2]:

- Self contamination of the plating baths due to the products of the redox reaction (phosphite ions,…) and ligands from the metal salts used for replenishment.
- Limited solution life
- Ecological disadvantages (wastes treatments)
- Too many surface conditioning steps
- The plating of large dimensions' parts requires the use of enormous baths and large quantities of reactants, which is neither easy nor economical.

So during these past five years several direct "wet" metallizing processes were developed[2,3] such as:

- Electroplating directly on conductive polymers such as poly(thiophene), poly(pyrrole),...
- Electroplating on polymers that have been made conductive using adsorbed conductive particles such as graphite or colloidal metals such as palladium.
- Electroplating on polymers that have been made conductive using conductive metal-filled paints especially in the field of Electromagnetic Interference Shielding (EMI/RFI).

The industrial demand and the need for new techniques which do not cause environmental problems have led us to develop a new direct metallizing procedure. We have called this process "Dynamic Chemical Plating" (DCP) [4-6]. To explain and clarify the dynamic aspect of the process we should compare the different conventional wet techniques.
Metal plating is based on the reduction of a metallic ion in aqueous or organic solutions. The electrons to reduce these ions could be supplied in different ways. In electroplating, baths function as a source of material (metal salt ions) while the necessary energy to reduce the metallic ions, at the substrate/solution interface, is ensured by an external current source. In the electroless deposition case, the baths are the common source of both material (the chelated metal ions) and energy. The reducing agent acts like an electron energy source in a stable bath. This energy is released in contact with a catalytic surface. The reaction will go on if the deposit has catalytic properties. In the Immersion or Displacement Plating process, the substrate itself is the source of electrons. The deposition is carried out by immersing a metallic substrate in a solution containing the ions of the metal to be plated. The metallic atoms at the surface are displaced by the metallic ions of the solution. These ions have a redox potential that is lower than the one corresponding to the substrate atoms and the species are displaced at the interface between the substrate and the solution.
In these three cases, baths are static. It means that when mechanical stirring or bubbling agitation is off, the diffusion of the different active species in solution is the main parameter that controls continuous substrate supply with metallic ions. The mechanisms of plating depend primarily on the oxidation-reduction potential of the electroactive species and their corresponding reactions kinetics on the surface. The solutions have minor influence on the plating procedure [6].

This paper describes the development of a procedure for metallizing surfaces especially of the nonconductor ones. This procedure may be considered as an evolution of the electroless plating method combined with the old spray procedure of preparing plastics for electroplating. This old method is well described in the literature [7-11] and is commonly called " the silver spray coating process" because silver was practically the only metal plated this way. Or, this old process may be considered as the first industrial aqueous plating process that does not use static liquid bath. Instead, the electroactive species are held in sprayed droplets that have a defined volume and a precise kinetic energy, according to the spraying air pressure.
Hence, the spraying procedure combined with the oxidation-reduction reaction constitutes the Dynamic Plating Concept.

It is also possible to obtain a metallic film on a catalytic substrate by spraying from an aqueous electroless copper or electroless silver bath onto the surface [12]. We have called this procedure Dynamic Electroless Plating (DEP) [6]. A limited aqueous film is formed at the surface and it constitutes a limited autocatalytic layer on the substrate. This layer is thermodynamically stable and the catalytic properties of the substrate allow the oxidation-reduction mechanism. The DEP and the silver spray coating processes have three common disadvantages :

- It is necessary to sensitize and activate the substrate to be plated. So there are two additional high cost conditioning steps in the process [8,9]

- The obtained deposits are very thin and are only used as pre-metallizing step for electroplating nonconductors [7,9,12]
- The metallic films are only "physisorbed" on the substrates, so there is a poor adhesion quality[13].

The main objective of this paper is to present the new concept of the dynamic metal deposition that can bypass these disadvantages and offers new applications possibilities.

EXPERIMENTAL

The treatment of the substrates prior to metal plating consisted of degreasing with an appropriate organic solvent depending on the nature of the used plastic. The samples had 40 mm x 40 mm surface area. All deposits were plated at room temperature. The surface morphologies of the substrates and the deposited films were observed by scanning electron microscopy using a "Philips XL20" micrograph coupled to an "EDAX" X-ray probe for qualitative and semi-quantitative elemental analyses. Inductively coupled plasma was used to determine the exact elemental composition of the deposits. A modified Maxtek PM 500 quartz microbalance was used to measure the growth of the metallic layers in the experimental conditions. A commercially available concentric double-nozzle spray gun was modified and used to spray the aqueous solutions for plating. The spraying apparatus was automated and was controlled using a home-made software . An air compressor was used as a spraying system during the experiments. It supplies compressed air to two different polyethylene coated stainless steel tanks that contain, respectively, the reducing agent solution and the oxidizing metal salt solution. The air pressure in the different compartments is independently adjustable using two different manometers. Both tanks are linked to the automated spray gun. The chemicals used were Sigma-Aldrich products of the reagent grade. The contact angle measurements were made using a "DIGIDROP" contact angle meter coupled to a microcomputer for image analysis (GBX Scientific Instruments-France). The liquid used was ultrapure water and all experiences were carried out at room temperature.

RESULTS AND DISCUSSION

Principle of the DCP Process

The novelty of this method is based on a sequentially highly controlled supply of the material (metallic ions) and the energy source (reducing agent) to the substrate. In other words it can be described as a metal plating process using sequential and simultaneous spraying of two or more solutions.

Using compressed air and the double nozzle spray gun, the reducing and the oxidizing agents are sprayed simultaneously and separately onto a nonconductor. Thus a very thin liquid film is formed. This film is adsorbed on the surface and contains the active species. This film is also thermodynamically unstable which means that the oxidation-reduction reaction will occur spontaneously. But under some specific experimental conditions, this spontaneous reaction can be controlled to avoid coarse precipitation. These conditions are the following :

- The sprayed liquid film, which is adsorbed on the substrate, must be thin enough. So the redox reaction occurs on the surface.
- The ratio of the sprayed oxidizing and reducing solutions must be adequate to give an optimal stoichiometric electrochemical reaction on the surface. This corresponds to the ratio of electrons exchanged between the oxidizing and the reducing agents for a fixed sprayed volume of solutions. We will call this the Ox/Red electron exchange proportion.

- The plastic substrate must have a uniform wettability and a relatively high surface energy ($\gamma_s >$ 40 mJ/m^2).

When these conditions are fulfilled, the redox reaction will be induced on the surface instead of a spontaneous precipitation of the metallic ions in the adsorbed liquid film. Then a compact, dense and adherent metallic film is formed on the substrate. Once the reaction is over in the liquid layer, the thin liquid film is refilled with active species by spraying again the solutions. The time separating the two successive refills is called "relaxation time" and during this time the spraying is off. In order to ensure the thickness growth of the metallic film, the spraying is carried out intermittently using successive liquid sprays. The procedure can then be described as a sequence of several spray/non-spray steps or in other words a sequence of several spraying/relaxation steps added.

Study of the Experimental Parameters

The conditions listed above should be fulfilled all the time during the whole plating operation. If a large quantity of liquid is sprayed, the aqueous film will flow and the redox reaction will occur in the liquid film and no metallic deposit will be observed on the substrate. So we define a threshold thickness of the adsorbed aqueous film and beyond this threshold the reaction will not occur on the surface but in the liquid layer independently of the surface characteristics (wettability,...). This threshold depends on the type of deposit and the nature of the polymer. Bypassing this threshold is probably one of the reasons why the old silver spray coating process, that uses continuous spraying, gave only very thin deposits (< 0.3 μm).[9]

Figure 1 shows the growth of Ni-B binary alloy on poly(ethylene terephthalate) (PET) as a function of the Ox/Red electron exchange ratio which is defined earlier. The thickness of the deposit was determined by weighing the substrate before and after plating, and also using the quartz microbalance to validate the measurements. The concentrations of the reducing agent (KBH_4) and the metallic ions Ni(II) were kept constant in the compartments. For a fixed flow value of KBH_4, the flow of the metallic ions solutions was varied by changing the air pressure. The sprayed quantity of liquid and the flow were adjusted not to exceed the threshold thickness of the aqueous adsorbed layer. The final sprayed volume was constant in the different experiments. We noticed the presence of two limits and beyond them no deposit occurred. On the other hand, we noticed the presence of an optimal area where the Ox/Red electron exchange ratio was between 0.07 and 0.09. This corresponds to the optimal conditions of plating. These values are calculated considering that the oxidation of KBH_4 is an irreversible 8-electron process [14].

Another important parameter is the surface wettability. We have measured the deposit weight per unit surface of Ni-B alloy on three different types of polymers: Acrylonitrile-Butadiene-Styrene (ABS), Poly(vinyl chloride) (PVC) and PET. The measurements were taken for fixed volumes of the sprayed reactants. The different polymer samples were chemically treated [15] to obtain different values of water contact angle. The results were classified by categories depending on the value of the contact angle on the different substrates. Figure 2 shows the variation of the deposit weight per unit surface as a function of the water contact angle on the different polymers. We notice that lower the contact angle, the higher the deposit weight. Although this tendency seems to be common to some other plating techniques also, it is important to notice its significant influence in this case. The deposits weights of the category 2 substrates show a fourfold increase in comparison to those of category 6 for the three types of polymers.

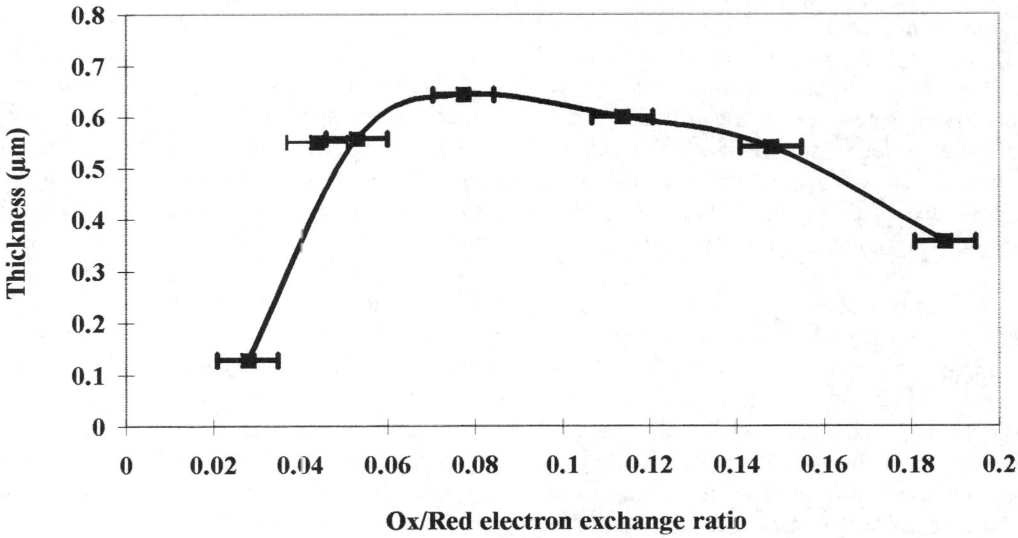

Figure 1. Growth of a Ni-B deposit on PET as a function of the Ox/Red electron exchange ratio for a constant volume of sprayed solutions.

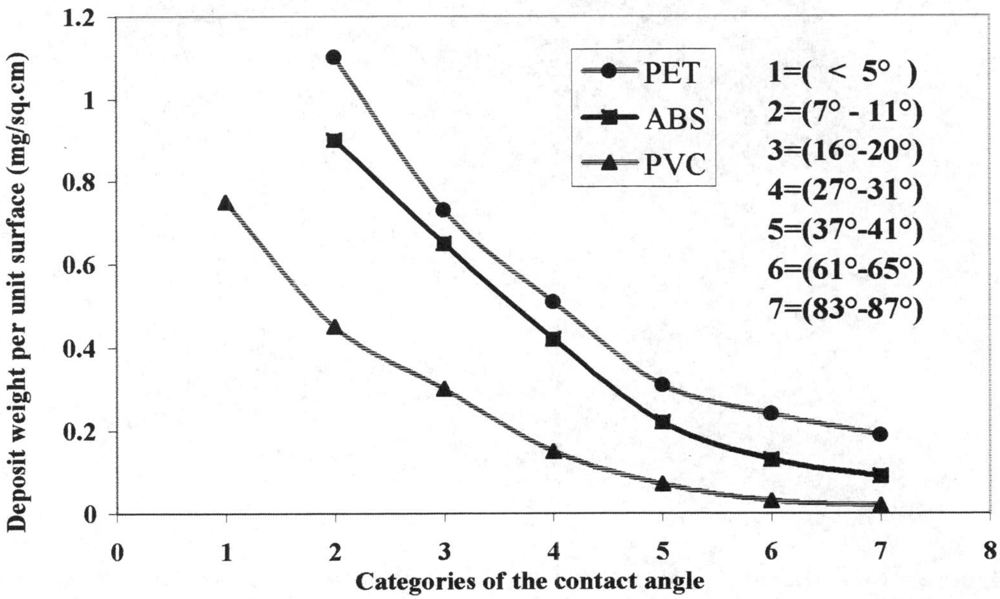

Figure 2. The deposit weight per unit surface of Ni-B film as a function of the contact angle of water on different polymers for constant volume of sprayed solutions.

It is striking to notice the difference in deposit weight per unit surface on the different substrates for the same category. The highest values are for the PET, then for ABS and finally for PVC where a very significant difference is seen. This means that the nature of the substrate itself has a considerable influence in addition to the wettability. This difference in the growth of the metallic film is probably due to the influence of the surface energy and the functional groups on the polymer surface. More precisely this difference occurs in the initial phase, because generally once the substrate is covered with enough thick deposit, the influence of the surface to be plated is eliminated [16]. In any case, these parameters should be examined and investigated in more detail in the future to clarify the role of the surface reactivity.

Fig.3 shows cross section of a 19 μm thick Ni-B deposit on PET polymer and Fig.4 shows its EDS spectrum. The deposit is very compact and its density is 7.3 g/cm^3. For comparison purpose, the average density of the electroless Ni-B is 8.2 g/cm^3. This difference between the two values is probably due to the high boron percentage in the DCP film. The EDS spectrum indicates an overall 11% content of boron. Figure 5 shows an SEM micrograph of this deposit showing a surface morphology with uniform fine grains. Except for some metallic agglomerates, the average grain size is about 32 nm. This is a confirmation of the compact character of our deposits. More crystallographic investigation will be made in the future to clarify this point. The existence of an internal nanoporosity should also be investigated.

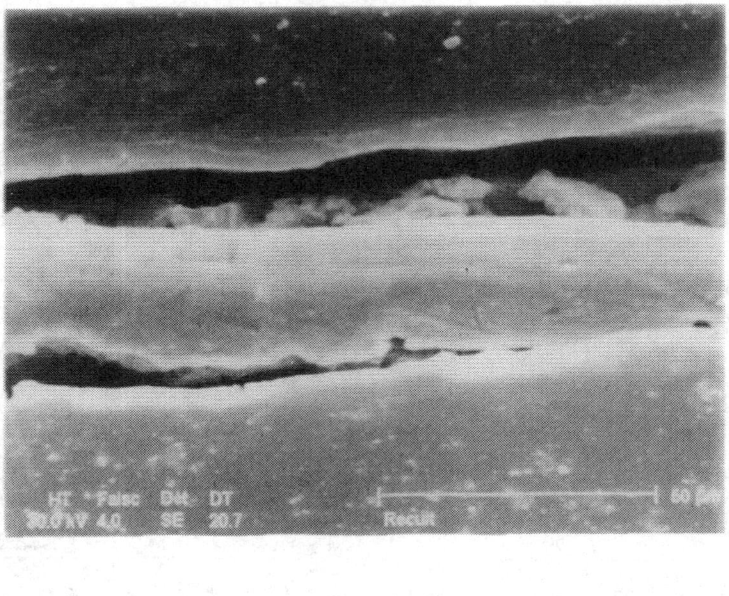

Figure 3. Cross section of a compact and dense Ni-B film deposited on PET polymer.

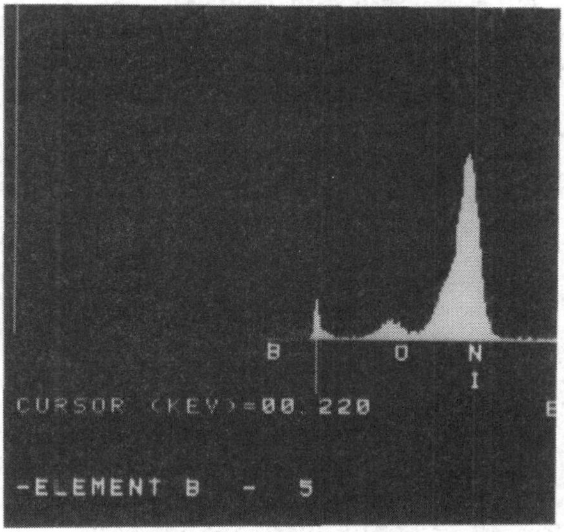

Figure 4. EDS spectrum of a Ni-B alloy film deposited on PET polymer.

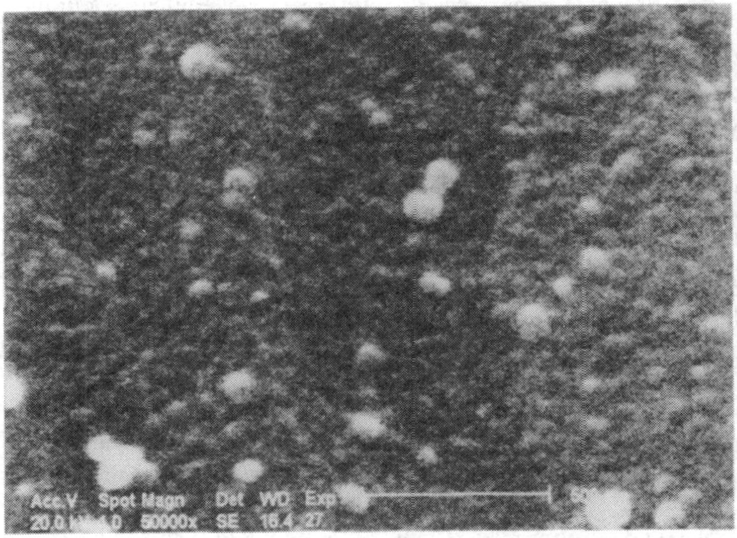

Figure 5. Scanning electron micrograph of Ni-B deposit on PET polymer.

Applications and Overall Characteristics

Using this process we have been able to metallize a wide range of conductor and nonconductor substrates. The examples given in this paper simply demonstrate and reveal some of its characteristics and potentialities. Uniform, adherent and dense films are obtained with nickel, cobalt, copper and their mixed binary and ternary alloys. Such alloys are very important now due to their interesting properties[17]. Their alloys with zinc, boron and tin were also prepared. This process shows a great flexibility and facility to metallize large parts without any dimensions limitations. It gives low cost deposits due the simplicity of the metallizing apparatus and the high stability of the chemicals used since the oxidizing and the reducing agents are separated. This is very important because it makes the preparation of composite deposits easier. We have incorporated Teflon and graphite particles into the metallic matrix by suspending these particles in the metallic ions solution. Finally it is important to mention the high deposition rate at room temperature. Under optimal conditions this rate reaches 40 μm/hour. This is very significant when we compare it to the nickel-phosphorus baths that work at 90°C and have a 25 μm/hour deposition rate [18,19]. It represents also an important energy saving compared to these baths when used for comparable applications.

CONCLUSION

The DCP method gives metallic deposits similar to those obtained by the static electroless technique. But it seems more useful and practical for many plating applications especially plating large dimension parts. We consider that this process is complementary to the other wet static techniques. The simplicity of DCP and its economical aspects make it a competitor process. (i) There is no need for sophisticated apparatus or equipment to perform plating, so DCP is a low cost process. (ii) There is no bath precipitation problems, as in the case of static autocatalytic technique. The DCP solutions are separated and are only used in adequate quantities when plating is needed. (iii) DCP is an environmentally friendly process and does not cause any environmental pollution. In fact, there is a full redox reaction on the surface and all the active species are consumed. Consequently the wastes generation is minimized. Finally, the spraying gun and its components are transportable, so the on-site plating is feasible making DCP a very useful and versatile method.

REFERENCES

1. G.Stremsdoerfer, J.R.Martin and P.Clechet, Plating Surface Finish., 81, No.1, 65-69 (1994).
2. J.Hajdu, Plating Surface Finish., 83, No.9, 29-33 (1996).
3. J.P. Langan, Plating Surface Finish., 82, No.9, 34 (1995).
4. G.Stremsdoerfer and A. Fares Karam, French patent 97-06834 (1997).
5. G.Stremsdoerfer and A. Fares Karam, paper presented at the 1997 International Colloquium onElectrochemistry (JE'97), Ecole Polytechnique de Montréal, Canada, Abstract No. CO10-7.
6. G.Stremsdoerfer, A.FaresKaram and J.M.Krafft, paper presented at the 2nd European Symposium on Composite Coatings, Ecole des Mines de Saint Etienne, France, 1996.
7. F.A.Lowenheim, "Modern Electroplating", Chap.31, Wiley-Interscience, New York, 1976.
8. G.Müller, "Les dépôts électrolytiques sur matières plastiques", pp.85-87, BPI Editions, Paris, 1967.
9. H.Narcus, "Metallizing of Plastics", Chap.2 & 5, Reinhold Publishing Co., New York, 1960.
10. S.Wein, Metal Finishing, 39, No.12, 666-671 (1941).
11. S.Wein, Metal Finishing, 40, No.1, 24-25 (1942).
12. S.Zhang, R.Han and C.Zheng, US Patent 5,492,613 (1992).
13. J.A.Lauchet, Surfaces, 159, No.6 , 60-63 (1983).
14. M.V.Mirkin, H.Yang and A.J.Bard, J.Electrochem.Soc., 139, 2212-2217 (1992).
15. N.V. Mandlich and G.A.Krulik, Plating Surface Finish., 80, No.12, 68-73 (1993).
16. C.D.Icovangelo, J.Electrochem.Soc., 139, 976-982 (1991).
17. M.Schlesinger and X.Y.Meng, J.Electrochem.Soc,137, 1858-1859 (1990).
18. G.O.Mallory and J.Hajdu, "Electroless Plating", Chap.1, AESF, Orlando, FL, 1990.
19. K.Parker, Plating Surface Finish., 79, No.3 , 29-33 (1992).

Part 2

SPECTROSCOPIC INVESTIGATION OF INTERFACIAL INTERACTIONS

Metallized Plastics 5&6: Fundamental and Applied Aspects, pp. 147-168.
K.L. Mittal (Ed.)

THE APPLICATION OF NEAR EDGE X-RAY ABSORPTION FINE STRUCTURE (NEXAFS) SPECTROSCOPY AND X-RAY PHOTOELECTRON SPECTROSCOPY (XPS) IN SURFACE ANALYSIS OF CHROMIUM-EVAPORATED POLYMERS AND ORGANIC FILMS

W.E.S. Unger, A. Lippitz, J. Friedrich and I. Koprinarov
Bundesanstalt für Materialforschung und -prüfung (BAM),
Unter den Eichen 87, D-12205 Berlin, Germany

K. Weiss and Ch. Wöll
Lehrstuhl für Physikalische Chemie I, Ruhr-Universität Bochum,
D-44780 Bochum, Germany

The formation of surface species resulting from chromium evaporation of organic materials was investigated by a multi-technique approach comprising X-ray photoelectron and X-ray absorption spectroscopies. In detail, poly(styrene), poly(bisphenol-A-carbonate), biaxially stretched poly(ethylene terephthalate), and octadecyltrichlorosilane self-assembling films, covered by chromium, were analysed. Direct indication for the existence of bisbenzene chromium (arene complexes) at the surface of the chromium evaporated polymers was found. By employing angle-resolved Near Edge X-ray Absorption Fine Structure (NEXAFS) spectroscopy it is possible to investigate not only surface chemical aspects of the interaction of chromium atoms with an organic substrate but also re-arrangements of molecular units in the investigated surface layer. It was revealed that well-oriented surfaces were re-arranged in course of chromium evaporation, step by step, to an isotropic state.

INTRODUCTION

Most often, X-ray photoelectron spectroscopy (XPS) has been applied for studying interfacial phenomena at metal evaporated organic (polymer) surfaces. However, there are cases where the capability of XPS is limited. One problem is, for instance, to differentiate between single and double bonded carbon species. Thus severe problems arise when aromatic rings, which are often components of polymers, have to be investigated in detail. Another open question is the interpretation of a low binding energy subpeak which usually occurs in C 1s spectra of polymers after deposition of certain amounts of a reactive metal. There are rather different approaches to interpret this spectral feature introduced by the workers in the field. The state of this matter will be briefly summarized in the *Discussion* section. As a consequence, it seems to be necessary to use alternative surface sensitive methods to provide new experimental data. Time-of-Flight Secondary Ion Mass Spectrometry (TOF-SIMS)[1], High Resolution Electron Energy Loss Spectroscopy (HREELS)[2], or Near Edge X-ray Absorption Fine Structure (NEXAFS) spectroscopy[3] are methods which may, in principle, contribute to this issue. In this study NEXAFS spectroscopy was employed because it had been successfully applied as a tool complementary to XPS in previous studies on modified polymer surfaces[4].

Another interesting aspect is the *effect of condensing metal atoms on the molecular orientation* at the organic surface subjected to metal evaporation. This aspect has not been considered very often in the past[5,6], however, it should be also relevant to a discussion of the complex adhesion phenomena between metals and organic materials. Angle-resolved NEXAFS spectroscopy has been found to be an analysis tool which is sensitive to alterations in the orientation of molecular units in a surface layer (cf. refs. 3, 4).

NEXAFS spectroscopy investigates both bound and continuum state resonances of molecules observed in soft x-ray photoabsorption with exciting photon energies close to an absorption edge of a selected constituent atom. While the resonance features above K-edges measured with organic materials are pretty broad due to short lifetimes, the experimental widths of the bound state resonances are rather small. In the case of polymers, sharp bound resonance features are most often correlated to C 1s, N 1s and O 1s $\rightarrow \pi^*$ transitions (when π bonds exist in the sample), thus, offering new possibilities for surface chemical analysis. Moreover, in the case of a preferential orientation of the polymer's segments and also with well-oriented self-assembled monolayers, NEXAFS spectra measured in the angle-resolved mode show a characteristic angle dependence[3]. Angle-resolved NEXAFS spectroscopy can be used here to exploit linear dichroism effects on non-isotropic surfaces. With the help of this special technique the effect of metal evaporation on the orientation of an organic substrate can be studied to some extent in selected cases.

In this communication, we present NEXAFS and XP spectra of polymers and organic films, metallized with chromium. These spectra were obtained at the synchrotron light source BESSY, Berlin, Germany, and with a VG ESCALAB electron spectrometer in our laboratory at BAM, Berlin, Germany. Examples which illustrate the analytical power of NEXAFS spectroscopy in the characterization of metallized polymers are presented. Model polymer systems used are poly(styrene), poly(bisphenol-A-carbonate), biaxially stretched poly(ethylene terephthalate) and well-oriented self-assembled octadecyltrichlorosilane (OTS) films. In the analysis of the polymers emphasis is placed on the effects on the aromatic rings.

EXPERIMENTAL

Samples

Poly(styrene) films were spin coated onto silicon wafers by dissolving non-commercial, atactic poly(styrene) (BASF AG, Germany) in toluene to a 1 wt.% solution. The resulting film thickness was around 100 nm. The oxygen content of the surfaces of these films was measured by XPS to be $\approx$ 1.5 at%.

200 nm thick self-supporting poly(bisphenol-A-carbonate) films were prepared from a methylene dichloride solution of a commercially available

polycarbonate sample (Macrolon 3200, Bayer AG, Germany) using the preparation procedure called "spontaneous film formation"[7].

Biaxially stretched 0.5 µm poly(ethylene terephthalate) foils (Mylar™, Du Pont de Nemours) were used. They were characterized to have an average molecular weight of 250000 Da, crystallinity of 60 - 65% (density measurement) and stretching ratios $\alpha_{||}$ and $\alpha_{\perp}$ of 3.5.

Octadecyltrichlorosilane ($CH_3(CH_2)_{17}SiCl_3$, *OTS*) films were deposited on oxidized silicon wafers as described elsewhere[8]. Analysis of the untreated OTS film revealed a rather smooth surface with a r. m. s. roughness of 1.5 Å (AFM), an ellipsometric thickness of 29.5 ± 1.5 Å and a C 1s spectrum consisting of a single peak at 285.0 eV binding energy with a FWHM of 1.4 eV.

Chromium Evaporation

Chromium was evaporated onto the polymer sample surfaces using a thermally shielded Knudsen cell (EFM 3, Omicron Vakuumphysik GmbH, Taunusstein, Germany) under UHV conditions. The flux control unit of this cell was calibrated by employing a quartz microbalance. 99.99 % pure chromium rods, supplied by BALTEC GmbH (material #B 8010 070 79), Schalksmühle, Germany, were used.

XPS Measurements

XPS analysis and chromium evaporation were carried out in a VG ESCALAB 200X electron spectrometer at BAM, Berlin. This instrument is equipped with a plasma reactor/metallization chamber, separated by a gate valve from the UHV system, where surface treatments can be carried out at 10^{-7} Pa. Mg K_{α} radiation was used with 15 kV and 20 mA settings. XP spectra were acquired in the CRR 40 mode at 20° take-off angle. Narrow scan spectra analysis was performed by routines which are part of the SCIENTA ESCA-300 data system (SCIENTA AB, Uppsala).

X-ray Absorption Spectroscopy Measurements

NEXAFS spectroscopy was carried out on the HE-TGM 2 and HE-PGM 3 monochromator beam lines (33.12 and 43.21, respectively) at the synchrotron light source BESSY (Berlin, Germany)[9]. In situ metal evaporation onto polymer samples was carried out again with the EFM 3 Knudsen cell. Spectra were acquired at the C K-edge and the O K-edge in the *partial electron yield* mode (PEY) at a retarding voltage of -150 V. Raw spectra were divided by the monochromator transmission function which was obtained with a freshly sputtered Au sample. Alignment of the energy scales was achieved by using the C 1s $\rightarrow \pi^*$ resonance, measured with a pyrolytic graphite sample (Advanced Ceramic Corp., Cleveland, USA), which was fixed at a value of 285.4 eV[10]. Characteristic features in the flux monitor signals were used to align the energy scales of the spectra. Reproducibility of the spectra was carefully checked by multiple scanning.

Angle-resolved NEXAFS spectra were recorded at angles of 20°, 55° and 90°, measured between the surface plane of the sample and the direction vector of the incident, linearly polarized light beam. An angle of 90° means that the electric field vector **E** lies in the surface plane, and 20° that **E** is nearly parallel to the surface normal. The reason for an angular dependence of spectra, taken from non-isotropic surfaces, is that the intensity of a selected resonance depends on the angle between the **E** vector of the linearly polarized synchrotron radiation and the main direction of the unoccupied molecular orbital. The geometric arrangement of the angle-resolved experiment is illustrated in figures 1a and b.

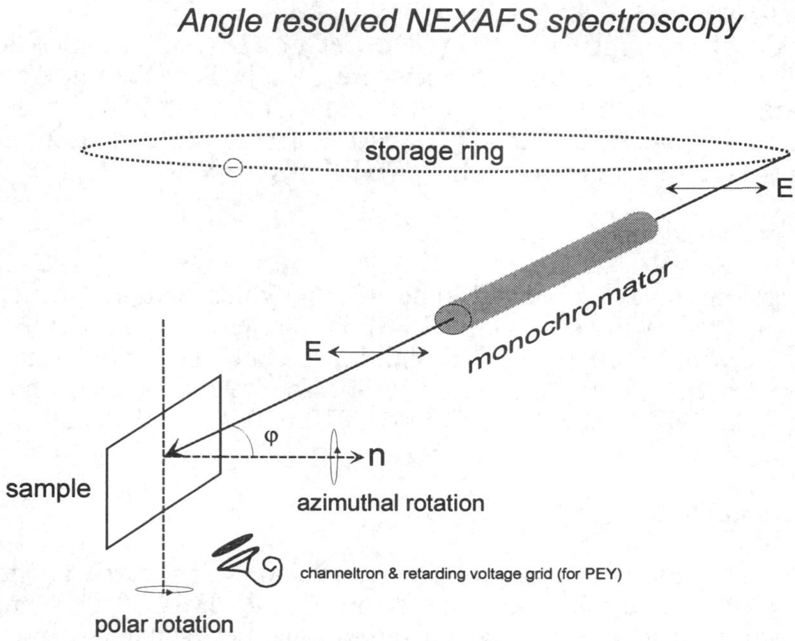

Figure 1a. Experimental setup for angle resolved NEXAFS spectroscopy. The angle 90°-φ is varied between 20° and 90°.

In principle, absorption spectra can be measured by monitoring the non-radiative decay (electron yield) of core holes, created in the sample by the photoabsorption process itself [3]. Photoelectrons as well as C- and O-KLL series electrons contribute to this electron yield signal measured by a channeltron detector. Surface sensitivity is achieved by using the *partial electron yield* mode because low energy electrons, which may originate from deeper layers, were suppressed from entering the detector by the pre-set threshold energy. The information depth, which is defined to be the depth from which 95% of the detected electrons originate[11], is estimated to be ≈ 3 nm for the C K-edge signal and ≈ 5 nm for the O K-edge signal[12]. Here it was assumed that the respective Auger electrons dominated the measured partial electron yield signal[3].

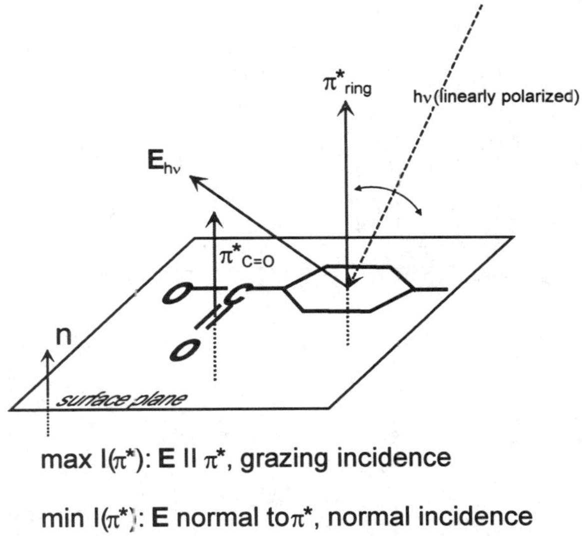

Figure 1b. Dependence of resonance intensities I vs. angle of incidence of linearly polarized synchrotron light to be expected for a polymer segment oriented parallel to the surface plane. The *vectors* 'π^*_{ring}' and '$\pi^*_{C=O}$' represent the directions of the maximum orbital amplitudes of the unoccupied π orbitals of the benzene ring and the carbonyl bond, respectively (cf. ref. [3]).

RESULTS

Cr on Poly(styrene)

XPS analysis of stepwise chromium evaporated poly(styrene) (cf. fig. 2) reveals a decrease of the intensity of the C 1s $\pi \rightarrow \pi^*$ shake-up feature. This is usually interpreted as a decrease in aromaticity of the organic substrate. At higher chromium exposures, a loss in aromaticity, i.e. a strong attenuation of the C 1s $\rightarrow \pi^*_{ring}$ resonance of poly(styrene), is also found by NEXAFS. Also at longer exposures, a new feature at lower binding energies occurs in the C 1s spectra. Full NEXAFS spectra of the C K-edge of poly(styrene) measured at medium resolution and for low chromium coverages are presented in fig. 3a. Inspection of the π* resonance region by high resolution NEXAFS spectroscopy reveals the formation of a new spectral feature at the high energy shoulder of the C 1s $\rightarrow \pi^*_{ring}$ resonance of poly(styrene) (cf. fig. 3b). This new feature should be correlated to another π* resonance resulting from chromium - arene hybrid orbitals which have been formed, for instance, in chromium-benzene complexes. To confirm this interpretation, we have measured the respective NEXAFS spectra of chromium complexes (cf. fig. 3c). As a result the intense C 1s $\rightarrow \pi^*_{ring}$ resonance of bisbenzene chromium, ($Cr(C_6H_6)_2$), was found exactly at 285.8 eV. This energy corresponds well to the one measured with the new feature observed in the high resolution spectra of chromium evaporated poly(styrene). We note here that, most probably, the low energy feature in the C 1s XP spectrum does not correspond to the new π* feature in the NEXAFS spectra.

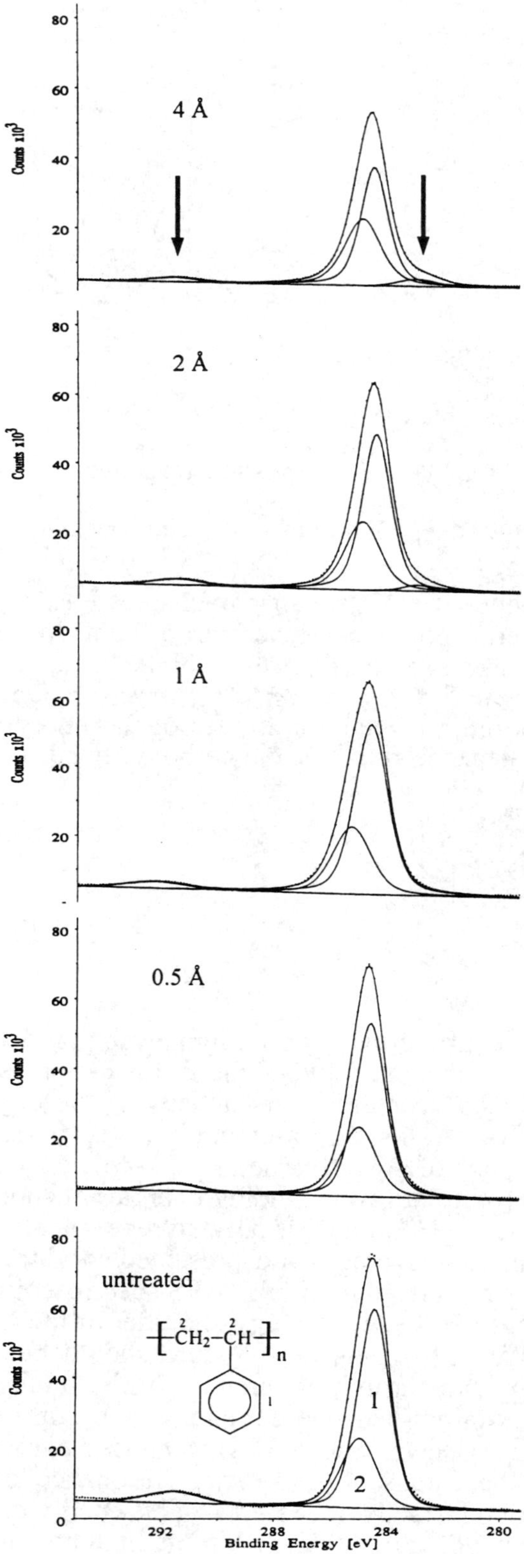

Figure 2. C 1s XP spectra of virgin and chromium evaporated poly(styrene).

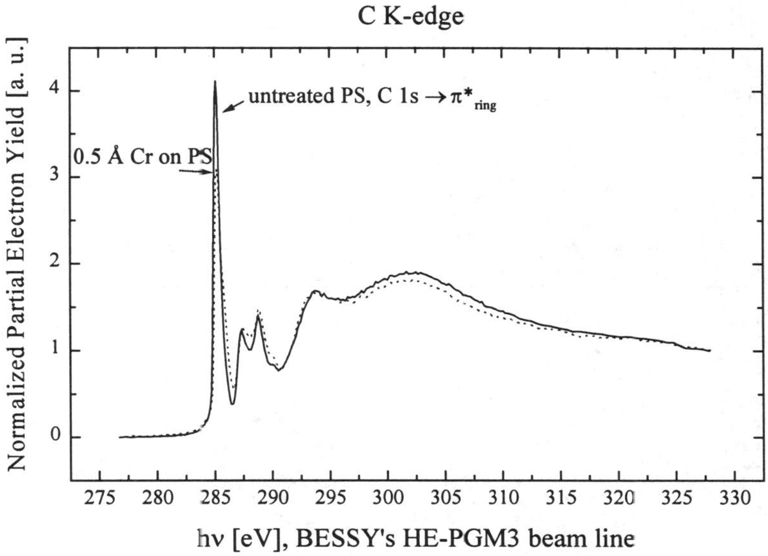

Figure 3a. NEXAFS spectra of the C K-edge of virgin and chromium evporated poly(styrene).

Additionally, the Cr L-edge was investigated (fig. 3d). Comparison of the spectra measured with chromium evaporated poly(styrene), thick chromium films, Cr_2O_3 and bisbenzene chromium supports the conclusion that chromium reacts with the poly(styrene). Only at higher chromium doses metallic Cr defines the L-edge spectrum. A chromium complex formation at low Cr doses is suggested by the occurrence of its finger-print in the respective Cr L-edge spectra (0.25 and 0.5 Å Cr).

Cr on Poly(bisphenol-A-carbonate)

Fig. 4 presents the XP spectra measured with chromium evaporated poly(bisphenol-A-carbonate). Again a loss in aromaticity and a new low binding energy feature are found in the C 1s spectra after chromium evaporation. Additionally, an intensity decrease of the carbonate related subpeak is observed. However, there is a superposition of the carbonate C 1s subpeak and the ring related C 1s $\pi \rightarrow \pi^*$ shake-up feature in the more complex spectrum of the poly(bisphenol-A-carbonate). Therefore, it is almost impossible to differentiate the effects of chromium evaporation on the phenylene rings and the carbonate group. Considering the respective O 1s spectra, a new low binding energy feature is observed after evaporation of chromium, which can be correlated to chromium oxide species. This contribution originates from both chromium - poly(bisphenol-A-carbonate) interaction (most probably a carbonate-chromium redox reaction) and chromium oxidation by residual gas components (less than 25 %). Because metallic chromium is an excellent getter this artifact cannot be fully avoided even at quite good vacuum conditions. The carbonyl bond related subpeak in the O 1s spectra is superposed by another new feature evolving in course of chromium evaporation. Usually this

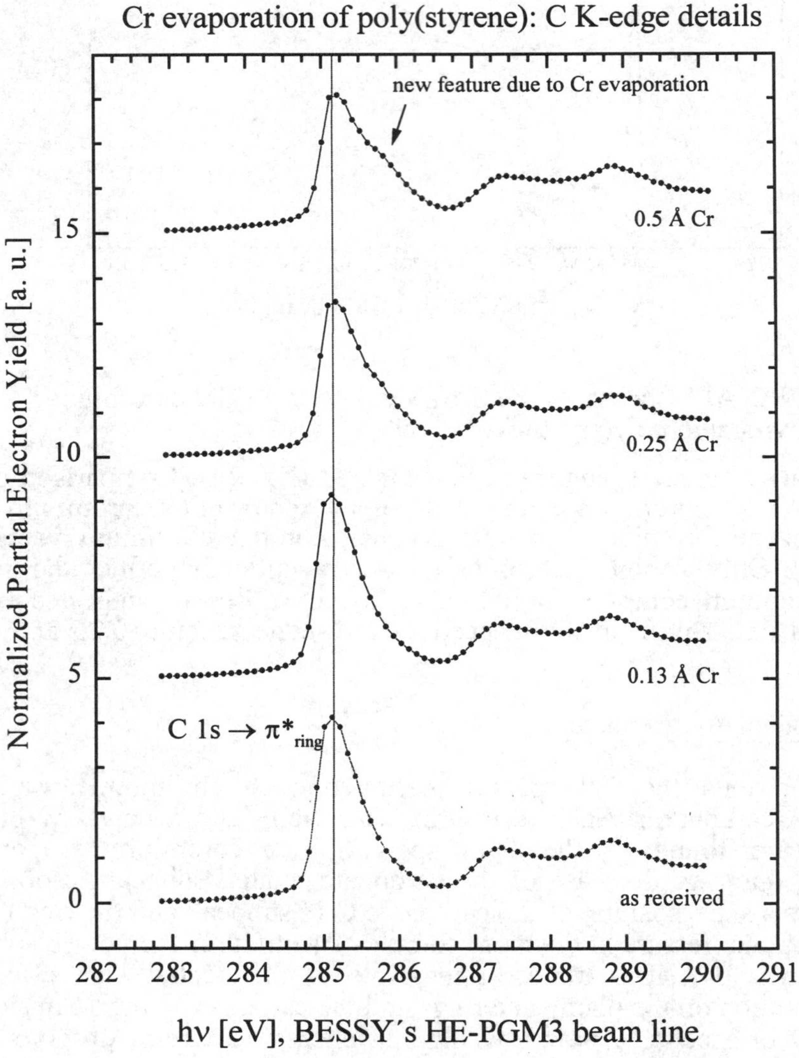

Figure 3b. High resolution NEXAFS spectra of the C K-edge region of poly(styrene).

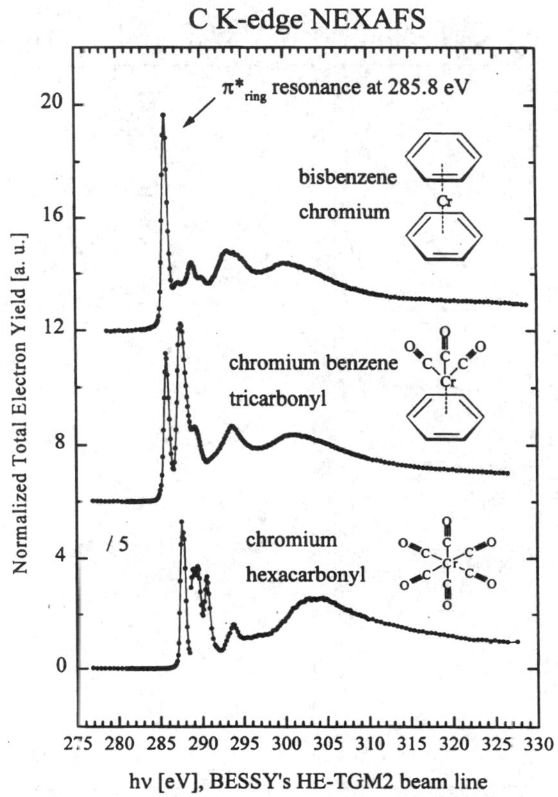

Figure 3c NEXAFS spectra of the C K-edge of chromium complexes condensed on a liquid nitrogen cooled Au covered Si wafer.

contribution is proposed to represent a metal-O-C complex formed by interaction with the carbonyl bond.

NEXAFS analysis of the C K-edge (cf. fig. 5a) reveals clear evidence for both chromium-phenylene and chromium-carbonate interactions, even at early stages of evaporation. A discussion of the spectral features found in the spectra of non-treated poly(bisphenol-A-carbonate) was given in a previous paper[13]. The carbonate related C 1s $\rightarrow \pi^*_{C=O}$ resonance gradually disappears with increasing chromium doses. Moreover, the intensity of the C 1s $\rightarrow \pi^*_{ring}$ resonance is diminished and broadened towards higher energy. This tailing is again (cf. the poly(styrene) case) attributed to the formation of a new C 1s $\rightarrow \pi^*_{ring}$ resonance occurring at an energy position also occupied by the respective π^*_{ring} resonance of bisarene (phenylene) chromium complexes. Inspection of the O K-edge and Cr L-edge regions (cf. fig. 5b) reveals strong indications for chromium-carbonyl (carbonate) interactions. Considering the "Cr multilayer" case in fig. 5b all carbonate related features disappear thus confirming the respective C K-edge results. This O K-edge spectrum covers chromium-oxygen species, Cr-O-C bonds or Cr...O...C complexes (as concluded from the respective XP O 1s spectra) and Cr oxide species. It cannot be exclusively explained by, for instance, Cr_2O_3 formation. The chromium L-edge spectrum is here dominated by features characteristic of elemental Cr.

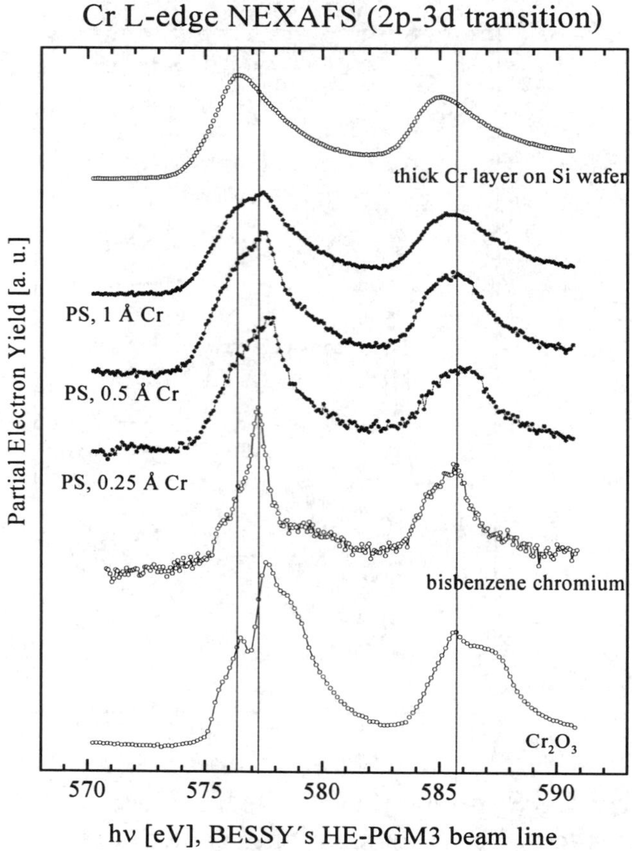

Figure 3d. High resolution NEXAFS spectra of the Cr L-edge region of chromium evaporated poly(styrene). For comparison the respective spectra of bisbenzene chromium, Cr_2O_3 and a thick Cr layer on a Si wafer are included.

Cr on Biaxially Stretched Poly(ethylene terephthalate) – Response of an Oriented Surface

The features in the C K-edge spectra (cf. figs. 6a) of pure poly(ethylene terephthalate) were discussed in detail in ref. 13. Essentials are repeated here briefly. The C K-edge spectrum is dominated by an intense C 1s $\rightarrow \pi^*_{ring}$ resonance at 284.9 eV which is characterized by a shoulder to higher energies. The shoulder is due to the phenylene ring symmetry perturbation induced by the polar ester group. The second intense resonance at 288.3 eV represents a C 1s $\rightarrow \pi^*_{C=O}$ transition characteristic of the carbonyl bond in the repeat unit. The respective O K-edge spectrum, shown in fig. 6b, is dominated by an intense O 1s $\rightarrow \pi^*_{C=O}$ resonance. Considering the angle dependence of the spectra, one should remember that the NEXAFS spectroscopy probes the amplitude and directionality of unoccupied MO's on the atom, characterized by the core hole. Only the spatial orientation (direction

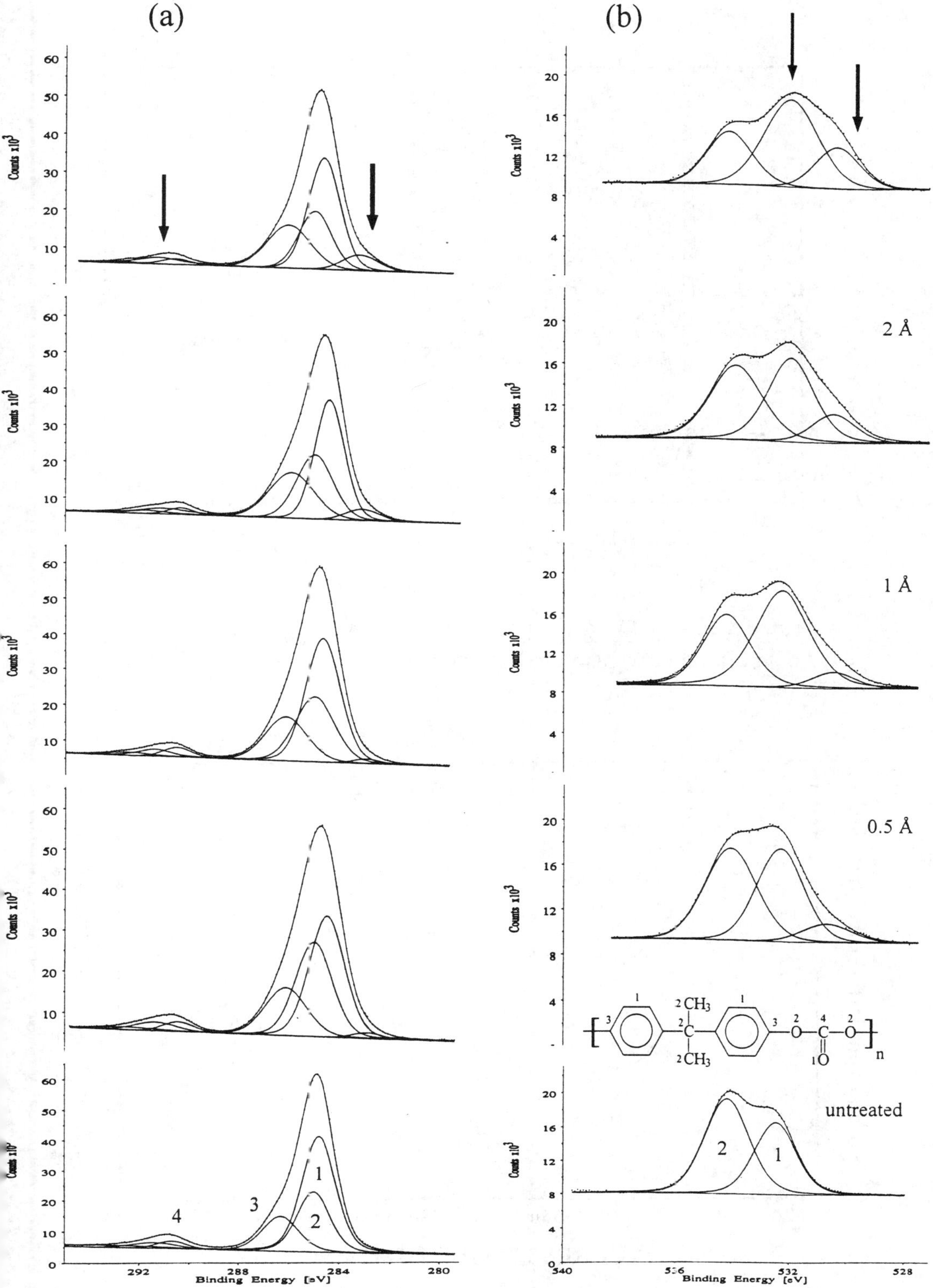

Figure 4. C 1s (a) and O 1s (b) XP spectra of virgin and chromium evaporated poly(bisphenol-A-carbonate).

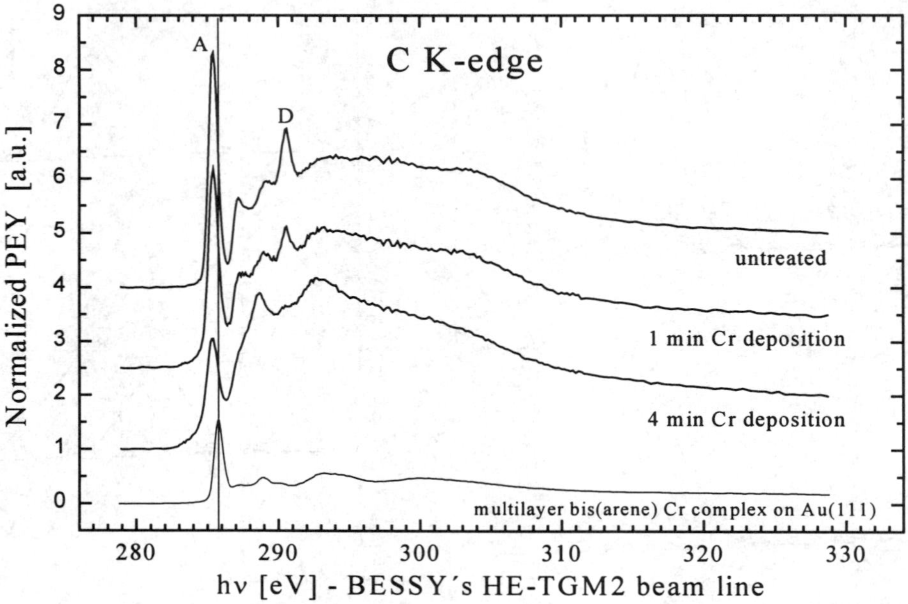

Figure 5a. NEXAFS spectra of the C K-edge of virgin and chromium evaporated poly(bisphenol-A-carbonate).

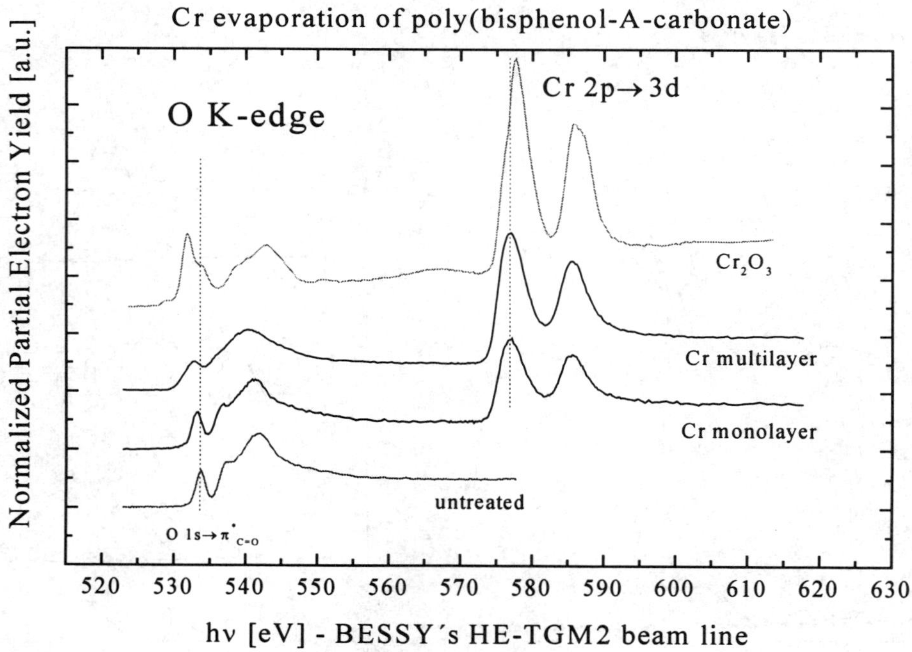

Figure 5b. NEXAFS spectra of the O K-edge and the Cr L-edge of virgin and chromium evaporated poly(bisphenol-A-carbonate).

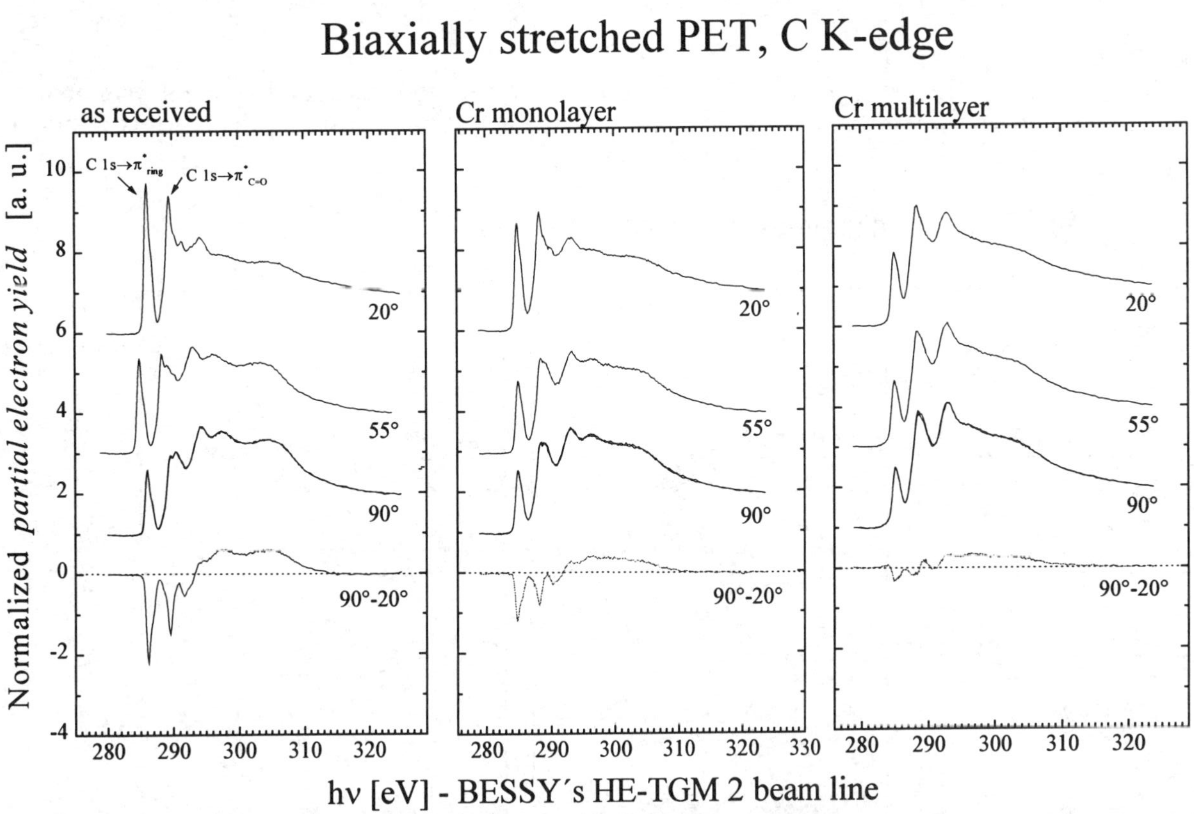

Figure 6a. Angle-resolved C K-edge NEXAFS spectra of biaxially stretched poly(ethylene terephthalate) foils in the "as received" state and after chromium deposition. '90° - 20°' difference spectra are evaluated and plotted at the bottom of each set. The 90° spectra were measured at the beginning and at the end of an experiment. Its reproducibility is illustrated by plotting both spectra together in each set of spectra.

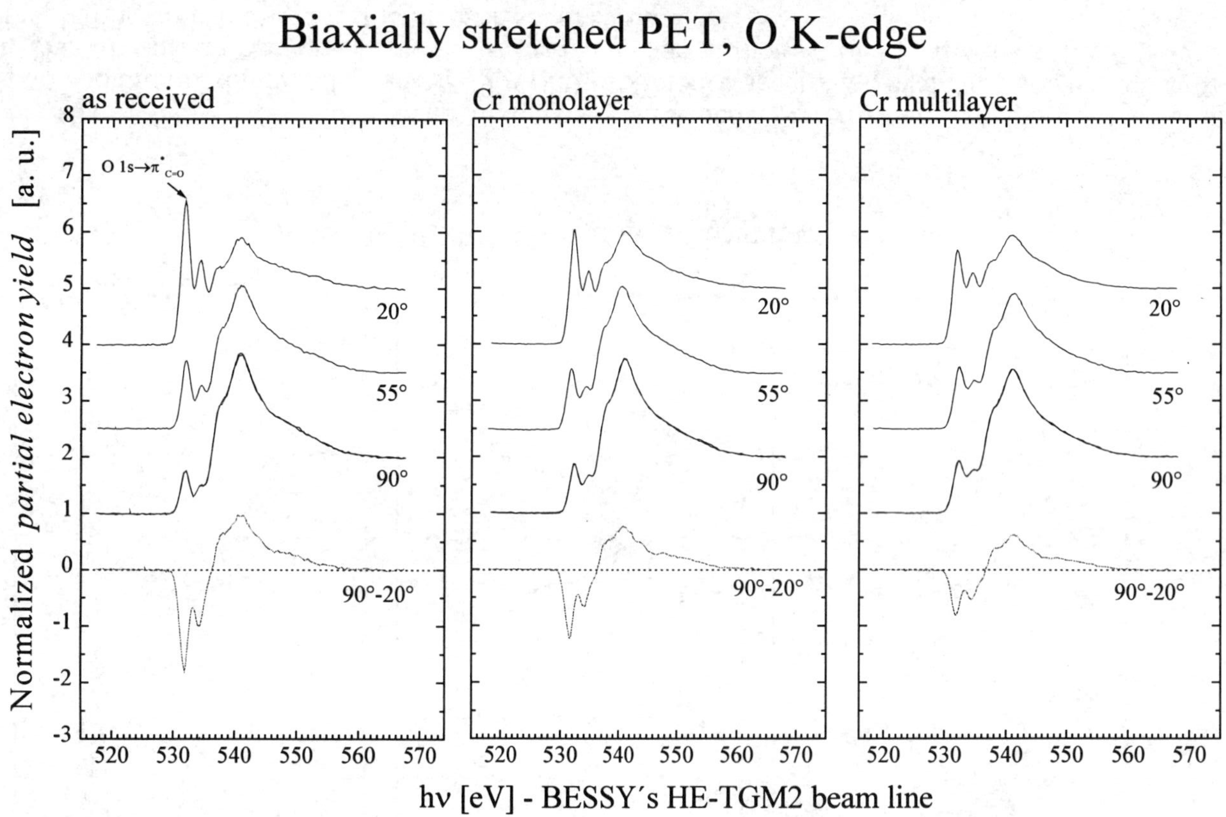

Figure 6b. Angle-resolved O K-edge NEXAFS spectra of biaxially stretched poly(ethylene terephthalate) foils in the 'as received' state and after chromium evaporation.

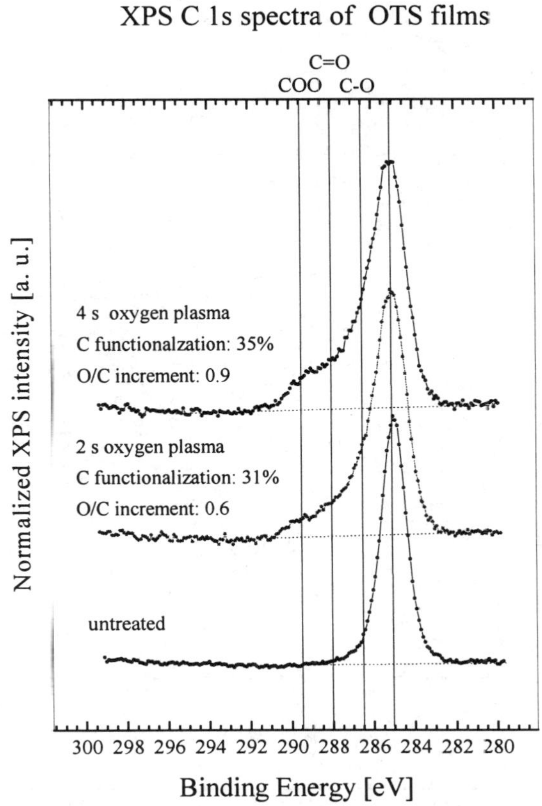

Figure 7. C 1s XP spectra of virgin and oxygen plasma treated OTS self-assembled films.

of maximal orbital amplitude) of a selected orbital determines the angle dependence of K-edge absorption spectra. In this picture, the π system of the poly(ethylene terephthalate) phenylene ring can be represented by a vector which is orthogonal to the ring plane. Therefore, the maximal intensity of a transition from C 1s into π^*_{ring} orbitals should be observed with the synchrotron light's **E** vector orthogonal to an aromatic ring plane. This situation is illustrated by fig. 1b. Because the poly(ethylene terephthalate) repeat unit is nearly planar[14], both intensity maxima of the C 1s and O 1s → $\pi^*_{C=O}$ transitions should occur for the same direction of the **E** vector. Considering the left columns of spectra presented in figs. 6a and b, the angle dependence of the key resonance features referred to above was consistently found to have their intensity maxima at glancing incidence of the synchrotron light beam (20°). These observations suggest that the planes of the poly(ethylene terephthalate) repeat units are preferentially oriented in a way that they are parallel to the surface plane of the sample. "90°-20°" difference spectra were evaluated to clarify the angle dependence of the respective spectra.

Recently, the area of the C 1s → π^*_{ring} resonance feature in the C K-edge difference spectra was successfully correlated to a laterally averaged degree of orientation of the molecular units in the investigated surface layer and its alterations due to surface treatments (oxygen plasma, cf. ref. 5).

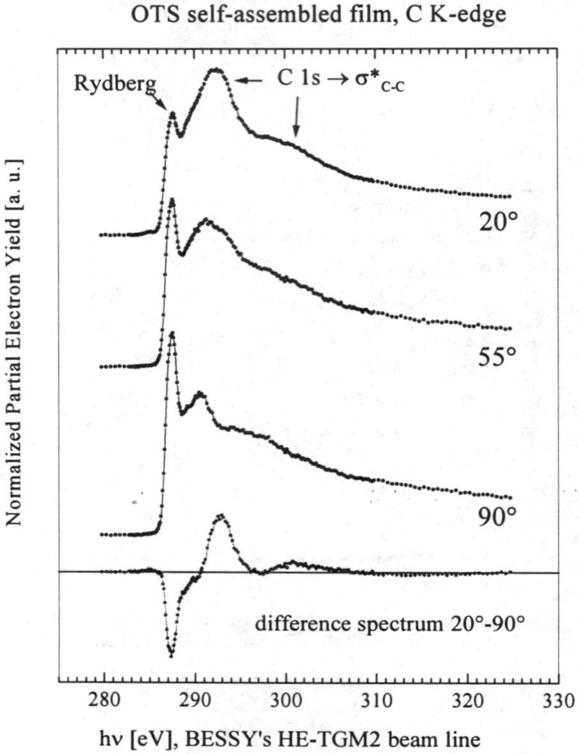

Figure 8. Angle resolved C K-edge NEXAFS spectra of an OTS self-assembling film in the 'as received' state. The '20°-90°' difference spectrum is evaluated and plotted at the bottom of the set.

The response of the biaxially stretched poly(ethylene terephthalate) foil t metallization with chromium is illustrated by the middle and right columns in figs 6a and b. Obviously, there are chemical effects, which alter the spectra in characteristic manner (and are not considered in this communication). Compariso of the difference spectra, presented in figs. 6a and b, suggests a significant *decreas in the preferential orientation,* even for chromium monolayer coverages of th sample. Multilayer coverage results in a rather strong decrease in orientation. In th latter case the C K-edge spectra reveal only a small residual orientation (≈ 10 % o the original one) whereas the O K-edge spectra are obviously much less affecte This is due to the enhanced information depth obtained when the O K-edge i analyzed.

<u>Cr on Highly Oriented OTS Self-assembled Layers - Response of an Oriente Surface</u>

In another experiment chromium evaporated, highly oriented self-assemble OTS films were investigated to study the response of a comparatively simpl

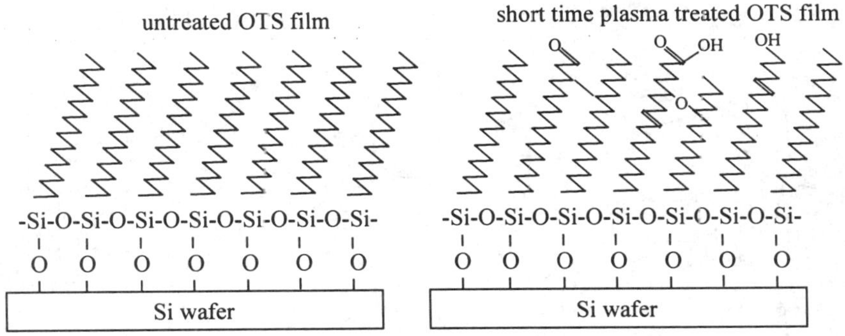

Figure 9. Idealized models of OTS self-assembled layers on oxidized Si wafers before and after mild D.C. oxygen plasma treatment.

system. XPS and NEXAFS spectroscopy analyses of the *untreated* OTS film revealed a single peak C 1s XP spectrum (fig. 7) and an angle dependent NEXAFS C K-edge spectra (fig. 8), respectively. The relevant features in the latter set of spectra are a sharp Rydberg resonance (cf. ref. 15) at 287.5 eV and two C 1s $\rightarrow \sigma^*_{C-C}$ contributions. The characteristic angle dependence of the spectra confirms the high degree of orientation of the alkyl chains forming the OTS film[3]. A model of this state of the OTS self-assembled film is given by fig. 9. Evaporation of chromium onto OTS films results in small alterations in the NEXAFS spectra due to chemical effects[16], however, the alkyl chains are still well oriented as verified by the respective difference spectrum (cf. fig. 10, upper right column). D.C. oxygen plasma treatment of the OTS films resulted, as expected, in oxygen functionalization of the alkyl chains, their entanglement and chain scissions. C 1s XP spectra, given in fig. 7, show the characteristic tailing of the main peak towards higher binding energies indicating the formation of C-O, C=O and O-C=O species. Only by applying a short plasma pulse, e.g. for a duration of 2 s, we found a treatment regime, where sufficient oxygen functionalization of the film can be observed by XPS (fig. 7) and a relatively high degree of orientation of the alkyl chains still exists. This was again verified by the respective difference in angle-resolved NEXAFS spectra which reveals indeed an angle dependence (cf. fig. 10, lower left column). Also in that case, the NEXAFS spectra show characteristic alterations due to chemical functionalization by plasma which correlate well with the XPS findings[16]. An idealized model of the state of the OTS film after rather mild plasma treatment is given by the right picture in fig. 9. NEXAFS analysis of the chromium evaporated *oxygen functionalized OTS film* verifies that its state is very different from that of the untreated OTS film after deposition of Cr. Actually, a plasma functionalized, oriented OTS film becomes fully entangled after chromium evaporation. Any angle dependence of the respective NEXAFS spectra is lost (cf. fig. 10, lower right column). On investigating the chemical state of the evaporated chromium sub-monolayers by XPS, the formation of Cr oxide species was found. The interpretation was that they are the result of an interfacial redox reaction[17].

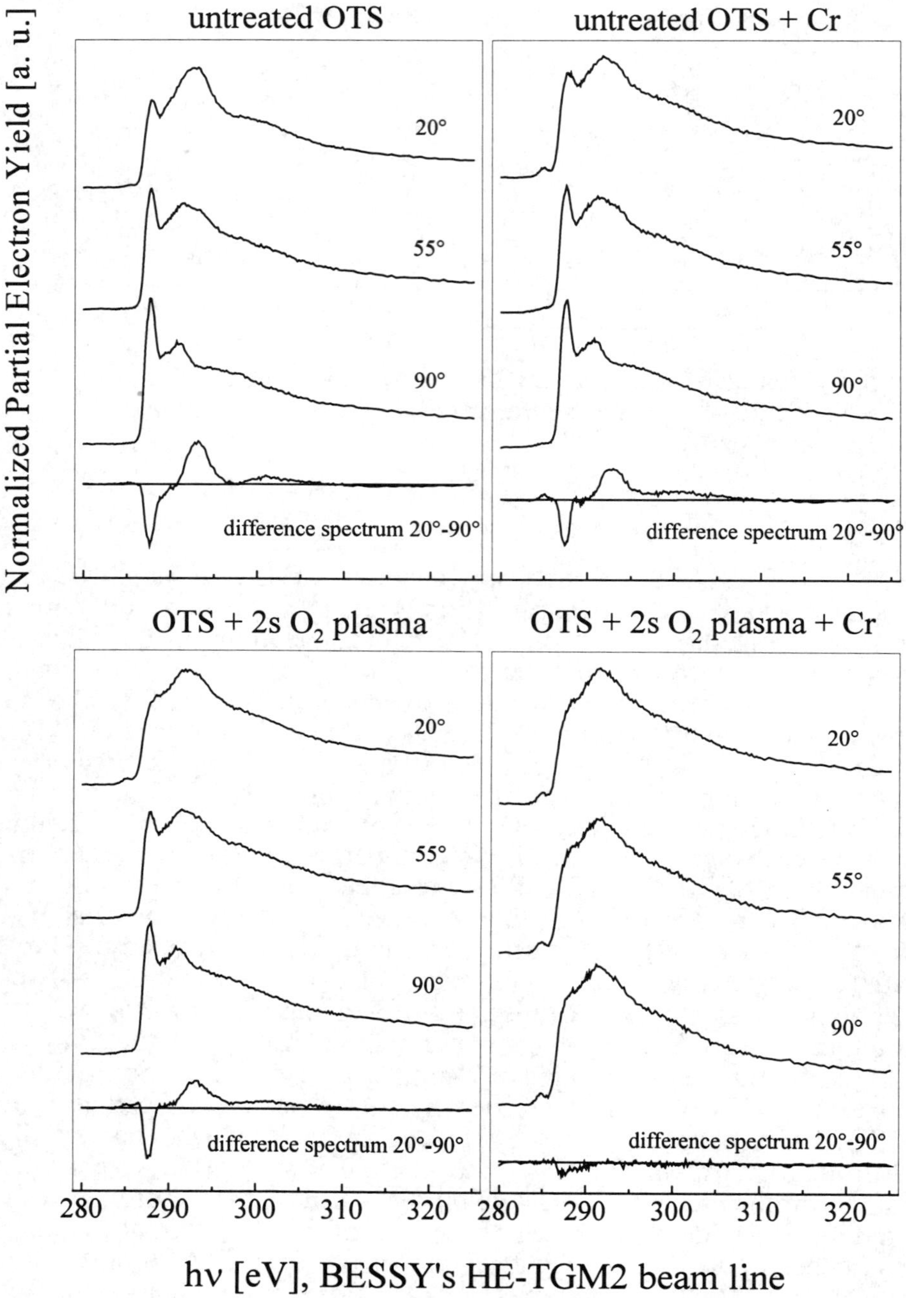

Figure 10. Angle resolved C K-edge NEXAFS spectra of OTS self-assembling films in the 'as received' state, after chromium evaporation, after mild D.C. oxygen plasma treatment and after plasma treatment and chromium evaporation. The '20°-90°' difference spectrum is evaluated and plotted at the bottom of the set of spectra in any case.

DISCUSSION

Concerning the problem of chromium species formed at the surface of aromatic polymers in course of chromium evaporation, very different approaches can be found in the literature. Several authors discuss the formation of a "π bonded arene chromium complex", where charge is transferred from Cr to the aromatic ring[18-24]. Alternatively, the formation of chromium carbide is claimed (cf. e.g. ref. 25). The latter model implicitly includes severe degradation of the polymer. By the way, Czanderna and collaborators reported also metal carbide formation when self-assembled monolayers of 16-mercaptohexadecanoate, which does not contain aromatic rings, were evaporated with chromium (or other metals)[26]. Often authors use the termini "electron-rich" or "carbide-like" species in order to express that *no real carbide formation is meant*[27-29]. Finally, a scheme was proposed which starts with chromium charge transfer complexes ("C-arene π-complex") at the early states of interaction and ends with a "σ-bonded C-Cr carbide-like species" upon further chromium evaporation[30]. Pireaux reviewed the state of the whole matter in 1994[2].

Here, we try to discuss the problem again considering our own results obtained by XPS and NEXAFS spectroscopies. Starting with the possibility of bisarene chromium complex formation, it is now clear that NEXAFS spectroscopy may confirm it by detection of the respective C 1s → π^*_{ring} resonance which is shifted with respect to the π^*_{ring} resonance of non-affected aromatic rings. In contrast, NEXAFS spectroscopy cannot confirm a "σ bonded C-Cr species" from an analysis of the C K-edge. It should be an extremely difficult task to extract features correlated to this species from the σ region of the C K-edge spectra of Cr evaporated polymers because in that NEXAFS region superposition by other σ* features originating from the substrate has to be expected.

XPS analysis of condensed bisarene chromium revealed partially inconsistent results in the past. A positive C 1s binding energy shift of about +0.4 eV relative to the C 1s binding energy of benzene itself was found[31]. Other authors found shifts of +0.7 eV[32] or -0.7 eV[33]. A theoretical treatment of the problem results in a rather small negative binding energy shift of about -0.05 eV for $Cr(C_6H_6)_2$ and -1.42 eV for $Cr(C_6H_6)$[34]. However, it is clear that all of the *experimental shifts* may not explain the new peaks observed in the C 1s spectra of chromium evaporated poly(styrene) and poly(bisphenol-A-carbonate), which were shifted by -1.46 eV or -1.43 eV, respectively (cf. figs. 2 and 4). Relying on the theoretical shifts, only the formation of a $Cr(C_6H_6)$ complex would fit the experimental data. On the other hand, the binding energy measured for the new peak in the C 1s spectra corresponds well to those usually found in XPS experiments with metal carbides or, as we can learn from ref. 30, to *other Cr-C species where sufficient negative charge is shifted to carbon atoms*. The latter interpretation is, from a more chemical point of view, to be preferred.

As discussed in ref. 2, bisarene chromium formation should have an effect on the binding energy and intensity of the π→π* shake-up feature because these parameters depend on the respective degree of π-electron delocalization. The problem is to derive reliable binding energy and intensity data for the shake-up feature from standard XP spectra. The π→π* shake-up feature is usually small in intensity and, e.g. in the case of poly(bisphenol-A-carbonate), is superposed by other spectral C 1s contributions. Furthermore, some authors (cf. literature quoted in ref. 2) point to the fact that in the case of polymers which have both aromatic rings and carbonyl sites, chromium preferentially interacts with the latter and that this interaction also influences the π-electron system of the aromatic ring. One consequence could be a correlated attenuation of C 1s $\pi\rightarrow\pi^*$ shake-up or C 1s $\rightarrow \pi^*_{ring}$ intensities observed in the respective XP or NEXAFS spectra. For poly(bisphenol-A-carbonate) we confirmed indeed a carbonyl site-chromium interaction both by NEXAFS spectroscopy and XPS. So a correlation as mentioned above cannot be fully excluded. However, an investigation of poly(styrene) should provide data *free* of this "cross-talk" problem. Considering the high resolution NEXAFS spectra of chromium evaporated poly(styrene), we found strong indications for the formation of chromium bisarene complex (new π^* resonance in the NEXAFS spectrum) for a 0.5 Å chromium layer but neither significant losses in aromaticity nor contributions by low binding energy components in the respective XP C 1s spectrum. This confirms a reaction scenario starting with a chromium aromatic ring complex (precursor?) and ending with *Cr-C species* characterized by a sufficient negative charge shifted to the carbon atom (cf. ref. 2, where the Al - poly(styrene) system is discussed). The loss of aromaticity observed by XPS with chromium evaporated poly(styrene) and, most probably, also for the poly(bisphenol-A-carbonate) samples at higher chromium doses could support, however, the assumption of a more severe degradation of the polymer structure, which is a prerequisite for Cr carbide formation. However, here we point again to the fact that XPS cannot differentiate between a real chromium carbide formation and the formation of other "σ-bonded C-Cr carbide-like species".

The study of the effect of chromium evaporation onto oriented organic substrates by angle-resolved NEXAFS spectroscopy revealed rather strong re-orientations of structural units in the surface layer towards an isotropic state. The effect can be exclusively observed when there are reactive sites at the surface (cf. the special case of OTS). Reactive sites in the given sense are oxygen functionalized sites which exist on both poly(ethylene terephthalate) and oxygen plasma treated OTS self-assembled films. Because strong indications for chromium-arene complex formation were found for poly(ethylene terephthalate) too, so its phenylene rings also have to be considered as potential reaction sites[35]. The experimental findings could be, at the one hand, understood assuming that chromium severely degraded the organic substrate. Chain scissions, carbide formation, oxide formation, etc. could occur. It is known for biaxially stretched poly(ethylene terephthalate) foils that plasma degradation may result in an isotropic state of the surface[5]. On the other hand, the effect could be (relying on a surface energy examination) due to a driving force for a re-orientation of an ordered organic surface. The surface energy should be lowered when chromium reaches an optimal coordination with its ligands, which can be aromatic rings and/or carbonyl groups in the investigated cases. Considering poly(ethylene terephthalate), we are once again faced with the problem that both the aromatic rings and the ester carbonyl units are reactive to form complexes with chromium[35]. Consequently, one cannot rate the individual potentials of these units to trigger re-orientation phenomena due to chromium evaporation.

CONCLUSIONS

It is now clear that due to the complexity (co-existence of species, competing reaction channels) of the problem, XPS alone is not able to give unique conclusions. Multitechnique approaches are required. Amongst other methods, NEXAFS spectroscopy has been shown to be qualified to provide complementary information. Employing this method, we found direct indications for the existence of bisarene chromium complexes at the surface of chromium evaporated poly(styrene) and poly(bisphenol-A-carbonate). Additionally, the degradation of the carbonate unit of poly(bisphenol-A-carbonate) by evaporated Cr was unequivocally evidenced by NEXAFS spectroscopy. Formation of a predominantly ionic chromium carbide as the dominating species as sometimes concluded from the low energy XPS C 1s feature is not very probable. This reaction pathway would be accompanied by a strong degradation of the polymer substrate itself. However, this is in contradiction to the adhesion promotion effect usually achieved by chromium interlayers. The overall conclusions are that the formation of surface species resulting from chromium evaporation of organic materials strongly depends on the substrate itself, so coexistence of species has to be taken into account and generalization of individual results should be made with extreme care.

By employing angle-resolved NEXAFS spectroscopy, it is possible to investigate not only selected chemical aspects of the interaction of chromium atoms with an organic substrate but also re-arrangement effects. Here we found that well-oriented surfaces are turned gradually in course of chromium evaporation to an isotropic state. The conclusion is that the re-orientation is induced by a driving force resulting from the reduction of surface energy, which can be obtained by an optimum co-ordination of chromium with its complexed ligands.

ACKNOWLEDGMENTS

The preparation of OTS films by Dr. K. Bierbaum is gratefully acknowledged. Special thanks are due to Dr. W. Braun, Dr. Ch. Jung, Ch. Hellwig and M. Mast at BESSY, Berlin, Germany.

REFERENCES

[1] P. Bertrand and A. Delcorte, SIMS XI, Orlando, USA, 1997, proceedings, submitted; P. Bertrand, P.A. Gollier and Y. Travaly, paper presented at the 1997 Joint International Meeting - the 192nd Meeting of The Electrochemical Society and the 48th Annual Meeting of the International Society of Electrochemistry, Paris, France, 1997.
[2] J. J. Pireaux, Synthetic Metals, 67, 39 (1994).
[3] J. Stöhr, "NEXAFS Spectroscopy", Springer Verlag, Heidelberg, 1992.
[4] W. E. S. Unger, A. Lippitz, C. Wöll and W. Heckmann, Fresenius J. Anal. Chem., 358, 89 (1997).
[5] I. Koprinarov, A. Lippitz, J. F. Friedrich, W. E. S. Unger and Ch. Wöll, Polymer, 38, 2005 (1997).
[6] Y. De Puydt, P. Bertrand and P. Lutgen, Surface Interface Anal., 12, 486 (1988).
[7] A. Hoffmann, K.-F. Becker, B. Pigur and G. Hinrichsen, Abstracts, 5. Berliner Polymeren-Tage, Potsdam, Germany, 1994, p. 173.
[8] K. Bierbaum, Thesis, 1995, Universität Heidelberg, Germany.
[9] Research at BESSY. A User's Handbook, Ch. Jung, editor, BESSY mbH, Berlin, Germany, 1993.
[10] P. E. Batson, Phys. Rev., B 48, 2608 (1993).
[11] Standard Terminology Relating to Surface Analysis, ASTM E 673-91c, Annual Book of ASTM Standards, Vol. 03.06 ASTM, Philadelphia, 1993), p. 525.
[12] M. P. Seah and W. A. Dench, Surface Interface Anal., 1, 2 (1979).

[13] L. G. M. Pettersson, H. Ågren, B. L. Schürmann, A. Lippitz and W. E. S. Unger, Intl. J. Quantum Chem., 63, 749 (1997).
[14] R. de P. Daubeny, C. W. Bunn and C. J. Brown, Proc. Roy. Soc. (London) A 226, 531 (1954).
[15] P. S. Bagus, K. Weiss, A. Schertel, Ch. Wöll, W. Braun, C. Hellwig and C. Jung, Chem. Phys. Lett., 248, 129 (1995).
[16] To be published.
[17] W. Unger, A. Lippitz, Ch. Wöll, J. Friedrich and L. Nick, Supplement a la Revue "Le Vide, les Couches Minces", 272, 95 (1994).
[18] P. S. Ho, B. D. Silverman, R. A. Haight, R. C. White, P. N. Sanda and A. R. Rossi, IBM J. Res. Develop., 32, 658 (1988).
[19] R. Haight, R. C. White, B. D. Silverman and P. S. Ho, J. Vac. Sci. Technol., A 6, 2188 (1988).
[20] R. C. White, R. Haight, B. D. Silverman and P. S. Ho, Appl. Phys. Lett., 51, 481 (1987).
[22] S. G. Anderson, J. Leu, B. D. Silverman and P. S. Ho, J. Vac. Sci. Technol., A 11, 368 (1993).
[23] P. N. Sanda, J. W. Bartha, J. G. Clabes, J. L. Jordan, C. Feger, B. D. Silverman and P. S. Ho, J. Vac. Sci. Technol., A 4, 1035 (1986).
[24] G. Polzonetti, M. V. Russo, A. Furlani and G. Infante, Chem. Phys. Lett., 214, 333 (1993).
[25] M. J. Goldberg, J. G. Clabes, A. Viehbeck and C. A. Kovac, Mater. Res. Soc. Symp. Proc., 108, 225 (1987).
[26] D. R. Jung and A. W. Czanderna, "Proc. ECASIA 95", H. J. Mathieu, B. Reihl and D. Briggs, editors, p. 513, Wiley & Sons, Chichester (1996); D. R. Jung and A. W. Czanderna, Appl. Surface Sci., 99, 161 (1996); D. R. Jung and A. W. Czanderna, J. Vac. Sci. Technol., A 13, 1337 (1995); D. R. Jung, A. W. Czanderna and G. C. Herdt, in "Polymer Surfaces and Interfaces: Characterization, Modification and Application", K. L. Mittal and K.-W. Lee, Eds., pp.189-221, VSP, Utrecht (1997).
[27] J. L. Jordan, C. A. Kovac, J. F. Morar and R. A. Pollak, Phys. Rev., B 36, 1369 (1987).
[28] C. A. Kovac, J. L. Jordan-Sweet, M. J. Goldberg, J. G. Clabes, A. Viehbeck and R. A. Pollak, IBM J. Res. Develop., 32, 603 (1988).
[29] J. L. Jordan, P. N. Sanda, J. F. Morar, C. A. Kovac, F. J. Himpsel and R. A. Pollak, J. Vac. Sci. Technol., A 4, 1046 (1986).
[30] G. Polzonetti, G. Iucci, C. Furlani, M. V. Russo, A. Furlani, G. Infante, G. Paolucci, B. Brena and D. Cocco, Chem. Phys. Lett., 267, 384 (1997).
[31] D. Wacker, K. Weiss, U. Kazmaier and Ch. Wöll, Langmuir, 13, 6689 (1997).
[32] F. Kinomura, T. Tamura, L. Watanabe and Y. Yokoyama, Bull. Chem. Soc. Japan, 49, 3544 (1976).
[33] S. Pignataro, A. Foffani and G. Distefano, Chem. Phys. Lett., 20, 350 (1973).
[34] A. R. Rossi, P. N. Sanda, B. D. Silverman and P. S. Ho, Organometallics, 6, 580 (1987).
[35] To be published.

Metallized Plastics 5&6: Fundamental and Applied Aspects, pp. 169-201.
K.L. Mittal (Ed.)

PENETRATION OF DEPOSITED Au, Cu, AND Ag OVERLAYERS THROUGH ALKANETHIOL SELF-ASSEMBLED MONOLAYERS ON GOLD OR SILVER*

G. C. Herdt, D. E. King, and A. W. Czanderna**

National Renewable Energy Laboratory
1617 Cole Boulevard, Golden, CO 80401 USA

ABSTRACT

The purpose of our research is to study the interactions or processes between vacuum-deposited metals (M) and the organic functional end groups (OFGs) of self-assembled monolayers (SAMs) on Au or Ag films under controlled conditions. Metal/SAM/Au or Ag systems are models for understanding bonding at M/organic interfaces and the concomitant adhesion between the different materials. In broad terms, the M/OFGs form interacting interfaces (e.g., Cr/COOH or Cu/COOH) in which the deposit resides on top of the OFGs or weakly interacting interfaces through which the overlayer penetrates and resides at the SAM/gold interface. We present results from weakly interacting systems (Au/CH_3, Cu/CH_3, Cu/$COOCH_3$, Ag/CH_3, Ag/COOH) and discuss the time-temperature dependence of the disappearance of the metal from the M/SAM interface following deposition using ion scattering spectroscopy (ISS) on these systems. X-ray photoelectron spectroscopy (XPS) and ISS were used to characterize alkanethiols terminated with CH_3 (C-6, -10, -12, -16, and -18 chain lengths), COOH (C-11 and C-16 chain lengths), and $COOCH_3$ (C-16) before and after depositing up to 1.0 nm Au, Ag, or Cu at ca. 10^{-7} torr. XPS spectra indicate that no strong interaction occurs between any of the deposited metals and the organic functional groups studied, although a stronger

---

*Performed under DOE contract DE-AC36-83CH10093.
**Performed as an Associated Western Universities Laboratory-Graduate Fellow as part of his doctoral dissertation at the University of Denver.

interaction is evident for Ag on a C-16 COOH than a C-11 COOH and a weak interaction results between Cu and $COOCH_3$. Although these interactions between the Cu or Ag and OFG are evident, metal penetration into these SAMs occurs. No interaction between Au, Ag, or Cu with CH_3 is evident. ISS compositional depth profiles indicate penetration occurs in less than 16 h in all cases and as short as 30 s for the M/CH_3 systems. The time to penetrate is in the following order: Ag, Cu, and Au/CH_3 (C-16) $<$Cu/$COOCH_3$ $<$Ag/COOH (C-16).

Abbreviations

BE	Binding Energy
CDP	Compositional Depth Profile
FT-IRRAS	Fourier Transform Infrared Reflection Absorption Spectroscopy
ISS	Ion Scattering Spectroscopy
M or MO	Metal or Metal Oxide
M/SAM	Metal/Self-Assembled Monolayer
MHA	Mercaptohexadecanoic acid, $HS(CH_2)_{15}COOH$
MMHD	Mercaptomethylhexadecanoate, $HS(CH_2)_{15}COOCH_3$
MUA	Mercaptoundecanoic acid, $HS(CH_2)_{10}COOH$
MUO	Mercaptoundecanol, $HS(CH_2)_{11}OH$
NREL	National Renewable Energy Laboratory
ODT	Octadecanethiol, $HS(CH_2)_{17}CH_3$
OFG	Organic Functional End Group
RAIRS	Reflection Absorption Infrared Spectroscopy
SAM	Self-Assembled Monolayer
XPS	X-ray Photoelectron Spectroscopy

INTRODUCTION

This paper is an extension of recent research[1-11] to elucidate penetration rates of metal overlayers through self-assembled monolayers (SAMs) using ion scattering spectroscopy (ISS)[6-7] or to identify metal/organic functional group interactions at metal/SAM interfaces, primarily by means of X-ray photoelectron spectroscopy (XPS).[1-5,8-11] As illustrated in Fig. 1a, SAMs have fulfilled a need for model systems with ordered organic surfaces of uniform chemical composition for studying interactions at metal (M) or metal oxide (MO)/organic interfaces.[1-11] In principle, atomically sharp interfaces can be formed between the deposited overlayer and the SAM surface. Chemical bonding at metal/organic interfaces plays an important role in the reliability and durability of many technological devices.[12] Bonding at organic/metal or organic/metal oxide interfaces typically has been studied either by surface analysis of organic species adsorbed onto single-crystal metal surfaces,[13,14] or by coating a metal oxide with a polymer, whereas metal/organic interfaces have been

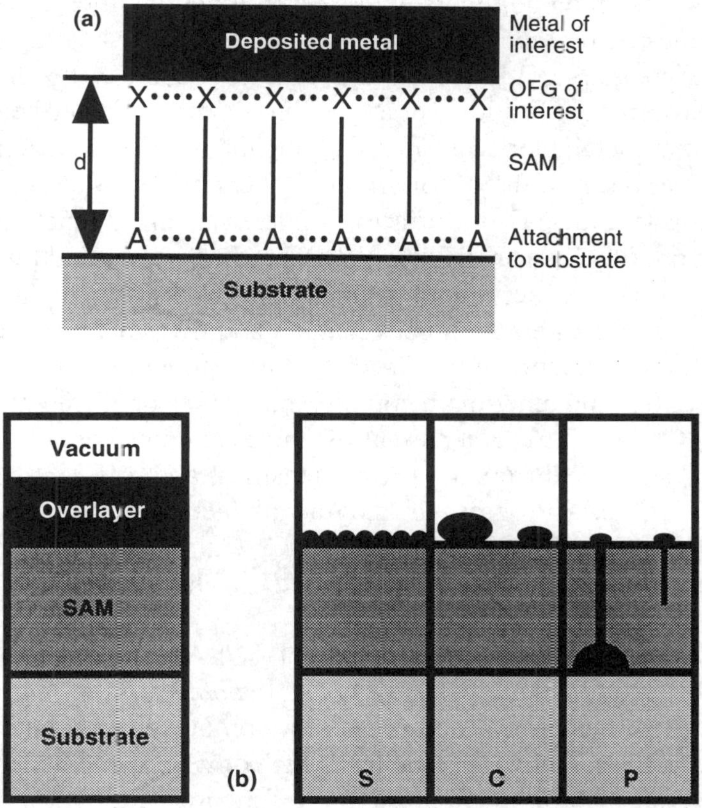

Fig. 1. (a) Idealized scheme for using an organized molecular assembly to form an interface between a metal overlayer and an organic functional group. The thickness *d* of the SAM is the length of the molecule at perpendicular attachment. For alkanethiol SAMs on gold, the attachment group is a thiolate, and the methylene chain is 28° from the normal at 295 K; (b) cross sections of the principally observed results to date in which there is a (S) strong metal-OFG interaction, (C) weak metal-OFG interaction, or (P) very weak metal-OFG interaction. The designations S, C, and P stand for smooth, clustered, and penetrating, respectively.

studied by surface analysis of metallized polymers.[15,16] The chemical environment studied in the first type of experiment bears little resemblance to that present at polymer/metal (oxide) interfaces. In contrast, the lack of structural and chemical order of polymer surfaces complicates the location and type of interactions with metal or oxide overlayers. MO or M/SAM studies have added a new approach, where the state of subdivision of the M or MO overlayer is different and selection of a highly ordered organic surface is possible.

Reactive transition metals (e.g., Cr and Ti) are frequently used to promote adhesion at metal/polymer interfaces, and they have been the subject of many studies.[5,11,17-20]

Good metal adhesion on polymers with oxygen functionalities is well known and can be 2 to 10 times better than on polymers without oxygen-bearing groups.[15,16] Less-reactive metals (e.g., Au, Ag, and Cu) are frequently used as metallization coatings of polymers, for mirror applications, or as grid lines in photovoltaic cells, which may also contain polymer/metal interfaces. For studying the stability or reactivity of interfaces, SAMs are ideal because they form ordered, thermally stable arrays of close-packed aliphatic chains terminated by a surface of uniform chemical functionality, as originally cited in the work of Allara and Nuzzo[21-22] and recently reviewed by others (Table I).[23] SAMs with a variety of terminating functional groups are now available, making it possible to study the interaction between a known metal species and an organic surface of a known functionality. SAMs can be formed from alkyl sulfides, alkanoic acids, alkanethiols, and other more complex molecules on different surfaces, e.g., Ag, Au, SiO_2, Al_2O_3, and GaAs and present tilt angles ranging from 0° to between 50° and 57° to the substrate,[24] different packing densities, and different contact angles with water or hexadecane.[10] Progress in studying alkanethiol SAMs terminated with CH_3, CH_2OH, COOH, $COOCH_3$, and CN with Ti, Cr, Cu, Ni, Al, K, Na, and Ag overlayers has been summarized and critically reviewed.[10] The specific M/OFG combinations studied to date are identified in Table II.

In support of the main goal of recent M/SAM research (understanding the interactions between M overlayers on SAMs), the *objectives* of recent work have been to: (a) identify new chemical compounds or complexes, and intermediate stages in compound formation, if any; (b) establish which or what combination of the three main configurations of metal may be present as a function of the ratio of deposited metal atoms to SAM organic functional groups (OFGs), as illustrated in Fig. 1b: S (smooth metal) that is chemically bonded or complexed with the SAM and probably spread out on top of the SAM, C (clustered) nonbonded metal that is on top of the SAM and probably in the form of large clusters, and P (penetrating) nonbonded metal that significantly penetrates the SAM and whose morphology could take many forms; (c) compare the quantities S, C, and P for deposition at different rates on surfaces at different temperatures, and, thereby, search for evidence of activated processes, metastable/nonequilibrium states, and results that depend on the degree of SAM ordering; (d) obtain images of clusters on the SAM (by scanning probe or electron microscopy), and use the distribution of cluster sizes, the rate of deposition, and the surface temperature as parameters to model the kinetics of cluster nucleation and growth; (e) characterize the degree of ordering and close packing of the SAM before and after metallization (e.g., are pinholes playing a role?); and (f) measure rates of penetration of metal through the SAM.

The lowest energy state of a M/SAM system is that for which all the deposited metal resides between the substrate and the SAM, because the surface free energies of metals are much higher than those of organics. Thus, we have two critical questions about M/SAM research: Are the configurations S and C of Fig. 1b stable at room temperature, and what factors determine that stability? The scheme depicted in Fig. 1b, and our corresponding identification of species S with chemical bonding and

Table I. Some Properties of Thiol SAMs/Au as Summarized by Dubois and Nuzzo.[23]

Strength of Au thiolate bond	~44 kcal/mol
Superlattice of S head groups on Au (111)	$(\sqrt{3}x\sqrt{3})R30°$
S-S nearest-neighbor spacing	0.499 nm
Tilt angle of methylene chain axis on Au (111) from surface normal	~25°-30°
Twist angle of the C-C-C plane from the plane containing the chain axis and the surface normal	~52°
Temperature at which significant chain disordering occurs	~380 K
Temperature at which a C22* thiolate monolayer decomposes/desorbs under vacuum conditions	~500 K

*C22 indicates the number of C atoms in each thiol molecule

Table II. Matrix of References to Metal/Organic Functional Group Systems $X-(CH_2)_n$ SH/Au, (X = OFG) Studied to Date.

OFG / Metal	CH_3	CH_2OH	$COOCH_3$	COOH	CN
Ag	6ab			6c	
Cu	58	3	8,27,11c	1,2,22	4,8
Ni				10	5
K			10		
Na	10				
Al	10		10		
Cr	11a		8,11c	11b	5,8,11a
Ti	9	9	9	9	

species C and P with nonbonding, are intended as idealizations (e.g., we neglect the role of metal carbide formation, which could occur along the entire chain).

Objectives (c) and (e) listed above are concerned with *how* a M/SAM system attains the observed chemical state and spatial distribution. The process of adlayer nucleation and growth, a competition that depends kinetically on surface temperature and deposition rate, may be considerably more complex than that found for two-dimensional systems if there is diffusion into (and not just over) the SAM. The mechanism of penetration may involve different possible rate-limiting steps, for example, lateral diffusion leading to penetration at defect sites vs. direct vertical diffusion that might be enabled by thermal motion (because of ambient thermal energy or transient thermal energy imparted by atom impact), chemical reaction with and saturation of all the possible bonding with the OFG, or disordering due to strong chemical effects.

The order, structure, and stability of alkanethiol monolayers on Au have been reviewed by Dubois and Nuzzo[23] and are briefly summarized in Table I. As implied, SAMs spontaneously self assemble but are also robust and stable. The characteristics of the ordered SAM include: (1) the presence of the desired OFG at the vacuum/SAM interface; (2) a low density of pinholes in the film; (3) uniform film thickness; (4) uniform average orientation of constituent molecules; (5) a low population of gauche bonds in the methylene chains; (6) lateral ordering (real-space Bravais lattice or translational symmetry vectors) and internal ordered structure of the unit cell of the SAM monolayer; and (7) a domain size of coherent lateral order. Details of experimental probes used to characterize the properties of SAMs have been critically reviewed.[10] As for stability, SAMs withstand extended atmospheric exposure, repeated washing, and temperatures from 60°C to 100°C. These and other film characteristics (e.g., their wettability by water and other liquids) are described further by Dubois and Nuzzo.[23] Only a few of the objectives (a)-(f) and control of the characteristics (1)-(7) are addressed in this summary of part of the work done to date.[10]

For prior studies *directly related* to the research in this *partial* review, we include the work by Tarlov[7] and us[6] on the Ag/ODT interface and by us on the Ag/COOH interface for SAMs bonded to Au films as the substrate. Using ion scattering spectroscopy (ISS), Tarlov showed that Ag deposited onto ODT at 300 K penetrates to the ODT/Au interface, but that Ag resides at the methyl end group when deposited at 90 K.[7] As we present below, Herdt and Czanderna studied the time and temperature dependence of Ag penetration through ODT, MUA, and MHA using ISS and showed that Ag penetrates rapidly through ODT above about 200 K, but penetrates less rapidly through MUA and MHA.[6]

The formation of Ag islands on Langmuir-Blodgett (LB) films made from cadmium arachidate, a polymeric phthalocyanine, and an alkyl-substituted polyglutamate statistical copolymer has been studied by Reiter et al. using X-ray reflectometry.[25] Their work indicates that 5 nm of Ag deposited at about 0.1 nm/s onto LB films formed islands that penetrated into the film to a depth of 1-2 monolayers. The interaction between vacuum-deposited Ag and poly(ethylene terephthalate) (PET) surfaces was studied by Gerenser,[26] who found evidence of charge transfer between deposited Ag atoms and the carbonyl O of PET at low Ag coverages. He also found evidence of additional bonding between vacuum-deposited Ag and plasma-induced COOH groups after treatment with an oxygen plasma. From his XPS results, he concluded that a unidentate Ag species was formed.

The interaction of Cu with the specific terminal OFGs on SAMs has been studied at NREL and elsewhere.[8,27] For a carboxylic acid end-group COOH, a unidentate copper complex is formed with the single-bonded oxygen atom.[1,2] The copper atoms of this complex exhibit a maximum charge of +1, with one charged Cu per COOH group. For a hydroxyl end-group CH_2OH, a similar copper oxide species is formed.[3] The possibility of penetration of Cu deposits into or through CH_2OH and CN SAMs was addressed following the reported penetration of Ag through ODT by Tarlov.[7] Czanderna et al.[1,2] saw no evidence of penetration for Cu on the MUA.

EXPERIMENTAL

A detailed list and critique of the experimental measurements used to characterize SAMs is available.[10] For the typical results presented in this article, the use of XPS and ISS will be described. The sample preparation methods described below are similar to those reported by other groups.

SAM Preparation

For the studies of Au deposited onto CH_3-terminated SAMs, substrates were prepared by evaporating 120 nm of silver onto polished Cr-primed (0.5 nm) silicon (100) wafers. The wafers were removed from the bell jar and cleaved into 10-mm x 30-mm strips. XPS analysis of the as-evaporated silver films revealed no contamination from either silicon or chromium. Methyl-terminated SAMs were self-assembled by immersing the freshly prepared silver films in 3 mM solutions of alkanethiols in ethanol. The thiols were obtained from Aldrich (typically 99% pure) and were used without further purification. The methyl-terminated alkanethiol chain lengths studied contained, 6, 10, 12, 16, and 18 carbons (designated as C-6 through C-18, respectively). To protect the films from air oxidation and contamination, the SAMs were stored in the thiol solutions until used. Samples were removed from solution and spin-rinsed with ethanol and dried under ultra-high-purity (UHP) nitrogen prior to analysis. The SAM film quality was assessed by static deionized water contact angles measured with a Rame-Hart goniometer, followed by XPS analysis.

For the studies of Cu or Ag deposited onto various SAMs, the SAMs were formed by immersing clean Au-coated Si <100> wafers in ethanolic solutions of octadecanethiol (ODT, $HS(CH_2)_{17}CH_3$), mercaptomethylhexadecanoate (MMHD, $HS(CH_2)_{15}COOCH_3$), 11-mercaptoundecanoic acid (MUA, $HS(CH_2)_{10}COOH$), or 16-mercaptohexadecanoic acid (MHA, $HS(CH_2)_{15}COOH$) using the well-known self-assembly process for alkanethiols on Au or Ag.[21,24,28] Some of the gold coatings were prepared by evaporating 80 nm of Au directly onto Ar^+ ion bombardment-cleaned Si wafer coupons,[29] whereas others were prepared by evaporating a 9-nm Cr adhesion layer followed by a 200-nm Au layer onto solvent-cleaned Si wafers. The hillock sizes of these two types of Au films have been measured by scanning tunneling (STM) and atomic force microscopies (AFM) to be 30-50 nm wide and 10 nm high for the Ar^+ ion sputtered substrates and about 100 nm wide and 10 nm high for the Cr-adhered Au films. In most cases, the Au surfaces were placed in a UV-ozone cleaner for 5 min to remove trace organic contaminants[30] prior to immersion in the (typically) 2-mM ethanolic alkanethiol solutions. The ozone treatment for 1 h produces a 1.7-nm-thick Au_2O_3 oxide on the gold that is subsequently reduced during the self-assembly process.[31] For the system Au/CH_3, 80-nm-thick Ag films were prepared by vacuum evaporation before self-assembly of the ODT. In all cases, the self-assembly was carried out in glass that was cleaned with piranha solution at 90°C and rinsed with absolute ethanol.[1,2] Water contact angles of less than 10° are observed on the Au surface only if such a cleaning procedure is used. Following incubation times of 2-4 days at 295 K, a sample was removed, rinsed in ethanol, dried with ultra-high-purity nitrogen, and checked with a Rame-Hart goniometer for a water contact angle to verify the presence of the SAM. Contact angles of 10°±2° to 20°±2°, 72°±2°, and 112°±2° observed in this study for the COOH-, $COOCH_3$ and CH_3-terminated SAMs, respectively, agree well with published results.[28]

Evaporative Deposition of the Metal Overlayers

Evaporative deposition of the metal onto the SAMs and subsequent XPS and ISS analyses were carried out using a Leybold-Heraeus LHS-10 surface-analysis system. The LHS-10 is equipped with an analysis chamber and a sample preparation chamber through which a sample introduction rod can be translated without breaking vacuum. The sample preparation chamber is equipped with two vacuum feedthroughs that provide power to crucibles used for evaporative deposition. Deposition rates are measured with a quartz- crystal oscillator using an IC 6000 deposition monitor. A negligible amount of sample heating occurs during deposition because the sample-to-source distance is about 38 cm. An X-ray source for XPS analysis, an ion gun for sputtering and ISS, and a concentric hemispherical energy analyzer (CHA) are attached to the analysis chamber. The CHA is used to determine the energies of scattered ions in ISS. A complete description of this apparatus and its use for XPS and ISS have been presented previously.[32]

XPS and ISS Measurements

The sample was mounted onto a hot/cold transfer rod (initially at 22°C) and inserted into the LHS-10 chamber maintained at a base pressure of 2-5x10^{-8} Pa. Deposition of the Cu or Ag (Metron, 99.999%) took place in the attached preparation chamber that permits *in situ* transfer of the sample to and from the XPS (or ISS) analysis chamber at constant temperature. The sample was allowed to outgas at room temperature for 1 h prior to heating or exposure to X rays. In our standard procedure, the sample was characterized by XPS at 295 K, heated or cooled as desired, recharacterized by XPS, exposed to the first incremental deposition of metal, and again characterized by XPS.[1-6] Repeated deposition/characterization cycles were performed in some cases, resulting in total coverages of up to 2 nm. Thus, the photoelectron escape depths permitted us to probe for M/OFG interface effects throughout the incremental deposition processes; the take-off angle was in the direction of the surface normal. Typically, a time period of up to 2 h was required for each cycle of XPS characterization involving a survey spectrum and narrow-scan spectra. The 2-h exposure is less than that known to damage SAMs.[11b] In this paper, we refer to the amount of metal deposited as a uniform coverage, e.g., 1.0 nm Ag or Cu, as though it is a smooth, continuous covering of the surface of the SAM. We are aware that clustering may occur, but we must provide a frame of reference. We address the attenuation of the photoelectron peak intensities in models used to estimate the degree of clustering or uniformity of the M overlayer. In all experiments, less than 3 min elapsed between the end of metal deposition and the beginning of the XPS or ISS analysis.

The photoelectron spectra were recorded using a non-monochromatic Mg $K\alpha$ X-ray source operated at 240 W. Narrow scans were taken using constant pass energies of 200 eV for N 1*s* (for coverages below 0.2 nm) and 50 eV for C 1*s*, N 1*s*, O 1*s*, Au 4*f*, Cu 2*p*, and Ag 3*d*. Survey scans were taken in the constant-transmission mode with a retardation ratio of 3. For determining XPS peak areas and binding energies, the C 1*s* narrow-scan line shapes were fitted using 85% Gaussian/15% Lorentzian (G/L) product functions of appropriate peak widths for each of the high-resolution scans,[33] and a linear background using a nonlinear least-squares package.[34] The binding energies (BEs) of all peaks are referenced to the Au $4f_{7/2}$ peak position at 84.0 eV to account for sample charging, which was found to be negligible.[35] The XPS data are used to evaluate the interface stoichiometry from adjusted peak intensities, the "chemical state" from BE shifts, and nonchemical BE shifts for partial to total penetration.

ISS compositional depth profiles were measured using 1 keV $^3He^+$ ions at a current density of 0.2 $\mu A/cm^2$ and temperatures from 113 to 295 K. The ca. 1-mm full-width-at-half-maximum (FWHM) ion beam was rastered over a 2-mm x 2-mm area with the detected signal gated at a 70% aperture to minimize the signal from the etch-pit walls. During ISS analysis, He from the ion gun caused the chamber pressure to increase from a base pressure below $2x10^{-9}$ Torr to a carefully maintained pressure of $4x10^{-7}$ Torr. Each ISS spectrum was acquired from 800 eV to 1000 eV (E/E_0 of 0.8 to 1.0 in which E_0 and E are the energies of the incident and scattered helium ions, respectively) over about 40 s. Detection of C with $^3He^+$ is limited by the number of H atoms bonded to the C. Carbon in CH_3 or CH_2 is difficult to detect with ISS, but C or CH exposed to the ion beam are detectable. Consequently, we did not attempt to detect C in our compositional depth profiles. Sulfur from the thiol has been observed by ISS in a previous work.[29] Accordingly, data were taken over an energy range E/E_0 of 0.80 to 1.0, to obtain depth profiles with good time resolution and good signal to noise.

RESULTS AND DISCUSSION

We present and discuss in this section a summary of the M/OFG combinations studied to date and rank them according to their degree of penetration into the SAM.[10] We then present and discuss our results for the penetration in the Au/CH_3, Cu/CH_3, Cu/$COOCH_3$, Ag/CH_3, and Ag/COOH systems.

A. Ranking of M/OFG Combinations According to Penetration

In Table III, based on the XPS and FT-IRRAS results, we have ranked the penetration in M/SAM systems according to the results reported for the different OFGs and different chain lengths of the SAMs. A dependence of the degree of penetration on the temperature of the sample--i.e., lower penetration for lower temperatures--has been found for the systems of Cr/CN, Ni/CN, and Cu/CN[5] T = 173 K, Ag/CH_3[6-7] T ≥ 90 K, and Ag/COOH.[6c] A dependence on the length of the alkanethiol molecule may also affect the XPS results shown in Table III; i.e., shorter molecules show evidence for permitting greater penetration.

We conclude from these rankings that the degree of penetration increases with decreasing reactivity. In order for the adsorbed metal atom or cluster to penetrate, it first must encounter the OFG. Penetration is unlikely if there is a reaction. If there is *no* reaction, then penetration can proceed in some cases. Rates of penetration have only been measured in the cases of Ag[6,7] and Na[36,37] on CH_3-terminated SAMs and for Ag on COOH-terminated SAMs.[38] The reactivity is highest for the oxygen-containing OFGs (except for Ag/COOH), intermediate for the CN OFG, and lowest for the CH_3 OFG.[10] For the metals, Ti is the most reactive, followed by Cr, Al, K, Ni, Cu, Na, and Ag, although the data for Al, K, Ni, and Na are scant. The decreasing reactivity within the transition metal series, Ti, Ni, and Cu, and the low reactivity of the noble metals, Cu and Ag, is in the same order as the filling of the d orbitals. The relative reactivity of trivalent Al and the monovalent alkali metals (Na, K) is not so clear at this point because so few of these M/SAM systems have been explored. Evaporating oxygen-free Al films has not been successful, as yet.[39] The M/SAM studies involving Al evaporation might have been affected by rapid oxidation of the deposit,[39] which could have had a severe effect on the chemical interactions at the interface.

There is some evidence that the reactivity of the incoming M with the SAM also depends on the SAM temperature.[5,11] Activation of M/SAM reactions may be because of electronic, steric, or structural factors and is not unexpected. Further temperature-dependent studies are warranted to learn more about activation energies in M/SAM reactions, as well as for rates of penetration.

In Table IV, we broadly summarize the penetration time for Au, Ag, and Cu through various SAMs based on our ISS results. The detailed results are discussed in Sections B, C, and D and also include XPS and contact-angle results. However, ISS has provided us with the most direct and consistent results for assessing the penetration times. Its limitations are discussed in Section D.

TABLE III. XPS Evidence for Penetration in M/SAM Systems (All M/OFG(CH_2)$_n$-S/Au) at Room Temperature (RT) and at Lower Temperatures (LT).

			Metal Shift+		C 1s attenuation+		Penetration+		
n	Metal	OFG	RT	LT	RT	LT	RT	LT	Ref.
15	Ti	COOH	HBE	?	?	?	Very low	?	9
15	Ti	$COOCH_3$	HBE	?	?	?	Very low	?	9
15	Ti	CH_2OH	HBE	?	?	?	Very low	?	9
15	Cr	$COOCH_3$	HBE	?	?	?	Very low	?	8, 11(c)
15	Ti, Cr	CH_3	HBE	?	?	?	Low	?	4,8,9
15	Cr	CN	HBE	?	?	?	Low	?	4,8
11	Cr	CN	HBE	HBE	Slight	OK	Low	Low	4
15	Cu	$COOCH_3$	HBE	?	?	?	Low/Mod.	?	8, 33
10	Cu	COOH	Zero	?	?	?	Low	?	1,2
10	Cu	CH_2OH	LBE	?	Low	?	Moderate	?	3
11	Ni	CN	Zero	HBE	Low	Slightly low	Moderate	Low	4
15	Cu	CN	HBE	?	?	?	Low	?	8
11	Cu	CN	LBE	HBE	Quite low	?	Moderate	Low	4
17	Ag, Cu	CH_3	LBE	HBE	Zero	?	High	Low	6, 7, 33
10	Ag	COOH	HBE	?	?	?	High	Low	6(c)
15	Ag	COOH	HBE	?.	?	?	Moder./High	Low	6(c)

+ *Notes:* The metal shift refers to whether a core level has been measured at low coverages to exhibit a shift to higher binding energy (HBE) or lower binding energy (LBE), respectively, with respect to the bulk metal core level. C 1s attenuation refers to the attenuation of the C 1s level compared to that expected for a uniform metal overlayer. References to the sources of these XPS results are given. Question marks (?) indicate that the information has not been reported. The degree of penetration is based on the interpretation by the authors of ref. 10 of the available XPS results, in which "very low" means the overlayer essentially resides on top (in the same plane) of the OFG and "high" means the overlayer penetrates to the SAM/Au interface.

Table IV. Time Required for 1.0-nm Metal Overlayers to Penetrate SAMs at 295 K as Determined by ISS

System	Penetration Time
$Au/CH_3(CH_2)_{17}S/Ag$	< 5 min
$Ag/CH_3(CH_2)_{17}S/Au$	< 5 min
$Ag/HOOC(CH_2)_{10}S/Au$	65 min
$Cu/CH_3(CH_2)_{17}S/Au$	135 min
$Cu/CH_3OOC(CH_2)_{15}S/Au$	ca. 440 min
$Ag/HOOC(CH_2)_{15}S/Au$	> 16 h

B. Penetration of Deposited Au Through C6 to C18 CH_3-terminated SAMs on Ag

ISS analysis. ISS compositional depth profiles (CDP) of the methyl-terminated SAMs studied are shown in Fig. 2. A marked difference in the silver ISS signal is seen as a function of the SAM chain length. For obtaining the plot in Fig. 2b, we used the raw data to determine to within ±2 min when the Ag ISS signal clearly emerged from the background noise. The data in Fig. 2a are presented to show the broad trend from sputtering the SAMs, but the raw data in the early stages of bombardment were used to produce Fig. 2b. ISS is found to be extremely sensitive to the number of carbons in the SAM chain, and, thus, to the thickness of the monolayer. In addition, the time to sputter through the SAMs to reach the silver substrate is not linear with carbon chain length, as shown by the increase in the erosion time per methylene group between 10 and 12 carbon atoms. Because the thickness of SAMs is roughly linear with increasing number of carbon atoms (about 0.14 nm/carbon atom),[28] we conclude that SAMs with more than 11 methylene groups have a greater packing density. This is consistent with longer-chain SAMs being more closely packed, ordered, having a greater film density, and an increased erosion time per methylene group from 11 to 18 carbon atoms in the SAMs. Thus, differences in film thickness and in the packing density in the organic monolayers are detectable using ISS.

ISS depth profiles were collected on a series of methyl-terminated SAMs with incubation times of 30 s and 42 min after deposition of 1 nm of gold (Fig. 3). The incubation time begins at the time metal deposition is complete; the sample was held in vacuum for the specified incubation time prior to the start of any analysis. For the 30-s incubation time, the data show that silver is detected before gold for the SAMs with chain lengths less than 16 carbon atoms. For SAMs with 16 carbons, both silver and gold are detected after 21 min. For the C18 SAM, gold is detected before silver. For silver to be detected before gold in the shorter-chain SAMs, the deposited gold must not only rapidly penetrate the SAM (in less than 30 s), but also, must diffuse

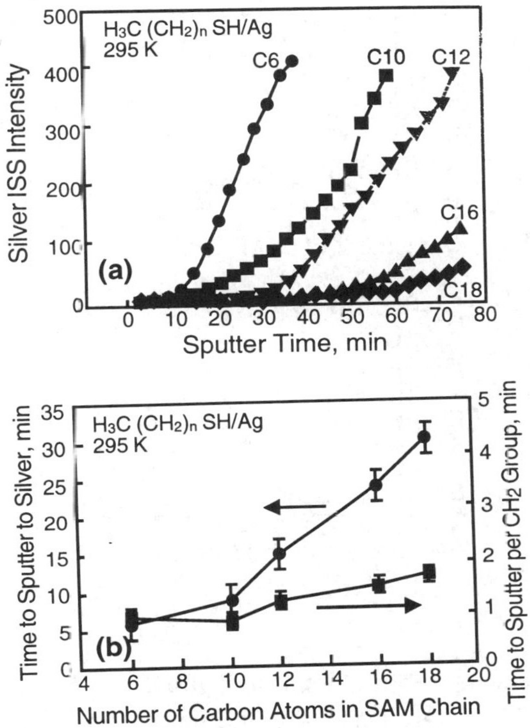

Fig. 2. (a) ISS compositional depth profiles (CDPs) of a series of methyl-terminated SAMs on silver substrates. Using the same ion beam as for ISS the remarkable sensitivity for the time to erode through SAMs with different number of carbon atoms is shown. SAMs differing in thickness by as few as 2 carbon atoms are clearly resolved from the erosion process. (b) The time to erode through SAMs from C-6 to C-18 increases nonlinearly as the number of carbon atoms in the SAM chain increases. Between SAMs with C-10 to C-12, the rate of damage by the ion beam decreases markedly from the increase in packing density or order of the longer SAM chains.

under the sulfur/silver interface and *under* the SAM. With a C16 SAM, some of the gold resides at the thiol/silver interface because both metals are detected simultaneously. For the C18 SAM, the gold is probably still in the process of diffusing through the SAM, so gold is detected earlier in the profile than silver.

Deposition of 1 nm of gold and incubating the samples for 42 min yields similar results, as shown in Fig. 3b. Silver is detected before gold in all cases, except for C16 and C18 SAMs, where gold remains at the thiol/silver interface and both are detected simultaneously. For C12 and shorter-chain SAMs, gold penetrates rapidly and diffuses below the SAM/silver interface; but for the longer-chain SAMs, the gold seems to reside longer at the SAM/silver interface. We conclude from the data that Au diffusion under Ag may be the rate-limiting step to reaching a steady state in these systems.

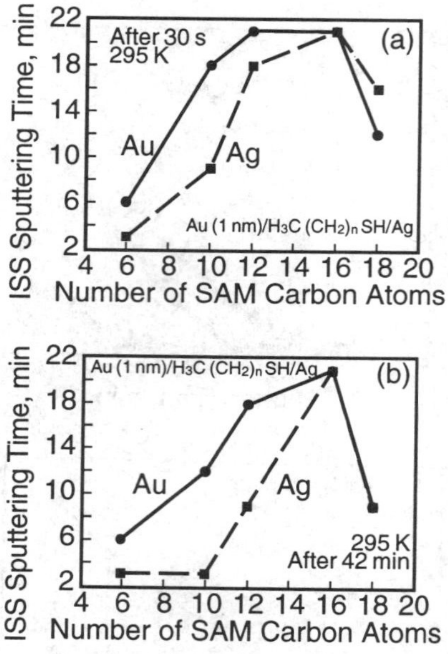

Fig. 3. (a) ISS CDPs were begun 30 s after 1 nm of gold was deposited. For SAMs with C-12 or less, there was enough time for the gold to completely penetrate the SAM and diffuse *under* the thiol SAM/silver interface. For the C-16 SAM, both gold and silver are detected at the same time, indicating that the diffusing gold is at the thiol SAM/silver interface. For the C-18 SAM, the gold penetration through the SAM is not complete after 30 s, and gold is detected before silver. (b) The CDPs were begun 42 min after 1 nm of gold was deposited. For C6-C12, silver is detected before gold, which is the same after a 30-s incubation time, but gold and silver are simultaneously detected for the C16 and C18 SAMs. This means that in 42 min, all of the gold has penetrated to the SAM/silver interface for the C16 and C18 SAMs.

To determine if the gold diffusion under a monolayer of silver is affected by gold/silver interactions, 1 nm of gold was deposited onto clean, silver-coated substrates. Samples were incubated for 30 s and 22 h as shown in Fig. 4. For both incubation times, ISS detected both the gold and the silver at the same time, indicating that a thiol monolayer clearly influences the diffusion of gold under the SAM into the silver film.

To study the metal penetration/diffusion process further, 1 nm of gold was evaporated on C18 SAMs, and then the samples were analyzed with ISS at incubation times of 30 s, 1 min, 39 min, and 960 min. Times to profile to gold and silver through the SAM are shown in Fig. 5. After 16 h of incubation time, silver is detected after sputtering for 21 min and gold is detected 6 min later. Thus, with enough time,

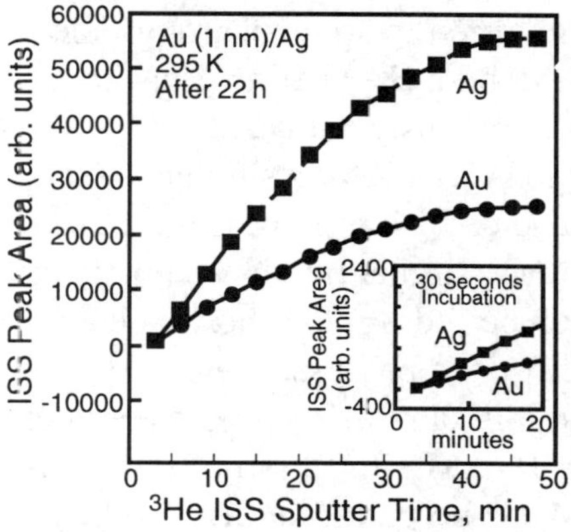

Fig. 4. For 1 nm of gold deposited on a clean silver substrate (without a SAM), gold is detected at the same time as silver after 30 s and also 22 h later. Therefore, gold diffusion to below the sulfur-silver interface is unique to the thiol-SAM systems.

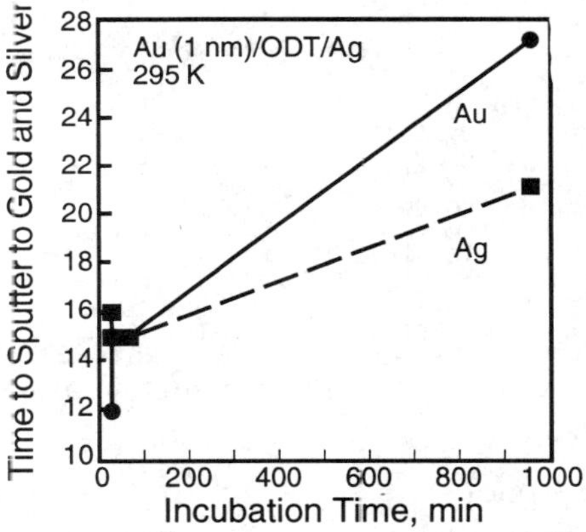

Fig. 5. The time to erode through a C18 SAM to reach the silver substrate or 1 nm of penetrating gold after different incubation times. After a 16-h incubation of 1 nm of gold on a C18 SAM, there is complete penetration through the SAM to below the SAM/silver interface. After 16 h, silver is detected several minutes before gold. In addition, the time to sputter through the SAM to the SAM/silver interface is considerably shorter than for the bare SAM, suggesting significant damage by the deposition and diffusion processes.

evaporated gold buries itself under the thiol/silver interface of the longer-chain SAMs From these data, we can assess the damage to SAMs resulting from the meta evaporation and diffusion process. As shown in Fig. 5, 21 min is the time needed t sputter through a C18 SAM to the SAM/silver interface 16 h after metal deposition Because 9 min less is required to sputter through the C18 SAM after Au penetratio (Fig. 5, 21 min) than the bare SAM film (Fig. 2b, 30 min), the Au deposition an diffusion processes have obviously damaged the SAM. The resulting thickness (orde or density) of the damaged C18 SAM is similar to that presented by a bare C18 SAM so the gold deposition and diffusion processes reduce by 33% the effective density o thickness of the C18 SAM. Similar trends are not detected for the shorter-chain SAM after a 42-min incubation period; so, the shorter-chain SAMS seem to be damaged les by the Au deposition/diffusion process.

XPS analysis. No significant differences are evident in the C 1*s* XPS spectra of bare SAM and that for a SAM after 2 nm of gold was evaporated (Fig. 6). A sligh increase in the FWHM of the photoline may indicate subtle changes in the film o changes in the surface chemistry. This result is consistent through the series of methy terminated SAMs studied, but contrasts with the ISS results, where significant damag to the film is evident after deposition of gold.

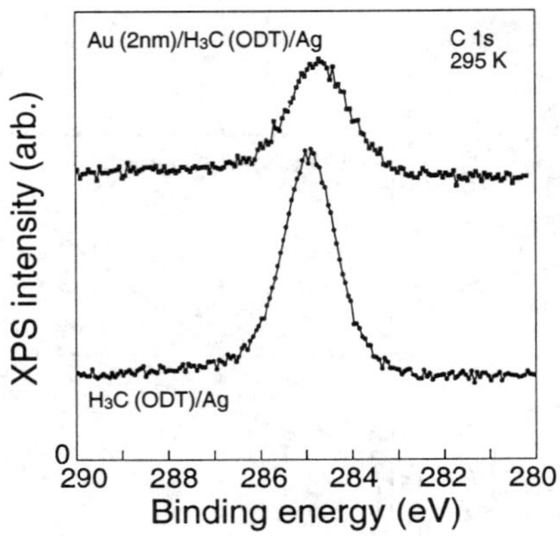

Fig. 6. The C 1*s* XPS photoline broadens slightly as gold is deposited from subt chemical or physical changes in the structure of the SAM. The result show is typical for all of the Au/SAM interactions for the C6 through C18 SAM studied. Because no significant changes are observed in either the Au 4*f* or 1*s* XPS spectra after depositing Au, no interaction appears to occur betwee Au and the methyl-terminated SAMs. The changes in XPS spectra before an after metal deposition are in marked contrast to the ISS results for a gol coated C18 SAM.

Damage to the SAMs is also assessed to be minimal from contact-angle data alone (Fig. 7). Water-contact angle measurements on different chain-length SAMs before and after deposition of 1 nm of gold onto methyl-terminated SAMS show a slight increase in contact angles after gold is deposited. However, within experimental error, the contact angle does not change appreciably after the metal deposition and incubation process. This indicates that some degree of film reordering or restabilizing occurs after the diffusion process is complete. The change in the increasing trend of the contact angle after metal evaporation for the C18 SAM may be real, and it indicates that the diffusion and reordering processes are not yet complete, or that the monolayer has been damaged by the process. It appears that the longer-chain SAMs are more subject to damage by metal evaporation and penetration than shorter ones. The contact-angle data alone do not provide the insight into the quality (order, density, and thickness) of the SAM that ISS does.

As a summary of the Au/H_3C-SAM systems, the erosion rate through a SAM with ISS is sensitive to the SAM thickness, and differences of two methylene groups or two bonded carbon atoms in length (less than 0.3 nm in thickness) are easily measured. Differences in the ordering (and thus, the density) of the SAM films are also detected from the time to sputter through SAMs with the ion beam used for ISS. Evaporated gold rapidly penetrates short-chain SAMs and diffuses under the thiol-SAM/Ag interface. As detected with ISS, the penetration and diffusion processes of the evaporated metal result in damage to a C18 SAM by decreasing its order and density. These structural changes result in a slight broadening the C 1*s* peak in XPS, but are not evident from contact-angle measurements. Finally, gold evaporated onto a bare silver film does not diffuse under the silver surface as detected in the SAM systems. The presence of the thiol/SAM has a direct influence on the positioning of evaporated gold under the silver surface.

C. Penetration of Deposited CU Through CH_3- and $COOCH_3$-terminated SAMs

XPS and ISS analyses of the Cu/CH_3 and $Cu/COOCH_3$ systems are discussed in this section. The $Cu/COOCH_3$ system was previously studied using reflection absorption infrared spectroscopy (RAIRS),[10] and some XPS characterization of this system was also reported by Zhang.[8] The objectives of work presented in this section are to compare and contrast penetration and interactions of Cu with CH_3- and $COOCH_3$-terminated SAMs and to extend prior studies of the $Cu/COOCH_3$ system.

XPS of Cu deposited on MMHD and ODT. Figure 8 compares the measured Cu $2p_{3/2}$, C 1*s*, O 1*s*, and Au $4f_{7/2}$ XPS peak intensities as a function of the amount of Cu deposited onto MMHD to values predicted by a layer-by-layer growth model. Although the C 1*s* and Au $4f_{7/2}$ intensities conform remarkably well to the predictions of the model, the measured Cu $2p_{3/2}$ intensity deviates from the model and is attributed to the formation of Cu clusters on the $COOCH_3$-terminated surface of MMHD. Comparable results were obtained for Cu deposited on methyl-terminated ODT. Because the mean free path for Cu 2*p* photoelectrons in a SAM is about 1.6 nm,[40] the sensitivity to different Cu depths in the SAM is relatively small at exit angles normal to the surface.

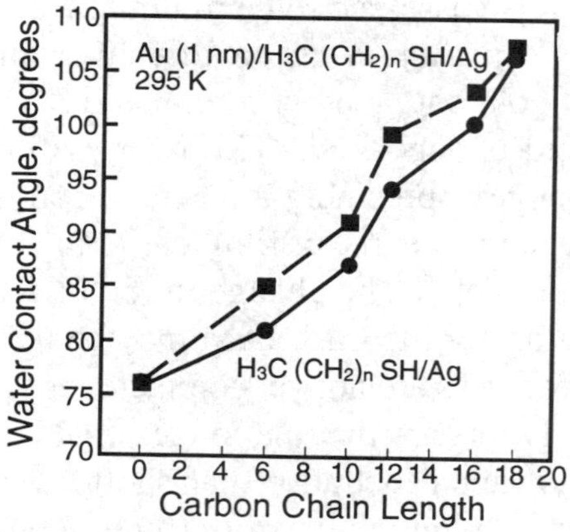

Fig. 7. The contact angles of water on C6 to C18 SAMs before and after depositing 1 nm of Au. Although contact angle is increased slightly after deposition of Au for shorter chains, little differences are measurable for C16 and C18. The minor differences in the contact angles for C16 and C18 SAMs before/after deposition indicate that SAMs have returned to their initial packing order after the deposition and diffusion processes have been completed.

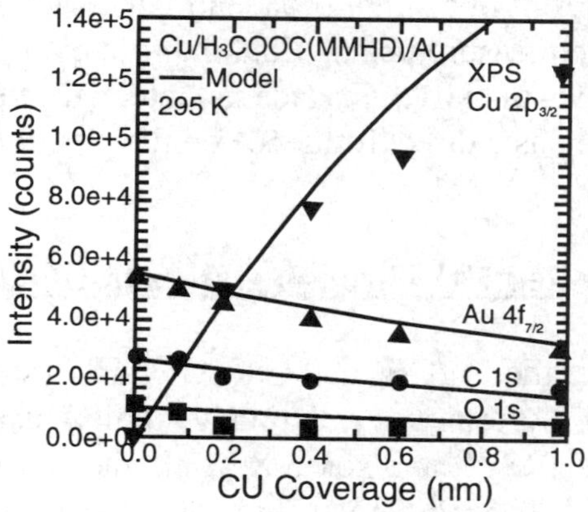

Fig. 8. Comparison of measured XPS intensities with a layer-by-layer growth model for incrementally deposited Cu on MMHD at 295 K.

Hence, obtaining repeated XPS scans is not a good means for determining the time required for metal overlayers to penetrate SAMs.

For incremental deposits of Cu onto ODT, the C 1*s* peak shifted to high binding energies (HBE) by about 0.3 eV with increasing Cu coverage, which is comparable to the C 1*s* peak shift observed for the Ag/ODT and Ag/MUA system.[6] The C 1*s* peak also exhibits slight broadening upon deposition of 0.1 nm of Cu, which is also

consistent with the observed behavior of the C 1*s* peaks in the Ag/ODT and Ag/MUA systems.[6]

For different Cu coverages on MMHD, XPS C 1*s* peaks were fitted using three Gaussian-Lorentzian product functions, corresponding to low, medium, and high binding energies (LBE, MBE, HBE), respectively, for carbon in methylenes (285.0 eV), the carbonyl carbon (289.3 eV), and the methyl carbon (287.3 eV) in the $COOCH_3$ OFG.[41] The signal-to-noise ratio of the MBE and HBE peaks is poor at Cu coverages above 0.1 nm, but the BE shifts in all three of these peaks are comparable, showing a 0.2-0.3 eV shift to HBE, which is consistent with BE shifts in other M/SAM systems discussed in this work.[10] The ratio of peak intensities for the LBE, MBE, and HBE C 1*s* components remains approximately constant with increasing Cu coverage, which agrees with Zhang and whose work is summarized in ref. 10.

The O 1*s* peaks for the Cu/MMHD system were fitted to two components corresponding to oxygen in C=O (LBE, 532.7 eV) and C-O-CH_3 (HBE, 534.1 eV).[41] At Cu coverages below 0.2 nm, the ratio of peak areas of the LBE and HBE O 1*s* components remains close to 1:1, and the peaks exhibit only a slight change in position. At coverages above 0.2 nm, a clear shift to lower BEs is evident in both components of the O 1*s* peak, and the O-CH_3 O 1*s* signal is even reduced more than the C=O component of the methyl ester. These results agree with the RAIRS results of Allara (as summarized in ref. 10), who reported a "weak interaction" between Cu and the O-CH_3 component of the $COOCH_3$ OFG of MMHD. However, this loss of O 1*s* intensity is not accompanied by a corresponding loss of intensity in the O-CH_3 C 1*s* peak; so, there is no evidence that the O-CH_3 bond is cleaved as Allara speculated, and whose work is summarized in ref. 10.

The XPS Cu 2*p* and X-ray-excited Cu LMM Auger peaks for the Cu/ODT and Cu/MMHD systems exhibited peak shifts amounting to a maximum of about 0.3 eV with increasing Cu coverage. These slight changes in the BEs as a function of Cu coverage are attributed to a combination of instrumental and peak-fitting errors and to deviations from bulk behavior at low coverages. It is also notable that the FWHM of both the Cu $2p_{3/2}$ and the Cu LMM peaks for these systems decreases markedly with increasing Cu coverage. The broader peaks may be attributed to the formation of clusters on the CH_3- terminated surface of ODT at low coverages, or to the dispersion of Cu as it penetrates into the alkanethiol chains; the narrower peaks are attributed to the formation of larger clusters at the highest Cu coverages. Auger parameters for both systems determined from the Cu $2p_{3/2}$ and X-ray excited Cu LMM peaks are more consistent with values reported for Cu(0) than for Cu(I).[42,43] Thus, the weak interaction that causes the BE shift in the O 1*s* results from only a partial electron transfer from (or sharing with) the Cu to the carbonyl and ester oxygens.

Time-dependence of Cu penetration through ODT and MMHD from XPS and ISS. Two experiments were carried out to characterize the amount of Cu resident on the OFG surfaces of ODT and MMHD as a function of time after deposition. In the first experiment, 1.0 nm of Cu was deposited onto the SAMs, and repeated XPS narrow scans of the Cu $2p_{3/2}$, C 1*s*, and Au 4*f* peaks were obtained so that changes in peak intensities could be used to monitor penetration of Cu through the SAM. Cu $2p_{3/2}$ peak intensities for the Cu/ODT and Cu/MMHD systems are shown in Fig. 9. The C 1*s* peaks for these systems are relatively insensitive to time-dependent changes in the system, so they were not analyzed further. A significant difference in the initial absolute intensity of this peak was observed in the Cu/ODT and Cu/MMHD systems, indicating a greater Cu coverage on MMHD than on ODT immediately after deposition, i.e., more penetration into ODT than MMDH occurred *during* deposition. From these data, the time for Cu to penetrate ODT appears to be between 60 and 200 min, but the time to penetrate MMHD is at least 270 min.

In the second set of experiments, ISS analysis was used to characterize penetration of Cu through ODT and MMHD after deposition of 1.0-nm Cu overlayers onto these SAMs after a chosen period of time. In these experiments, the ISS signal is used to determine the Cu coverage on SAMs as a function of time after deposition (about 100 s were required to transfer the sample to the analysis chamber after deposition was terminated, and 40 s were needed to complete the first ISS spectrum). A new sample was used for each measurement to minimize the ion-beam damage effects on penetration of Cu through the SAMs. The results of these experiments are shown in Fig. 10. It is clear that Cu penetrates through ODT in 135 min and through MMHD in about 440 min. The disappearance of the initial Cu ISS signals results from the continued penetration of Cu into the SAM.

The ISS Cu peak intensity for 1.0 nm of Cu (i.e., 140 s after depositing 1.0 nm Cu on MMHD) is about six times more intense than for Cu on ODT. Thus, a greater Cu coverage is formed on MMHD than on ODT under identical deposition conditions and amounts of deposition. However, quantification of the actual surface coverage of Cu on these SAMs is not straightforward; differences between the electronic states of "bulk" Cu and Cu clusters or islands could substantially change the neutralization probability of ions scattered from the clusters versus a Cu bulk metallic standard.

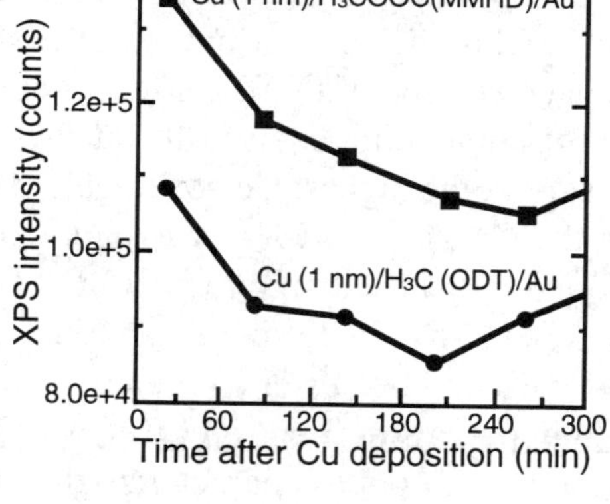

Fig. 9. XPS Cu $2p_{3/2}$ peak intensity as a function of time after Cu deposition for 1.0 nm Cu depositions onto ODT and MMHD SAMs. The uncertainty in the XPS counts is less than the size of the circles or squares used in the plots.

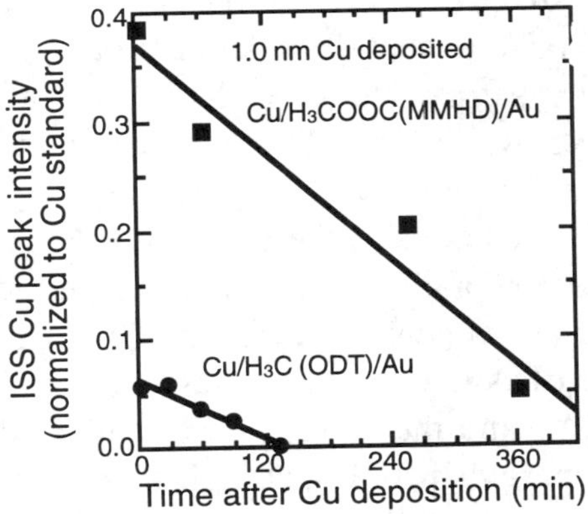

Fig. 10. ISS Cu peak intensity as a function of time after depositing 1.0 nm Cu onto ODT and MMHD SAMs.

Discussion of the Cu/ODT and Cu/MMHD systems. Evidence from the present study is not sufficient to elucidate the nature of the interaction at the Cu/$COOCH_3$ interface, beyond stating that it is weak and appears to occur preferentially with the C-O-CH_3 oxygen of the OFG. The relative weakness of the interaction is clear from the lack of change in the Cu 2*p* and Cu LMM peaks when compared with those measured for Cu/CH_3 and because the Cu 2*p* peak position is more consistent with Cu(0) than the expected values for Cu(I) or Cu(II). Although Cu is known to form both unidentate and bidentate complexes,[44] the interaction of Cu with the singly-bonded oxygen in $COOCH_3$ is more consistent with the unidentate formed when Cu is deposited onto COOH to form O=C-OCu.[1] For the $COOCH_3$ group, Cu-Cu interactions can dominate rather than Cu-O interactions, because the methyl group shields the more-active oxygen and scission of the O-CH_3 bond is not favored. Shielding could also explain why the Cu/$COOCH_3$ interaction is observed only at Cu coverages above 0.4 nm; at lower coverages, the Cu probably forms small clusters on the $COOCH_3$ surface and does not interact as metallic Cu.

The difference in the rate of penetration of Cu through ODT and through MMHD can result from the difference in interactions between Cu and CH_3 (weak, if any) or with $COOCH_3$ (some); it is also possible that lateral interactions between $COOCH_3$ OFGs inhibit transport pathways in the Cu/$COOCH_3$ system. The difference in the time that Cu and Ag take to penetrate ODT (135 min vs < 5 min) can be attributed to differences in the nucleation behavior and/or mobilities of these species on the CH_3-terminated surface.

Penetration of Deposited Ag Through CH_3- and COOH-terminated SAMs

XPS and ISS of Ag/COOH and Ag/CH_3. The Ag/MUA, Ag/MHA, and Ag/ODT systems were studied with XPS and ISS at incremental Ag coverages. The typical results reported here are for Ag/MUA, for which we also have the most complete data.

For the range of Ag coverages studied, results for the Ag/MHA and Ag/ODT systems agree well with those for Ag/MUA, except O 1*s* data are obviously not obtained from XPS studies of the Ag/ODT system. The O 1*s* XPS spectra of Ag on MUA as a function of Ag coverage are shown in Fig. 11a. We determined from "peakfits" that these O 1*s* spectra could all be fitted best with an LBE peak at 531.9 eV and an HBE peak at 532.9 eV and that the positions of the LBE and HBE components remained constant within experimental error for all Ag coverages studied for both Ag/MUA and Ag/MHA. Furthermore, the ratio of peak areas for the LBE and HBE peaks remains constant at 1:1 as a function of Ag coverage. These results do not indicate that a complex is formed between Ag and the COOH OFG, unless the interaction occurs without a detectable binding-energy shift. If a complex were formed, we would have expected to observe the O 1*s* feature at 531.3 eV, as previously observed in compounds with Ag bonded to O.[26,46]

XPS C 1*s* and Ag 3*d* spectra corresponding to the data in Fig. 11a are shown in Figs. 11b and 11c. With increasing Ag coverage, the Ag 3*d* peak position shifted to LBE by about 0.4 eV for Ag on MUA; for corresponding coverages, the Ag 3*d* shifted to HBE by about 0.1 eV for Ag on ODT.[6b] The XPS binding-energy shift of the Ag $3d_{5/2}$ peak for Ag/ODT and Ag/MUA as a function of Ag coverage is shown in Fig. 12. The binding-energy shift differences probably result from initial- or final-state effects for Ag on a metal (Ag/ODT, in which Ag resides at the ODT/Au interface) and Ag on an insulator (Ag/MUA).[47-50] The binding-energy shift to HBE in the C 1*s* peak for Ag on MUA is about 0.2 eV (Fig. 11b), which is smaller than the value of about 0.4 eV for similar coverages of Ag on ODT.[6b] Coverage, temperature, and time-dependent factors could result in the observed XPS data, because final-state screening of the core holes depends on the metallic properties, and hence, the cluster size. Other possible origins for the shifts in the C 1*s* peaks have been discussed by Tarlov.[7]

Figure 13 shows the absolute XPS peak intensities as a function of Ag coverage on MUA. The solid lines in the figure correspond to intensities calculated assuming layer-by-layer growth of the Ag on the MUA. An unambiguous interpretation of these results is not possible. The XPS intensities in the experimental data are found to be less than that predicted by the model in all cases, which might be explained by X-ray damage to the SAM during the course of the experiments.[51,52] Similar behavior was observed for the Cu deposited onto CN for coverages below 0.6 nm; but above 0.6 nm, the attenuation supported the hypothesis of overlayer clustering.[4] Fluctuation in the C 1*s* HBE data (corresponding to C=O bonded C) results from the relatively low count rate measured for this peak. In contrast, Tarlov found no attenuation of the C 1*s* peak with Ag deposited to 1.0 nm at 295 K, which is consistent with rapid, complete penetration of Ag through the ODT he studied.[7]

ISS CDP of Ag/CH_3 and Ag/COOH from 113 K to 293 K. ISS compositional depth profiles (CDPs) were obtained from 113 to 293 K to study the penetration of Ag into ODT, MUA, and MHA, as a function of temperature. For all of these experiments, 1.0 nm of Ag was deposited at a rate of about 0.01 nm/s onto the SAM prior to depth profiling. The elapsed time after starting metal deposition and completion of the first ISS spectrum was about 300 s (100 s for deposition; 160 s for

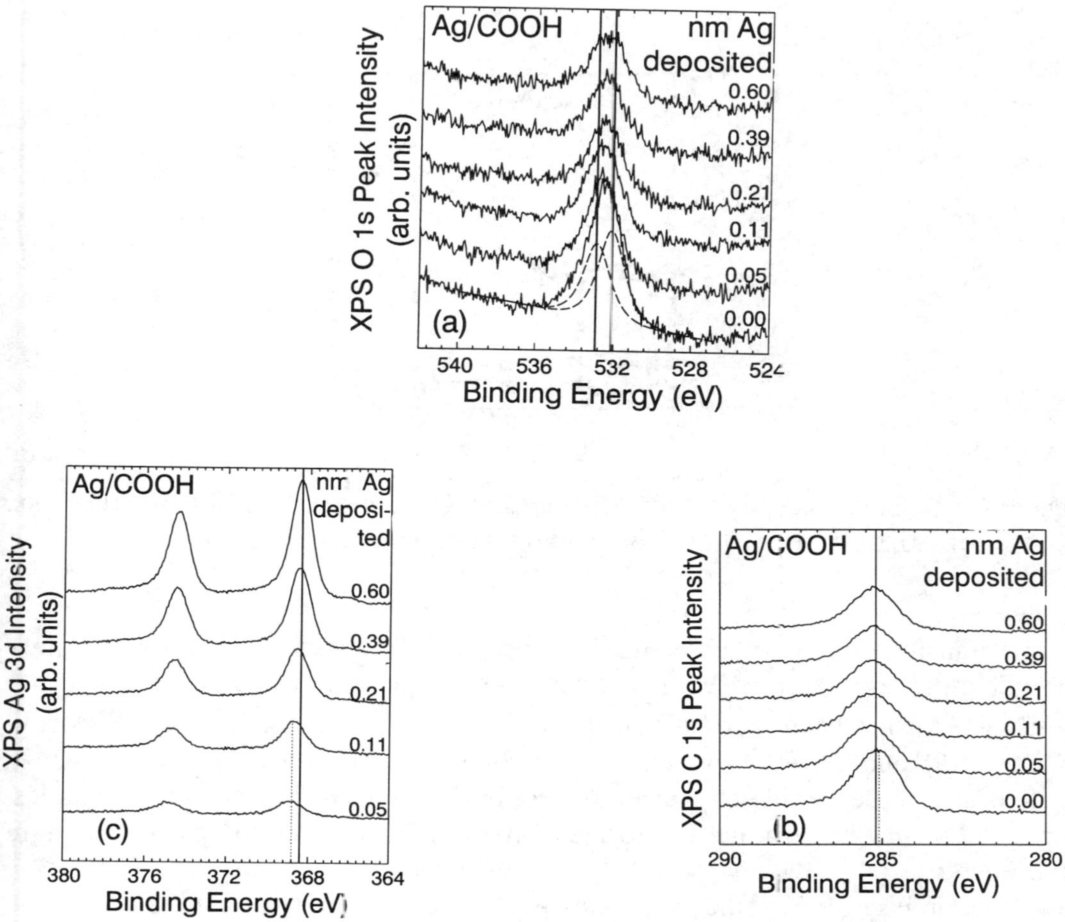

Fig. 11. XPS line shapes for Ag coverages of 0.0, 0.05, 0.11, 0.21, 0.39, and 0.60 nm on COOH of MUA at 295 K for (a) O 1*s*, (b) C 1*s*, and (c) Ag 3*d*.

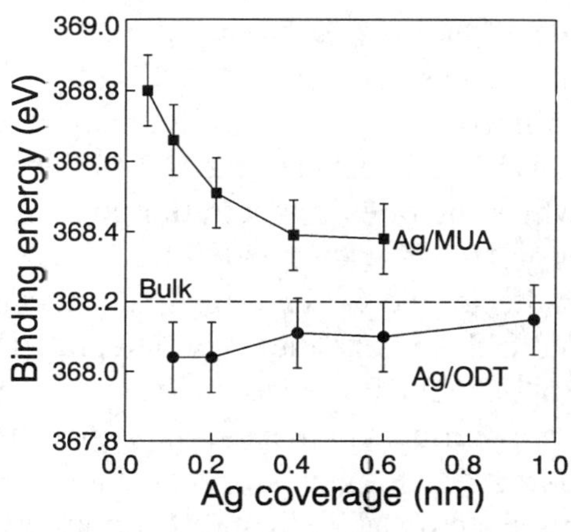

Fig. 12. XPS Ag $3d_{5/2}$ peak binding-energy shifts for Ag/ODT and Ag/MUA as a function of Ag coverage at 295 K.

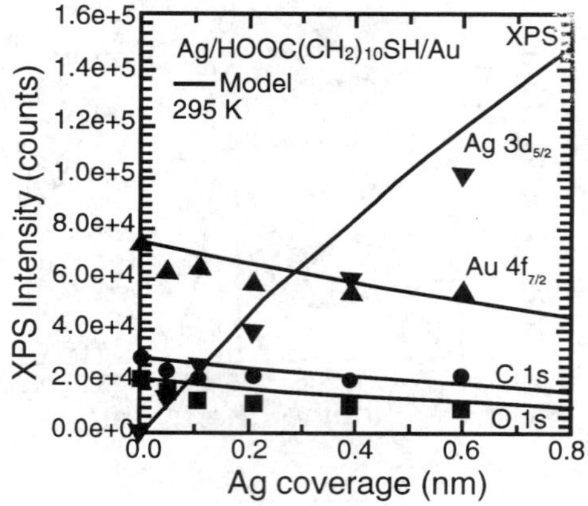

Fig. 13. XPS intensities (areas) for Au $4f_{7/2}$, O $1s$, C $1s$, and Ag $3d_{5/2}$ for Ag coverages of 0.0, 0.05, 0.11, 0.21, 0.39, and 0.60 nm on MUA at 295 K.

sample transfer, and 40 s for the first ISS spectrum). Thus, 5 min elapsed for time-dependent processes to proceed before our "initial" spectrum was taken during depth profiling. An ion-beam current density of 0.2 $\mu A/cm^2$ was used for all ISS depth profiles. XPS and ISS measurements were carried out on several samples at 193 K and verified that no detectable oxygen signal could be obtained from the ODT samples. These data indicate that no ice formed on the samples during low-temperature experiments.

The plots in Fig. 14 show the *fraction* of Ag (Ag intensity divided by the sum of the Ag and Au intensities) in the ISS depth profiles of Ag/ODT as a function of erosion time at temperatures from 113 to 293 K. The data are consistent with an increasing rate of Ag penetration to the ODT/Au interface at progressively higher temperatures, which results in greater Au intensities that are detected concomitantly with Ag when the Ag is at the ODT/Au interface. The time to erode through a "bare" ODT SAM to the ODT/Au interface[53] is about 1000 s at 0.2 $\mu A/cm^2$. The plots in Fig. 14 are consistent with a model of Ag residing on the methyl end-group of ODT at liquid nitrogen temperatures and at the ODT/Au interface at 293 K.[6] The data also elucidate the temperature regime over which the rate of penetration increases rapidly. The residual Ag signal for erosion times greater than about 2000 s, which corresponds to the location of the ODT/Au interface, indicates the presence of Ag clusters at the ODT/Au interface.

Figure 15 shows the actual ISS Ag peak intensity (integrated peak area) as a function of erosion time for 113 K to 293 K. Differences in the slopes of these CDPs for erosion times greater than 2000 s *may* reflect changes in the Ag cluster size or distribution at the ODT/Au interface (discussed below), but other interpretations are also possible given the complexity of this system. More than three monolayers of Ag were initially deposited, so substantial amounts of Ag must be in clusters at the ODT/Au interface because the Au signal is such a large fraction when the interface is reached, which is about 2000 s for eroding ODT when all the Ag penetrated to the ODT/Au interface. The amount of Ag present in the SAM likely depends on temperature and is difficult

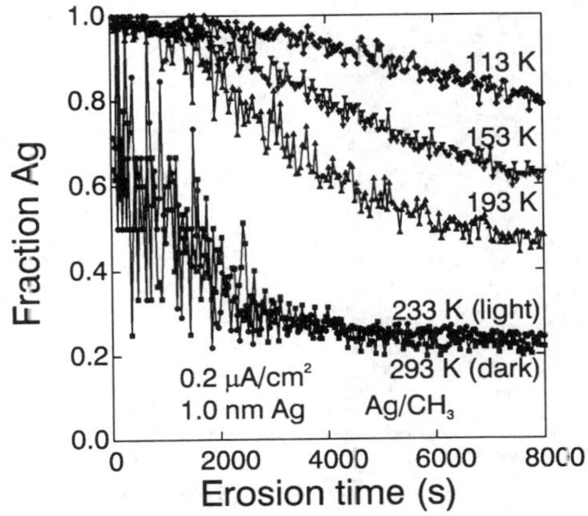

Fig. 14. Fraction of Ag signal (see the text) as a function of erosion time at 0.2 μA/cm² for 1.0 nm Ag on ODT/Au at temperatures of 113, 153, 193, 233 (solid circles), and 293 K (solid squares).

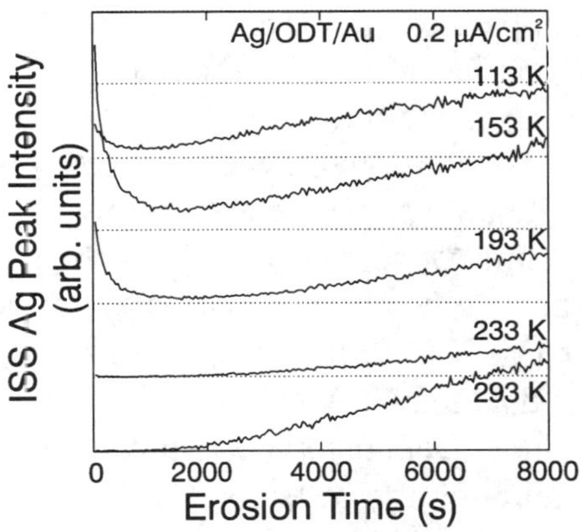

Fig. 15. ISS Ag peak intensities as a function of erosion time at 0.2 μA/cm² for 1.0 nm Ag on ODT/Au at 113, 153, 193, 233, and 293 K. Each horizontal grid line represents the "zero" for the CDP above it. The distance between grid lines corresponds to the same area of integrated peak intensities. The overlap of the data taken at 233 K and 293 K is the important point; identifying each point for the two temperatures is not important. The initial spectrum for all ISS depth profiles is about 300 s after the onset of time-dependent processes (see the text).

to quantify from the ISS experiments because of possible knock-in effects. The variations of the relative Ag and Au peak intensities as a function of temperature at erosion times of 2000 s indicate that 37% to 98% of the ODT/Au interface is covered by Ag.

The integrated initial ISS Ag peak intensities for Ag/ODT and Ag/MHA are shown as a function of temperature in Fig. 16 and provide a measure of the maximum ISS Ag

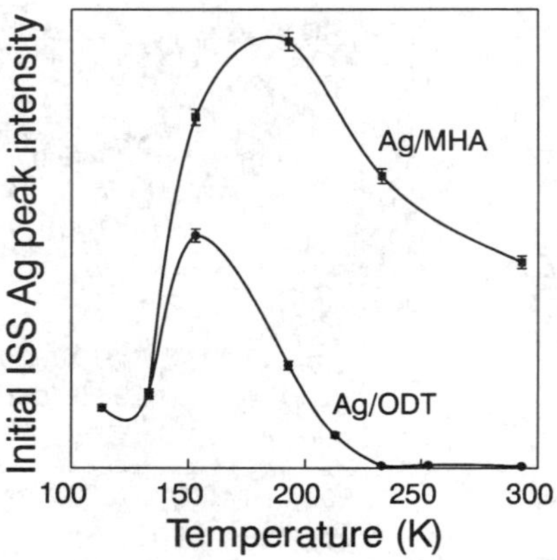

Fig. 16. Initial ISS peak intensities for 1.0 nm Ag on MHA and ODT after deposition at temperatures from 113 K to 295 K.

signal obtainable with our apparatus for 1.0 nm of Ag before extensive penetration into the SAM has occurred. In contrast to Fig. 14, the initial ISS Ag signal decreases above and below 153 K. The reproducibility of these data was carefully checked for coverages of 1.0-nm Ag. Two additional ISS CDPs were taken for a coverage of 5.0-nm Ag at 113 K to clarify the unexpected results below 153 K. The initial ISS signal intensity at 113 K and 5.0-nm Ag coverage is about the same magnitude as the data taken at 153 K with 1.0-nm Ag coverage. We have not determined if the Ag peak intensity is a maximum at 153 K (1.0-nm Ag) and 113 K (5.0-nm Ag).

Compared with ODT, the data for Ag/MUA (not shown) and Ag/MHA (Fig. 16) also show a larger initial ISS signal. Data taken (not shown) that are similar to that in Fig. 14 show the rate of Ag penetration is slower for MUA than for ODT, and even slower for MHA. For example, the ISS peak area was measured at four different positions on the same sample for 1.0 nm of Ag on MUA and MHA as a function of time after deposition at 295 K, after waiting for about 5, 15, 35, and 65 min after deposition; a fifth ISS Ag peak intensity was measured after 900 min for both MUA and MHA. As plotted in Fig. 17, the data show that after deposition, Ag remains on the surface of MUA for at least 5 min, and some Ag is on the MUA surface for more than 1 h. For MHA, Ag remains on the surface for more than 1 h (i.e., the Ag ISS intensity is unchanged), and the ISS intensity is only reduced to 68% of its initial value after 15 h, which shows that complete penetration to the SAM/Au interface is greatly retarded by the longer-chain (C_{16} vs C_{11}) MHA. By comparison, 1.0 nm of Ag penetrates ODT in less than 5 min at 295 K (Fig. 17). Ag on MUA is an intermediate case, in which Ag remains on the surface for 5 to 15 min but then penetrates into the MUA; but 15% of the deposit is still on the surface after 1 h.

The slower Ag penetration through MUA and MHA compared with that for ODT at 295 K provides some indication of the mechanism of metal penetration through SAMs. First, weak interactions between Ag and the COOH group may retard the onset of Ag transport through MUA and MHA. We did not detect evidence for formation of a Ag

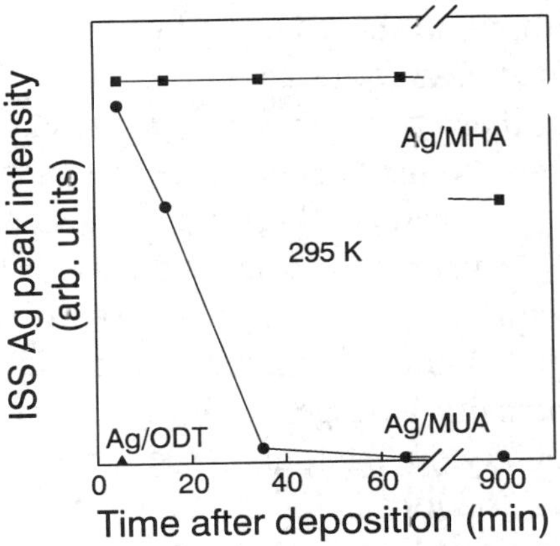

Fig. 17. ISS peak intensities for 1.0 nm Ag on MUA and MHA as a function of time after Ag deposition at 295 K.

unidentate complex with O, even though Ag-O interactions of 18 kcal/mol to 69 kcal/mol are known to exist.[54] Secondly, we speculate that at least partial hydrogen bonding between surface COOH groups, possibly having the "sideways dimeric" structure reported by Smith et al.,[27] restricts atom transport through the COOH-terminated SAMs. However, the dimers do not necessarily result in an "ordered" surface network, because their interactions disturb the potential for maximum packing density. In situ infrared studies of the COOH surface are needed to clarify the type of hydrogen bonding, in contrast to those[27] that permitted ex situ copper oxidation.[55] In addition to the likely absence of a Ag/CH_3 interaction, the CH_3 group on ODT experiences large rotational oscillations at 295 K,[56] and so, transport of Ag to the ODT/Au interface is not inhibited by interactions with the OFG. The increased rate of penetration of Ag through MUA relative to that for MHA is attributed to a greater movement of the OFGs in the less tightly packed shorter-chain alkanethiol.[57] A third speculative possibility is that the presence of the COOH group on MUA and MHA reduces the number of defects in the SAM, so that fewer transport pathways are available in these films, because the MHA is more defect-free than the MUA.

In Fig. 16, the initial ISS Ag peak intensities from CDPs of 1.0-nm Ag on MHA and ODT are compared for temperatures from 113 K to 295 K. The decrease in intensity below 150 K for ODT, which is challenging to interpret as transport to the interface, is now discussed. In the simplest case, the time-dependent processes relevant to understanding the results above and below 150 K include nucleation and growth into clusters of deposited Ag, the cluster size distribution formed, and diffusion through the ODT or ODT domain boundaries to the ODT/Au interface. The last step depends on the integrity and reproducibility of the ODT SAM, e.g., the number of pinhole defects in the SAM, gauche defects between chains above 200 K,[23] the nature of domain boundaries in the SAM, and other diffusion paths between the CH_3 end-group and the ODT interface. We *speculate* that the temperature for obtaining a maximum initial Ag signal depends on the amount and location of Ag deposited. For example, the rate of

Ag lost from the clusters formed on the ODT surface is likely to increase at higher temperatures (diffusion) and decrease for larger clusters (less ODT defects available). The cluster size formed during the deposition typically decreases at lower temperatures and approaches a single or a few atoms. It is possible that such small "cluster" sizes at 113 K permit faster penetration than the larger cluster size formed at 153 K. At temperatures above 153 K, we suspect the rate of diffusion through the ODT is the dominant transport mechanism. The cluster size at 5.0-nm Ag coverage at 113 K is expected to be larger than for 1.0 nm, and more Ag is initially present, which could explain why the Ag peak intensity is larger at 5.0 nm than at 1.0 nm of coverage. Obviously, changes in the mobility of Ag atoms at the surface and through the ODT as a function of temperature and variations in the structure of the Ag/ODT/Au interfaces play a role, but cannot be determined from our work.

As seen in Fig. 16, a similar decrease occurs for MHA, though below 200 K, and we think the reasons for the decrease are the same as presented above for Ag on ODT. The initial ISS Ag peak at 295 K for 1.0-nm Ag on MHA is more intense than the most-intense initial ISS peak observed for 1.0-nm Ag on ODT. This result indicates that the Ag coverage of the COOH surface is higher than on ODT, i.e., the cluster size is smaller. Because the Ag initially remains longer on the COOH surface of MHA, the ISS signal can be a maximum for any given temperature, depending on lateral diffusion kinetics. Differences in Ag coverage on ODT and MHA probably result from differences in surface energy (and thus clustering), as well as from penetration of Ag through the two SAMs.

In summary, both the Ag/ODT and Ag/MHA systems show a marked decrease in ISS Ag signal intensity at temperatures below about 200 K, and the Ag ISS peak intensity for MHA remains consistently larger than that for ODT, even at lower temperatures. We interpret these results to mean that 1.0-nm Ag on MHA has more (and smaller) clusters that cover a greater surface area than 1.0-nm Ag on ODT. We suggest that penetration of metal overlayers on SAMs may be linked to the available free volume between alkanethiol chains that constitute the SAM or at domain boundaries within the SAM. From this perspective, Ag penetrates MHA slowly at 295 K because partial hydrogen bonding between COOH OFGs inhibits motion of the alkanethiol chains, thus limiting the access of surface Ag atoms into the free volume between the chains.[10] Furthermore, the ISS experiments show the penetration processes are complex, and the models used for the XPS intensity plots do not adequately reflect the complexity. Further work is clearly needed to reveal the details of the Cu/OH[3] or Cu/CN[4] systems and should include using ISS.

Our research on the Ag/ODT and Ag/COOH systems shows that much work remains to be done before the behavior of weakly or noninteracting metal overlayers on model organic surfaces can be well understood. Using in situ AFM to understand the temperature-dependent clustering of Ag overlayers on SAMs and IR spectroscopy to probe changes in (weak) bonding at these interfaces should be especially valuable. Elucidation of metal nucleation and growth behavior on SAMs as a function of temperature by use of AFM could play a key role in understanding the competition between the kinetics of metal cluster growth and metal diffusion through SAMs as a function of initial cluster size. The time evolution of the IR spectra of an Ag-metallized SAM could provide valuable information about specific penetration mechanisms. Detailed studies of Ag/SAM systems with selected OFGs may provide a basis for predicting the behavior of other metal/SAM systems with weak metal/OFG interactions.

CONCLUDING REMARKS

In the research to date, the interactions in several M/SAM systems have been *identified*, along with *qualitative* relationships among the effects of chemical reactivity, temperature, and penetration at a metal/organic interface. With further research, the task remains to understand *quantitatively* how M/SAM interfaces are formed. This will require more-complete characterization of the reactants, the structures, and the kinetics of the reactants and the structures of M/SAM systems.

The initial research on M/SAM systems, which was begun at the National Renewable Energy Laboratory and initially reported in 1989 for Cu/COOH, Cu/CH_3, Cu/CN, and Cu/CH_2OH,[58] has served as a test of their use as model systems, and of whether M/SAM research will have benefits in applied areas, (e.g., real M/polymer interfaces). The results described above are clearly more readily interpreted and more detailed than *most* of their counterparts in the M/polymer interface literature. A notable exception is the more completely interpreted and detailed work by Grunze et al. on metals deposited onto polyimide.[59] Their work also escapes the complications of SAM quality and detect concentrations that have not been assessed in our work. The surface order of COOH OFG SAMs has also been questioned,[57] which may reduce its value as a model system. Based on the presence of similar reactants at a M/polymer interface, such as OFGs and metal atoms, our research approach succeeds in identifying factors that affect the initial bond between a metal and the polymer as the interface is formed in vacuum.

From the initial results, a continued broad effort is warranted for studying a matrix of 10 to 15 metals and metal oxides with about 10 different OFGs to evaluate the full potential of the approach and of this subfield of surface science. This should lead to the identification of submatrices targeted at specific classes of materials and applications.

Future characterization needs to include obtaining (1) detailed, molecular orbital descriptions of new metal-organic bonds in a unique environment (because of the difference between bringing individual molecules to a metal surface compared with bringing individual metal atoms to an organic surface), (2) diffusion processes and rates for metals on organic surfaces, (3) penetration processes and rates for metals into organic surfaces, and (4) characterization for the quality of SAMs and defect concentrations prior to M/SAM studies.[10] These advances will be based on further work with valence-band probes, lateral imaging probes, and probes that can measure rates of penetration, as discussed above. Control of defects and a capability for the characterization of defects will be necessary to understand the relative role of intrinsic vs. extrinsic effects in chemical and physical processes in M/SAM systems.

Finally, the results of M/SAM studies may be applied to the understanding of M/polymer interfaces, and for this reason, M/SAM interfaces are often described as "model" M/polymer interfaces. Studies of metal oxides, semiconductors, and other inorganics deposited onto SAM substrates should provide results than can be applied to the understanding of other types of inorganic/organic interfaces. Further relevance to inorganic/organic interfaces actually used in practice and to interfacial reaction mechanisms under ambient conditions can be gained by exposure of inorganic/SAM interfaces to different atmospheres, temperature extremes, radiation, and chemical reagents. Inorganic/organic interfaces with well-defined and stable properties are needed in advanced devices.[60-62] Research to identify and modify reactions at inorganic/SAM interfaces will provide the fundamental understanding to fill that need.

ACKNOWLEDGMENTS

The authors are pleased to acknowledge support by the U.S. Department of Energy under Contract DE-AC36-83CH10093 through NREL's Director's Development Fund. We are especially grateful for the helpful discussions with D.L. Allara, P. Zhang, and R.W. Collins, and to D.L. Allara and G. Whitesides for providing most of the alkane thiols we used in our experimental work. We thank D.L. Allara, P. Zhang, R.L. Opila, K. Konstadinidis, M.D. Porter, M. Grunze, H.G. Rubahn, L.H. Dubois, R.G. Nuzzo, M.J. Tarlov, and P. Zhang for permission to use figures from their work in ref. 10, and R.W. Collins, G. Scoles, A.J. Bard, J.E. Houston, K.H. Gray, R.W. Linton, S.M. Lindsay, G.N. Robinson, A.N. Parikh, K. Edinger, and F. Eisert for providing copies of their publications.

REFERENCES

1. A.W. Czanderna, D.E. King, and D. Spaulding, J. Vac Sci. Technol. **A 9**, 2607 (1991).
2. D.E. King, A.W. Czanderna, and D. Spaulding, in "Metallized Plastics 3 Fundamental and Applied Aspects," K.L. Mittal (Ed.), pp. 149-161, Plenum, NewYork, 1992.
3. D.R. Jung, D.E. King, and A.W. Czanderna, Appl. Surface Sci. **70/71**, 127 (1993).
4. D.R. Jung, D.E. King,, and A.W. Czanderna, J. Vac. Sci. Technol. **A 11**, 2382 (1993).
5. D.R. Jung and A.W. Czanderna, Mater. Res. Soc. Symp. Proc. **304**, 131 (1993).
6. (a) G. Herdt and A.W. Czanderna, Surface Sci. Lett. **297**, L109 (1993); (b) J. Vac. Sci. Technol. **A 12**, 2410 (1994); (c) J. Vac. Sci. Technol. **A 13**, 1275 (1995).
7. M.J. Tarlov, Langmuir **8**, 80 (1992).

8. P. Zhang, Ph.D. dissertation, Department of Materials Science and Engineering, Pennsylvania State University, 1993.
9. R.L. Opila, K. Konstadinidis, D.L. Allara, and P. Zhang, Surface Sci. **338**, 300 (1995).
10. D.R. Jung and A.W. Czanderna, Crit. Rev. Solid State Mater. Sci. **19**, 1 (1994).
11. D.R. Jung and A.W. Czanderna, J. Vac. Sci. Technol. (a) **A 12**, 2402 (1994); (b) **A 13**, 1337 (1995); (c) Appl. Surface Sci. **99**, 161 (1996).
12. E. Sacher, J.J. Pireaux, and S.P. Kowalczyk, (Eds.), "Metallization of Polymers," ACS Symposium Series, No. 440, American Chemical Society, Washington, D.C., 1990.
13. F.P. Netzer and M.G. Ransey, Crit. Rev. Solid State Mater. Sci. **17**, 397 (1992).
14. M.R. Albert and J.T. Yates, "The Surface Scientist's Guide to Organometallic Chemistry," American Chemical Society, Washington, D.C., 1987.
15. J. M. Burkstrand, J. Vac. Sci. Technol. **20**, 440 (1982).
16. J.M. Burkstrand, J. Appl. Phys. **50**, 1152 (1978); **52**, 4795 (1982).
17. K. Konstadinidis, R.L. Opila, J.A. Taylor, and A.C. Miller, Mater. Res. Soc. Symp. Proc., **304**, 83 (1993).
18. S.G. Anderson, J. Leu, B.D. Silverman, and P.S. Ho, J. Vac. Sci. Technol. **A 11**, 368 (1993).
19. M.J. Goldberg, J.G. Clabes, and C.A. Kovac, J. Vac. Sci. Technol. **A 6**, 991 (1988).
20. J.L. Jordan, C.A. Kovac, J.F. Morar, and R.A. Pollack, Phys. Rev. **B 36**, 1369 (1987).
21. R.G. Nuzzo and D.L. Allara, J. Am. Chem. Soc. **105**, 4481 (1983).
22. D.L. Allara and R.G. Nuzzo, Langmuir **1**, 45 (1985); 52 (1985).
23. A. Ulman, "An Introduction to Ultrathin Organic Films," Academic, New York, 1991; L.H. Dubois and R.G. Nuzzo, Annu. Rev. Phys. Chem. **43**, 437 (1992).
24. D.L. Allara, Crit. Rev. Surface Chem. **2**, 199 (1993).
25. G. Reiter, C. Bubeck, and M. Stamm, Langmuir **8**, 1808 (1992).
26. L.J. Gerenser, J. Vac. Sci. Technol. **A 6**, 5 (1988).
27. E.L. Smith, C.A. Alves, J.W. Anderegg, M.D. Porter, and L.M. Siperko, Langmuir **8**, 2707 (1992).
28. C.D. Bain, E.B. Troughton, Y.-T. Tao, J. Evall, G.M. Whitesides, and R.G. Nuzzo, J. Am. Chem. Soc. **111**, 321 (1989).
29. D.E. King and A.W. Czanderna, Surface Sci. Lett. **235**, L329 (1990).
30. J.R. Vig, J. Vac. Sci. Technol. **A 3**, 1027 (1985).
31. D.E. King, J. Vac. Sci. Technol. **A 13**, 1247 (1995).
32. J.R. Pitts, Ph.D. dissertation, Department of Physics, University of Denver, CO, 1985.
33. **P.M.A. Sherwood, in "Practical Surface Analysis," D. Briggs and M.P. Seah (Eds), Volume 1, 2nd ed., John Wiley, New York, NY, 1990.**
34. Using Peakfit,® Jandel software, Jandel Scientific, San Rafael, CA.
35. D.E. King, A.W. Czanderna, and D. Spaulding, J. Vac. Sci. Technol. **A 11**, 180 (1993).

36. K. Bammel, J. Ellis, and H.-G. Rubahn, Chem. Phys. Lett. **201**, 101 (1993).
37. F. Balzer, K. Bammel, and H.-G. Rubahn, J. Chem. Phys. **98**, 7625 (1993).
38. G. Herdt and A.W. Czanderna, J. Vac. Sci. Technol. **A 13**, 1275 (1995).
39. Independent observations by D.L. Allara et al. at Penn State U., R.L. Opila et al. at AT&T-Bell Labs, and A.W. Czanderna et al. at NREL.
40. S. Tanuma, C.J. Powell, and D.R. Penn, Surface Interface Anal. **17**, 911 (1991); P.E. Laibinis, C.D. Bain, and G.M. Whitesides, J. Phys. Chem. **95**, 7017 (1991).
41. A. Dilks, in "Electron Spectroscopy: Theory, Techniques, and Applications," C.R. Brundle and A.D. Baker (Eds.), Vol. 4, p. 278, Academic Press, New York 1981.
42. J.F. Moulder, W.F. Stickle, P.E. Sobol, and K.D. Bomben, in "Handbook of X-Electron Spectroscopy," J. Chastain (Ed.), Perkin-Elmer corporation, Eden Prai Minnesota, 1992.
43. C.D. Wagner, in "Practical Surface Analysis," D. Briggs and M.P. Seah (Eds), Volume 1, p. 595, Auger and X-ray Photoelectron Spectroscopy, 2nd ed., John Wiley, New York, NY, 1990.
44. F.A. Cotton and G. Wilkinson, "Advanced Inorganic Chemistry," 5th ed., Wiley, New York, 1988.
45. D. Briggs and M.P. Seah, (Eds.), "Practical Surface Analysis," Wiley, Chichester, 1983.
46. P.J. Trotter, M.G. Mason, and L.J. Gerenser, J. Phys. Chem. **81**, 1325 (1977).
47. G.K. Wertheim, Z. Phys. **B 66**, 53 (1987); **D12**, 319 (1989).
48. G.K. Wertheim, S.B. DiCenzo, D.N.E. Buchanan, and P.A. Bennett, Solid State Comm. **53**, 377 (1985), and references therein. The sign of the surface core-level shift resulting from band narrowing is not easily predicted for Cr because the 3*d* band is half-filled.
49. S.B. DiCenzo, and G.K. Wertheim, J. Elect. Spect. Rel. Phen. **43**, c7, (1987).
50. I. Jirka, Surface Sci. **232**, 307 (1990).
51. P.E. Laibinis, R.L. Graham, H.A. Biebuyck, and G.M. Whitesides, Science **254**, 981 (1991).
52. D.R. Baer, M.H. Engelhard, D.W. Schulte, D.E. Guenther, Li-Qiong Wang, and P.C. Rieke, J. Vac. Sci. Technol. **A 12**, 2478 (1994).
53. D.E. King and A.W. Czanderna, to be published.
54. A.W. Czanderna, J. Vac. Sci. Technol. **16**, 408 (1977).
55. D.E. King, and A.W. Czanderna, Langmuir **10**, 1630 (1994).
56. A. Ulman, Adv. Mater. **3**, 298 (1991).
57. M. Grunze, private communication.
58. D. Spaulding, M.S. thesis, Materials Science Dept., University of Denver, Denver CO, 1989.
59. T. Strunskus, M. Grunze, G. Kochendoerfer, and Ch. Woll Langmuir **12**, 2712 (1996).
60. A.W. Czanderna and R.J. Gottschall, (Guest Eds.), Mater. Sci. Engr. **53**, 1-168 (1982).

61. J.D. Swalen, D.L. Allara, J.D. Andrade, E.A. Chandross, S. Garoff, J. Israelachvili, T.J. McCarthy, R. Murray, R.F. Pease, J.F. Rabolt, K.J. Wynne, and H. Yu, Langmuir **3**, 932 (1987).
62. A.W. Czanderna and A.R. Landgrebe, (Guest Eds.), Crit. Rev. Surf. Chem. **2** (Nos. 1-4) and **3** (No. 1), 1993.

Metallized Plastics 5&6: Fundamental and Applied Aspects, pp. 203-210.
K.L. Mittal (Ed.).

AN IN-SITU X-RAY PHOTOELECTRON STUDY OF THE INTERACTION BETWEEN VAPOR-DEPOSITED Al ATOMS AND HYDROXYL OR AMINE TERMINATED SELF-ASSEMBLED MONOLAYERS

A. Hooper and D.L. Allara

Department of Materials Science and Engineering and Department of Chemistry
The Pennsylvania State University
University Park, PA 16802

K. Konstadinidis and R. L. Opila

Bell Labs, Lucent Technologies
600 Mountain Avenue
Murray Hill, NJ 07974

The vapor deposition of aluminum metal onto self-assembled monolayers (SAMs) was monitored using x-ray photoelectron spectroscopy (XPS). The functionalized SAMs had the structure X-(CH2)$_{15}$-SH where X = CH_2OH and CH_2NH_2. The Al was found to be highly reactive with the nitrogen and oxygen containing functional groups. No chemical reaction was observed between Al and carbon (-CH_2-), up to metal coverages of 1.6 nm.

INTRODUCTION

The reactions of metals with organic films has wide application in the development of electronic devices, design of new materials and coatings, and fundamental surface chemistry. Many studies involving organic films suffer from the inability to reproducibly prepare or to characterize surfaces fully. In the study of polymer surfaces, for example, it is difficult to eliminate surface contamination and to fully know the density of specific organic functional groups at the surface due to the conformation of the polymer. Self-assembled monolayers (SAMs) provide an ideal opportunity to study clean, well characterized organic surfaces[1,2]. In the past they have served as model systems for the study of the interaction of metals deposited on organic films[3].

Much of the previous work related to metals on SAMs has been summarized in two review papers[3,4]. Aluminum evaporated on polymers has been studied previously including polyimides[5], poly(acrylic acid) and polyethylene[6], poly(ethylenenterephthalate)[7], and poly(vinyl alcohol)[8]. These aluminum on polymer studies show that the aluminum always reacts with oxygen atoms contained in carbonyl, ether, or hydroxyl groups present in the polymers. For the Al on polyimide studies, no reaction was observed between the Al and N atom in the imide ring. Al has not been observed to react with the carbon atoms in organic films (with the exception of carbon atoms directly bonded to oxygen atoms). It has been suggested that interaction between Al and C can occur through the formation of an Al-O-C complex[7,9,10].

In this work, we have conducted a study of the interaction of evaporated Al atoms with two alkanethiolate SAMs on gold surfaces. The SAMs are terminally functionalized: $X(CH_2)_{15}SH$, where X = $-CH_2OH$ and - CH_2NH_2. The chemical state of the SAM and the deposited Al film was monitored using high resolution XPS in situ after deposition. The results will be compared to previous studies of titanium on SAMs[11].

EXPERIMENTAL

The preparation of alkanethiol films has been described in detail previously[11]. The SAMs were prepared on gold films that had been deposited on chromium coated single crystal silicon wafers. The films were characterized with ellipsometry, infrared spectroscopy, and contact angle measurements to ensure that the films were densely packed and well-ordered[1,12]. The synthesis of the $HS(CH_2)_{15}CH_2NH_2$ amine thiol compound will be detailed elsewhere[13].

The XPS analyses were performed on a Scienta ESCA 3000 spectrometer equipped with a monochromatic Al Kα source, and is described in detail elsewhere [14,15]. A pass energy of 75 eV and an energy step of 0.05 eV was used for the analysis, with a resulting full width at half maximum for the Au $4f_{7/2}$ of 0.52 eV. The binding energy of the Au $4f_{7/2}$ was set at 84.0 eV as a reference for all spectra. Due to differences in x-ray beam power for SAM samples analyzed on different days, intensities were normalized to the Au $4f_{7/2}$ intensities when relative areas of different types of SAMs were compared.

Following initial XPS analysis of the SAM, the sample was transferred under vacuum to the preparation chamber, which was isolated from the analysis chamber by a gate valve. The pressure in the preparation chamber remained below 1×10^{-8} Torr during deposition. Al was deposited at a rate of 0. 05 nm/min by evaporation from a graphite crucible heated to approximately 975 °C. The deposition rate was checked periodically with Rutherford backscattering spectroscopy. After each deposition, the Al/SAM specimen was transferred without exposure to air to the analysis chamber.

RESULTS

Analysis of Unreacted SAMs

A comparison of the C 1s, O 1s and N 1s core-level spectra taken at 10° take-off angle for the SAMs is shown in Figures 1-3. Improved signal-to-noise ratio and higher surface selectivity is obtained at this angle. Peak areas and energies were determined by curve-fitting using a mixed 90% Gaussian- 10% Lorentzian shape (with the notable exception of the Au 4f peaks which had a larger Lorentzian component). In addition, the binding energies were all determined with respect to the Au $4f_{7/2}$ at 84.0 eV.

Hydroxyl SAM. Two carbon peaks are observed at 285.14 eV and 286.95 eV corresponding to the main alkyl chain and the carbon directly bonded to the hydroxyl group, respectively. A single oxygen peak appears at 533.35 eV corresponding to the hydroxyl functional group.

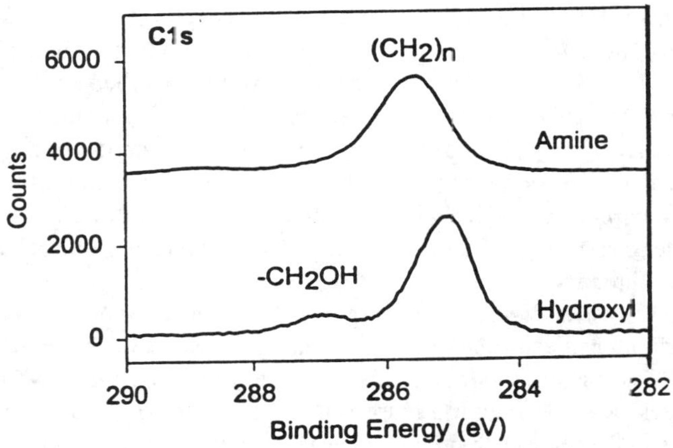

Figure 1. Comparison of C 1s core-level spectra of $HS(CH_2)_{15}CH_2NH_2$ and $HS(CH_2)_{15}CH_2OH$ self-assembled monolayers deposited on Au.

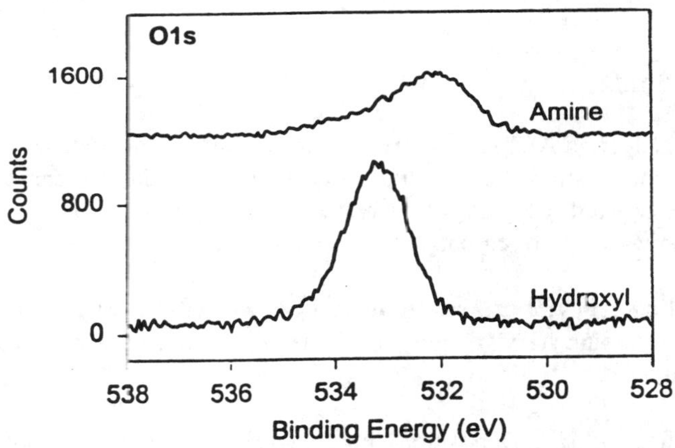

Figure 2. Comparison of O 1s core-level spectra of $HS(CH_2)_{15}CH_2NH_2$ and $HS(CH_2)_{15}CH_2OH$ self-assembled monolayers.

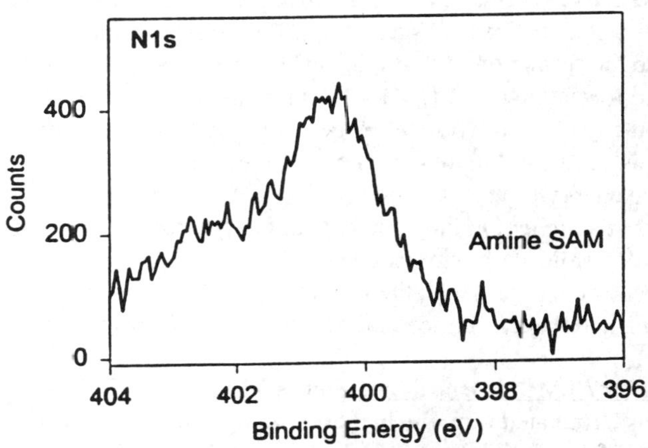

Figure 3. N 1s core-level spectrum of $HS(CH_2)_{15}$ CH_2NH_2 self-assembled monolayers deposited on Au.

Amine SAM. Inspection of the amine O 1s core level spectrum in Figure 2 reveals that the amine monolayer contains impurities. The appearance of two peaks at 531.15 and 533.77 eV is not expected in a pure amine monolayer. This impurity could be water adsorbed on the surface of the SAM (the O 1s transition for water is expected at approximately 533 eV). The source of this impurity could also be the reaction of the amine during or after the formation of the SAM (possibly after the SAM was exposed to air). A suspected impurity is due to the reaction of the amine with ambient CO_2 to produce a carbamate ($NHCO_2H$). The integrated area of these oxygen peaks is approximately 50% of the area of the hydroxyl terminated monolayer. For this reason these results are to be considered as preliminary.

The N 1s region shown in Figure 3 shows two peaks observed at 400.46 and 402.36 eV. The ratio of the integrated intensities of the high to low energy peaks is 0.32. Hence, if the peak at 400.46 corresponds to a NH_2 species, then the sample is predominately amine with some impurity.

Three carbon peaks are observed in Figure 1 at 285.61 , 286.93, and 288.84 eV. The peak at 285.61 eV corresponds to the methyl chain, however the assignment of the other two peaks is not as obvious. Previous studies with nitrile terminated thiols have observed the (CN) C 1s peak at 286.86 at a 10° take-off angle[11] so the peak at 286.93 is probably the carbon atom bonded to the amine. The peak at 288.84 eV corresponds to a COOH species and is consistent with a carbamate impurity.

Deposition of Al on SAMs

C 1s, O 1s, N 1s, and Al 2p spectra were recorded after depositions of Al on the SAM surfaces. Figures 4a and 5a show the C 1s spectra taken at a 10° take-off angle for each Al exposure for the $-CH_2OH$ and $-CH_2NH_2$ films, respectively. Figures 4b and 5b show the O 1s peaks. Figures 4c and 5c show the Al 2p regions for the SAMs. Figure 5d shows the N1s core level for the amine monolayer.

One interesting behavior is the shifting of the C 1s methylene and O 1s peaks with respect to the Au $4f_{7/2}$ with increasing Al coverage. This behavior has been observed in previous studies and remains unexplained[16].

Hydroxyl after Al Metal Deposition. Figure 4a shows the C 1s region for the hydroxyl terminated SAM. After deposition of 0.1 nm of Al, the higher binding energy peak in the C 1s spectrum associated with $-CH_2OH$ is completely gone. Attenuation of the C 1s intensity of the methylene chain is observed with higher Al coverage. No aluminum carbide, expected at approximately 282.5 eV, is observed. Figure 4b shows the O 1s region for the SAM. After 0.1 nm of Al has been deposited, the oxygen peak drops in intensity, shifts to lower energy, and becomes very broad (due to the formation of multiple peaks). This behavior is likely due to the loss of the CH_2-OH bond and formation of a CH_2-O-Al or aluminum oxide species. Some aluminum oxide may form due to the reaction of Al with oxygen impurity contained within the analysis chamber. The Al 2p region is shown in Figure 4c. The Al 2p region has a small underlying peak centered at 74 eV which is the energy of the Au $5p_{1/2}$ core level. At 0.1 nm Al coverage, the Al 2p peak is very broad and appears at a higher binding energy than is expected for aluminum metal. Metallic Al is expected at 72.8 eV, while aluminum oxide (Al_2O_3) would appear at 74.2 eV. The broad peak that is initially observed is consistent with the formation of aluminum oxide, or Al-O-C. Metallic Al eventually appears, and this can be seen in the 1.6 nm Al coverage spectrum.

Amine after Al Metal Deposition. Figure 5a-5d shows the spectra for the amine terminated SAM. For the NH_2 terminated film, the highest binding energy C 1s peak disappears after the addition of 0.1 nm of Al. The higher energy oxygen, carbon, and nitrogen peaks attributed to the impurity (at 533.77, 288.84, and 402.36 eV respectively) disappear by 0.1 nm coverage. After the 0.1 nm deposition the nitrogen peak becomes broad, and shifts approximately 1 eV to

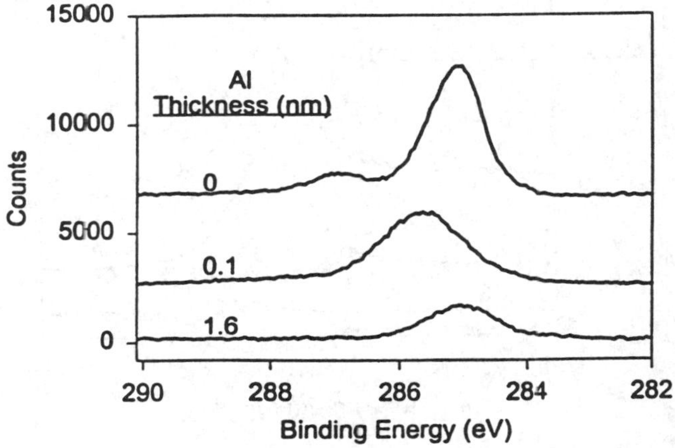

Figure 4a. Changes in the C 1s core-level spectrum of a $HS(CH_2)_{15}CH_2OH$ self- assembled monolayer deposited on Au after exposure to evaporated Al.

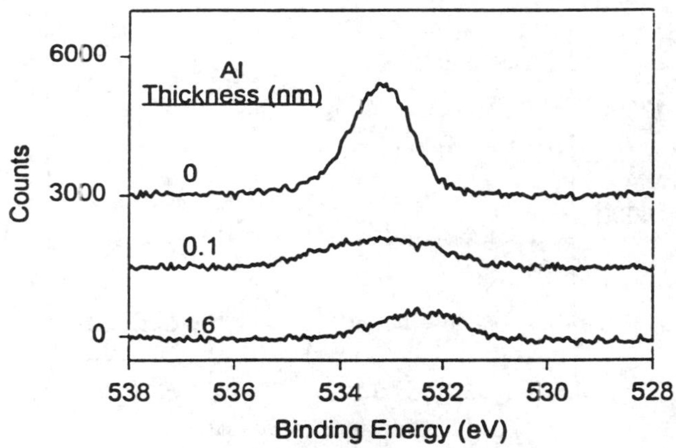

Figure 4b. Changes in the O 1s core-level spectrum of a $HS(CH_2)_{15}CH_2OH$ self-assembled monolayer deposited on Au after exposure to evaporated Al.

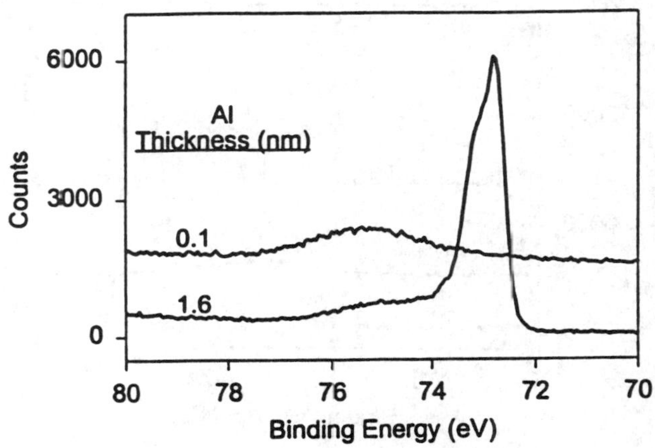

Figure 4c. Changes in the Al 2p core-level spectrrum of a $HS(CH_2)_{15}CH_2OH$ self-assembled monolayer deposited on Au after exposure to evaporated Al.

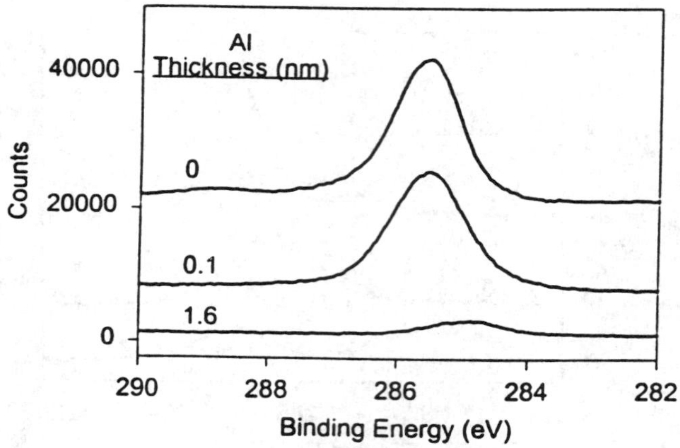

Figure 5a. Changes in the C 1s core-level spectrum of a $HS(CH_2)_{15}CH_2NH_2$ self-assembled monolayer deposited on Au after exposure to evaporated Al.

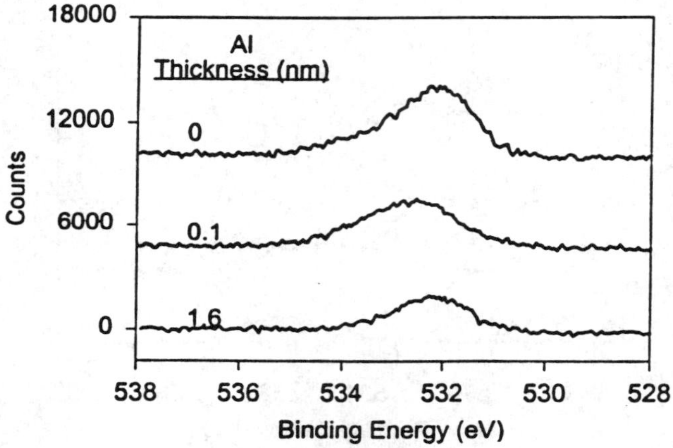

Figure 5b. Changes in the O 1s core-level spectrum of $HS(CH_2)_{15}CH_2NH_2$ self-assembled monolayer deposited on Au after exposure to evaporated Al.

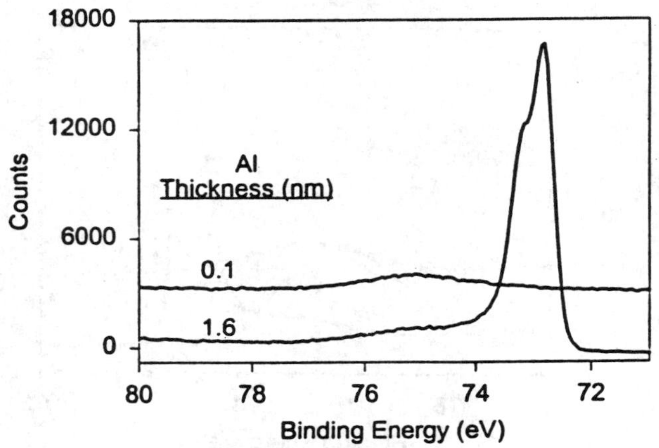

Figure 5c. Changes in the Al 2p core-level spectrum of a $HS(CH_2)_{15}CH_2NH_2$ self-assembled monolayer deposited on Au after exposure to evaporated Al.

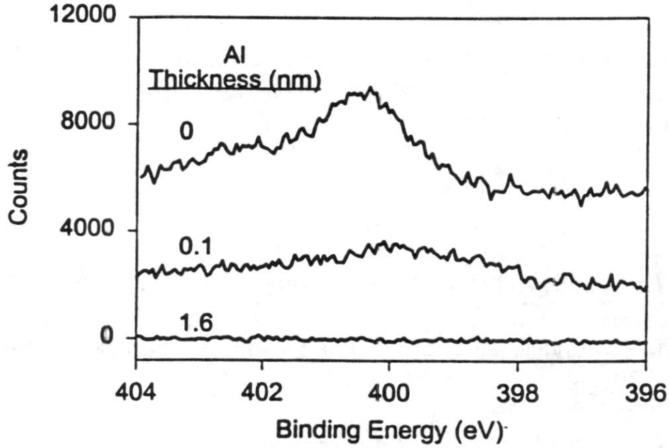

Figure 5d. Changes in the N 1s core-level spectrum on a $HS(CH_2)_{15}CH_2NH_2$ self-assembled monolayer deposited on Au after exposure to evaporated Al.

lower energy. This is consistent with the formation of an Al-N bond. One interesting effect is the rapid loss of the amine signal, and no nitrogen is observed after deposition of 1.6 nm Al. It is unlikely that this is due to attenuation by the Al. Another possibility is that the N core-level peak has become so broad that it is undetectable. Oxygen is observed at both coverages. No metallic Al is seen at 0.1 nm coverage, but metallic Al occurs before 1.6 nm Al coverage.

For SAMs, the monolayer functional group coverage is approximately $4.6x10^{14}$ functional groups per square cm[17]. By using the bulk density of aluminum, a 1:1 ratio of Al atoms to SAM functional groups will occur at 0.076 nm aluminum coverage. However, two or more atoms of Al can react with a single terminal functional group (for example, in Al_2O_3 each Al has reacted with 1.5 O's). In contrast, a single layer of aluminum atoms arranged in a cubic lattice on the SAM surface has $4.89x10^{15}$ Al atoms per cm^2. Thus a single atomic layer of Al has approximately eleven times the number of SAM surface functional groups.

CONCLUSION

Al binds to the O and N containing functional groups. This is determined by the rapid loss of C 1s peaks associated with the functional groups. No interaction between aluminum and carbon is observed. Changes in the O 1s and N 1s binding energies suggest that chemical bonding occurs between the Al and the oxygen on nitrogen atoms.

This result can be compared to results obtained earlier for Ti on SAMs [11]. For both systems the metal is observed to react rapidly with the surface functional groups while leaving the alkyl chain portion of the monolayer intact. The primary difference between Ti and Al on SAMs is that Ti reacts with the SAMs to produce titanium carbide, while no such interaction is observed with Al. Both Ti and Al depositions result in a gradual attenuation of the C 1s peak associated with the methylene chain with coverage.

ACKNOWLEDGMENTS

The authors acknowledge the allocation of time and services in the SCIENTA ESCA laboratory of Lehigh University. Professional and technical assistance of Dr. Alfred C. Miller is greatly appreciated.

REFERENCES

1. R. G. Nuzzo, L. H. Dubois, and D. L. Allara, J. Amer. Chem. Soc., 112, 558 (1990).

2. C. D. Bain, G. M. Whitesides, Y.-T. Tao, J. Evall, and R. G.. Nuzzo, J. Amer. Chem. Soc., 111, 321 (1989).

3. D. R. Jung and A. W. Czanderna, Crit. Rev. Solid State Mater. Sci., 19, 1 (1994).

4. G. C. Herdt, D. R. Jung, and A. W. Czanderna, Prog. Surface Sci., 50, 103 (1995).

5. Lj. Atanasoska, S. G. Anderson, H. M. Meyer III, L. Zhangda, and J. H. Weaver, J. Vac. Sci. Technol., A5, 3325 (1987).

6. B. M. DeKoven and P. L. Hagans, Appl. Surface Sci., 27, 199 (1986).

7. M. Bou, J. M. Martin, and Th. Le Mogne, Appl. Surface Sci., 47, 149 (1991).

8. P. Stoyanov, S. Akhter, and J. M. White, Surface Interface Anal., 15, 509 (1990).

9. S. Akhter, X.-L. Zhou, and J. M. White, Appl. Surface Sci., 37, 201 (1989).

10. J. J. Pireaux, M. Vermeersch, C. Gregoire, P. A. Thiry, and R. Caudano, J. Chem. Phys, 88, 3353 (1988).

11. K. Konstadinidis, P. Zhang, R. L. Opila, and D. L. Allara, Surface Sci., 833, 300 (1995).

12. P. E. Laibnis, G. M. Whitesides, D. L. Allara, Y. T. Tao, A. N. Parikh, and R. G. Nuzzo, J. Amer. Chem. Soc., 113, 7152 (1991).

13. H. Skriba and D. L. Allara, to be published.

14. U. Gelius, B. Wannberg, P. Baltzer, H. Felliner-Feldbergg,G. Carlson, C. G. Johansson, J. Larsson, P. Munger, and G. Vegerfors, J. Electron Spectrosc. Relat. Phenom., 52, 747 (1990).

15. G. Beamson, D. Briggs, S. F. Davies, I. W. Fletcher, D. T. Clark, J. Howard, U. Gelius, B. Wannberg, and P. Balzar, Surface Interface Anal., 15, 541 (1990).

16. D. R. Jung and A. W. Czanderna, J. Vac. Sci. Technol., A13, 1337 (1995).

17. L. H. Dubois and R. G. Nuzzo, Ann. Rev. Phys. Chem., 43, 437 (1992).

Metallized Plastics 5&6: Fundamental and Applied Aspects, pp. 211-229.
K.L. Mittal (Ed.)

EARLY STAGES OF COPPER/POLYIMIDE INTERFACE FORMATION

M. Kiene, T. Strunskus and F. Faupel

Technische Fakultät der CAU Kiel
Kaiserstr. 2
24143 Kiel, Germany

In this paper we use a multi-technique approach to characterize the initial stages of copper/polyimide interface formation. Copper was vapor-deposited onto fully cured PMDA-ODA polyimide (PI). Deposits from submonolayer thickness up to 1.6 nm were investigated with x-ray photoelectron spectroscopy (XPS), transmission electron microscopy (TEM) and atomic force microscopy (AFM). Copper was found to form clusters growing with increasing coverage at the polymer surface, even at room temperature (RT) deposition. Annealing of the RT deposits up to 350 °C, still well below the glass transition temperature of PI, leads to significant reduction of the copper XPS signal and is attributed to further cluster growth, confirmed by TEM and preliminary AFM results. No major Cu diffusion into the polymer occurs under these conditions. This suggests that no significant diffusion is to be expected upon annealing continuous Cu films. Changes in the lineshapes and positions in the x-ray induced copper Auger spectra were attributed to cluster size effects indicating that these data cannot be taken as evidence for chemical bond formation at the Cu/PI interface.

INTRODUCTION

Metal coating of polymers is of considerable technological importance and is used for a variety of applications ranging from low-cost foils for food packaging, reflectors for car lights, compact discs and electrically shielded computer cases to packaging in microelectronics. Besides a fundamental interest in metal-polymer interaction and adhesion these applications have stimulated intensive research during the last decade[1,2]. Multilayers incorporating high-temperature polymers such as polyimides and copper wiring have substantially enhanced speed and density of packaging structures[3], and polyimides are also of interest as interlevel dielectrics in very large scale integrated circuits (VLSI)[4], because they provide a unique combination of properties such as low dielectric constant, ease of planarization and patterning by lithographic techniques, chemical resistance, and thermal stability. Accordingly, the copper-polyimide interface has been investigated using a variety of techniques (For recent reviews see, for example Matienzo and Unertl[5] or Pireaux[6]) including XPS[7-12], ultraviolet photoelectron spectroscopy (UPS)[8], cross-sectional TEM[13-14], medium energy ion scattering (MEIS)[15], Rutherford backscattering (RBS)[16-17], Fourier transform infrared reflection-absorption spectroscopy (FT-IRAS)[18-19], and radio-tracer diffusion studies[20].

It turned out that the extent of chemical interaction between a metal and a polymer has a strong influence on metal/polymer interface formation. Strong chemical interactions between the metal atoms and the polymer functional groups lead to bond formation already in early deposition stages and results in the formation of a relatively sharp interface. This is, for example, observed for the deposition of very reactive metals like chromium and titanium on polyimide films[8,16,21].

Noble metals deposited at room temperature onto a polyimide film show Volmer-Weber type of growth, i.e., they do not wet the polyimide surface but form three-dimensional metal clusters and islands[11,22-24]. For low copper coverages, a preferential interaction of copper with the PMDA-part was observed in XPS, UPS and FT-IRAS measurements[8,11,19]. These observations indicate that only a weak chemical interaction occurs between copper and the polyimide and that the nucleation of the copper clusters occurs near the PMDA-units of the polymer. In radio-tracer diffusion studies[20], MEIS[15] and TEM[13] measurements it was observed that thermally evaporated copper atoms diffused into the polyimide film when deposited at elevated temperatures and low deposition rates to form metallic copper clusters inside the polymer matrix. A strong dependence on deposition rate and substrate temperature was found in the radio-tracer and TEM experiments for various noble metals deposited onto polyimides and polycarbonates[13-15,20]. Pronounced diffusion took place at high substrate temperatures and low deposition rates, while only little diffusion occurred at low substrate temperatures and high deposition rates. These results were confirmed by computer simulations[25-26]. **In light of this, a question of great practical importance** arises, as to whether diffusion also occurs from a continuous film as has often, at least implicitly, been suggested in the literature (see, e.g. refs. 16 and 27).

In this paper we report on XPS, TEM and AFM investigations of room temperature Cu deposits on polyimide and samples that were subsequently annealed below the glass transition temperature of the polymer. Due to the short inelastic mean free path of the photoelectrons[28-29] XPS particularly lends itself to studies of the growth mode and diffusion behavior in the near surface region where radio-tracer and other techniques for diffusion measurements are not sensitive. As cluster formation proved to play an important role in the interpretation of the XPS data, complementary investigations with TEM and AFM were carried out. The results suggest a modification of well-established interpretations of surface spectroscopy data with respect to diffusion and allow one to rule out significant diffusion from a continuous Cu film into polyimide. Moreover, it is shown that x-ray induced Auger data for Cu on PI can be interpreted without resorting to chemical bond formation.

EXPERIMENTAL

The polyimide films were prepared by spin-coating poly(amic acid) onto a Si wafer and curing in the usual way[30] with a final curing step of 60 min at 400 °C in a nitrogen atmosphere. These films are expected to exhibit a preferred molecular orientation parallel to the surface[31]. Polyimide films were degassed in ultra high vacuum (UHV) prior to Cu deposition. Cu evaporation and heat treatment were performed in UHV in the preparation chamber of the photoelectron spectrometer. Prior to deposition, the Cu flux was measured with a quartz crystal microbalance, which was calibrated gravimetrically. XPS investigations were performed in situ using a hemispherical electron analyzer (VG MKII) and a non-monochromatized Al K_α x-ray source. The pressure in the vacuum system was 5x 10^{-9} Pa during XPS analysis, and $<5\times10^{-8}$ Pa during evaporation and heating. All XPS measurements were done with electron emission normal to the sample surface. The spectra were taken at 20 eV pass energy with an energy resolution of 1.2 eV measured at the $Ag3d_{5/2}$ line. The peak areas used for the quantitative analysis were determined after a Shirley-type background removal.

For the TEM investigations an electron transparent film of about 0.1 µm thickness was removed from the wafer and transferred to TEM grids prior to the metal deposition, which was performed in the aforementioned UHV preparation chamber. Due to the relatively thick polymer samples the contrast is weak, resulting in a detection limit for the copper clusters of about 2 nm (see fig. 1). Best images in our microscope (Phillips CM30) were observed with 200 kV electron energy.

AFM measurements were carried out with a Digital Instruments Nanoscope II in tapping and lateral force modes under ambient conditions.

RESULTS AND DISCUSSION

Copper/Polyimide Interface Formation during Room Temperature Deposition

Copper was deposited at room temperature onto spin-cast polyimide films, and the evolving copper-polyimide interface was investigated using transmission electron microscopy (TEM), x-ray photoelectron spectroscopy (XPS), and atomic force microscopy (AFM).

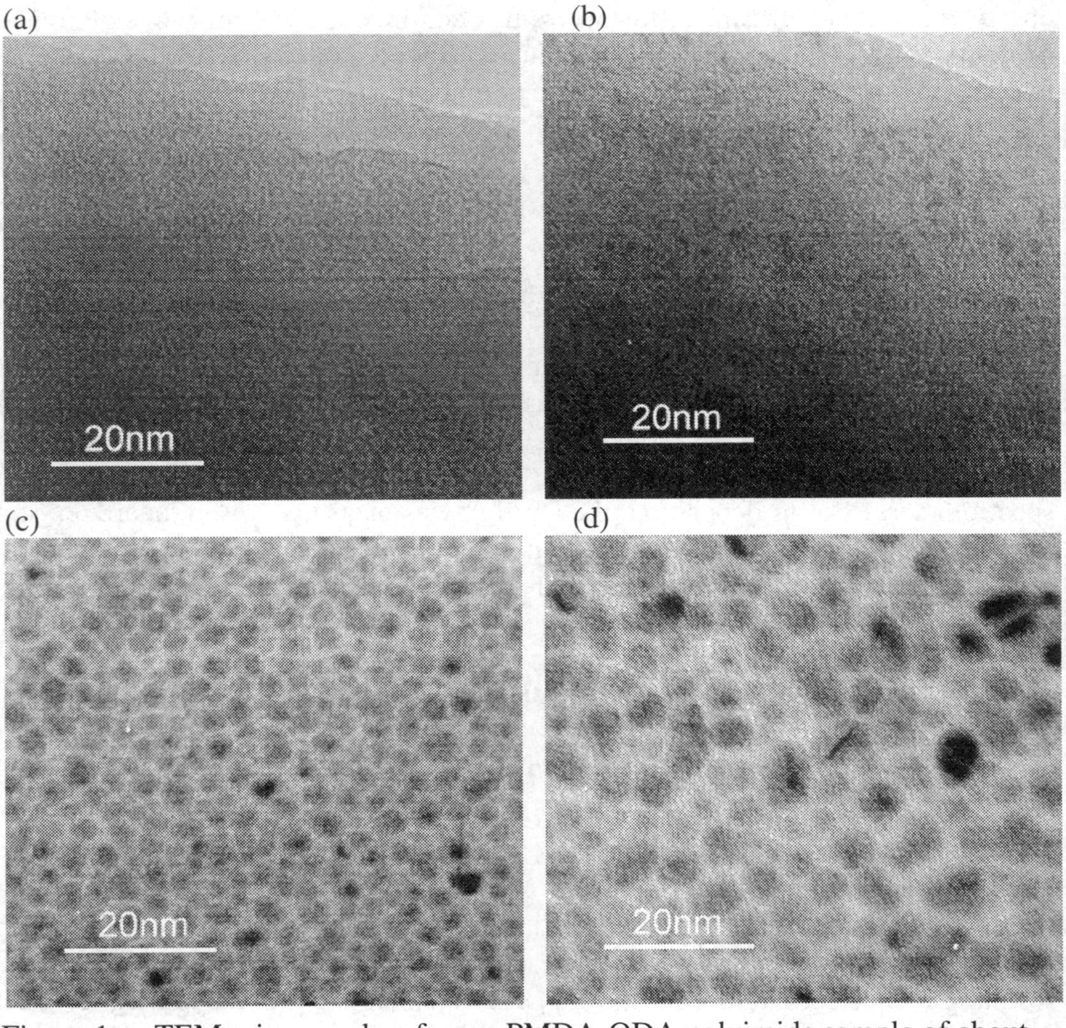

Figure 1 TEM micrographs of pure PMDA-ODA polyimide sample of about 0.1μm thickness (a) and after room temperature deposition of Cu. Nominal coverages are (b) 0.8 nm, (c) 2.4 nm, and (d) 6.4 nm. (The bright areas are due to thinner PI regions, originating from the cutting process).

Fig. 1 shows a selection of TEM micrographs of a) the clean PI film which exhibits the speckle pattern typical of amorphous polymers and b)-d) of the evolving copper/polyimide interface. For the lowest copper coverages, the contrast in the TEM pictures is not sufficient to observe distinct copper clusters. The detection limit is about 2 nm. Below a nominal copper coverage of about 0.8 nm, the copper clusters are too small to be identified in the TEM images. The low number density at 0.8 nm coverage indicates that only a small fraction of the copper clusters is bigger than 2 nm. This is supported by the TEM images for higher coverages, where more and larger clusters are visible.

The $Cu2p_{3/2}/C1s$ peak ratio was used to quantitatively monitor the processes occurring at the interface. Fig 2. shows the evolution of the $Cu2p_{3/2}/C1s$ peak ratio for nominal coverages from 0.03 nm up to 1.6 nm, the highest coverage investigated with XPS. For comparison, the $Cu2p_{3/2}/C1s$ peak ratio calculated for layer-by-layer growth is also shown. Radio-tracer measurements show that the condensation coefficient for copper deposited onto polyimide at RT is close to unity[32], i.e., all copper atoms arriving at the surface remain on the sample.

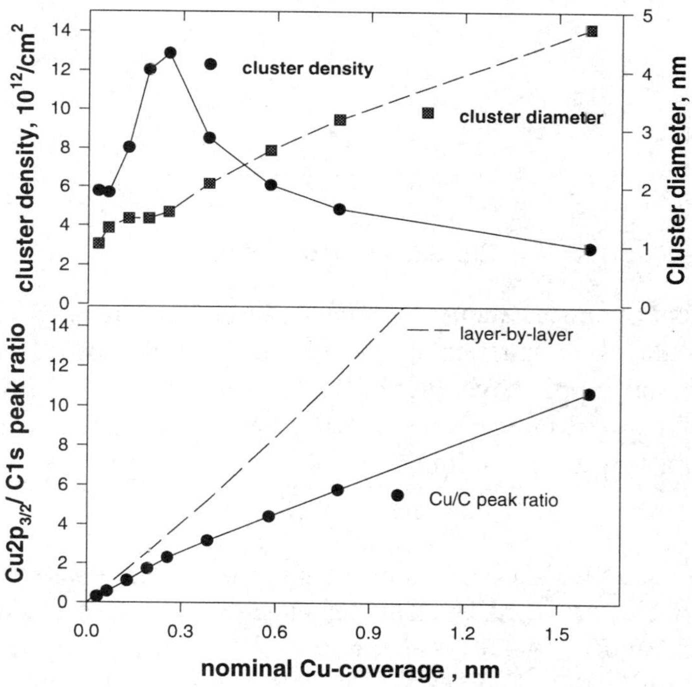

Figure 2 Top: Cluster size and areal density calculated from XPS data (see text). Bottom: Experimentally determined $Cu2p_{3/2}/C1s$ peak ratio connected by solid line to guide the eye, and $Cu2p_{3/2}/C1s$ peak ratio calculated for layer-by-layer growth (dashed line), versus nominal copper coverage.

Starting already at low coverages there is a deviation from layer-by-layer growth. The deviation from layer-by-layer growth is indicative of island formation and/or intermixing of the copper and the polyimide. To estimate the effect of cluster formation on the $Cu2p_{3/2}/C1s$ intensity ratio a single layer of spherical clusters with uniform size on top of the PI film was assumed. Following Lörincik[33] the $Cu2p_{3/2}$ and C1s intensities were calculated as a function of cluster size for different nominal Cu coverages. In Fig. 3 the calculated $Cu2p_{3/2}/C1s$ intensity ratio is plotted as a function of cluster diameter for two nominal Cu coverages. For comparison the $Cu2p_{3/2}/C1s$ intensity ratio obtained for a uniform copper layer is indicated by the symbols on the y-axis.

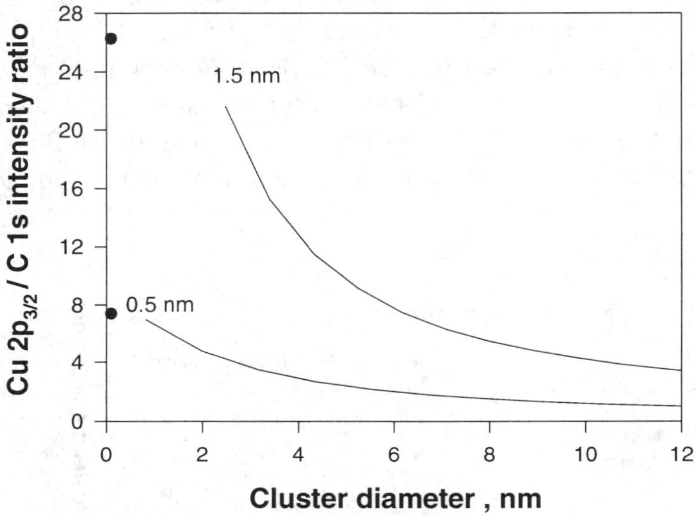

Figure 3 Effect of cluster diameter on the $Cu2p_{3/2}/C1s$ intensity ratio calculated for spherical copper clusters on a polyimide surface for nominal copper coverages of 0.5 nm and 1.5 nm. The filled circles on the y-axis indicate the intensity ratio obtained for a uniform copper layer on polyimide

The lines start with a data point for a close-packed 2-dimensional layer of spherical copper clusters. Smaller diameter clusters cannot be packed in a single layer as was assumed for the calculation. With increasing cluster size the $Cu2p_{3/2}/C1s$ intensity ratio decreases. The decrease is most pronounced for high nominal copper coverages and for smaller cluster diameters. Cluster growth leads to a decrease of the $Cu2p_{3/2}$ signal due to the stronger self attenuation of the copper signal in the larger copper clusters. On the other hand, the C1s intensity increases with increasing cluster size because the total area covered by the clusters decreases, overcompensating the effect of the stronger attenuation due to the larger clusters on the covered area of the surface.

When the cluster size is large compared to the inelastic mean free path of the photoelectrons in copper, which is 1.07 nm for $Cu2p_{3/2}$ and 1.78 nm for C1s photoelectrons, respectively, the $Cu2p_{3/2}$ signal is to a good approximation proportional to the area covered by the copper clusters and the carbon C1s signal is proportional to the area not covered by the clusters. Thus, the change in the $Cu2p_{3/2}$/C1s intensity ratio can be estimated by simply considering the change in the ratio of the covered to the uncovered area. As the cluster size decreases one has to correct this simple picture by considering the exponential increase or decrease of the signal with thickness, respectively. The change in the $Cu2p_{3/2}$ signal upon cluster growth is smaller than estimated from the change in areas, because normalized to its area a larger cluster will give a higher $Cu2p_{3/2}$ signal than a smaller cluster. In the C 1s signal the change is also smaller than estimated from the change in the areas, because the area covered by the copper clusters will also contribute to the C1s signal. This contribution decreases with increasing cluster size. Taken together the change in the $Cu2p_{3/2}$/C1s intensity ratio is shallower than estimated from the change in the covered to the uncovered area.

If the copper coverage is known one can use the measured $Cu2p_{3/2}$/C1s intensity ratio to estimate the cluster size and the areal density of the clusters. In the top part of Fig. 2 cluster sizes and cluster densities calculated from the measured $Cu2p_{3/2}$/C1s peak ratios are shown. Up to a coverage of about 0.3 nm there is an increase in the number of clusters, indicating that new nuclei are formed upon deposition of copper. Above 0.3 nm the number of clusters decreases by agglomeration, and the clusters grow in size. This general behavior has been confirmed in independent XPS measurements, where the Cu2p/Cu3p intensity ratio has been used to determine the cluster size[32], and is also supported by the TEM measurements shown in Fig. 1. Cluster sizes and densities from XPS and TEM measurements are summarized in Table I.

Table I. Cluster sizes and densities as determined from TEM micrographs and XPS analysis for different nominal Cu coverages.

nom. Cu-coverage, nm	**Cluster diameter from XPS, nm**	**Cluster diameter from TEM, nm**	**Cluster density XPS/TEM, $10^{12}/cm^2$**	**Comments**
0.8	3.2	≤ 2	4.8 / n.a.	most clusters not visible in TEM
1.6	4.7	2-3	2.9 / 3.9	clusters not clearly resolved in TEM
2.4		5-6	- / 1.9	spherical clusters
3.2		6-8	- / 1.2	most clusters spherical
6.4		n.a.	- / 0.8	irregular cluster shapes
9.6		n.a.	- / 0.32	irregular cluster shapes

Apparently, surface diffusion already leads to appreciable cluster formation even at temperatures near room temperature. This indicates a significant surface mobility of Cu atoms or small clusters. (The maximum temperature of the polymer surface during metal deposition was always below 50 °C) Up to a coverage of 3.2 nm most of the clusters observed are of spherical shape (Fig. 1b,c), but the number density of clusters decreases indicating cluster aggregation in addition to growth due to newly deposited material. At higher coverages the clusters become more and more irregularly shaped (Fig. 1d), indicating a (surface) diffusion limited mode of cluster growth and aggregation. Even at 13 nm Cu coverage (not shown), the highest coverage investigated, the Cu layer is not continuous. Always more and smaller clusters were observed in the TEM measurements than calculated from the XPS intensities (e.g. see Table I), but the observable clusters were not sufficient to account for all of the deposited copper. Therefore a fraction of the order of 1/3 of the deposited copper is not visible as clusters in the TEM micrographs and thus present as atoms or clusters smaller than 2 nm. This is also supported by the Auger spectra (see the section on Chemical Interaction and fig. 8) which indicate different cluster sizes to be present in the near-surface region. If all of the copper would be present on the polyimide surface and in the form of atoms and clusters smaller than 3 nm (as suggested by the TEM micrograph) it should show a much higher $Cu2p_{3/2}/C1s$ intensity ratio than is experimentally observed. We, therefore, conclude that a fraction of the order of some ten percent of the copper is located beneath the polymer surface after room temperature deposition and shows only a diminished XPS signal.

Influence of Subsequent Temperature Treatments

Room temperature deposits were heated to elevated temperatures, and the influence of the different temperature treatments on the interface morphology was investigated with XPS, TEM and AFM. The initial copper coverages and deposition rates were used as additional parameters.

The evolution of the $Cu2p_{3/2}/C1s$ peak ratio with time at 350 °C is shown in fig. 4 for two different deposition rates. The peak ratio obtained after deposition at room temperature is larger for the higher deposition rate indicating that smaller clusters have been formed (see fig.3) and/or more copper stays on the surface due to enhanced nucleation at the surface by the higher Cu flux. For each of the subsequent data points the sample was heated to 350 °C at 200 °C/min, kept there for a given annealing time and cooled down again at 60 °C/min for the XPS measurement. Only the dwell time at 350 °C is used on the x-axis, while the changes occurring during heating and cooling have been neglected.

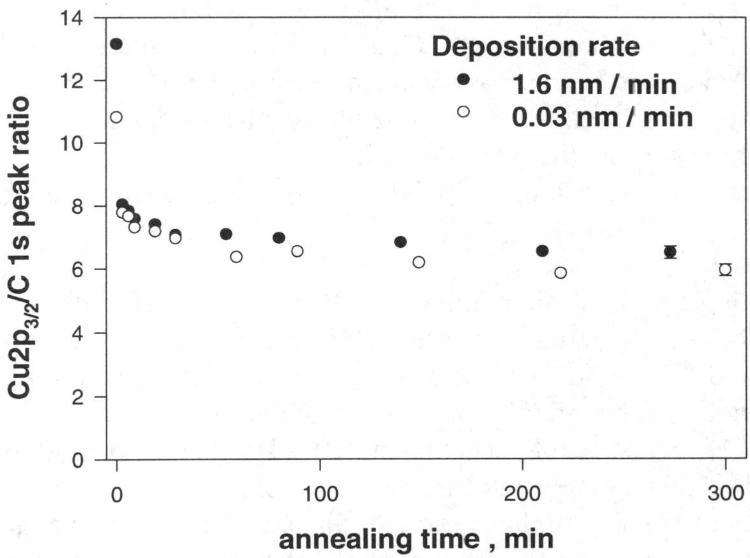

Figure 4 $Cu2p_{3/2}$/C1s peak ratio versus annealing time at 350 °C for 1.6 nm of Cu deposited onto PMDA-ODA polyimide at room temperature at rates of 1.6 nm/min and 0.03 nm/min.

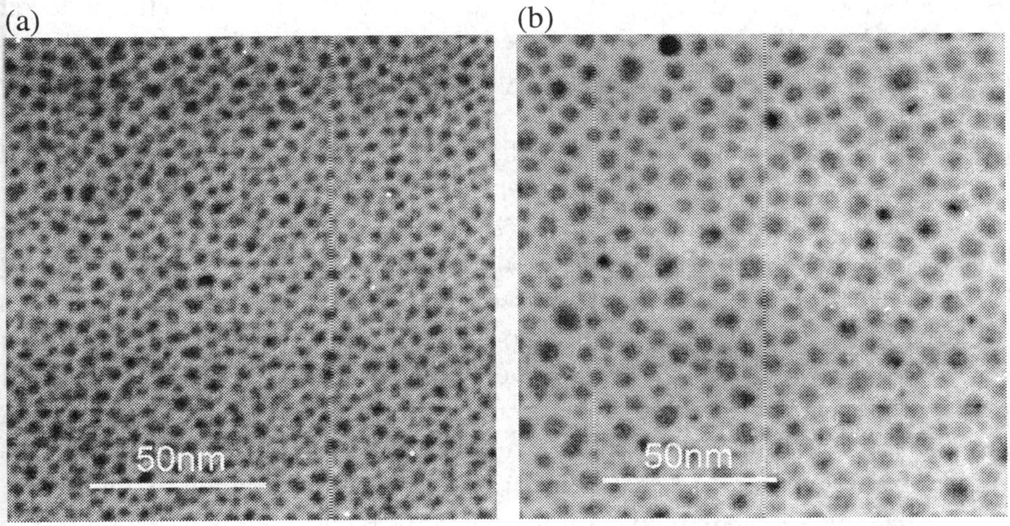

Figure 5 TEM micrographs of a PMDA-ODA polyimide film of about 0.1 μm thickness (a) after deposition of 1.6 nm of Cu at room temperature at a rate of 0.16 nm/min and (b) after subsequent annealing at 350 °C for 30 min. (The white bars measure 50 nm each).

The biggest decrease in the $Cu2p_{3/2}/C1s$ peak ratio occurs after the first annealing cycle for 3 minutes at 350 °C, and beyond 50 minutes the changes are within the error bars of our intensity ratio measurement. There are at least three distinct processes which can be responsible for the observed decrease:
1. Diffusion of copper into the polyimide film
2. Redistribution of copper into the polyimide film, e.g. by embedding of copper clusters
3. Cluster growth

Fig. 5 depicts TEM micrographs of a PI sample (a) after deposition of 1.6 nm of Cu at room temperature at a rate of 0.16 nm/min and (b) after subsequent annealing at 350 °C for 30 min. After room temperature deposition, very small size metal particles of the order of the TEM resolution limit of about 2 nm are just visible (cf. fig. 1). Subsequent annealing at 350 °C gives rise to substantial cluster growth resulting in relatively uniform size clusters of about 6 nm. Moreover, smaller clusters of about 3 nm have formed. Since smaller clusters exhibit a much higher mobility than larger ones the bimodal cluster size distribution can only be understood if one assumes that the smaller clusters are located within the polyimide film just below the surface, whereas the larger clusters are on the surface. Since the decrease in the $Cu2p_{3/2}/C1s$ peak ratio upon heating is too small to allow significant redistribution of the copper into the polyimide film, we conclude that most of the copper, which is not accumulated in the larger clusters at the surface, was already migrated below the polyimide film surface after deposition at room temperature. The smaller size originates from the reduced diffusivity in the bulk. It has to be stressed that the smaller clusters are located very close to the surface, otherwise they would not contribute to the XPS intensity.

Figure 6 shows the first results of the ongoing AFM investigations. By scanning the surface in the tapping mode the morphology of the surface is imaged. Only minor differences are observed between the clean surface (fig. 6a) and the surface covered with 1.6 nm of copper (fig. 6c). There is a slight increase in roughness from 0.3 nm (rms), being in good agreement with Unertl and Mack[34], to 0.68 nm. Unlike Unertl and Mack, who used much higher coverages, we were not able to resolve distinct clusters in the tapping mode. In the lateral force mode, a variety of spherical particles 8 - 12 nm in diameter can be seen on the metallized sample (fig. 6d) which we assume to be the larger metal clusters mentioned above, broadened by the tip geometry. The clean PI, for comparison shown in fig. 6b, shows only weak features on a scale similar to the one in the tapping mode. As the copper clusters are visible in the lateral force mode we conclude that these clusters are at the surface or are only partially embedded. This is also confirmed by the XPS data. If the clusters would be embedded completely inside the polymer, this would cause a much stronger reduction in the $Cu2p_{3/2}/C1s$ intensity ratio than is experimentally observed. Apparently, motion of the Cu particles into the polymer bulk is simply blocked because of the high bulk viscosity at temperatures up to 350 °C, still well below the bulk glass transition temperature of polyimide.

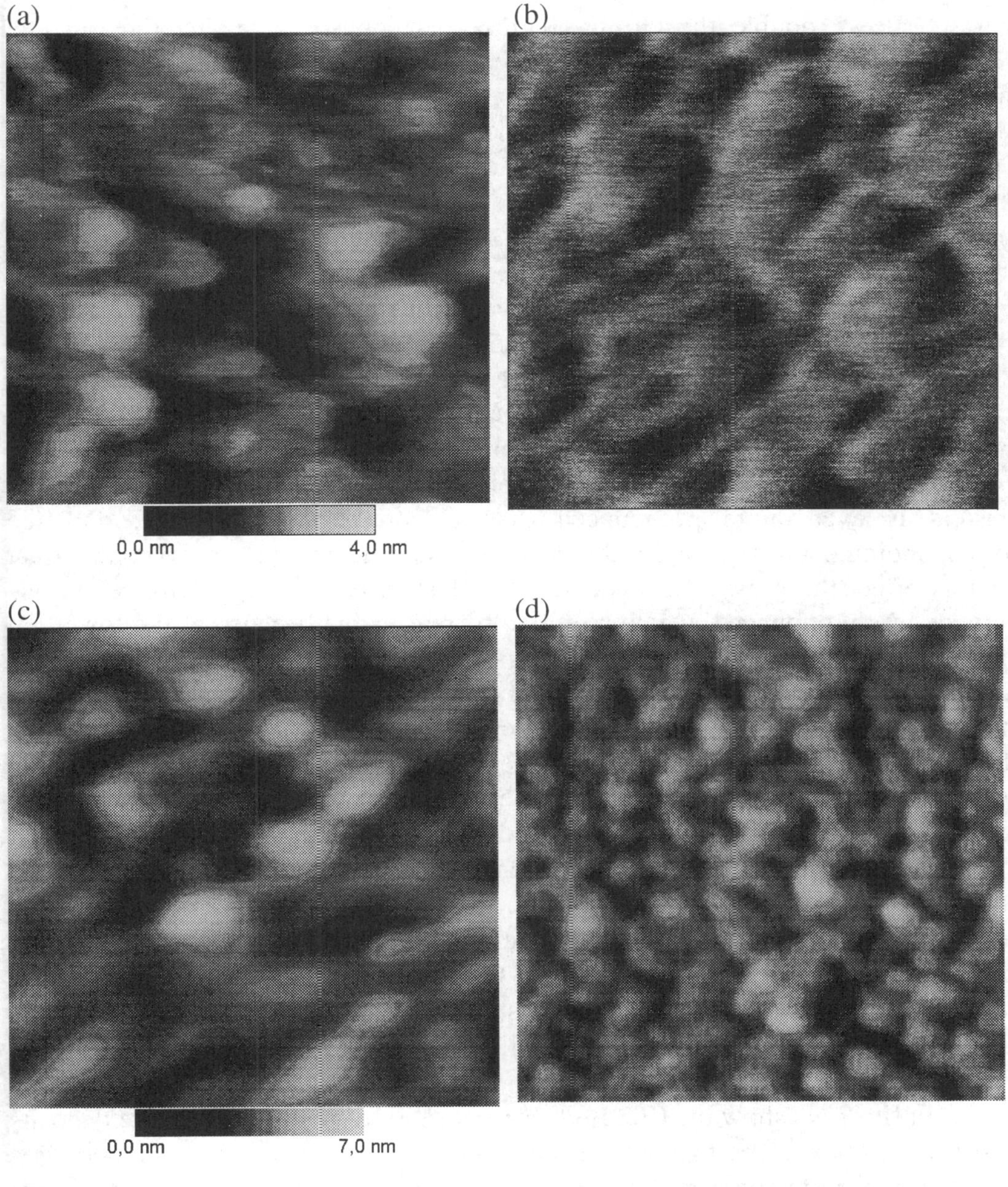

Figure 6 AFM images of pure PI in (a) tapping mode (b) lateral force mode and after deposition of 1.6 nm Cu at room temperature and subsequent annealing at 350 °C for 5h, (c) in tapping mode and (d) in lateral force mode. (The areas are 200x200 nm^2 each).

A rough estimate of the cluster size effect on the $Cu2p_{3/2}/C1s$ peak ratio is obtained when one considers the fraction of the surface area covered by the clusters. If we double the cluster diameter (which seems to be a reasonable estimate based on our TEM results) the area covered by each cluster increases by a factor of four, but the number of clusters decreases by a factor of eight. So the total area covered by all the clusters decreases by a factor of two, and the $Cu2p_{3/2}$/C 1s peak ratio decreases by more than 50%. This rough estimate is already in very good agreement with the experimentally observed XPS results. A more detailed calculation for a doubling of the cluster size from 3 nm to 6 nm showed a drop in the $Cu2p_{3/2}$/C 1s peak ratio by more than 55% (see Fig. 3), which is more than experimentally observed. Thus an increase of the cluster size by a factor of two can already account entirely for the observed decrease of the $Cu2p_{3/2}$/C1s peak ratio. It should be noted that a decrease in the metal intensity in surface-sensitive techniques is not necessarily indicative of metal diffusion into the bulk as has often been stated in the literature (e.g. ref. 16).

Judging from the TEM micrographs the growth of the larger observable clusters is even more pronounced than a doubling of the size, and the corresponding intensity drop for this process is significantly larger. To compensate for this effect the growth of some smaller clusters below the polyimide surface must be less pronounced, which seems to be reasonable because of the lower Cu diffusion coefficient in the PI bulk in comparison to the surface. Recent computer simulations of the interplay of metal diffusion and aggregation in an amorphous matrix indicate that there is also significant trapping of diffusing atoms by the clusters at or near the surface[26]. If these atoms were located deeper in the polyimide film after room temperature deposition and diffused back towards the surface this would also compensate for the intensity drop upon cluster growth.

Another important conclusion can be drawn from the saturation in the intensity drop in figure 4. After 50 min. the clusters have apparently grown so large that they become stable in size. It was not possible to fit the time dependence of the $Cu2p_{3/2}$/C1s peak ratio in terms of Ostwald ripening to the law of Lifshitz-Solyozov-Wagner[35]. This strongly suggests that cluster growth does not occur by dissociation of isolated atoms from smaller clusters and diffusion to larger ones but by diffusion of whole clusters.

In fig. 7 we show the $Cu2p_{3/2}$/C1s peak ratio for room temperature deposits of different thicknesses that were heated subsequently to different temperatures and kept at each temperature for 30 minutes. The ratios were normalized to their room temperature values to allow comparison of the relative intensity drops. Up to 200 °C we observe only minor changes in the peak ratio. This indicates that only little cluster growth occurs. Above 200 °C, the peak ratio decreases after each heating step indicating increased Cu mobility and larger cluster growth. The smaller drop in the $Cu2p_{3/2}$/C1s peak ratio in Fig. 7 at higher coverages can be attributed to the larger initial cluster size at higher coverages, which leads to reduced mobility.

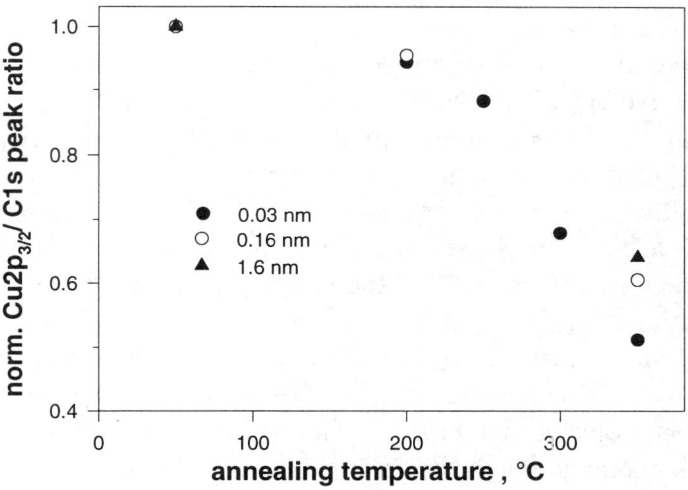

Figure 7 Normalized $Cu2p_{3/2}/C1s$ peak ratio versus annealing temperature for copper deposited onto polyimide at room temperature. Coverages are 0.03 nm (solid circles), 0.16 nm (open circles) and 1.6 nm (solid triangles). The annealing time was 30 minutes at each temperature.

Although we cannot rule out that other processes also contribute to the observed intensity changes, we conclude from our TEM and XPS measurements that cluster growth by aggregation is by far the dominant effect. If diffusion of copper into the bulk or Ostwald ripening would contribute significantly, we would expect these processes to be much faster and more pronounced for single copper atoms and very small copper clusters. Our annealing experiments, however, show only a very small dependence on the copper coverage. Most importantly, the occurrence of a plateau in the intensity ratio for prolonged heating at 350 °C (fig. 4) is in sharp contrast to ordinary diffusion of Cu into the polymer, which would lead to a continuous drop of the Cu intensity. This indicates that the clusters are stable on the polyimide surface and that no significant diffusion of copper occurs even from small (< 2 nm) copper clusters. We, therefore, can rule out significant diffusion from a continuous copper film, e.g. a conduction line, into the polymer.

Chemical Interaction of Copper Atoms

In the following section we address the question as to which chemical interactions can occur between copper atoms and polyimide and how they affect the interface formation.

A general consensus exists that thermally evaporated copper interacts only weakly with polyimide, but the details of the interaction are still controversial. In particular, it has been proposed that the interaction is essentially van der Waals like[5,19,23]. There have also been reports suggesting chemical bond formation between copper atoms and polyimide carbonyl groups or polyimide endgroups at low copper coverages[9-11]. The different chemical interaction models for the copper/polyimide interface are essentially based on different interpretations of the same XPS results. Changes in the C1s and O1s polyimide core level spectra occurring at low copper coverages were either taken as evidence for chemical bond formation between copper atoms and polyimide[9-11] or were interpreted as final state screening effects[19]. Correspondingly, the rather broad $Cu2p_{3/2}$ emission and the additional Auger transition separated by about 2 eV from the main transition in x-ray induced Cu $L_3M_{4,5}M_{4,5}$ Auger spectra at low copper coverages were taken as evidence for the formation of a Cu^{+1} species[11] or were explained alternatively as a cluster size effect[19]. The following results show that only very weak, if at all, chemical bonds are formed between copper and polyimide.

In fig. 8 we show $Cu2p_{3/2}$ and Cu $L_3M_{4,5}M_{4,5}$ spectra for a 0.04 nm copper deposit obtained on a cooled (T = -108 °C) polyimide substrate directly after deposition and after warming to 50°C, for a 0.03 nm copper deposit obtained after deposition at room temperature and after subsequent annealing to 300°C, and for a bulk copper reference sample. The modified Auger parameter $\alpha' = E_b(Cu2p_{3/2}) + E_k(Cu\ L_3M_{4,5}M_{4,5})$, which is independent of differential charging and spectrometer calibration, is listed for each spectrum in Table II together with reference data on Cu, Cu_2O and of Cu clusters of known size. The cooled sample shows a single but very broad Cu $L_3M_{4,5}M_{4,5}$ Auger transition with an α' of 1847.4 eV. After warming to 50 °C the $L_3M_{4,5}M_{4,5}$ Auger transition is shifted to higher kinetic energy, and α' increases to 1849.0 eV. Heating to 200 °C and 300 °C leads to a further increase of the modified Auger parameter to 1850.0 eV and 1850.6 eV, respectively (see Table II).

The sample obtained after room temperature deposition of 0.03 nm copper exhibits two $L_3M_{4,5}M_{4,5}$ Auger transitions separated by about 2.2 eV in good agreement with the data of Pertsin and Pashunin[11]. The shift of the transition at higher kinetic energy to a lower Auger parameter compared to copper bulk material was explained as a cluster size effect, whereas the Auger transition at lower kinetic energy was assigned to copper in the Cu^{+1} state bonded covalently to the polyimide carbonyl groups[11]. Note, the separation of 2.2 eV matches closely the difference between the Auger parameters for Cu and Cu_2O(see Table II).But the data obtained for the low temperature deposit show that this separation is just coincidence. A similar low kinetic energy Auger transition is observed directly after deposition at -108 °C, but upon warming it shifts continuously to higher kinetic energies(see Table II).Such a continuous shift is not possible if it would be a transition from Cu^{+1} to Cu^{0}, but is exactly what would be expected for cluster growth. We, therefore, assign the transition at lower kinetic energy not to Cu^{+1} but to very small copper clusters or even to single copper atoms.

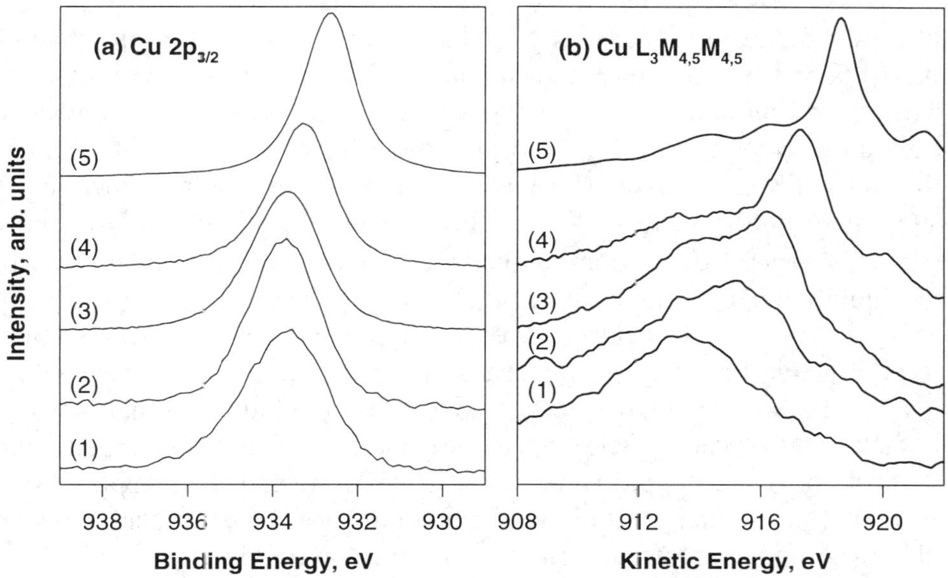

Figure 8. $Cu2p_{3/2}$ XPS spectra and X-ray induced Auger electron spectra of **0.04nm Cu on PI deposited at -108 °C (1) and after subsequent** warming to 50 °C (2), of 0.03 nm Cu on PI deposited at RT (3) and after subsequent annealing to 300 °C (4), and of Cu-bulk sample as a reference (5).

Table II. Auger parameters obtained for small copper deposits on PI at different annealing steps, and of some reference systems.

Annealing Temperature °C	**Auger Parameter** $\alpha' = E_B(Cu2p) + E_{kin}(Cu\ LMM)$	**Spectra in Fig. 8**
-108 (dep. temp.)	1847.4	(1)
50	1849.0	(2)
200	1850.0	
300	1850.6	
RT (dep. temp.)	1847.7 / 1849.9	(3)
200	1850.1	
250	1850.2.	
300	1850.6	(4)
350	1850.7	
Cu-Bulk (Cu^0) [36]	1851.3	(5)
Cu_2O (Cu^+) [36]	1849.1	
Cu clusters/Graphite diameter 1.1nm [37]	1849.9	
Cu clusters/Graphite diameter 2.4nm [37]	1851.1	
Cu atoms in SiO_2 [36]	1847.9	

As just stated the changes occurring in the Auger $L_3M_{4,5}M_{4,5}$ spectrum after warming from -108 °C to 50 °C are indicative of cluster growth and can only be explained in terms of a high mobility of the copper on the polyimide surface. This is in contradiction to the view that copper atoms bond covalently to the polyimide carbonyl groups or to polyimide endgroups and corroborates the fact that the XPS and Auger data have to be interpreted as non-chemical effects as suggested by Strunskus et al.[19]. In these terms the additional $L_3M_{4,5}M_{4,5}$ Auger transition observed for room temperature deposits at low copper coverages is due to single copper atoms or very small copper clusters coexisting with the larger clusters. Since smaller clusters are very mobile on the polyimide surface it seems to be unlikely that they are located on the polymer surface. Note, XPS and TEM results obtained for higher copper coverages also indicated that a fraction of the copper was located below the polyimide surface after room temperature deposition. If the copper is deposited at -108 °C and the sample is subsequently warmed to room temperature we observe only one Auger transition, indicating that all of the copper remained on the polymer surface during deposition and forms larger similar sized clusters upon warming.

Heating the room temperature deposit leads to a decrease in intensity of the Auger transition at lower kinetic energy and to a shift of the main transition to higher Auger parameters (see Fig. 8 and Table II). Comparison with results obtained by DeCrescenzi et al.[37] (see Table II) for copper clusters on graphite leads to a cluster size of ≈1 nm after room temperature deposition. This is in good agreement with the cluster size obtained in fig. 2 for a 0.03 nm room temperature deposit. After annealing to 300 °C the size has increased to ≈2 nm consistent with the postulated doubling of cluster sizes as concluded in the Section on Influence of Subsequent Temperature Treatment. The disappearance of the Auger transition at lower kinetic energy indicates that the very small copper clusters or copper atoms located just below the polyimide surface have become mobile upon heating to 300 °C and diffused deeper into the polyimide, agglomerated to larger clusters below the surface, or diffused back to the surface and were trapped by larger clusters. Considering the observed cluster growth and the rather small change in the $Cu2p_{3/2}/C1s$ intensity ratio upon heating (see Fig. 7) the latter two must be by far the dominant effects, but radio-tracer diffusion studies showed that traces of copper also diffused deeper into the polymer under these conditions[20].

CONCLUSIONS

Copper deposits from 0.03 to 1.6 nm nominal thickness on PMDA-ODA polyimide have been investigated by XPS, AFM, and TEM. Growth and agglomeration of small Cu clusters takes place even at room temperature. Annealing of room temperature deposits leads to significant cluster growth at temperatures above 200 °C but not to any major diffusion into the polymer bulk.

The occurrence of a plateau in the $Cu2p_{3/2}/C1s$ intensity ratio allows one to conclude that no significant diffusion is to be expected upon annealing of a continuous Cu film on PMDA-ODA polyimide. Cluster growth at the surface does not occur by Ostwald ripening, involving dissociation and association of Cu atoms from or to clusters, but is mediated by diffusion and aggregation of whole clusters.

We note that the absence of significant diffusion into the polymer and of cluster formation at larger penetration depth in the present study is not at all in conflict with the aforementioned earlier reports of pronounced diffusion and aggregation in the bulk at high temperatures and low deposition rates, or with radio-tracer experiments. The latter are sensitive enough to detect a very low concentration of atoms that diffuse deeply into the polymer even after RT deposition and subsequent annealing.

Models suggesting chemical bond formation between copper atoms and the polyimide carbonyl groups were based on XPS results obtained for room temperature deposition. In the preceding section we demonstrated conclusively that there is no evidence for copper atoms in a stable chemical bound state. Our observations indicate that copper atoms are very mobile on and in the polyimide and agglomerate to clusters already at room temperature.

ACKNOWLEDGMENTS

The AFM investigations were performed at the GKSS Forschungszentrum Geesthacht, Abt. für Membranforschung, Teltow together with A. Keller and H. Kamusewitz. We thank K. Mosig and C. v. Bechtolsheim for help with the TEM and A.Thran for helpful discussions. Financial support of our research into the field of metal-polymer interfaces throughout the years by the Volkswagen-Stiftung is gratefully acknowledged. The equipment for TEM sample preparation was, in part, financed by the Technologiestiftung Schleswig-Holstein.

REFERENCES

1. (a) K.L. Mittal and J.R. Susko, (Eds.), "Metallized Plastics 1: Fundamental and Applied Aspects," Plenum Press, New York 1989.
(b) K.L. Mittal, (Ed.), "Metallized Plastics 2: Fundamental and Applied Aspects," Plenum Press, New York 1991.
(c) K.L. Mittal, (Ed.), "Metallized Plastics 3: Fundamental and Applied Aspects," Plenum Press, New York 1992.
(d) K.L. Mittal, (Ed.), "Metallized Plastics: Fundamentals and Applications," Marcel Dekker, New York 1998.

2. E. Sacher, J.J. Pireaux, and S.P. Kowalczyk, (Eds.), "Metallization of Polymers," ACS Symposium Series No. 440; American Chemical Society, Washington D.C., 1990.
3. R. R. Tummala and E. J. Rymaszewski (Eds.), "Microelectronics Packaging Handbook," Van Nostrand Reinhold, New York, 1989.
4. P. Springer, Semiconductor Intl. 52 (Nov. 1994).
5. L. J. Matienzo and W.N. Unertl, in "Polyimides: Fundamentals and Applications," M. Ghosh and K.L. Mittal (Eds.), pp. 629-696, Marcel Dekker 1996.
6. J.J. Pireaux, Synth Met., 67, 39 (1994).
7. N.J. Chou and C.H. Tang, J. Vac. Sci. Technol. A, 2, 751 (1984).
8. R.C. White, R. Haight, , B.D. Silverman, and P.S. Ho, Appl. Phys. Lett., 51, 481 (1987).
9. P.N. Sanda, J.W. Bartha, J.G. Clabes, J.L. Jordan, and C. Feger, J. Vac. Sci. Technol A, 6, 2188 (1988).
10. R.G. Mack, E. Grossman, and W.N. Unertl, J. Vac. Sci. Technol. A, 8, 3827 (1990).
11. A.J. Pertsin and Yu.M. Pashunin, Appl. Surf. Sci., 47, 115 (1991).
12. G.D. Davis, B.J. Rees, and P.L. Whisnant, J. Vac. Sci. Technol. A, 12, 2378 (1994).
13. F.K. LeGoues, B.D. Silverman, and P.S. Ho, J. Vac. Sci. Technol. A, 6, 2200 (1988).
14. F. Faupel, R. Willecke, A. Thran, C. v. Bechtolsheim, M. Kiene, and Th. Strunskus, Mittal Festschrift Volume, in press.
15. R.M. Tromp, F. LeGoues, and P.S. Ho, J. Vac. Sci. Technol. A, 3, 782 (1985).
16. F.S. Ohuchi and S.C. Freilich, J. Vac. Sci. Technol A, 6, 1004 (1988).
17. J.H. Das and J.E. Morris, IEEE Trans. Components, Packaging, Manuf. Technol. B, 17, 620-625 (1994).
18. D.S. Dunn and J.L. Grant, J. Vac. Sci. Technol. A, 7, 253 (1989).
19. T. Strunskus, C. Hahn, D. Frankel, M. Grunze, J. Vac. Sci. Technol. A, 9, 1272 (1991).
20. a) F. Faupel, Phys. stat. sol. a, 134, 9 (1992).
b) F. Faupel, D. Gupta, B.D. Silverman, and P.S. Ho, Appl. Phys. Lett., 55, 357 (1989).
21. P. O. Hahn, G. W. Rubloff, J. W. Bartha, F. K. LeGoues, and P. S. Ho, Mater. Res. Soc. Symp. Proc., 40, 251 (1985).
22. J.T. Wetzel, D. A. Smith, G. Appelby-Mougham, Mater. Res. Soc. Symp. Proc., 40, 271 (1985).
23. T. Strunskus, M. Grunze, G. Kochendoerfer and Ch. Wöll, Langmuir, 12, 2712 (1996).
24. H. M. Meyer III, S. G. Anderson, Lj. Atanasoska and J. H. Weaver, J. Vac. Sci. Technol. A, 6, 30 (1988).

25. B.D. Silverman, Macromolecules, 24, 2467 (1991).
26. A. Thran, and F. Faupel, Defect and Diffusion Forum, 143-147, 903 (1997).
27. K. Shanker and J.R. MacDonald, J. Vac. Sci. Technol. A, 5, 2894 (1987).
28. G. Gergley, M. Menyhard, K. Pentek, A. Sulyok, A. Jablonski, B. Lesiak, and Cs. Daroczi, Vacuum, 46, 591 (1995).
29. S. Tanuma, C. J. Powell, and D. R. Penn, Surf. Interf. Anal., 21, 165 (1993).
30. C. Feger, Polym. Eng. Sci., 29, 347 (1989).
31. T. P. Russell, J. Polym. Sci. Polym. Phys. Ed., 22, 1105 (1984).
32. M. Kiene, Ph.D. Thesis, University of Kiel, Kiel, Germany, 1998.
33. J. Lörincik, Appl. Surf. Sci., 62, 89 (1992).
34. W.N. Unertl and R.G. Mack, in "Metallized Plastics 3: Fundamental and Applied Aspects," K.L. Mittal (Ed.), pp. 85-108, Plenum Press, New York, 1992.
35. J.W. Christian, "The Theory of Transformations in Metals and Alloys," Pergamon Press, Oxford 1965, 2nd Ed., Part I. 1975
36 C.D. Wagner, in "Practical Surface Analysis,"D. Briggs, M.P. Seah (eds.), Page Bros., Norwich, 1984.
37. a) M. DeCrescenzi, M. Diociauti, L. Lozzi, P. Picozzi, and S. Santucci, Solid State Commun., 74, 115 (1990).
b) M. DeCrescenzi, M. Diociauti, L. Lozzi, P. Picozzi, S. Santucci, C. Battistoni, and G. Mattongo, Surf. Sci., 178, 282 (1986).

Metallized Plastics 5&6: Fundamental and Applied Aspects, pp. 231-236.
K.L. Mittal (Ed.)

THE SPONTANEOUS REACTION OF Cu (hfac) (TMVS) VAPOR WITH DUPONT TEFLON AF1600

D. Popovici, E. Sacher and M. Meunier

Groupe des Couches Minces and Département de Génie Physique, École Polytechnique de Montréal, C. P. 6079, Succursale Centre-Ville, Montréal (Québec) H3C 3A7, Canada

and D. Léonard and P. Bertrand

Unité de Physico-Chimie et de Physique des Matériaux, Université Catholique de Louvain, 1 Place Croix du Sud, B-1349 Louvain-la-Neuve, Belgique

Preliminary X-ray photoelectron spectroscopic (XPS) and time-of-flight secondary ion mass spectroscopic (TOF-SIMS) measurements have confirmed the reaction of Teflon AF1600 with vapors of copper (hexafluoroacetylacetonate)(trimethyl vinyl silane) [Cu(hfac)(TMVS)], when exposed in the dark. XPS reveals the deposited copper to be in the Cu^{II} state, while TOF-SIMS has shown the organometallic to have lost its ligands in bonding to the surface.

INTRODUCTION

Future microelectronic multilayer devices will demand signal transmission times so short that the RC contribution must amount to no more than a small fraction of the overall delay time[1]. Here, R is the metal line resistance and C the capacitance of the interlayer insulator. There are three ways of accomplishing this: 1) reduce R with a more conductive metal; 2) reduce C with an insulator of lower permittivity; 3) reduce R by shortening the metal lines. R may be reduced by replacing the presently used aluminum with copper, which has a resistivity ~50% lower. C may be reduced by replacing the presently used polyimide insulation with fluoropolymers, having dielectric constants ~40% lower. A recent addition to the Teflon® family of fluoropolymers, Teflon AF, is soluble in perfluorinated solvents, permitting its use in spin-on applications; it is available as AF1600 (T_g = 160°C) and AF2400 (T_g = 240°C).

The Teflon AF series differs from others Teflons in that the fluorocarbon chain, $(CF_2)_n$, is a minor component[2]. The major component is the 2,2-bistrifluoromethyl-4,5-difluoro-1,3-dioxole ring, containing two ether oxygens. The Teflon AF1600 repeat unit contains two dioxole rings to one C_2F_4 moiety while the AF 2400 repeat unit contains seven dioxole rings to one C_2F_4.

Recent work in our group[3] has demonstrated the feasibility of the laser-induced deposition of Cu from the copper(I) precursor, copper (hexafluoroacetylacetonate) (trimethyl vinyl silane) [Cu(hfac)(TMVS)]. Deposition was successful on both TiN and Teflon PFA substrates. Cu(hfac)(TMVS) is extremely stable and has a high vapor pressure at slightly elevated temperatures, making it suitable for chemical vapor deposition (CVD). A recent book[4] discusses the mechanism of copper deposition from Cu^I(hfac)(TMVS). It begins with an activated reaction at the SiO_2 substrate surface, leading to the loss of TMVS and the bonding of Cu^I(hfac) to the surface. Subsequent reaction with Cu^I(hfac)(TMVS) leads to a disproportionation and the production of bonded Cu^0 and $Cu^{II}(hfac)_2$, which is subsequently lost as vapor. No explanation is offered as how elemental copper can bond and still retain its zero valency. Their overall mechanism is:

$$Cu^I(hfac)(TMVS) \xrightarrow{heat} substrate\text{-}Cu^I(hfac) + TMVS\uparrow \quad (1)$$

$$2[substrate\text{-}Cu^I(hfac)] \xrightarrow{heat} substrate\text{-}Cu^0 + Cu^{II}(hfac)_2\uparrow \quad (2)$$

The present study was provoked by our finding that Cu(hfac)(TMVS) reacted with Teflon AF1600 (and Teflon PFA) surfaces at low temperatures, in the absence of laser light, giving a monolayer or less of Cu-containing product. Here, we report our initial findings on this unanticipated reaction between a non-reactive fluoropolymer and a stable organometallic. We use (1) X-ray photoelectron spectroscopy (XPS), comparing the spectra with those obtained on unreacted Teflon AF 1600[5], and (2) time-of-flight secondary ion mass spectroscopy (TOF-SIMS), where the presence of fragments is used to indicate the state of the bonded copper.

EXPERIMENTAL

As reported earlier[5], Teflon AF1600, supplied by DuPont, was dissolved in Sigma Fluorinert FC-77 and spun onto a freshly cleaned Si wafer; it was then dried at 170°C for 3 hours and, finally, at 330°C for 10 minutes. Before exposure to Cu(hfac)(TMVS), purchased from the Schumacher Corporation as CupraSelect, the Teflon AF1600 was heated to 110°C for 10 minutes

Exposure was carried out in the following manner: a 100 sccm flow of H_2 carrier gas was passed through a bubbler of CupraSelect, held at 45°C. This vapor, containing CupraSelect at a partial pressure of 0.5 Torr, was passed into a chamber containing the Teflon AF1600 sample at 55°C; the exposure lasted 10 minutes. Samples were then submitted for XPS and TOF-SIMS analyses, with an unavoidable exposure to air: several minute in the case of XPS and several days in the case of TOF-SIMS.

Both XPS and TOF-SIMS techniques are well adapted for this type of study. Similar applications have recently been reviewed[2]. XPS is based on the photoelectric effect: irradiation by high energy monoenergetic X-rays causes the emission of electrons, whose binding energies indicate the orbitals from which they were emitted and whose chemical shifts, caused by the electron density at the emitting atom as influenced by its neighbors, indicate the chemical environment at the atom. TOF-SIMS is a form of SIMS carried out under "static" conditions, in which less than a monolayer of surface is removed during the course of the experiment; the use of a TOF spectrometer assures a high mass resolution ($M/\Delta M \approx 10^4$ at lower masses), a high sensitivity due to its high transmission, and a high detection efficiency.

XPS analysis was carried out on a VG ESCALAB Mark II apparatus, using Mg K_α radiation at 1253.6 eV. Survey scans were obtained at an analyzer pass energy of 100 eV and at a grazing angle of incidence. TOF-SIMS analysis was carried out on a Charles Evans & Associates TRIFT mass spectrometer, using a pulsed Ga ion beam, as recently described[6]. Secondary ions were extracted at an acceleration voltage of 3 kV, electrostatically deflected 270° and mass analyzed by measuring their time-of flight distributions from the sample surface to the position-sensitive detector.

RESULTS AND DISCUSSION

Cu(hfac)(TMVS) was found to react with Teflon AF1600 after a short exposure, in the absence of laser light, and at a temperature of 55°C, substantially below the conventional CVD temperature range of 120-130°C. This unexpected reaction is shown by the XPS survey scan in Figure 1 and the TOF-SIMS spectra in Figure 2.

The XPS scan in Figure 1 indicates the presence of C, O, F and Cu. Satellites in the Cu2p region (930-970 eV) indicates the presence of divalent Cu, although Cu(hfac)(TMVS) contains monovalent Cu. A high resolution scan of the Cu2p region, not shown, indicates the presence of only one divalent species, although zerovalent Cu is expected[4]; while air oxidation is a possible reason for the presence of divalent Cu, it is our experience that such oxidation is slow and, were it to have taken place, both monovalent and divalent Cu would have been in evidence. High resolution scans of the C1s, O1s and F1s regions showed no obvious changes when compared to untreated spectra[5]; they are not shown. A blank was also run: Teflon AF1600 was exposed under the same conditions but in the absence of Cu(hfac)(TMVS).

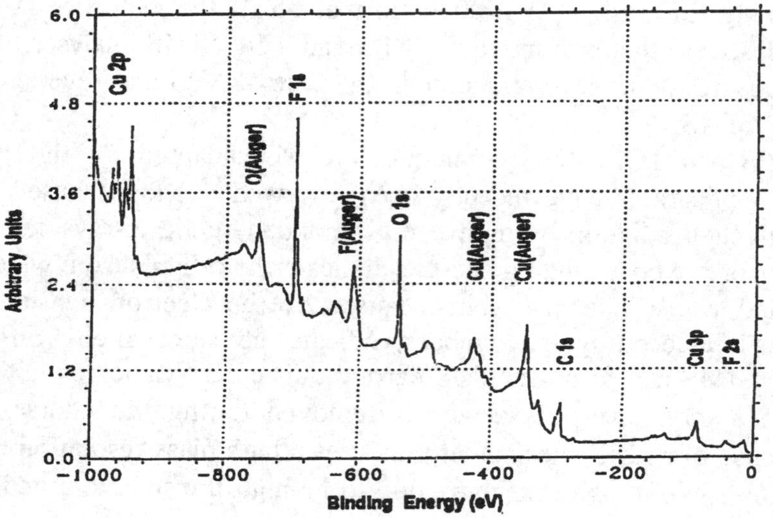

Figure 1. An XPS survey scan of the Teflon AF1600 surface after reaction with Cu(hfac)(TMVS).

Its spectrum was identical to that in Figure 1, except for the absence of all Cu peaks. It, too, is not shown.

The TOF-SIMS scans are found in Figure 2, where treated (a) and untreated (b) Teflon AF1600 spectra are shown; both general (0-200 amu) and Cu-specific (62.05-65.65 amu positive scans are shown for each. Both treated and untreated Teflon AF samples are characterized by the same high intensity peaks , relatable to the AF structure: C^+ (12 amu), CF^+ (31 amu), CF_2^+ (50 amu), CF_3^+ (69 amu), $C_2F_3^+$ (81 amu), $C_2OF_3^+$ (97 amu), $C_3OF_3^+$ (109 amu), $C_3F_5^+$ (131 amu) and $C_3F_7^+$ (169 amu). In the Cu specific spectra, ^{63}Cu and ^{65}Cu, in the natural isotopic ratio of 70:30, are observable only on the treated Teflon AF; peaks at 63 and 65 amu, observable on both treated and untreated Teflon AF, are due to $C_2HF_2^+$ and $C_5F_5^+$, respectively. No other copper-containing fragments were found.

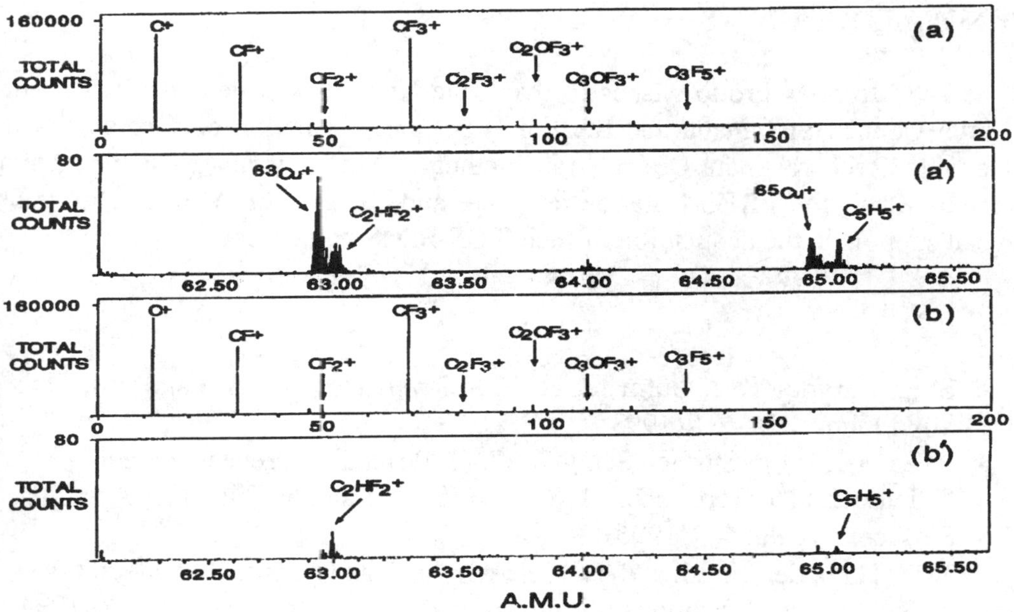

Figure 2. General and Cu-specific, respectively, TOF-SIMS scans of (a,a') treated and (b,b') untreated Teflon AF1600.

It is important to note that only $^{63}Cu^{+}$ and $^{65}Cu^{+}$ are found in the TOF-SIMS spectra, despite the XPS identification of Cu^{II}. This suggests, first, a bond between Cu^{II} and the Teflon AF surface which, when broken, causes an increased electron density at the Cu. Second, it is clear that the Cu^{II}, on bonding to the substrate, has also lost its hfac ligand; the observation of such intact fragments is expected[7] if the ligand were still bonded to the Cu.

The fact that the spectra remain unchanged for long periods of time under UHV condition in the spectrometers, and that Cu^{I} has been oxidized to Cu^{II} on reaction with Teflon AF1600, clearly shows the *chemical* nature of this reaction.

The mechanism for this chemical reaction is unclear. While the Cu^{I} has clearly undergone an oxidation to Cu^{II}, the reduction portion of the couple is not yet identified. It may involve the lost hexafluoroacetylacetonate fragment.

CONCLUSIONS

The present results indicate that Cu(hfac)(TMVS) reacts spontaneously with Teflon AF1600. Although the site of the reaction is unidentified, it is most probably the oxygen of the dioxole rings. In reacting, the copper loses both ligands and is oxidized to Cu^{II}. This is in contradistinction to the mechanism proposed in the literature[4], based on transmission FTIR studies carried out on high surface area SiO_2 substrates[8]. It may well be that both mechanisms are substrate-dependent.

ACKNOWLEDGEMENTS

The Montreal group wishes to thank the Natural Science and Engineering Research Council of Canada and the Fonds pour la Formation de Chercheurs et l'Aide à la Recherche du Québec for funding. The Louvain-la-Neuve group wishes to thank the FRFC-Loterie Nationale and the Région Wallone for their financial support in the acquisition of their TOF-SIMS equipment.

REFERENCES

1. S. P. Murarka, R. J. Gutmann, A. E. Kaloyeros and W. A. Landford, Thin Solid Films, **236,** 257(1993).
2. E. Sacher, Prog. Surface Sci., **47,** 273(1994) and references therein.
3. R. Izquierdo, J. Bertomeu, M. Suys, E. Sacher and M. Meunier, Appl. Surface Sci., **86,** 509(1995).
4. M. J. Hampden-Smith and T. T. Kodas, in "The Chemistry of Metal CVD," T. Kodas and M. Hampden-Smith (Eds.), chap.5, VCH, New York, 1994.
5. E. Sacher and J. E. Klemberg-Sapieha, J. Vac. Sci. Technol., **A15**, 2143 (1997).
6. K. Piyakis, E. Sacher, A. Domongue, J. J. Pireaux, G. Leclerc, P. Bertrand and J. B. Lhoest, Appl. Surface Sci., **84,** 227(1995).
7. L. T. Weng, P. Bertrand, G. Lalande, D. Guay and J. P. Dodelet, Appl. Surface Sci., **84,** 9(1995).
8. J. Farkas, M. J. Hampden-Smith and T. T. Kodas, J. Electrochem. Soc., **141,** 3539, 3547(1994).

Metallized Plastics 5&6: Fundamental and Applied Aspects, pp. 237-253.
K.L. Mittal (Ed.)

THE DEPOSITION OF COPPER ONTO TEFLON AF1600: AN XPS COMPARISON OF VAPOR DEPOSITION AND SPUTTERING

D. Popovici, J. E. Klemberg-Sapieha, G. Czeremuszkin,
A. Alptekin, L. Martinu, M. Meunier and E. Sacher

Département de Génie Physique and
Groupe de Couches Minces
École Polytechnique de Montréal
C.P. 6079, Succursale Centre-Ville
Montréal, Québec H3X 3A7, Canada

Electron beam evaporation and sputtering were used to deposit copper onto Teflon AF1600. Cu/polymer interfacial interactions were analyzed by X-ray photoelectron spectroscopy (XPS) and a scratch test was used to determine the so-called practical adhesion. Both evaporation and sputtering led to the loss of fluorine, as well as the formation of C-O$^{\bullet}$ free radicals and Cu-C interfacial bonding. Additionally, the higher energies of sputter deposition led to the loss of oxygen and the formation of inorganic fluorides. The adhesion of copper was greater in the case of sputtering.

INTRODUCTION

The need of the semiconductor industry to increase the signal transmission speed beyond the present limit of 300 MHz requires faster transmission lines. This is because the overall speed of large, high-density chips is presently not limited by the transistor speed but, rather, by the on-chip circuits that connect them[1]. Specifically, the resistance (R) of the conductors and the capacitance (C) of the insulators cause what is known as an “RC” time delay, which limits the chip speed when it is larger than the inherent switching time of the transistor. The goal of the microelectronics industry of reducing the RC time delay must be achieved for smaller dimensions and larger die sizes. This is a difficult challenge, for interconnects with smaller cross sections have higher resistances, as do the longer lines found in larger dies. Further, thinner dielectrics have higher capacitances. Thus, one is limited to any of three ways to reduce the RC delay time: (1) decrease R by using a higher conductivity metal, (2) decrease C by using dielectrics having lower dielectric constants and (3) decrease the length of the metal lines by using multilevel interconnect devices.

Copper, for example, having a resistivity 65% of the presently used aluminum[1], is seen as the most promising alternative metal, while fluoropolymers such as the Teflons, which have dielectric constants near 2, are considered as potential replacements for polyimides and SiO_2. The well-known chemical inertness of the fluoropolymers has created a strong interest in the metal-fluoropolymer interface and any specific reactions between them, in situations in which strong interfacial adhesion is required. This is the case for multilevel interconnect devices, which are made of alternating layers of polymer and metal. Thermal and hygroscopic expansion coefficient mismatches introduce interfacial stresses during the fabrication process. Any tendency to breakdown of the device due to these stresses must be prevented by strong interfacial adhesion[2].

Diffusion barriers, such as TiN or Ta, were not considered at this stage because the extensive diffusion of copper into Teflon AF1600 has not yet been demonstrated. In the present work, we focused on the copper/Teflon AF1600 interface. Copper provides benefits from a reduced resistivity and a relatively good electromigration resistance. In addition, it can be deposited with excellent step coverage and is capable of filling narrow, high aspect ratio contacts and vias. It can also be deposited at relatively low temperature, which is a technological advantage[3]. Teflon AF1600, the Teflon we have chosen to consider, is a copolymer of 35% tetrafluoroethylene (TFE) and 65% 2.2-bistrifluoromethyl-4,5-difluoro-1,3-dioxole (TDD) and is manufactured by DuPont; it has the repeat unit structure $-(TFE)-(TDD)_2-$. Teflon AF1600 also has the distinct advantage of being soluble in perfluorinated solvents, permitting its deposition by spin-on techniques[4].

The copper metallization of different types of Teflon, such as Teflon PFA and FEP, has been studied in our laboratory in an attempt to elucidate chemical interactions at the interface[5,6]. We showed that these surfaces are not easily attacked by copper. In our previous study of copper deposition onto Teflon AF1600[7], we provisionally related adhesion to the presumed formation of Cu-O bonds at the interface. Here, we consider specific reactions, based on the separated peaks of XPS spectra.

EXPERIMENTAL

The copper/Teflon AF1600 interface was studied using X-ray photoelectron spectroscopy (XPS), which is a convenient method because of its high surface sensitivity. The base pressure in the analysis chamber of our VG ESCALAB 3 Mark II spectrometer was 10^{-9} Torr. No electron flood gun was used to reduce sample charging during the data accumulation. The spectral resolution for our Mg K_α source was measured to be 0.8 eV. The XPS spectra were accumulated under conditions of minimal exposure to the X-ray source (10 mA and 6 kV), in order to avoid degradation of the Teflon surface during analysis[8].

Teflon AF1600 was purchased from DuPont as a clear solution containing 12% solids and was further diluted 50% in Sigma Fluorinert FC-77 before application. For both vapor deposition and sputtering, the Teflon AF1600 was spun onto a freshly cleaned Si wafer at a spin speed of 3000 rpm. After air drying at room temperature, the samples were baked at 115^{0}C for 15 minutes and then annealed at 165^{0}C for 3 hours. To obtain maximum uniformity of coating thickness and enhanced adhesion, a supplementary heating was performed at 300^{0}C for 15 minutes, as recommended by the manufacturer[4]. A final Teflon thickness of 1-2 μm was obtained.

The spectrometer possesses a preparation chamber where vapor deposition and sputtering were carried out without exposing the samples to air. Our evaporation unit is an electron beam evaporator similar to the one developed by Waldrop and Grant[9,10]. A thin wire of pure copper was positioned 9 cm from the sample and biased at +6 kV. A nearby wire filament acted as an electron source. Evaporation of the target took place at electron bombardment currents of 15 mA, at a pressure of $2x10^{-8}$ Torr. An evaporation rate of 0.2 nm/sec. and vapor deposition times of 2, 4, 6 and 12 seconds gave nominal copper layer thicknesses of about 0.4, 0.8, 1.2 and 2.4 nm, respectively, as measured by our quartz crystal monitor.

Our sputtering unit was also mounted in the preparation chamber of the spectrometer. The target was a 99.9% pure copper plate and was sputtered by Ar^+ ions at an energy of 6 kV. The intensity of the Ar beam was controlled by a focusing electrostatic lens, using an excitation potential of 4 kV. The Teflon samples were positioned at a distance of 5 cm from the sputter source at an angle of 65^0. A sputtering rate of 0.4 nm/min and sputtering times of 1, 2, 4 and 10 minutes gave nominal copper layer thicknesses of 0.4, 0.8, 1.6 and 4 nm, respectively (in the case of sputtering, our angle-resolved studies showed that there was no metallic copper layer on the Teflon AF1600 surface but, rather, the reacted copper was uniformly distributed throughout the observable depth).

For evaporated copper, two sorts of samples were used for the measurement of practical adhesion: after the evaporation of 0.15 μm of high purity copper onto 0.7 μm Teflon substrates on glass, one sample was post-annealed in vacuum at 200^{0}C for 30 minutes. Practical adhesion was determined by a microscratch tester (MST) developed at CSEM, Switzerland[11], using a Rockwell C hemispherical diamond stylus of 0.8 mm radius. At a table speed of 5 mm/min, a progressively increasing load ranging from 0.01 to 3 N was applied at a rate of 1 N/min. The critical load failure criterion was arbitrarily considered to be the moment when the copper layer began to be completely and consistently removed from the Teflon surface. It was determined by optical examination, using an optical microscope and a scanning electron microscope (SEM). For both non-annealed and annealed samples, five scratch tests were carried out and the results were averaged.

Spectral peaks were separated using an in-house nonlinear least-mean-squares program. The Gaussian / Lorentzian ratios and the Full Width at Half Maximum (FWHM) values used were those determined by us for other fluoropolymers[12], as it is seen in Table 1.

Table 1. Gaussian/Lorentzian (G/L) ratios and FWHM for C, O and F XPS spectra.

	C1s	O1s	F1s
G/L	40/60	60/40	30/70
FWHM (eV)	1.8	2.0	2.0

A Shirley background subtraction was made prior to peak separation, and charge compensation was carried out by peak alignment: the $Cu2p_{3/2}$ peak for the thickest vapor-deposited copper layer was arbitrarily placed at a binding energy of 931 eV, which is a value attributed to bulk copper[13]. This places the major F1s peak, to be discussed below, at 687.3 eV. This shift, subsequently carried out for all the samples, gave C1s, O1s and F1s peak alignments for all the spectra to within ±0.1 eV.

The relative concentration of each element of interest was determined from the peak area divided by the atomic sensitivity factor, also determined in our laboratory. To compare elements having different electronic mean free paths, the concentrations were corrected using the ratios of the mean free paths of the element being compared[14].

RESULTS

The repeat unit of the pure Teflon AF1600 and its XPS spectrum was discussed in a previous paper[12]; it is seen in Fig.1. As it can be noticed in this figure, the five peaks under the envelope are attributed to the five carbon environments shown.

Vapor Deposition

1) C1s spectra. A typical C1s spectrum is presented in Fig. 2a. The spectrum of the pure Teflon AF1600 is easily recognizable in the spectrum of the

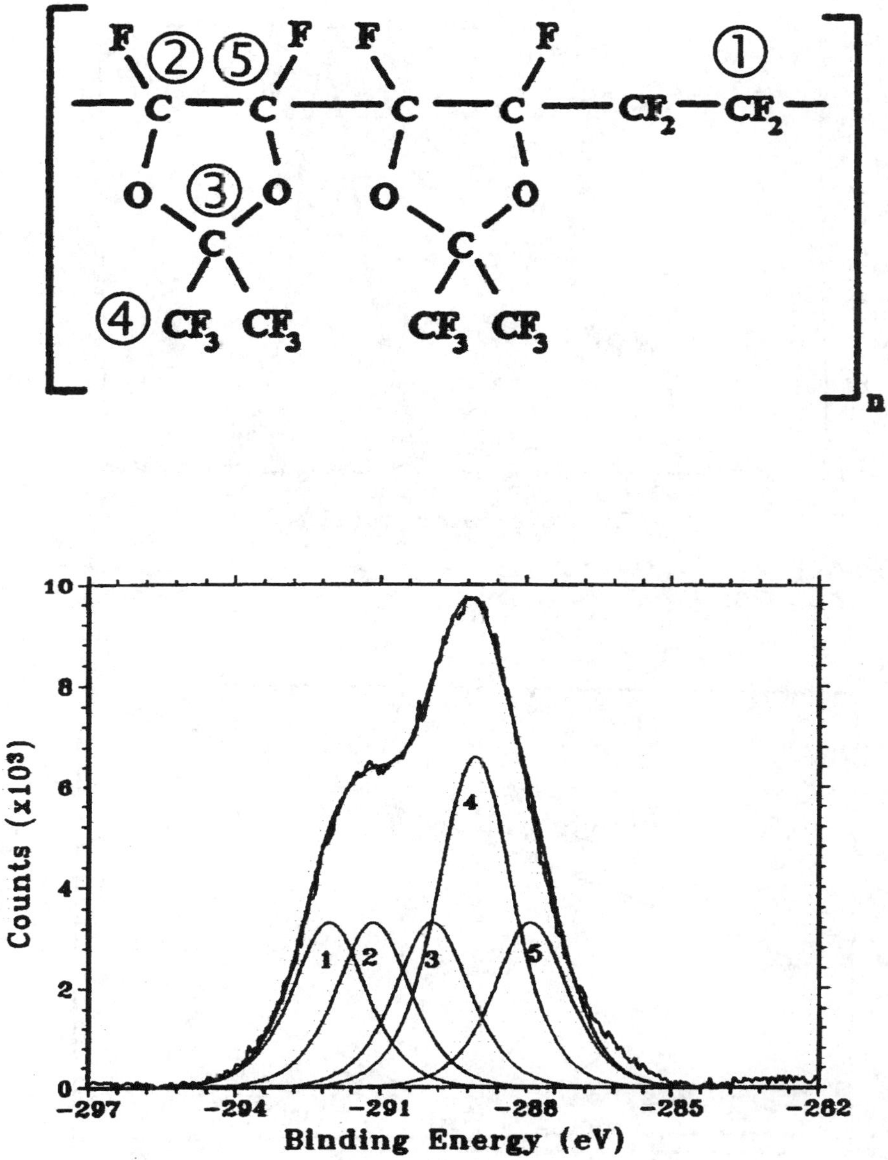

Figure 1. The structure of the Teflon AF1600 repeat unit and its XPS spectrum.

metallized sample above a binding energy of 287 eV. The new contribution arising at lower binding energy was too wide to accommodate one peak. Three supplementary peaks, placed at 283.7, 285.4 and 287.3 eV were necessary to fit the envelope. A fourth peak was infrequently required at 282.0 eV but there are

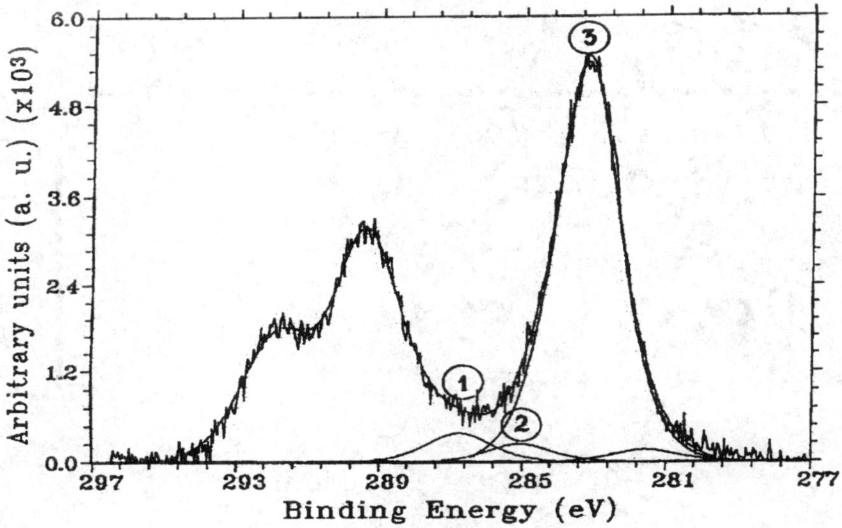

(a)

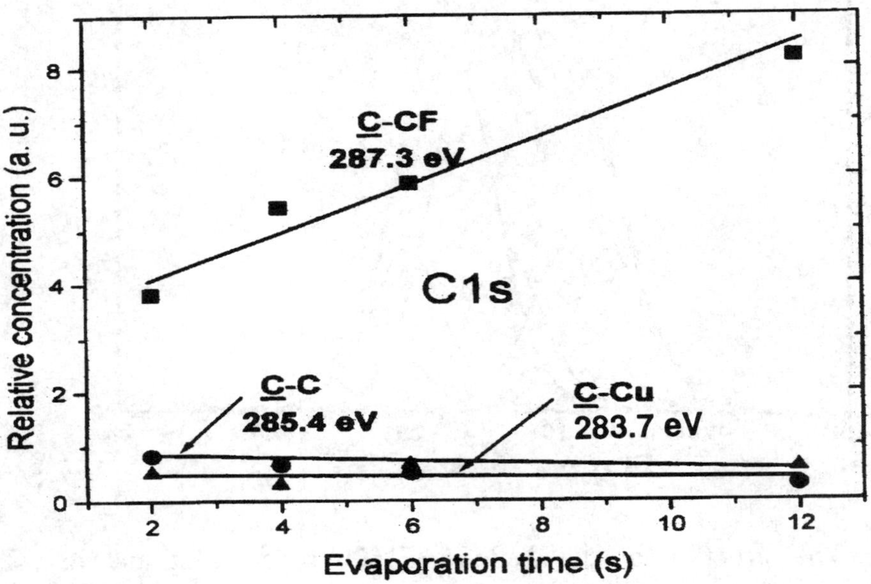

(b)

Figure 2. a) C1s spectrum for 12 seconds of vapor deposition. Peaks 1, 2 and 3 are attributed to the formation of C-CF, C-C and Cu-C, respectively; b) Evolution of the newly formed C1s peaks on vapor deposition.

serious doubts about its existence; its relative area represents 4.6% or less of the total area. Peaks 1, 2 and 3 are attributed to C-CF, C-C and C- Cu, respectively[15].

The evolution of these three supplementary peaks is shown in Fig. 2b. Peak 3, which is associated with carbide formation, increases linearly. Because it does not extrapolate to zero, there is evidently a high initial rate of carbide formation. Peaks 1 and 2 are due to the partial and total loss of fluorine, respectively. These processes occur at the onset of deposition and remain constant thereafter. Their similar evolution suggests that both processes are triggered by the initial deposition of copper.

2) F1s spectra. A typical F1s spectrum is presented in Fig. 3a. The area under the envelope decreases slowly, showing a total loss of about 10% fluorine, which is in good agreement with the degradation of C1s spectra. As reported in an earlier paper[14], the fluorine spectrum of pure Teflon AF1600 could be separated into three peaks assigned to CF, CF_2 and CF_3 in the ratio 1:1:3. The evolution of F1s peaks is shown in Fig 3b. The original three peaks decrease essentially in the same ratio of 1:1:3, suggesting that the fluorine is lost uniformly.

3) O1s spectra Fig. 4a shows a typical O1s spectrum. It was separated into four peaks, the two original peaks of Teflon AF1600 at 532.6 and 533.4 eV, and two newer ones at 529.6 and 530.4 eV appearing on copper deposition. Pairs of peaks at identical chemical shifts indicate that the two slightly different oxygen environments of Teflon AF1600 have equal probabilities of attack by copper.

The chemical shift of the newer peaks toward lower binding energies indicates a more electronegative environment, suggesting the possible generation of oxygen free radicals via O-C bond breaking. Thus, the formation of $-O^{\bullet}$ is expected to produce a shift to lower binding energies; the $-C^{\bullet}$ simultaneously produced may react with another similar free radical to form C-C bonds, contributing to peak 2 in the previously discussed C1s spectrum (Fig. 2a), or it may be obscured by other peaks. Another possible attribution of the new peaks, the formation of C-O-Cu, is discounted because associated peaks in the C1s and $Cu2p_{3/2}$ spectra do not exist.

The evolution of the O1s peaks is shown in Fig. 4b. Vapor deposition occurs with little loss of oxygen. The rate of formation of oxygen free radicals is higher at the beginning before falling to a constant value.

4) $Cu2p_{3/2}$ spectra. A typical $Cu2p_{3/2}$ spectrum is presented in Fig. 5a. As the amount deposited increases, the envelope becomes narrower, with FWHM values decreasing from 2.48 to 1.81 eV, indicating the noticeable presence of Cu^0. As already mentioned, this latter peak was arbitrarily placed at 931 eV.
The relative concentration of carbide present at the interface was determined from the ratio between the total amount of copper and the concentration of C-Cu measured from the C1s spectra. This ratio remained essentially unchanged,

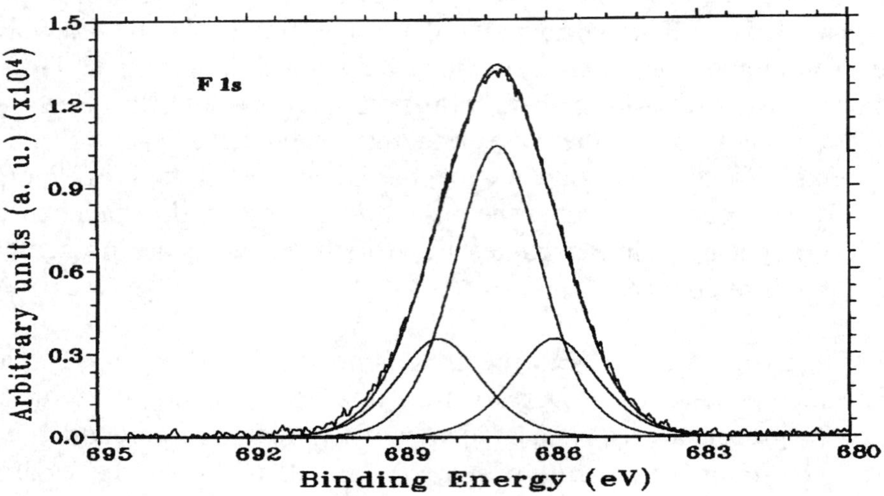

(a)

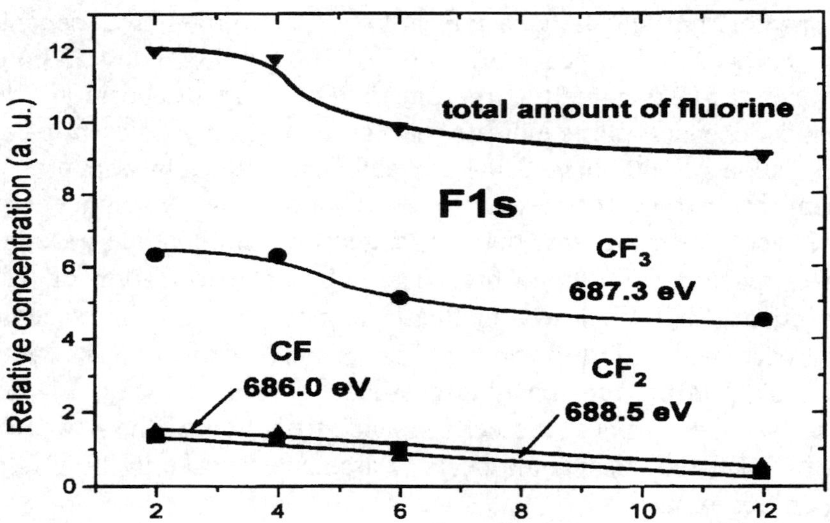

(b)

Figure 3. a) F1s spectrum for 2 seconds of vapor deposition; b) Evolution of the F1s peaks, as well as the total change of fluorine on vapor deposition.

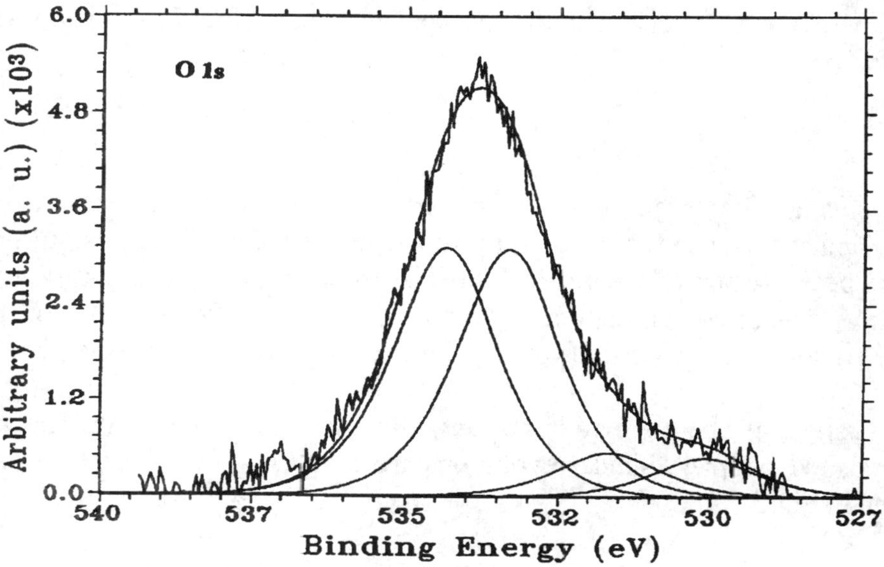

(a)

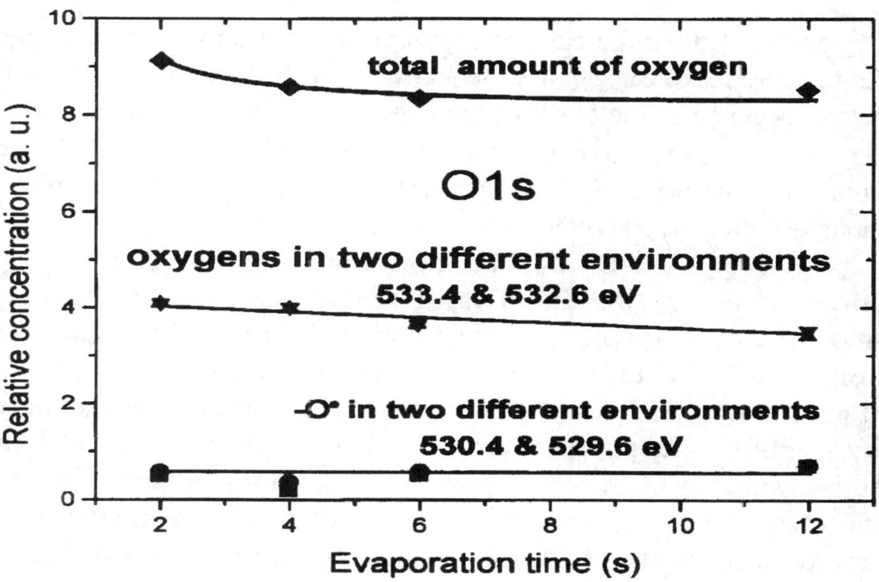

(b)

Figure 4. a) O1s spectrum for 6 seconds of vapor deposition; b) Evolution of the O1s peaks, as well as the total change of oxygen on vapor deposition.

showing that about 50% of the deposited copper remained in the metallic state. The evolution of the Cu^{I} peak and the carbide peak from the C1s spectra are shown in Fig. 5b.

Sputtering

In general, the spectra of sputtered samples show characteristics qualitatively similar to those found for vapor deposited samples. The positions of the separated peaks remained identical to those in the case of vapor deposition to within ±0.1 eV. Nevertheless, there are several differences in the evolution of the peaks. Further, some new peaks indicate the formation of additional species.

1) C1s spectra There are no important differences compared to the spectra of vapor deposited samples. Similar evolutions are found for the C-CF, C-C and C-Cu, as seen in Fig.6.

2) F1s spectra As shown in Fig. 7a, the spectrum indicates a new peak at a binding energy of 682.5 eV. Its evolution, presented in Fig. 7b, indicates it increases with the amount of coverage. It is attributed to F^- ion[15]. The other peaks are the ones normally found in pure Teflon AF1600 in a constant 1:1:3 ratio.

3) O1s spectra Unlike the case of evaporation, where the loss of oxygen was negligible, the amount of oxygen lost on sputtering was about 30%. The total amount of oxygen declined at the onset of the deposition before stabilizing after 4 minutes. The loss of bonded oxygen from the continually forming $C\text{-}O^{\bullet}$ free radicals accounts for this behavior. An equilibrium between these two oxygen species is established after several minutes, as seen in Fig. 8.

4) $Cu2p_{3/2}$ spectra The $Cu2p_{3/2}$ spectrum, which is presented in Fig. 9a, differs qualitatively from the spectrum of vapor deposited copper sample in Fig. 5a. The position of the major peak during sputtering was established, relative to the fluorine peak at 687.3 eV, to be at 932.0 eV; this indicates that there is no Cu^{0} left unreacted, nor is any found visually. A new structure develops at the higher binding energy side of the $Cu2p_{3/2}$ spectrum. Two additional peaks, at 934.3 and 936.0 eV, are necessary to fit the envelope. These positions suggest a second, more electropositive, Cu^{I} environment and the presence of Cu^{II}, respectively. In addition, the presence of a shake-up (not shown) is clear proof for the existence of the Cu^{II} state. Figure 9b shows the correlation between the concentration of elements for different species after the sensitivity and probe depth corrections. Thus, the continuous lines represent the evolution of the three copper peaks shown in Fig. 9a. The continuous thicker line represents the sum of fluoride peaks ($Cu^{I}F$ (934.3) eV and $Cu^{II}F_2$ (936.0 eV) and it is easily seen to correlate very well with the evolution of the fluoride peak F^- (682.5 eV), represented as a dashed line.

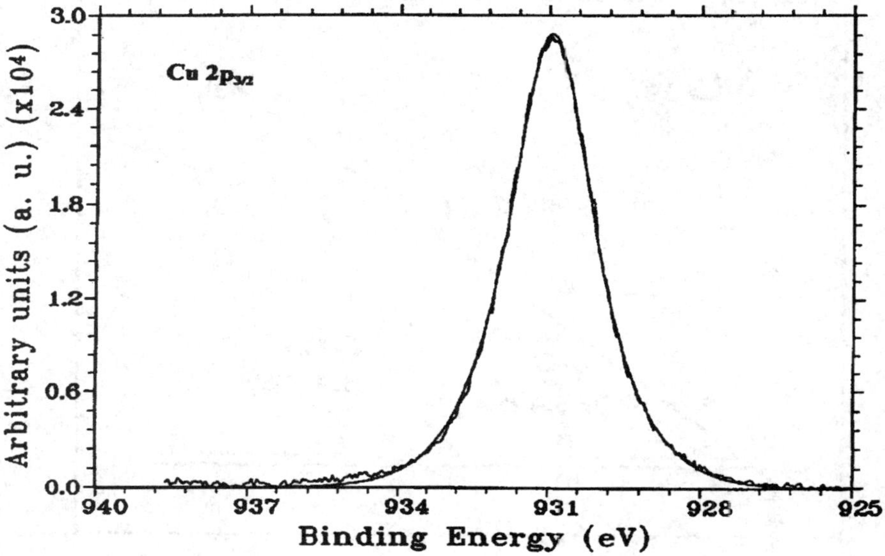

(a)

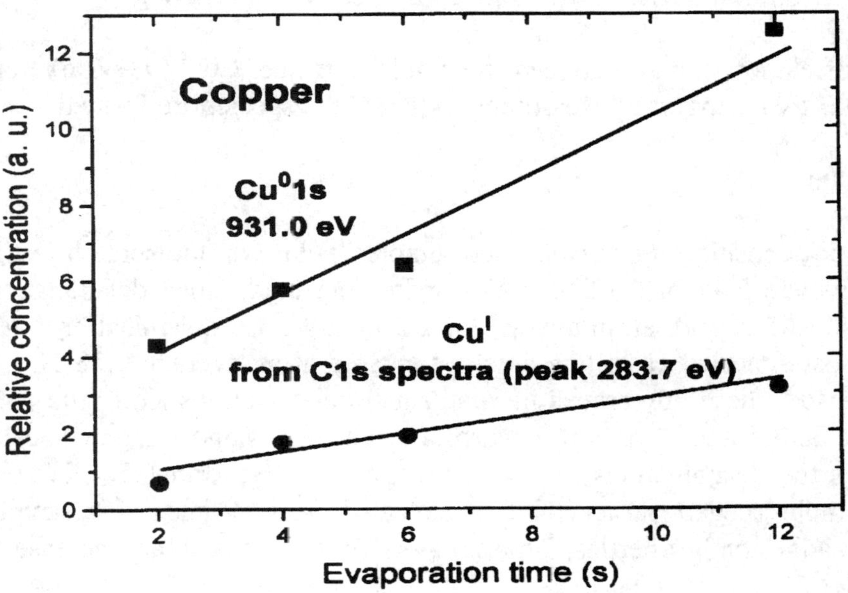

(b)

Figure 5 a) $Cu2p_{3/2}$ spectra for 6 seconds of vapor deposition; b) Evolution of Cu^0 and Cu^I (estimated from peak 3 of the C1s spectrum, assuming CuC formation).

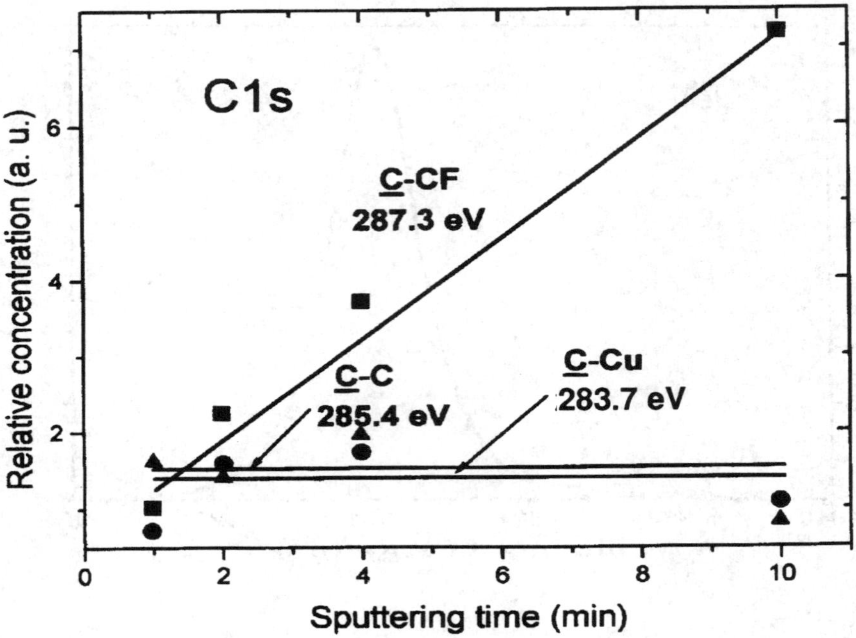

Figure 6. Evolution of the newly formed C1s peaks on sputtering.

The carbide correlation found between the continuous line, CuC (932.0 eV) and line, CuC (287.3 eV), supports the assumption that Cu_nC species are formed.

Adhesion Results

a) Vapor deposition For unannealed samples, failure during scratch testing began at an average load of 0.12 N. With increasing load, small delaminations were evident at 1.65 N and, at an average of 2.2 N, larger Cu delaminations were seen. In the post-annealed case, tiny cracks appeared at an average 0.28 N. As the force increased, these tiny cracks became small delaminations along the path. It appears that failure under tension, occurring at an early stage for nonannealed samples, delays the delaminations to higher force values. Nevertheless, failure in unannealed samples began earlier than in annealed ones. While both samples showed weak adhesion properties, annealing to 200^0C caused an increase in adhesion.

b) Sputtering For unannealed samples, failure manifested itself as the sudden onset of delaminations at a load of 0.56 N. Annealing at 200^0C caused some cracking of the copper coating without delamination, suggesting a marked

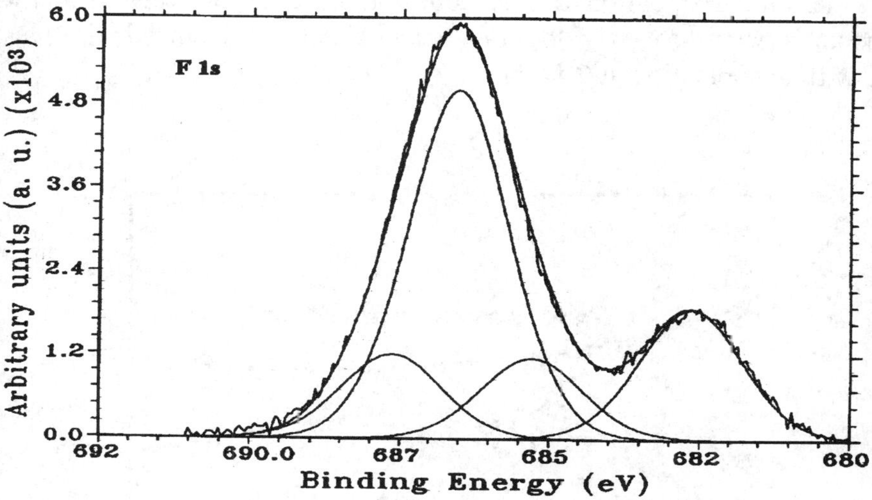

(a)

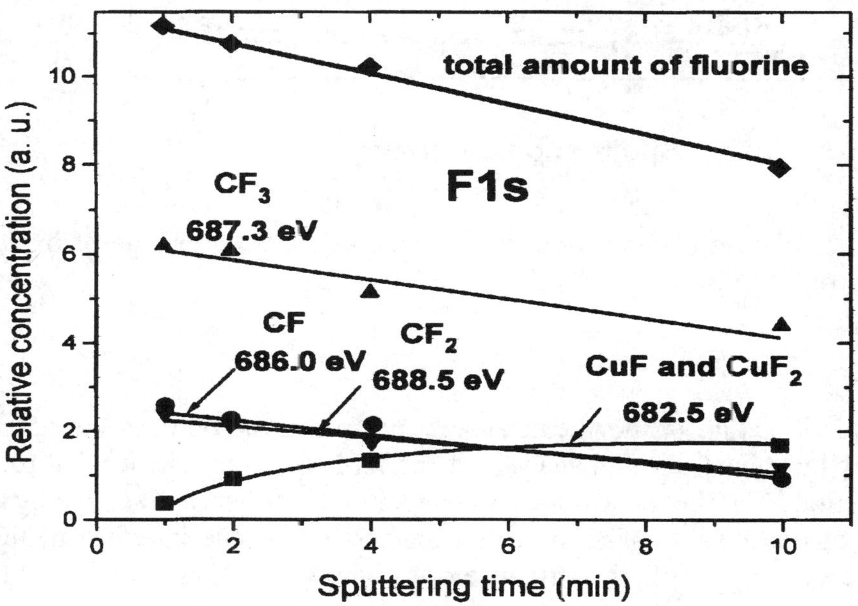

(b)

Figure 7 a) F1s spectrum for 10 minutes of sputtering; b) Evolution of the F1s peaks, as well as the total change of fluorine on sputtering.

increase in adhesion. Evaluation of the adhesion in areas between the cracks showed substantially improved adhesion values: failure occurred as plastic deformation, with delamination at 2.04 N.

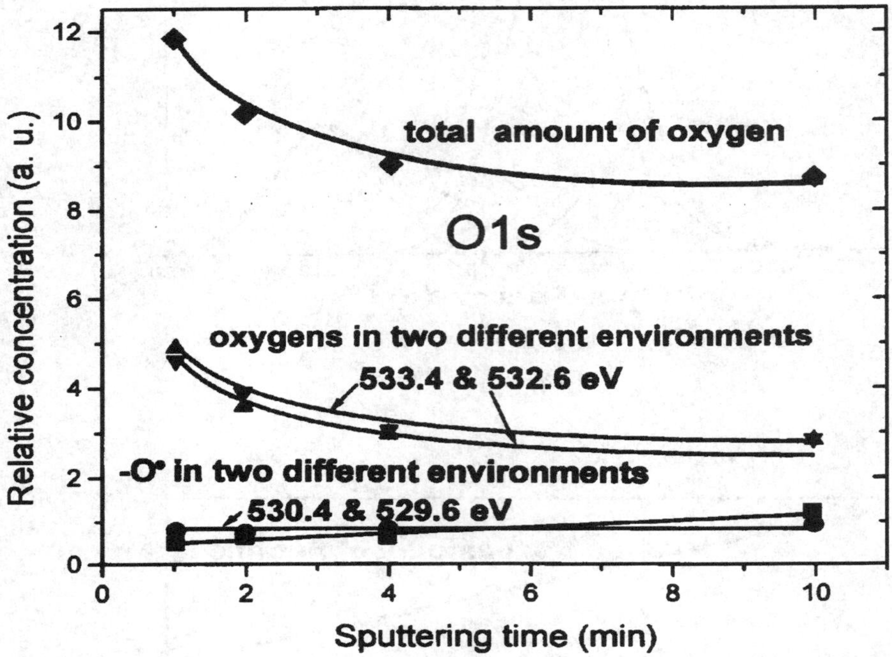

Figure 8. Evolution of the O1s peaks, as well as the total change of oxygen on sputtering.

DISCUSSION

An overall picture of the chemical reactions provoked by evaporation and sputtering may by gained from an analysis of the XPS spectra. The driving force for such reactions on the deposition of evaporated copper is the energy of condensation of the metal[16]. The increased temperature at the interface induces degradation reactions in the initial structure of Teflon AF1600, as revealed by the XPS spectra; these processes occur only on initial deposition. Degradation is due to the uniform loss of fluorine, as revealed in Fig. 3b. The copper bonds uniquely to carbon, forming carbide throughout the deposition, although approximately half of the depositing copper atoms remain in the metallic state in the thickness range used. There is no important oxygen loss, although modifications of the pure Teflon AF1600 O1s spectrum indicate single bond breaking of the oxygen of the dioxole rings, to produce free radicals. No evidence of C-O-Cu bond formation was found in this analysis.

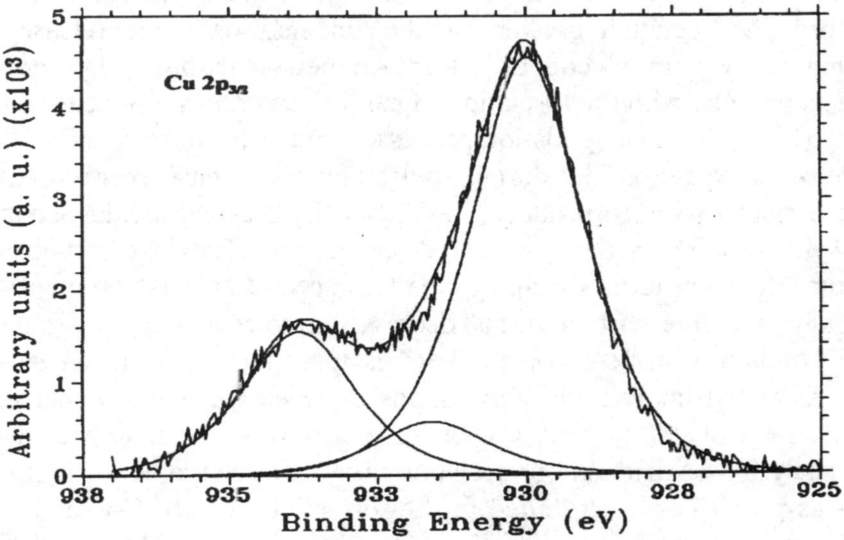

(a)

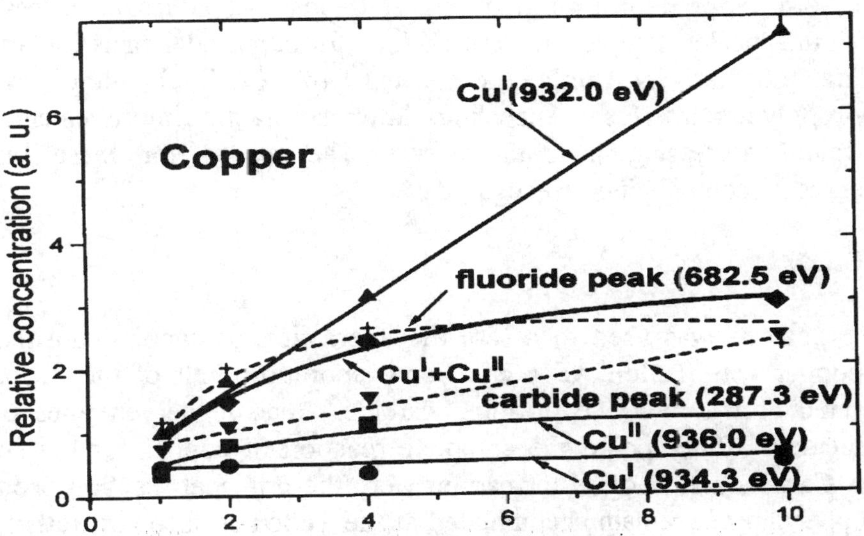

(b)

Figure 9. a) $Cu2p_{3/2}$ spectrum for 1 minute of sputtering; b) Correlation between concentration of elements for different species. Carbide (Cu-C) and fluoride ($Cu^{2+}F^-_2$ and Cu^+F^-) formation correlations support the assumption of Cu_nC formation.

In the case of sputtering, the kinetic energy of the copper atoms, which is at least an order of magnitude greater than the condensation energy released by the condensation of evaporated copper[17], leads to new reactions. The depositing copper reacts entirely, with the formation of carbide and fluoride species, while the carbon spectrum indicates degradation processes similar to those found in the case of evaporation. Oxygen is lost during sputtering: the intense energy causes not only single bond breaking to produce -C-O•, but the breaking of the second C-O bond, as well, resulting in the loss of oxygen atoms. The loss of fluorine and oxygen atoms does not radically change the C1s spectra because both species are lost uniformly: the five carbon peaks decrease uniformly with the same ratio. While C-F bonds are broken, not all the F is lost: some is retained as copper fluorides. As shown in Fig. 9b, correlations between Cu and F concentrations reveal a consistent overall picture of chemical reactions at the interface. No such attempt at analysis was made in our previous study[7], where the new feature in C1s spectrum was provisionally attributed to the formation of both C-C and Cu-O-C bonds when a very small amount of Cu was deposited. The present, more complete and accurate study, using a range of Cu thicknesses, gives no indication of Cu-O-C formation. Rather, as we show for the first time, copper reacts exclusively with carbon during both evaporation and sputtering.

For both vapor deposition and sputter deposition, adhesion is less than desired. For unannealed samples, the critical load for copper delamination appears to correlate qualitatively with the extent of Cu-C bonding at the copper/fluoropolymer interface. Annealing shows substantial improvement, with adhesion values increasing in both cases. The reason for these marked improvement is presently being investigated.

CONCLUSIONS

XPS analysis was used to assess the interaction of vapor deposited and sputtered copper with Teflon AF1600. For evaporation, half of the depositing copper reacted at the interface, forming carbide. The higher energies of the sputtered incident atoms led to the complete reaction of copper for the sputter deposition. Carbide and fluoride formations were the principal reaction products. Reacted copper did not remain accumulated at the Teflon surface but, rather, was uniformly distributed throughout the observation depth. Metal adhesion was stronger in the case of sputtering, increasing on annealing at 200^0C.

ACKNOWLEDGMENTS

This work was supported by the Natural Sciences and Engineering Research Council of Canada and by the Fonds pour la Formation de Chercheurs et l'Aide à la Recherche du Québec.

REFERENCES

1. S. P. Murarka, R. T. Gutmann, A. E. Kaloyeros and W. A. Lanford, Thin Solid Films, **236**,275 (1993).
2. E. Sacher, Prog. Surface Sci., **47**, 273 (1994).
3. P. Singer, Semiconductor Intl., 52 (Nov. 1994).
4. DuPont Specialty Polymers Technical Information Bulletins H-16577-1 and H-52454.
5. M. K. Shi, B. Lamontagne, A. Selmani and L. Martinu, J. Vac. Sci. Technol., **A12**, 44 (1994).
6. M. K. Shi, B. Lamontagne, L. Martinu, E. Sacher, M. R. Wertheimer and A. Yelon, J. Vac. Sci. Technol., **A12**, 804 (1994).
7. M. K. Shi, J. E. Klemberg-Sapieha, G. Czeremuszkin, L. Martinu and E. Sacher, "Proc. International Symposium of Plasma Chemistry-12," University of Minnesota Press, Minneapolis, MN, 1995, p. 51.
8. D. Popovici, E. Sacher, L. Martinu and M. Meunier, J. Appl. Polym. Sci., accepted for publication.
9. J. R. Waldrop and R. W. Grant, J. Vac. Sci. Technol., **A1**,1553 (1983).
10. J. R. Waldrop, P. Kowalczyk and R. W Grant, J. Vac. Sci. Technol., **21**,607 (1982).
11. C. Julia-Schmutz and H. E. Hintermann, Surface Coatings Technol., **48**, 1 (1991); P. Leroux, A. Raveh, J. E. Klemberg-Sapieha and L. Martinu, "Proc. 1993 Ann. Tech. Conf.", Society of Vacuum Coaters, Albuquerque, NM, p. 472.
12. E. Sacher and J. E. Klemberg-Sapieha, J. Vac. Sci. Technol.,**A15**, 2143 (1997).
13. Vacuum Generators Scientific Technical Bulletin (1990).
14. R. Ke, R. T. Haasch, N. Finnegan, L. E. Dottl, R. C. Alkire and H. H. Farell, J. Vac. Sci. Technol. **A14**, 80 (1996).
15. C. D. Wagner, W. M. Riggs, L. F. Davis, J. F. Moulder and G. E. Muilenberg, "Handbook of X-Ray Photoelectron Spectroscopy", p. 44, Perkin-Elmer Corporation, Eden Prairie, MN, 1979.
16. E Honig, RCA Rev., **18**, 195 (1957).
17. C. I. Maissel and R. Glang, "The Handbook of Thin Film Technology", p. 3.23, McGraw-Hill, New York, 1970.

Metallized Plastics 5&6: Fundamental and Applied Aspects, pp. 255-268.
K.L. Mittal (Ed.)

BULK AND INTERFACE DEGRADATION PROCESSES IN POLYMER ELECTROLUMINESCENT DEVICES

Brian H. Cumpston and Klavs F. Jensen

Department of Chemical Engineering
Massachusetts Institute of Technology
77 Massachusetts Avenue, Cambridge, MA, 02139-4307

Significant improvements in efficiencies and brightness of polymer-based electroluminescent (EL) devices have been achieved in the few years since the discovery that conjugated polymers could be used in such devices. Short operating lifetime, however, remains a key problem for the technology. Electrode-polymer interactions, oxidation during polymer processing, and photo-oxidation have been shown to adversely impact photoluminescence efficiency and device lifetime. We present *in situ* Fourier transform infrared reflection absorption spectroscopy (FT-IRRAS) and *ex situ* X-ray photoelectron spectroscopy (XPS) investigations of some of these chemical degradation processes.

INTRODUCTION

Recently, there has been increasing interest in applications of polymeric materials in electro-optic devices. Electroluminescent polymers constitute a particularly promising application. Over the course of the last five years, several groups around the world have demonstrated the great flexibility of these materials in producing light-emitting devices that are easy to assemble, are relatively efficient, and produce light across the full visible spectrum.[1-7] Many envision these polymers revolutionizing the field of light-emitting diodes (LEDs), which is currently dominated by inorganic semiconductors. Polymers offer the opportunity to produce lightweight, large area, full color display devices that can be built on flexible substrates that roll up when not in use.[8] Significant progress has been made to synthesize polymers that are more efficient, easier to process, and are compatible with flexible polymeric substrates.[9]

Two important classes of polymers for LED applications are the derivatives of poly(*p*-phenylene vinylene) (PPV) and polythiophene (PT). Sidegroups are added to these conjugated polymer backbones to increase solubility and improve the ability to process the polymers by commercial dip- and spin-coating techniques. The addition of sidegroups further allows engineering of the band gap, that is, tuning the emission color of the polymer. Typically, light-emitting devices are constructed in a layered structure with the polymer layer sandwiched between two electrodes - a low workfunction metal such as Ca for electron injection and a transparent film of indium tin oxide (ITO) for hole injection (Fig. 1). DC drive voltages are typically less than 10 volts with current densities less that 1 mA/mm^2. Most devices operate with a luminance of 100-500 cd/m^2, one to five times brighter than a typical color television set. Luminescence in these systems is achieved through the radiative recombination of electron-hole pairs that exist as excitons.[6] Since the radiative excitons are stabilized by Coulombic attraction, the emitted light has an energy typically 0.1-0.2 eV less than the bandgap of the polymer.

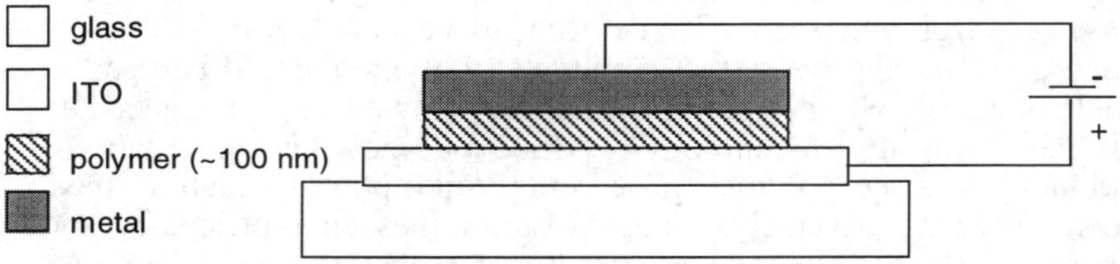

Figure 1. Typical sandwich structure for polymer-based LEDs.

Having demonstrated light emission over the entire visible spectrum, attention has begun to focus on improving the stability of polymer light-emitting devices. In the past, the interface between the polymer and the metal electrode that is used to pass electrical current through the active polymer layer has been problematic because highly reactive metals such as Mg or Ca had to be used to increase the efficiency of the device. Interfacial interactions apparently lead to the formation of 'dark spot' defects, areas of the LED that become non-emissive and increase the overall impedance of the device.[10] It remains unclear whether non-uniformities in the polymer film result in localized hot spots, which lead to short circuits, or chemical interactions produce non-emissive areas, perhaps by impeding charge injection into the polymer layer. Researchers have addressed the interface problem by using less reactive metals[11] in conjunction with charge-transporting layers,[12] and through the development of electrochemical cells[7] that operate under lower applied voltages, use less reactive metal electrodes, and do not exhibit dark spot defect formation. Relatively little work has considered the degradation of the active polymer layer itself. This paper reviews recent work that has addressed the effects of polymer oxidation in electroluminescent devices, the mechanisms for oxidation of the polymers and possible avenues to improve their stability.

EXPERIMENTAL

Idealized structures of the PPV derivatives studied are shown in Figure 2. MEH-PPV, MEH-PPV-CN, and PPV-DP were supplied by Uniax Corporation and A-PPV[13] was supplied by Philips Research Labs. Regioregular poly(3-octyl thiophene) (P3OT) was provided by Richard McCullough of Carnegie Mellon University. Polymer solutions were spin coated in a nitrogen drybox onto Au (100 nm thick) and Al (300 nm thick) metallized Si wafers. A 100 nm Ti interlayer was used to promote the adhesion of Au to the Si wafer. All polymers had a film thickness of 30-40 nm as measured by profilometry.

Infrared reflection absorption spectroscopy (IRRAS) was used to probe both the bulk polymer and the polymer/metal interface. This is a single reflection technique in which the incident IR beam strikes the sample at a grazing angle of about 10° from the surface. For films that are about 50 nm and less, the ratio of interfacial material to bulk material is sufficiently high that IRRAS can provide information from the polymer/metal interface.[14,15] It is important to note that the IRRAS technique is only sensitive to chemical groups with dipoles perpendicular to the surface (parallel to the plane of incidence) due to the selection rule at metal surfaces.[16]

X-ray photoelectron spectroscopy (XPS), which is highly surface sensitive due to the shallow escape depth for electrons, was used *ex situ* to characterize the top 5-10 nm of the polymer film.[17] XPS complements the IR results and aids in their interpretation. XPS spectra were referenced to the C1s peak at 285.0 eV and were fitted with 100% Gaussian peaks.

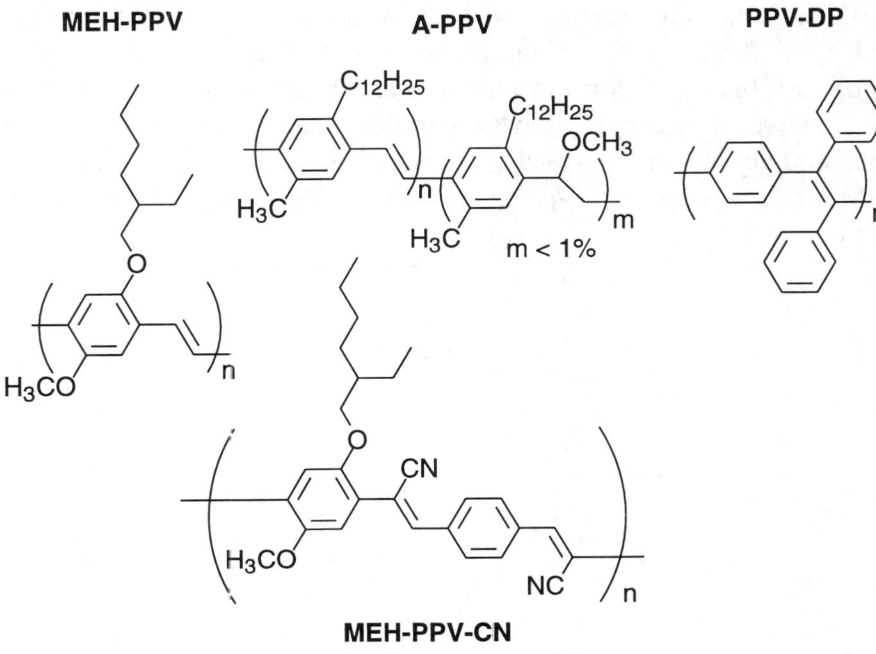

Figure 2. Idealized structures of PPV derivatives under study.

In situ photo-oxidation studies were performed using IRRAS. Films were UV irradiated using an unfiltered low pressure mercury lamp. The 254 nm line was the strongest with an irradiance of approximately 144 $\mu W/cm^2$. The second most intense line at436 nm was about 15 $\mu W/cm^2$.Visible irradiation was performed using an Ar^+ laser. Both the 514 nm and 457 nm lines were used, each with an irradiance of 10 mW/cm^2. During irradiation, samples were contained in a stainless steel cell under flowing dry air.

RESULTS

Photo-Oxidation of PPV Derivatives

As-coated films of MEH-PPV were identical on both Au and Al substrates as shown by IRRAS and XPS. The major IR vibrational bands for MEH-PPV are shown in Figure 3 (Au substrate): CH_2 and CH_3 stretching (2850-2950 cm^{-1}), phenyl ring modes (1597, 1506, and 1415 cm^{-1}), asymmetric and symmetric CH_2 and CH_3 deformations (1465 and 1380 cm^{-1}), out-of-plane CH wag (860 cm^{-1}), ether C-O-C stretching (1350, 1255, 1208 and 1040 cm^{-1}), and *trans* double bond CH wagging (969 cm^{-1}). Some as-coated samples show impurity oxidation in the form of a carbonyl group at 1725 cm^{-1}. Although it is unclear which portion of the molecule is oxidized, the position of the carbonyl band indicates the presence of an aliphatic aldehyde. The X-ray photoelectron C1s spectra of as-coated MEH-PPV on Au and Al show peaks at 285.0 eV for aliphatic and aromatic carbons and at 286.3 eV for carbon atoms attached to ether oxygens.

When films were exposed to dry air in the absence of light, no change was observed by IRRAS. Also, when films were exposed to UV irradiation in a vacuum of less than 10^{-6} torr no change was observed after up to 48 hours of exposure. However, when films were simultaneously exposed to UV or visible irradiation and dry air, the polymer oxidized. The evolution of the spectrum for a film of MEH-PPV cast onto Au and photo-oxidized by 514 nm irradiation is shown in Figure 3.

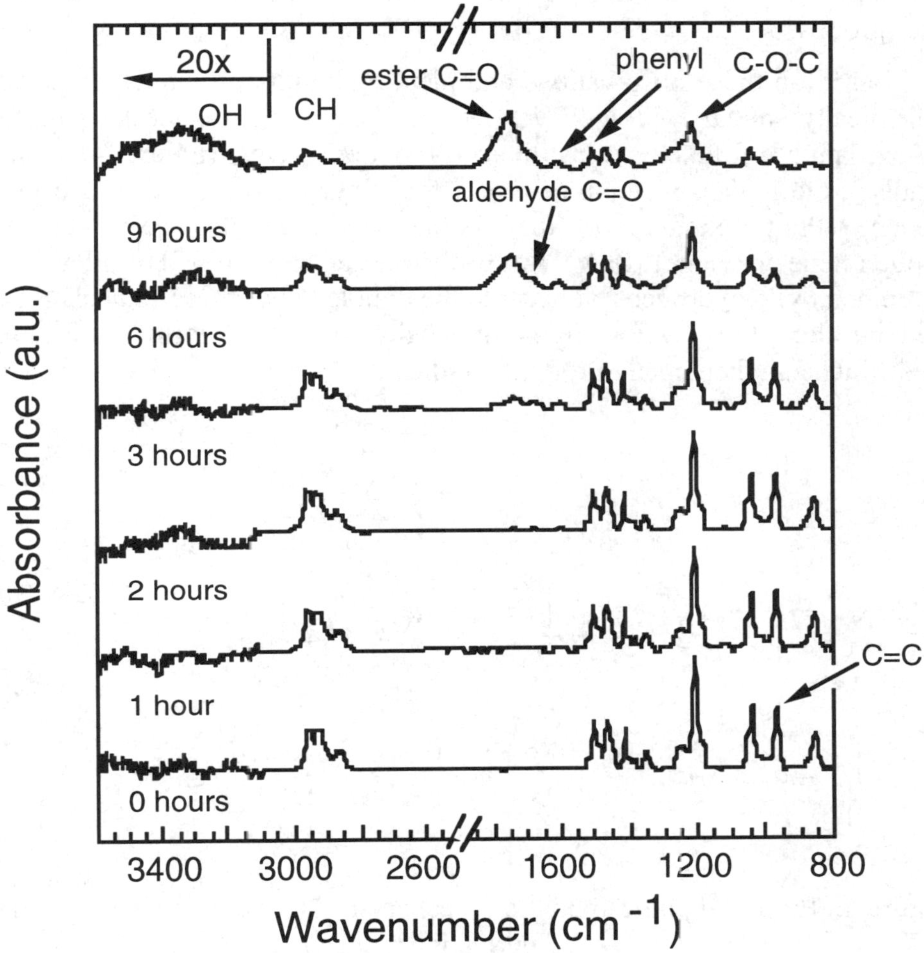

Figure 3. IRRAS spectra after photo-oxidation of MEH-PPV by 514 nm irradiation: (a) as coated and after (b) 1h, (c) 2h, (d) 3h, (e) 6h, and (f) 9h.

The spectra are taken at various times during the exposure to dry air and 514 nm irradiation: as-coated, after 1h, 2h, 3h, 6h, and 9h. The most obvious change is the growth of a broad carbonyl peak centered around 1740 cm^{-1} with a smaller shoulder at 1682 cm^{-1}. The peak at 969 cm^{-1} that is associated with the vinyl double bond in the polymer backbone is attenuated but not fully destroyed during irradiation indicating that the conjugation length of the polymer is shortened. Further evidence for chain scission at the vinyl double bond is the growth of the phenyl ring band at 1597 cm^{-1}, which is normally IR inactive in symmetric molecules such as MEH-PPV. The position of the carbonyl peak at 1740 cm^{-1} is indicative of an ester, and peaks due to the ether linkage around 1200 cm^{-1} are attenuated but remain relatively strong after 9 h of exposure in support of ester formation. There is a weak, broad band centered around 3400 cm^{-1} which is due to the formation of hydroxyl groups. These appear in very low concentration and may further react to form an aromatic aldehyde; this is consistent with the observation of the shoulder at 1682 cm^{-1}. XPS results also indicate the formation of ester groups.

Based on these observations and identical results previously reported for the chemically similar BCHA-PPV,[18] we propose a general mechanism for the photo-oxidation of alkoxy substituted PPV derivatives (Figure 4). In this mechanism, the initial oxidation occurs at the vinyl double bond in the polymer backbone. Photo-oxidation of the double bond has also been observed in poly(phenylene vinylene) (PPV), the backbone segment of BCHA-PPV.[19] Since there are no allylic hydrogens adjacent to the double bond, these radicals are likely formed by the [2+2] cycloaddition of singlet oxygen to form a 1,2-dioxetane species which can then open to form a biradical.

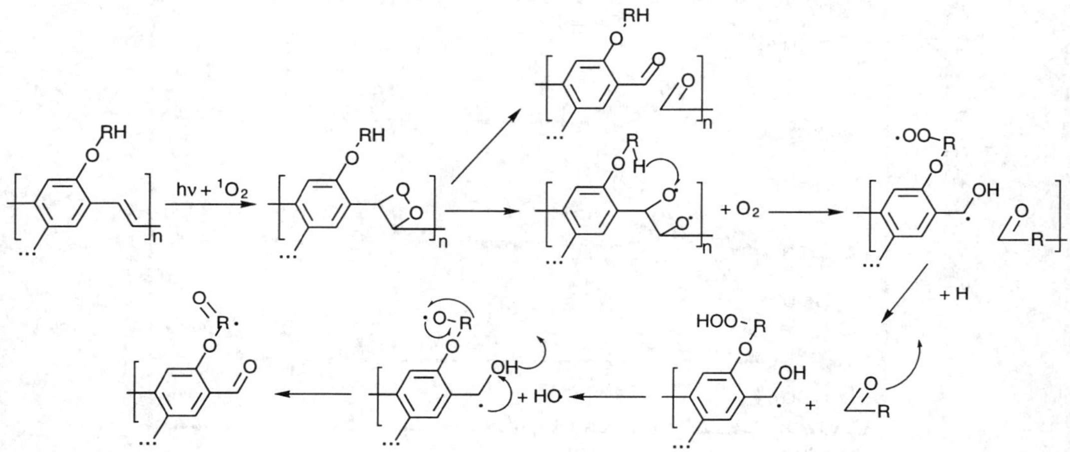

Figure 4. Proposed photo-oxidation mechanism for alkoxy substituted PPV derivatives.

The singlet exciton of MEH-PPV has an energy of about 2.2 eV and a lifetime of 145 ps,[20] too short-lived to transfer energy to molecular oxygen. However, the MEH-PPV triplet level should be sufficiently energetic (the BCHA-PPV triplet lies at 1.49 eV[21]) and long-lived to produce $^1\Delta_g$ O_2 which has an energy of 0.98 eV. The BCHA-PPV triplet exciton has been shown to generate singlet oxygen with unit efficiency in solution and with a maximum efficiency of 0.02 in the solid state.[21] The energy and lifetime of the MEH-PPV triplet exciton have not been reported. Once the biradical has formed it may abstract hydrogens from the aliphatic sidechains to initiate the formation of hydroperoxides which then decompose to give ester and hydroxyl groups. The abstraction of a hydrogen from the carbon adjacent to the ether oxygen is favored because the resulting radical is stabilized by the oxygen lone pairs. The mechanism also indicates the reaction of hydroxyl groups to form aldehydes, but it should be pointed out that the formation of a biradical species is in direct competition with the cleavage of the dioxetane to directly form two aromatic aldehydes.

To confirm the role of alkoxy oxygen in the formation of esters, we performed irradiation experiments using the alkyl substituted polymer A-PPV (see Fig. 2). IRRAS spectra for the photo-oxidation of A-PPV on a Au substrate using 514 nm light show that substantial oxidation does occur, but the predominant carbonyl feature forms at 1685 cm^{-1} indicating the presence of an aromatic aldehyde. There is no evidence of ester formation in the case of the alkyl substituted PPV. This result implies that the ester forms at the alkoxy group in MEH-PPV and is not formed via oxidation of aldehyde or ketone groups.

The photo-oxidation of MEH-PPV-CN and PPV-DP was also studied to assess the effect of vinyl substitutions on the reactivity of PPV derivatives. The electron withdrawing and/or steric effect of the substitutions was very pronounced. For photo-oxidation experiments performed on Al substrates using 514 nm light and a total exposure of approximately 380 J, we find that both of these derivatives are much more stable relative to MEH-PPV.

Computational Chemistry Studies of PPV Reactivity Towards Oxygen

All of the above results indicate that the vinyl double bond in these PPV derivatives is the most reactive segment in the chain and that the predominant mechanism for oxidation at this site is the [2+2] cycloaddition of singlet oxygen. In order to investigate the role of conjugation length and sidegroups on the stability of this bond, computational chemistry calculations were performed at the semi-empirical level using MOPAC. The Austin Model 1 (AM1) parameterization set was used to minimize the geometry of dimeric and oligomeric segments of the polymers. Mulliken overlap population analysis was used to determine electron densities around particular atoms and at specific bonds. These methods are not as rigorous as *ab initio* or density functional methods, but they are computationally inexpensive and should give reasonably good results for making relative comparisons.

Calculations on dimer, trimer, and tetramer segments of PPV indicate that conjugation length has a minimal effect on electron density at the vinyl double bond. The bond order is a way of viewing the electron density of the bond: the higher the bond order, the greater the density of electrons between the two atoms. The order of every vinyl bond in the dimer, trimer, and tetramer is 1.83. Although it seems from these results that conjugation length should have no effect on the reactivity of PPV, higher molecular weight oligomers must be investigated before this conclusion can be generalized.

As one might expect, substitution of electron withdrawing groups on the ring structure is not as effective as direct substitution on the double bond as a means of decreasing the reactivity of the molecule. The results of the calculations show that substitution of nitrile groups onto the vinylene segment reduces the bond order of the double bond, thus reducing its reactivity in agreement with the experimental results discussed above for MEH-PPV-CN. The substitution of two phenyl rings in a *trans* configuration across the double bond is successful in increasing the atomic charge of the two carbon atoms, but also increases the bond order relative to unsubstituted PPV. This may be related to the high degree of non-planarity in this structure caused by the bulky phenyl substitutions. The fact that PPV-DP performs so well when irradiated with visible light may be related to steric protection of the double bond from singlet oxygen.

Characterization of Poly(3-octyl thiophene) Photo-oxidation

Recognizing that the vinyl double bond and ether linkage present in MEH-PPV and other PPV derivatives may enhance photo-oxidation, poly(3-octyl thiophene) (P3OT) provides an interesting comparison because it does not contain these chemical groups (Figure 5). This polymer was also subjected to simultaneous exposure to dry air and UV irradiation, the same treatment as discussed above for the alkoxy PPV derivatives. After 72 hours of exposure, the polymer remains unchanged. XPS has also confirmed that P3OT does not oxidize as a result of this exposure. The behavior of P3OT is identical on both Au and Al. In addition, neither of these films exhibited degradation when exposed to visible irradiation at 457 or 514 nm. In all cases, no carbonyl groups form and there is no loss of signal intensity, particularly in the CH stretching region near 3000 cm^{-1}.

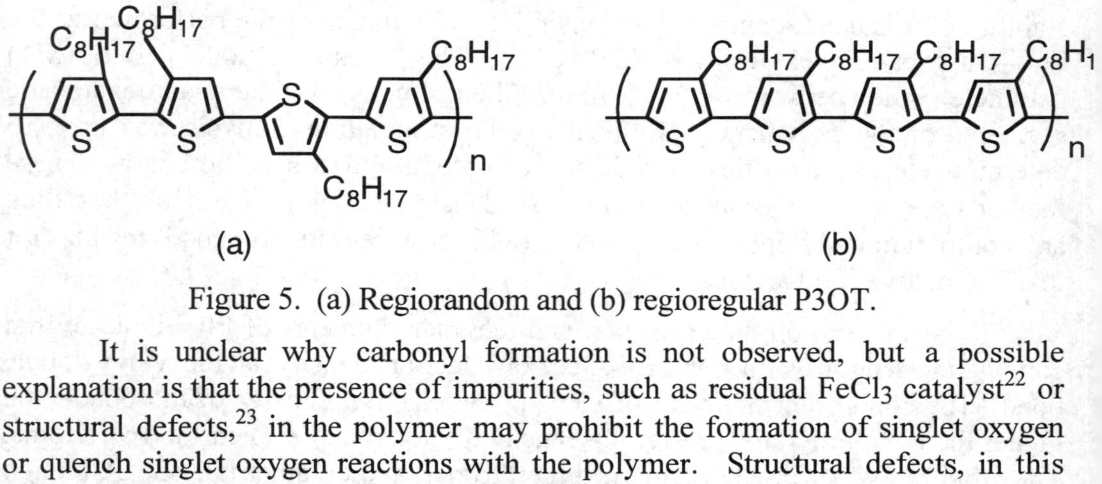

Figure 5. (a) Regiorandom and (b) regioregular P3OT.

It is unclear why carbonyl formation is not observed, but a possible explanation is that the presence of impurities, such as residual $FeCl_3$ catalyst[22] or structural defects,[23] in the polymer may prohibit the formation of singlet oxygen or quench singlet oxygen reactions with the polymer. Structural defects, in this case, refer to the presence of head-to-head and tail-to-tail linkages in the polymer backbone. The level of iron impurity in this batch of P3OT was determined by atomic absorption spectroscopy to be less than 200 ppm. This suggests that even low levels of impurity may quench reactive species in the polymer. These impurities may efficiently quench luminescence as well.

When highly purified and regioregular P3OT (99% head-to-tail structure) (Fig. 5b)[24] was irradiated in air under conditions of either UV or visible irradiation, films coated on Al wafers showed some degree of oxidation, while those coated on

Au wafers did not oxidize. A series of other metal substrates (Ag, Cr, Cu, and Ni) was studied in order to assess the role of the metal in the photo-oxidation of regioregular P3OT. The relative degradation of P3OT coated on these metals was quantified by the loss of CH stretching intensity as measured by IRRAS. This measurement correlates to either oxidation of the alkyl sidechains of P3OT and/or thinning of the film due to oxidation. These experiments showed that only the films coated on the Al substrate underwent substantial degradation.

Density functional theory calculations were performed to try to understand this effect. The DMol package of Biosym was used to optimize geometries for a single thiophene ring in the vicinity of a single metal atom. The local density approximation was used to characterize polythiophene/metal interactions. A double numeric basis set with polarization functions for all elements except hydrogen was employed. This model will not accurately predict the behavior of polymer films on a metal substrate, but may hint at the interactions that exist between the two materials. For all metals, it was found that two equilibrium geometries exist: one in which the metal atom is positioned above one of the α-carbons of the thiophene ring, and one where the metal atom is centered directly above the ring. In general the latter configuration, that is, with the metal atom centered above the ring, is only slightly energetically favorable. The structure of thiophene itself is very planar; however, when an Al atom is allowed to equilibrate above the thiophene ring, the sulfur atom is pushed down out of the plane by about 20°. Also the C_α-C_β bonds become longer than the C_β-C_β bond. That is, the bond length alternation pattern of the thiophene ring is reversed, resulting in a quinoid structure instead of the energetically favorable benzoid structure. The quinoid form is suspected to be more susceptible to singlet oxygen addition across the exposed double bond between the ring structures. The issue of metal/polythiophene interactions has been discussed in greater details by other authors who make similar conclusions.[25-27]

Bulk Degradation of MEH-PPV in LEDs

MEH-PPV based LEDs were fabricated as demonstrated in Fig. 1 on either KCl or ZnSe substrates to allow direct probing through the backside of the device by IR spectroscopy. The experimental apparatus is illustrated in Figure 6a. These devices were operated in constant voltage mode from 10-20 VDC. Devices were initially operated in a vacuum of less than 10^{-6} torr and emitted light without any observable change in the IR spectrum for 16 days. Over this time, current densities were increased from 1 to 165 mA/cm^2 to test the operational limits of the device, ultimately operating at greater than 70 mA/cm^2 for 120 hours.

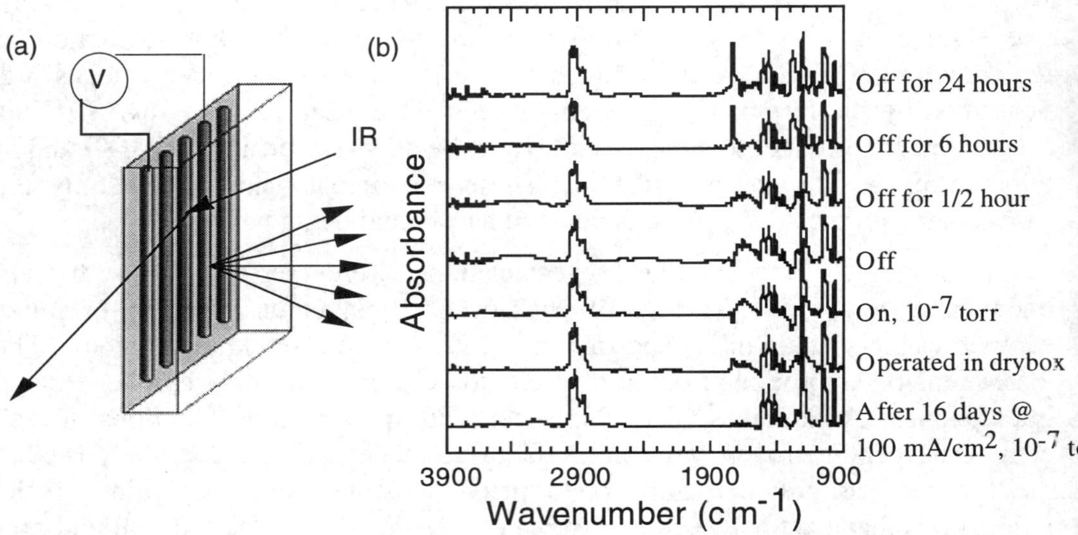

Figure 6. (a) Experimental apparatus for *in situ* LED characterization and (b) IRRAS results.

After 16 days of continuous operation in vacuum, the device was transferred under vacuum to a nitrogen drybox for visual inspection. At this time, there was no FTIR or visual evidence of polymer degradation and no dark spot defects were apparent on the metal electrode. The device was then operated for approximately two minutes in the drybox before returning it, under vacuum, to the spectrometer compartment. The first spectrum taken, prior to turning the LED on again, indicated the presence of carbonyl groups in the polymer film (Figure 6b). These carbonyl groups apparently formed during operation in the drybox which contains low parts per million levels of oxygen. Two carbonyl peaks are observed: an intense band at 1740 cm^{-1}, typical of an ester, and a smaller shoulder at 1685 cm^{-1}, indicative of an aromatic aldehyde. Attenuation of the *trans*-hydrogen wagging band at 965 cm^{-1} indicates that the conjugation length of the polymer is being shortened by scission of the vinyl double bond in the PPV backbone. These observations are consistent with changes previously reported for thermal and photo-oxidation of PPV and its derivatives.[18,21,28-32]

The IR spectra also reveal that while the device is operating, certain peaks become negative, indicating that IR emission is taking place from these groups. Emission peaks are present at 1740 cm^{-1}, 1290 cm^{-1}, and 1130 cm^{-1}; the position of these bands is typical for ester compounds.[33] Infrared emission from MEH-PPV LEDs is due to high temperatures caused by high current densities in the device. At steady state operating temperatures, IR absorption is swamped by the emission signal which reaches a magnitude that appears to be a function of the temperature of the device but not of the concentration of carbonyl groups at this level of oxidation. Infrared emission from ester compounds heated to 30 °C has been reported elsewhere.[34,35] When the LED was turned off, emission peaks became less negative, passed through zero intensity, and increased in a positive (absorption) sense for approximately 24 hours before reaching a saturation absorption value. This result indicates that radiative cooling of the device under high vacuum requires about one day before the temperature drops below a threshold value for the detection of IR radiation emitted from the LED. The device temperature will be significantly lower when not operated under high vacuum conditions because of conductive and convective cooling.

Visual inspection of the LED after failure revealed the presence of "dark spot" defects on the metal electrode. Optical microscopy revealed dark spots along with holes and apparent blisters on the metal surface. These defects were only observed in active regions of the device, not in areas of the device structure where ITO had been removed. This suggests that degradation of the electrode may be induced by areas of high current density localized around defects in the polymer film. These defects may be structural or may be created by electrochemical redox reactions or as a result of polymer oxidation which occurs in the vicinity of excited-state polymer segments. The origin of the defects is unknown, but it has been reported that high current density in these regions results in localized heating which leads to the eventual failure of the device.[36] Also, some of these failures may be due to electromigration of the metallic cathode.[37] Fluorescence microscopy images show that there is significant degradation of MEH-PPV around the sites of these defects. The blue-shift in photoluminescent emission is due to oxidation which results in shortened conjugation length along the MEH-PPV chain. Oxidation in these defect regions may be enhanced by higher current densities near these defects or by higher temperatures.

CONCLUSIONS

We have found that alkoxy derivatives of PPV undergo photo-induced degradation in the presence of light and oxygen. The mechanism for this process has been established in which excited state singlet oxygen is formed by photosensitization *via* the non-radiative triplet exciton species of the polymer. Singlet oxygen undergoes [2+2] cycloaddition across the electron-rich vinyl double bond to form a 1,2-dioxetane species. This dioxetane initiates free radical processes that ultimately lead to the formation of esters and aromatic aldehydes. Identical irradiation of the dialkyl substituted A-PPV was performed and only aromatic aldehydes were formed. This observation supports our conclusion that ether oxygen in the alkoxy sidegroup participates in the formation of the ester.

The addition of electron withdrawing groups to PPV-like backbones was shown to decrease the overall reactivity of the polymer. These observations were supported by AM1 semi-empirical calculations which showed that substitution of highly electronegative groups directly onto the vinyl double bond proves effective in improving the oxidative stability of these polymers. The substitution that had the largest effect on photo-oxidative stability was the strongly electron-withdrawing nitrile group. The computational results take into account only stability and do not consider ease of synthesis or efficiency of light emission.

The photo-oxidation of the polythiophene derivative P3OT has been shown to be very sensitive to both the chemical purity and structural regularity of the polymer as well as the nature of the metal substrate on which it is coated. Two effects seem to be at work. First, very pure polymer is more likely to oxidize because fewer defects and impurities can quench either the excited states of the polymer or singlet oxygen. Second, the nature of the metallic substrate influences the degree to which the polymer will photodegrade under irradiation. The interface chemistry remains to be fully delineated, but it is known that reactive metals such as Al can interact with the thiophene rings and cause a change in the bond-length alternation along the chain.[27] The resulting quinoid structure with double bond character between the rings could make the molecule more susceptible to 1,2-cycloaddition by singlet oxygen. The increased reactivity of P3OT coated onto Al versus Au could also be explained in terms of charge transfer from the metal to the polymer.[27]

In situ results on the degradation of MEH-PPV in operating LEDs show that photo-oxidation and oxidation during electroluminescent operation involve identical processes, that is, the reaction between excited-state singlet oxygen and MEH-PPV. This is to be expected since the generation of singlet oxygen occurs *via* photosensitization by the triplet excited state of the polymer. Photo-oxidation studies, which are easier to perform, can be expected to give meaningful results when used to screen electroluminescent polymers for their chemical stability. It is apparent that by carefully fabricating devices using highly purified materials and by operating in well-controlled atmospheres, devices made with currently known polymers can last for long times without excessive loss of performance. ITO does not appear to be a source of oxygen during device performance provided it is deposited under very clean conditions. Since only minute quantities of contaminants can initiate degradation of the active polymer though, it is important to search for polymers that have inherently greater chemical stability yet still have high luminescence efficiency.

ACKNOWLEDGMENTS

The authors would like to thank Ian Parker of Uniax Corporation for fabricating LEDs for *in situ* studies. Financial support was provided by the Office of Naval Research, Grant # N00014-95-1-0693. The MOPAC and DMol computational chemistry packages were provided by Biosym Corporation.

REFERENCES

1. J.H. Burroughes, D.D.C. Bradley, A.R. Brown, R.N. Marks, K. Mackay, R.H. Friend, P.L. Burns, and A.B. Holmes, *Nature* **347**, 539 (1990).
2. D.R. Baigent, N.C. Greenham, J. Grüner, R.N. Marks, R.H. Friend, S.C. Moratti, and A.B. Holmes, *Synth. Met.* **67**, 3 (1994).
3. D. Braun, A. Brown, E. Staring, and E.W. Meijer, *Synth. Met.* **65**, 85 (1994).
4. D. Braun and A.J. Heeger, *Appl. Phys. Lett.* **58**, 1982 (1991).
5. D. Braun, G. Gustafsson, D. McBranch, and A.J. Heeger, *J. Appl. Phys.* **72**, 564 (1992).
6. D.D.C. Bradley, *Synth. Met.* **54**, 401 (1993).
7. Q. Pei, G. Yu, C. Zhang, Y. Yang, and A.J. Heeger, *Science* **269**, 1086 (1995).
8. P. Yam, *Scientific American* **273** (1), 82 (1995).
9. G. Gustafsson, G.M. Treacy, Y. Cao, F. Klavetter, N. Colaneri, and A.J. Heeger, *Synth. Met.* **55-57**, 4123 (1993).
10. P.E. Burrows, V. Bulovic, S.R. Forrest, L.S. Sapochak, D.M. McCarty, and M.E. Thompson, *Appl. Phys. Lett.* **65**, 2922 (1994).
11. N.C. Greenham, S.C. Moratti, D.D.C. Bradley, R.H. Friend, and A.B. Holmes, *Nature* **365**, 628 (1993).
12. S. Aratani, C. Zhang, K. Pakbaz, S. Höger, F. Wudl, and A.J. Heeger, *J. Electron. Mater.* **22**, 745 (1993).
13. E.G.J. Staring, R.C.J.E. Demandt, D. Braun, G.L.J. Rikken, Y.A.R.R. Kessener, T.H.J. Venhuizen, H. Wynberg, W. ten Hoeve, and K.J. Spoelstra, *Adv. Mater.* **6**, 934 (1994).
14. J.P. Lu, P.V. Nagarkar, D. Volfson, F. Trusell, K.F. Jensen, and S.D. Senturia, *J. Vac. Sci. Technol. A* **in press**,
15. R.O. Carter, C.A. Gierczak, and R.A. Dickie, *Appl. Spectr.* **40**, 649 (1986).
16. R.G. Greenler, *J. Chem. Phys.* **44**, 310 (1966).
17. A. Dilks, in "Photon, Electron, and Ion Properties of Polymer Structure and Properties," D. W. Dwight, T. J. Fabish, and H. R. Thomas, Editors, p. 293, American Chemical Society, Washington, D.C. (1981).
18. B.H. Cumpston and K.F. Jensen, *Synth. Met.* **73**, 195 (1995).
19. M. Yan, L.J. Rothberg, F. Papadimitrakopoulos, M.E. Galvin, and T.M. Miller, *Phys. Rev. Lett.* **73**, 744 (1994).
20. G.R. Hayes, I.D.W. Samuel, and R.T. Phillips, *Phys. Rev. B* **52**, R11569 (1995).
21. R.D. Scurlock, B. Wang, P.R. Ogilby, J.R. Sheats, and R.L. Clough, *J. Am. Chem. Soc.* **117**, 10194 (1995).
22. M.S.A. Abdou and S. Holdcroft, *Macromolecules* **26**, 2954 (1993).
23. R.D. McCullough, R.D. Lowe, M. Jayaraman, and D.L. Anderson, *J. Org. Chem.* **58**, 904 (1993).
24. R.D. McCullough, S. Tristram-Nagle, S.P. Williams, R.D. Lowe, and M. Jayaraman, *J. Am. Chem. Soc.* **115**, 4910 (1993).
25. M. Boman, S. Stafstrom, and J.L. Brédas, *J. Chem. Phys.* **97**, 9144 (1992).

26. P. Dannetun, M. Boman, S. Stafström, W.R. Salaneck, R. Lazzaroni, C. Fredriksson, J.L. Brédas, R. Zamboni, and C. Taliani, *J. Chem. Phys.* **99**, 664 (1993).
27. V. Parenté, R. Lazzaroni, A. Selmani, and J.L. Brédas, *Synth. Met.* **67**, 147 (1994).
28. I. Murase, T. Ohnishi, T. Noguchi, and M. Hirooka, *Synth. Met.* **17**, 639 (1987).
29. K. Yoshino, T. Kuwabara, T. Iwasa, T. Kawai, and M. Onoda, *Jpn. J. Appl. Phys.* **29**, L1514 (1990).
30. F. Papadimitrakopoulos, K. Konstadinidis, T. Miller, R. Opila, E.A. Chandross, and M.E. Galvin, *Chem. Mater.* **6**, 1563 (1994).
31. F. Papadimitrakopoulos, M. Yan, L.J. Rothberg, H.E. Katz, E.A. Chandross, and M.E. Galvin, *Mol. Cryst. Liq. Cryst.* **256**, 663 (1994).
32. B.H. Cumpston, K.F. Jensen, F. Klavetter, E.G.J. Staring, and R.C.J.E. Demandt, *Mater. Res. Soc. Symp. Proc.* **413**, 35 (1995).
33. D. Lin-Vien, N.B. Colthup, W.G. Fateley, and J.G. Grasselli, "The Handbook of Infrared and Raman Characteristic Frequencies of Organic Molecules," Academic Press, Boston (1991).
34. Y. Terui and K. Hirokawa, *Vib. Spectr.* **6**, 309 (1994).
35. Y. Terui and K. Hirokawa, *Vib. Spectr.* **6**, 315 (1994).
36. J.C. Scott, J.H. Kaufman, P.J. Brock, R. DiPietro, J. Salem, and J.A. Goitia, *J. Appl. Phys.* **79**, 2745 (1996).
37. B.H. Cumpston and K.F. Jensen, *Appl. Phys. Lett.* **69**, 3941 (1996).

Part 3

SURFACE MODIFICATION AND ADHESION ASPECTS

Metallized Plastics 5&6: Fundamental and Applied Aspects, pp. 271-293.
K.L. Mittal (Ed.)

PLASMA PRETREATMENT OF POLYMER SURFACES - RELEVANCE TO ADHESION OF METALS

Jörg F. Friedrich, Wolfgang E. S. Unger, Andreas Lippitz, Ivaylo Koprinarov, Gerhard Kühn, Steffen Weidner, and Lydia Vogel

Bundesanstalt für Materialforschung und -prüfung
Unter den Eichen 87
D-12200 Berlin, Germany

The application of selected surface sensitive analytical methods and the measurements of molecular weights of polymers give some new insight into chemical processes which occur at a polymer surface during the interaction with an oxygen plasma. The chemical and morphological stabilities of polymer segments in the near-surface layer were investigated by XPS (X-ray Photoelectron Spectroscopy), NEXAFS (Near Edge X-ray Absorption Fine Structure) spectroscopy and SEIRA (Surface Enhanced Infrared Absorption spectroscopy). The results were correlated with those obtained with LB (Langmuir-Blodgett), SAM (Self-Assembled Monolayers) and oligomer model films. For thin polymer layers (30 to 500 nm) the changes in molecular weight distributions of some polymers were systematically investigated by SEC (Size Exclusion Chromatography), MALDI-TOF MS (Matrix-Assisted Laser Desorption Ionisation Time-Of-Flight Mass Spectrometry) and ThFFF (Thermal Field Flow Fractionation) for oxygen and helium plasma exposures. Atmospheric corona discharge and plasma jet treated polymer surfaces were included into these investigations. Generally, the polymer surfaces were relatively stable on exposures to an oxygen low-pressure plasma up to ca. 2 s. This is an important information for the establishment of maximum adhesion to metals or polymer coatings. At longer treatment times a fast destruction of the polymer structure, decrease in the molecular weight, and anisotropy were observed. Some important chemical oxidation pathways were identified, especially for PET. Polymers can be classified in groups of different degradation behaviours on exposure to an oxygen plasma.

INTRODUCTION

The plasma modification of polymer surfaces has been intensively investigated in the last four decades[1]. However, as early as in 1939 glow discharge modification of paper and polymer sheets was used to improve wetting with insulating oils in electrical condensers[2]. The first scientific work was published by Rossmann[3], who modified PE (polyethylene) in a low-pressure plasma. As a chemical tool for the deliberate functionalization with amino groups Hollahan et al. described the modification of PP (polypropylene) in a NH_3 plasma and proposed a reaction mechanism[4]. Now, this technique is well established and many of the complicated chemical and physical processes occurring during the interaction of plasma particles and radiation with the polymer surface and the near-surface bulk are known. The interpretation of some chemical processes leading to the incorporation of plasma gas atoms or groups and the formation of functional groups during the plasma exposure was possible by the application of XPS (X-ray Photoelectron Spectroscopy) and SIMS (Secondary Ion Mass Spectrometry) techniques by Clark et al.[5], Briggs et al.[6], Amouroux[7] and others[8,9] especially for polyolefins and PET (poly(ethylene terephthalate)) beginning in the 1970´s.

However, a detailed knowledge of the chemical reactions at the surface, the spatial rearrangement of segments or of the whole macromolecule and exact information on changes in molecular weight distribution (MWD) of polymers occurring during plasma exposure is lacking. The oxidation pathways of polymer surfaces in an oxygen plasma were commonly explained by means of photo oxidative-like reactions on the basis of XPS, SIMS or IR-ATR (Attenuated Total Reflection) data[8-14]. In 1966 the process of Crosslinking by Activated Species of INert Gases (CASING) was introduced by Hansen and Schonhorn[15]. Hudis found that the gel formation during exposure of polymers to hydrogen plasma was due to the action of intense plasma vacuum UV radiation (VUV)[16]. Some efforts were made to use the plasma-induced crosslinking as a polymer surface modification technique with applications in adhesion improvement and barrier formation. These investigations brought more insight into degradation reactions of PET[8]. In a series of papers we have shown by XPS, NEXAFS (Near Edge X-ray Absorption Fine Structure spectroscopy), SEIRA (Surface Enhanced Infrared Absorption spectroscopy) and adhesion tests that optimal surface modification can be obtained with 0.1 to 2 s short plasma pulses[17-24]. Longer treatment leads to a strong degradation of the polymer structure and orientation and often to a decrease in bondability[21,25]. It is assumed that at longer exposure times low-molecular weight degradation products cover the surface and form a Weak Boundary Layer (WBL) which is detrimental for good adhesion. Therefore, the effects of different plasma treatments on the molecular weight of polymers were also examined in this study.

In addition to the investigation of the plasma surface modification in low-pressure plasmas of glow discharges excited with microwaves (MW), radio-frequency (RF) or direct current (DC) power we have also applied the atmospheric corona discharge and an atmospheric plasma jet working with air or oxygen. Here our interest is focussed on the comparison of reaction products formed at the polymer surfaces in low-pressure and atmospheric-pressure discharges, the results in improving adhesion properties, the possible acceleration of processes by using atmospheric-pressure plasmas, and the handling of such a plasma-jet.

The intention of this work was to collect more information and details on processes at a polymer surface during plasma exposure to obtain a more general idea on chemical reaction pathways. So, we have analyzed the topmost atomic layer at polymer surfaces after complete oxidation in an oxygen plasma in order to answer the relevant questions: What is the optimum treatment time in terms of functionalization, preserving the molecular structure and improving the adhesion properties ? Which new adhesion-relevant species are formed ? What is the effect of a plasma treatment on the orientation of macromolecular segments at the surface ? How stable are aromatic and carbonyl groups under plasma exposure ? What is the influence of an addition of helium to the oxygen plasma ? How long does the modification in oxygen and helium plasmas reach into the polymer bulk ? What is the influence of oxygen and helium plasmas on molecular weight distributions ? Are there crosslinking reactions on oxygen plasma modification ? Which post-plasma reactions are possible ? Are C=C double bonds formed at the surface of PP? etc.

Recently many surface analytical methods have been described, which can be employed to obtain the chemical composition. We have used the XPS, NEXAFS[26-31], and SEIRA spectroscopy[17,18]. However, none of these methods provides any information on changes in the molecular masses and their distribution in polymers. Especially Matrix-Assisted Laser Desorption Ionisation Time-of-Flight Mass Spectrometry (MALDI-TOF MS)[32-34] and Thermal Field Flow Fractionation (ThFFF)[22,35] represent new powerful methods for the determination of molecular masses and functionalities of oligomers and polymers as well as of slightly-crosslinked polymers (gels). MALDI-TOF MS allows simultaneous determination of molecular masses and mass distributions of synthetic polymers up to a mass range (M_W) of *ca.* 10.000 Da and more (up to 500,000 Da depending on the type of spectrometer and polymer). Its principle is well-described[36-38], whereas the principle of ThFFF is fairly unknown[39-41]. It is a special type of chromatography without a solid stationary phase and utilizes the different hydrodynamic behaviours of monomers, oligomers, polymers and gels in a flow channel with differentially heated walls. As a standard method for measuring the molecular weight distribution, Size Exclusion Chromatography (SEC) was used.

EXPERIMENTAL

Sample Preparation

100 nm thick layers of poly(ethylene terephthalate) (PET), aromatic polycarbonate (PC), polystyrene (PS), poly(ethylene glycol) (PEG), poly(propylene glycol) (PPG), and poly(methyl methacrylate) (PMMA) were spin coated. Biaxially stretched foils of PET (DuPont, Mylar, 0.5 to 300 µm thick), PP (5 to 70 µm thick, Trespaphan, Hoechst), and PE (50 µm thick, Lupolen, BASF) were also used. Furthermore, for MALDI-TOF MS analysis PET oligomers were synthesized as models after the procedure described in ref. 33. The M_w was ca. 1,700 g/mol and the end groups were exclusively OH groups. No macrocycles were present.

For the SEC and ThFFF analyses PET, PC (poly(bisphenol-A-carbonate)) and PS films were prepared with thicknesses from 30 to 500 nm by spin coating or dipping into the polymer solutions.

Plasma Treatment

The polymers were treated in oxygen or helium low-pressure plasmas with MW, RF, and DC excitation[29,32]. The plasmatron consists of a cylindrical plasma reactor with a volume of ca. 50 dm^3. The sample was fixed on a grounded sample holder. The reactor was equipped with a MW source, coupled to a quartz window (applied power ca. 150 to 300 W) and an RF linear electrode (25 cm; applied power ca. 150 W). The exposure time was varied between 0.1 to 600 s, the gas pressure was 6 Pa and the gas flow ca. 5 dm^3/h. With DC and RF plasmas also considerably soft plasma conditions (low Wattage per Flow rate and Molecular Mass, W/FM) were used to minimise the destructive plasma attack, and, furthermore, to provide similar energy inputs into low- and atmospheric-pressure plasmas for correlation purposes.

Atmospheric-pressure plasmas were generated using a corona discharge in air or a spark jet discharge using either air or an oxygen:nitrogen = 1:1 mixture (for details see refs. 18, 21,42).

XPS Analysis

XPS analysis and O_2 DC plasma treatments were carried out in a VG ESCALAB 200X electron spectrometer (VG Scientific, East Grinstead, UK) equipped with a plasma reactor chamber separated by a gate valve. DC plasma treatments were carried out at 6 Pa pressure of 99.999% grade oxygen. During the treatment, the sample was 25 mm away from the luminous plasma region. Spectra were acquired using Mg Kα excitation (15 kV, 20 mA) in the CRR mode (Constant Retarding Ratio 40, i.e. at constant relative resolution) at a take-off angle of 20°. XP spectra analysis and quantification were performed by routines which are part of the SCIENTA ESCA-300 (SCIENTA AB, Uppsala, Sweden) and VG`s VGS 5250 data system, respectively (see also refs. 30,43-46).

NEXAFS Analysis

The polymers treated in the XPS preparation chamber were exposed to laboratory atmosphere during the transfer into the NEXAFS apparatus. NEXAFS spectroscopy was carried out on the HE-TGM2 monochromator beam line (33.12) at the synchrotron light source BESSY (Berlin, Germany). Spectra were acquired at the C_K and the O_K edges in the *partial electron yield* mode (PEY) at a retarding voltage of -150 V. The monochromator resolution at the C 1s edge was better than 0.8 eV. Raw spectra were divided by the monochromator transmission function which was obtained with a freshly sputtered Au sample. Alignment of the energy scale was achieved by using the C 1s $\rightarrow \pi^*$ resonance measured with a pyrolytic graphite sample (Advanced Ceramic Corp., Cleveland, USA) which was fixed at 285.4 eV. The absolute energy calibration of the monochromators gave a resolution no better than about ± 0.2 eV. NEXAFS spectra were recorded at angles of 20, 55, and 90°, measured between the surface plane of the sample and the direction vector of the incident, linearly polarized light beam. The changed degree of molecular orientation after plasma treatment was referenced to the unmodified self-assembled monolayer (as prepared = 100 % molecular orientation) or unmodified polymer surface (as received = 100 % molecular orientation) evaluated by the respective 20-90° difference spectra[17], see for example Fig. 7. The information depth was estimated to be ≈ 3 nm for the C_K edge signal and ≈ 5 nm for the O_K edge signals[44,48].

SEIRA Analysis

The principle of SEIRA as applied to monomolecular layers or polymer surfaces was described in refs. 49-51 and will only be briefly repeated here. SEIRA requires the existence of an Ag or Au island film (3 to 10 nm effective thickness) on the surface of an IR target or on the plasma modified polymer. The organic sample must be in contact with the surface of the elliptically shaped metal islands. The metal islands show diameters from 15 to 45 nm (measured by scanning electron microscopy) which are $\ll \lambda_{IR}$. The enhancement of vibration intensity is similar to that of Surface Enhanced Raman Scattering (SERS) and is due to surface plasmon polariton or collective electron resonance associated with the nature of the metal ellipsoids[50]. With these particles electrical field line density is strongest at the tip and is weakest at the flat side of ellipsoids. The integrated intensity (I_i) of a vibrational absorption band *i* depends on the magnitude and orientation of the transition moment $\mathbf{M_i}$ with respect to the electrical field vector $\mathbf{E}$ (eq. 1), where Θ_i is the angle between $\mathbf{M_i}$ and $\mathbf{E_i}$ as[51]:

$$I_i \sim |\mathbf{M_i} \cdot \mathbf{E}|^2 \sim \cos^2 \Theta_i \qquad (1)$$

The electrical field lines are aligned perpendicularly to the surface at every point; therefore, optical selection rules are valid[50]. The enhancement of the IR signals is in the range of 10 to 1000 compared to standard transmission IR[51].

All fatty acids used for SEIRA were prepared as Self-Assembled Monolayers (SAM) and were used as short-chain models for PE. The respective IR spectra did not show free carboxylic group vibrations (≈1700 cm^{-1}), only carboxylate group related signals ($Ag^{+}COO^{-}$, ν_s ≈ 1395 cm^{-1} and ≈ 1550 cm^{-1}) were detected. This is in contrast to LB layers where both carboxyl and carboxylate group signals were found in the spectra.

MALDI-TOF MS Analysis

Lower molecular-weight fractions or oligomers as models for polymers up to 10,000 Da were characterised by MALDI-TOF MS. To obtain information on the penetration depth of the plasma effects, spin coated polymer films with thicknesses from 30 to 500 nm were used. The MALDI III spectrometer from Kratos (Manchester; UK) works with a N_2 UV laser (λ=337 nm) in a linear or in the reflection mode. The spin coated films on the sample slide were plasma treated and subsequently impregnated with a solution of trihydroxyacetophenone or dihydroxybenzoic acid, the so called "matrix", which is necessary to ionize the macromolecules without fragmentation. The addition of traces of $AgNO_3$ increases the ion yield.

SEC Analysis

A commercial Waters 150C chromatograph (Manchester, UK) was used. Before analysis the plasma-treated spin coated films of thicknesses from 30 to 500 nm deposited on Si wafers or on aluminium foil were dissolved, for example, PET in chloroform or hexafluoroisopropanol and filtered to hinder the obstruction of SEC columns by gel particles. PET foils of 1.5 µm thickness were also plasma treated and analyzed.

ThFFF Analysis

Molecular weight distributions of PET could also be determined by ThFFF. This method allows to analyze slightly crosslinked polymer particles (gels) from some nanometers up to ca. 100 µm, resp. from 10^3 to 10^{18} Da, by separation in a solvent flow passing a flow channel. This separation technique utilises the different hydrodynamic properties of oligomers, polymer molecules, and gel particles. One side of the flow channel can be heated up to 80 °C. It is possible to analyze without filtering the polymer solution to remove gel particles; furthermore, also the size of gel particles as well as those of additives or mineral fillers can be measured by ThFFF and a multi-angle light scattering detector (MALLS). A ThFFF apparatus supplied by Wyatt Technology, Hamburg, including RI (refractive index), UV, and MALLS detectors was used in our experiments.

RESULTS

Modification of Polymer Surfaces

Time Dependence. To determine the chemical and structural changes at polymer surfaces in dependence on the time of exposure to an O_2 plasma, the introduction of oxygen onto the surface, the changes in the supermolecular structure and the correlation to adhesion properties were investigated. Starting with monomolecular SAM and LB layers of C_{18-22} fatty acids and trichlorosilanes on Si wafers (here, stearic acid and octadecyltrichlorosilane (OTS), see Experimental) as models for polymer surfaces and continuing with PET surfaces XPS and NEXAFS analyses were performed (Fig. 1). Additionally, the results of peel strength measurements of 100 nm thick thermally evaporated Al films on oxygen plasma treated PET (peel test measurements at DuPont, Luxembourg) also show a correlation between macroscopic and molecular properties.

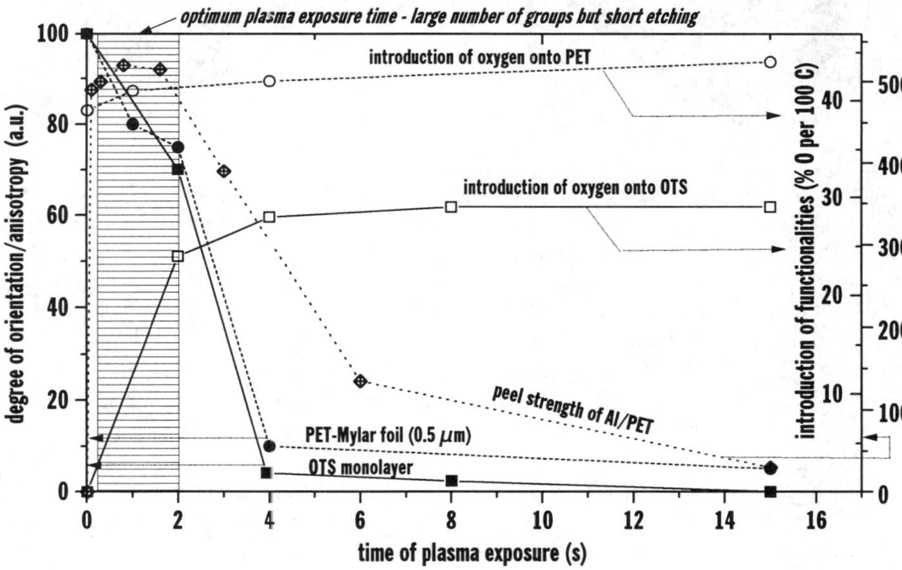

Figure 1 Assessment of optimum treatment time of OTS and PET in an O_2 plasma, assigned by nearly complete preservation of molecular orientation, high density of functional groups and maximum peel strength (see also ref. 18) of thermally evaporated Al (100 nm) on O_2 plasma treated PET.

The degree of molecular orientation was referenced to the unmodified SAM layer or PET surface as mentioned before (= 100 %). From Fig. 1 it can be concluded that the optimum time interval of exposure to an O_2 plasma is from 0.1 to 2 s as for the functionalization degree (4 to 24% O for OTS, PE and PP), the preservation of the orientation/anisotropy (95-70% for OTS and PET) and the adhesion properties (Al-PET). For all MW, RF and DC O_2 plasmas this optimum range is valid. Changes in the supermolecular structure of PET and PP are also visible in SFM (Scanning Force Microscopy) topographic pictures after ca. 1 s exposure to an O_2 plasma which is interpreted as molecular relaxation of stretched and frozen macromolecules as well as the beginning of etching (Fig. 2).

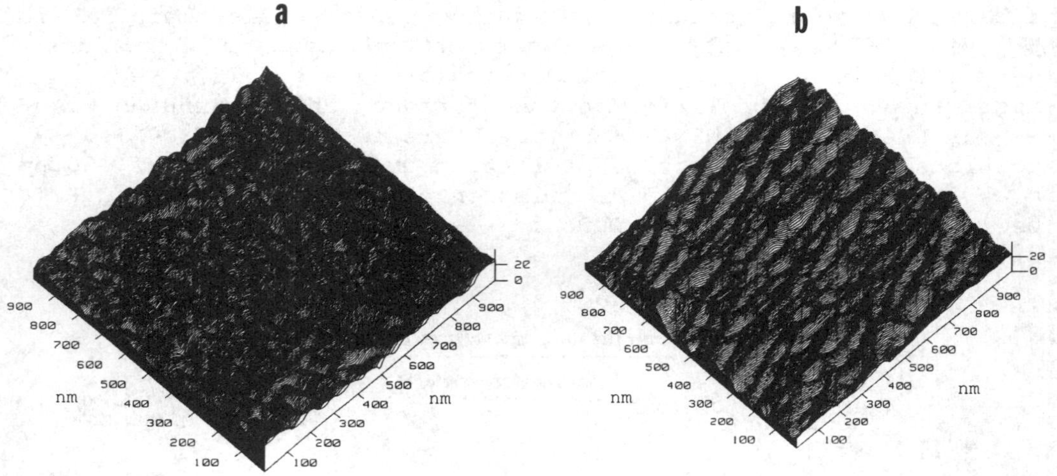

Figure 2 Scanning Force Microscopic pictures of untreated biaxially stretched PP (Trespaphan) (a) and after 1 s DC O_2 plasma treatment (b).

For the untreated biaxially stretched PP the ripple topography (r.m.s. roughness ≈ 4 nm for a 500 x 500 nm area) is characteristic [19]. Short exposure to an O_2 plasma (1 s, XPS O/C ≈ 5 %) results in a smoothing of the surface (r.m.s. ≈ 1.3 nm).

Plasmas at Atmospheric Pressure. In comparison to the RF low-pressure plasma treatment, the O introduction onto the surface of biaxially stretched PP with atmospheric-pressure plasmas is much faster. Referenced to the exposure time needed to reach 20% O per 100 C atoms (XPS) at the surface of PP for all types of plasmas the plasma jet with the O:N = 1:1 gas mixture (see Experimental) was about 33 times and the corona treatment in air was ca. 11 times faster than the RF O_2 low-pressure plasma. The species produced and their concentration relations on the surface differ only slightly in dependence on the type of plasma applied.

In Fig. 3 the formation of C=O groups estimated from FTIR-ATR measurements (1700 cm^{-1}), the polar component of the surface energy (for experimental details see ref. 20), and the O introduction (XPS) onto the surface of biaxially stretched PP are shown in dependence on treatment time. At the beginning of the corona discharge treatment the introduction of O functional groups is very fast and after t > 0.1 s the reaction rate is much slower. The same effect can be observed for the O_2 low-pressure plasma treatment after ca. 2 to 3 s. The time period of fast introduction of O functional groups was interpreted as modification and saturation of the uppermost surface layer with O followed by oxidation of deeper layers.

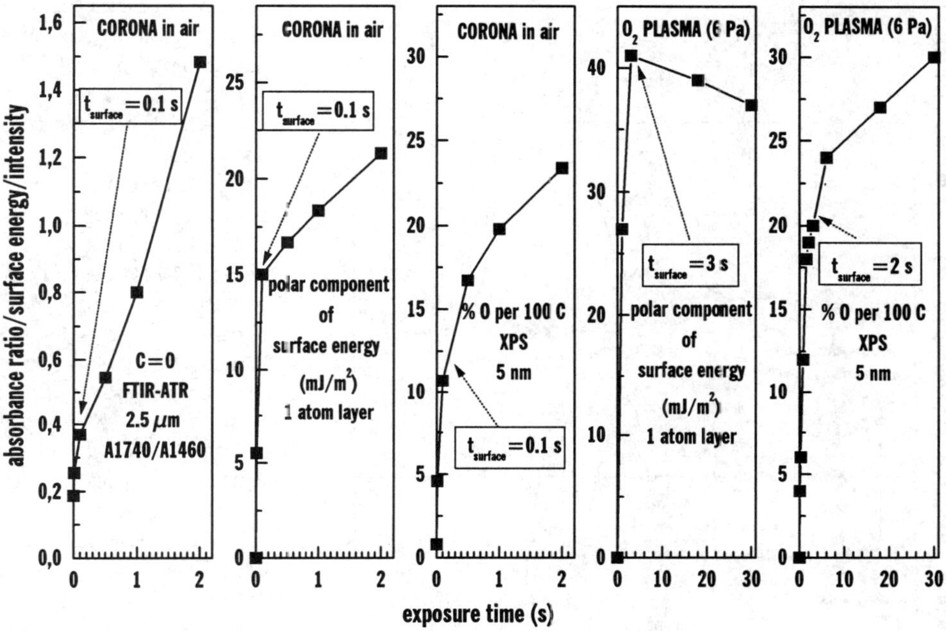

Figure 3 Increase in absorbance of C=O groups (ATR, ratio of 1740/1460 cm^{-1}), in O concentration (XPS), and in the polar component of surface energy with time of oxidative plasma treatments at the surface of biaxially stretched PP.

Types of Functional Groups. The C1s peak of PET treated for 15 s in a DC O_2 low-pressure plasma is shown in Fig. 4. In the difference spectrum (treated-untreated) the chemical composition is clearly demonstrated. Besides peaks 1-5, which can be assigned to original structural components of PET, new subpeaks appear due to the formation of peroxyacids or carbonates (290-291 eV), acids (289.5 eV, superposed by the original ester peak), ketones and aldehydes or possibly phenols (287.5-288.0 eV), alcohols, peroxides (near 286 eV, strongly superposed), polyenes or quarternary carbon (near 284.0-285.5 eV)[30,43]. For the interpretation of spectra it has to be noted that all subpeaks are broadened after each plasma treatment and additionally the β-shift must be taken into consideration.

Poly(bisphenol-A-carbonate) shows similar behaviour, however, the carbonate group is very fragile towards the oxidative attack of a plasma. A new C1s peak at 288.0 eV can also be assigned to phenolic groups which can be easily formed by the Photo-Fries

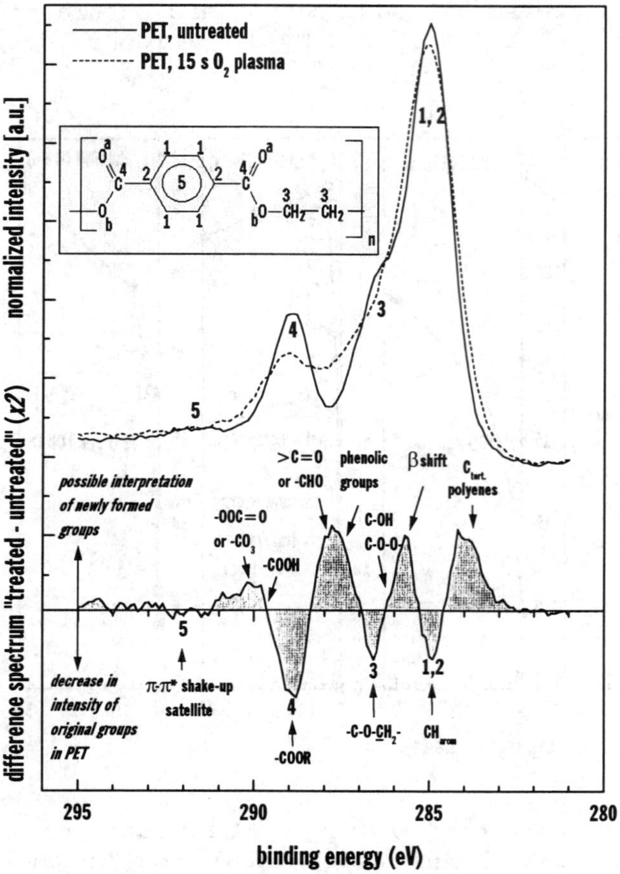

Figure 4
XPS C1s peak of PET before and after 15s exposure to a DC O_2 low-pressure plasma and the theoretically expected peak positions of possible reaction products including their assignment on the binding energy scale.

rearrangement[3] besides ketones or aldehydes. To form ketones or aldehydes the backbone of the PET and PC macromolecules must be scissioned by destruction of ester/carbonate or aromatic structures. This will be explained in more detail in connection with the discussion of NEXAFS results. PP shows a strong decrease in CH_3 group concentration due to the treatment in the O_2 plasma (XPS C2s valence band contribution at ca. 17 eV)[27]. Peak fitting of the C1s signal obtained with plasma oxidized PP or C_{18} alkane layers (OTS) reveals the existence of single, double and triple C to O bonded groups with an intensity ratio of about 5:3:2 [19] and changes in the orientation (see Fig. 1 for OTS). The formation of high concentration of C=C bonds with OTS and also with PP was verified in the NEXAFS C_K edge spectra by a π^* resonance at low $h\nu$ values (for PP see Fig. 5). In the difference spectrum of the spectra of treated and untreated PP a splitting of the π^* resonance indicates the existence of conjugated double bonds. This behaviour corresponds to the decrease in CH_3 group concentration. Furthermore, a large number of newly formed C=O species were detected in the NEXAFS O_K edge spectrum, besides the σ^* resonance of C-O single bonded groups.

Hexatriacontane (HTC, $C_{36}H_{74}$) exposed to the oxygen plasma shows the formation of one C=C double bond and at longer treatment times of two C=C bonds per molecule (see Fig. 5). HTC as prototype for PE is the only aliphatic hydrocarbon which shows traces of di and trimerization providing probably precursors for crosslinking in an O_2 plasma (see Fig. 5). With other polymers it was also possible to detect crosslinking products after exposure to an O_2 plasma.

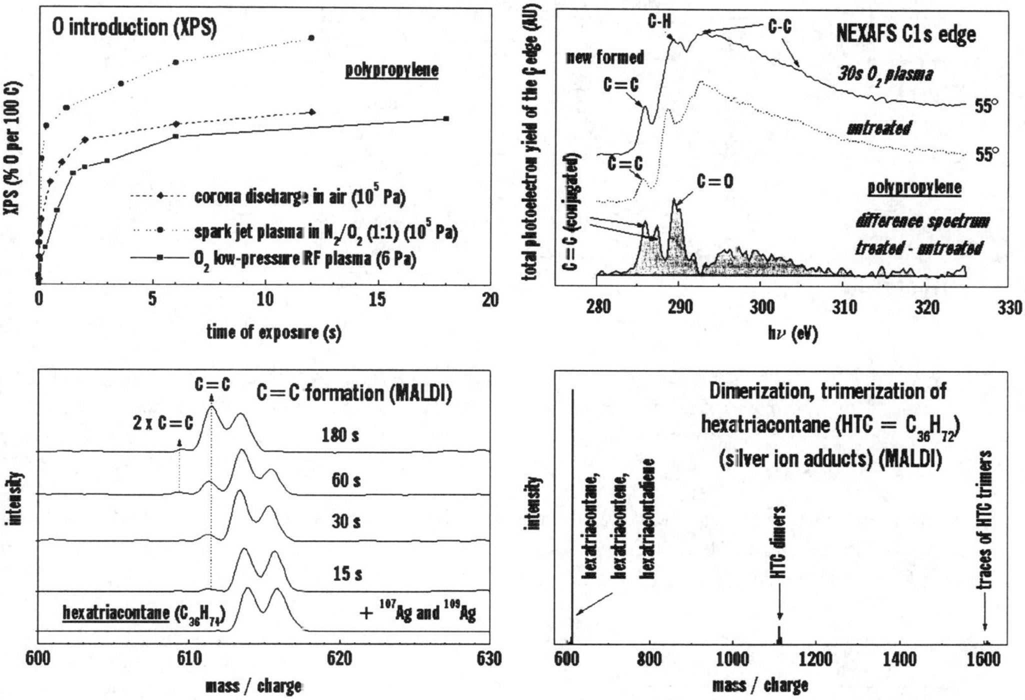

Figure 5 NEXAFS and XPS (oxygen introduction) spectra of the PP surface as well as MALDI spectra of hexatriacontane (HTC) films after exposure to an O_2 low-pressure plasma.

C_{18} SAM and LB layers (2.9 nm) were exposed to an oxygen plasma as exemplified in the SEIRA spectra in Fig. 6. CH_3 end groups of the aliphatic molecules could be detected (2963 cm^{-1}). The spectrum of an untreated stearic acid LB film (see Fig. 6) shows free COOH groups which are located in the gaps between silver islands. The rapid oxidation as detected by XPS could not be measured; however, the free COOH groups were completely destroyed within 2 s of O_2 plasma exposure but after 10 s a significant concentration of newly formed C=O groups appears.

<u>Orientation Effects</u>. Angle-dependent NEXAFS spectra of biaxially stretched PET-Mylar foils ($\alpha_\perp$=3.5 and $\alpha_\parallel$=3.5) reflect a significant anisotropy of π^* resonances (aromatic rings, C=O) (Fig. 7). The π^*_{ring} and $\pi^*_{C=O}$ resonances were most intense at grazing incidence (20°), therefore, they should be in the plane of the surface. Thus,

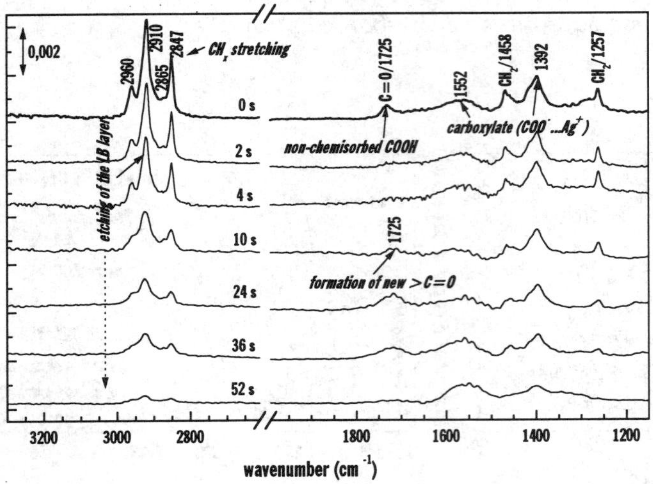

the electrical field vector of the synchrotron radiation **E** is nearly parallel to the surface normal. The reason for an angle dependence of spectra taken from non-isotropic surfaces is that the intensity of a selected resonance depends on the angle between the **E** vector of the linearly polarised

Figure 6 SEIRA spectra of a Langmuir-Blodgett monolayer of stearic acid (C_{18}) prepared on silver islands (ca. 30 nm in diameter) in dependence on its degradation on exposure to a low-pressure O_2 DC glow discharge plasma.

synchrotron radiation and the main direction of the unoccupied molecular orbital. The NEXAFS spectra in Fig. 7 show that the original preferential orientation of both π electron containing groups (π^*_{ring} and $\pi^*_{C=O}$) is only slightly diminished after 1 s exposure to the O_2 plasma (see difference spectra, compare with Fig. 1). The preferential orientation totally collapses after treatment times ≥ 4 s.

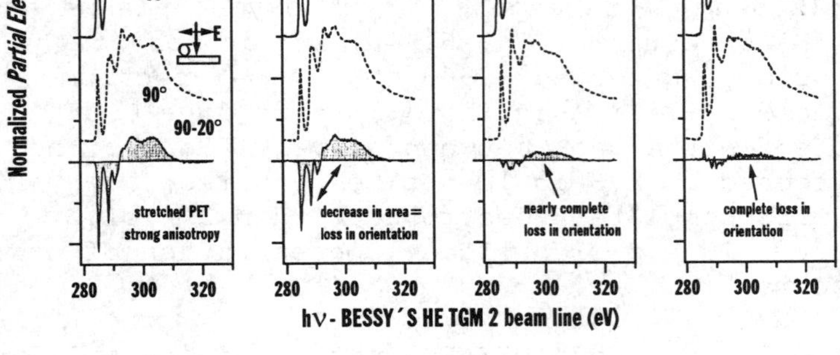

Figure 7 Angle-dependent NEXAFS C_K edge spectra of PET-MYLAR foils before and after exposure to DC O_2 low-pressure plasma.

This result is in agreement with that obtained with OTS (see Fig. 1 and ref. 54). The orientation of the C-C chain segments was also disturbed as seen by the broader σ* resonances at higher hν values (300 to 310 eV, see Fig. 7).

Polymer Degradation at the Surface. It is well known that organic polymers exposed to an O_2 plasma show different degradation mechanisms and etching rates[56,57]. Polymers with O functional groups (e.g. CO_3, COOH, COOR), easily forming stable degradation products (CO_2, H_2O, CO), are quickly degraded (PC, cellulose, poly(vinyl acetate), poly(vinyl alcohol) etc.). Polyolefins (PE, polybutadiene, also PP) tend to crosslink and, therefore, they are more stable. Aromatic rings act also as stabilising elements against the O_2 plasma attack. However, their large cross section at plasma particle impacts leads to the destruction of aromatic rings at the topmost layer. The very short wavelength plasma UV radiation has enough energy to produce ring destruction, presumably also in deeper layers.

Details on the degradation of aromatic rings and carbonyl groups and the appearance of a newly formed carbonyl species are shown by the NEXAFS spectra in Fig. 8 for PC, PS and PET spin coated films as well as for a 1.5 µm thick PET foil. Additionally,

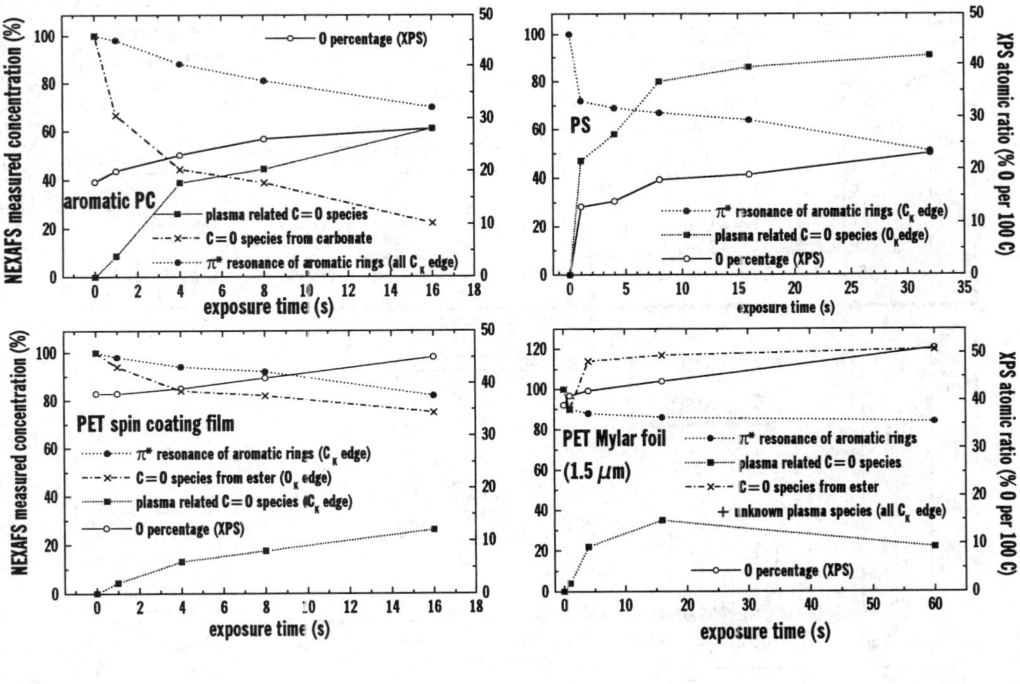

Figure 8 Dependence of the C1s π* resonance intensities (NEXAFS) and the introduction of oxygen functional groups (XPS) for PET, PC, and PS on the time of exposure to a DC O_2 low-pressure plasma.

the O introduction after exposure to O_2 low-pressure plasma is plotted (XPS). The concentration of the new C=O species which must possess a different chemical environment compared with the original ester or carbonate groups increases simultaneously with the O introduction (XPS). The C1s binding energies in the XP spectra suggest the existence of ketones or aldehydes which is supported by the NEXAFS data. Differences in the degradation behaviour of aromatic rings and original C=O groups can be identified in dependence on the type of polymer. PC shows a fast decrease in the C=O resonance intensity caused by decarboxylation and Photo-Fries rearrangement. PS loses twofold faster its aromatic rings than PC and PET which can be explained by the single-bonded side group position and splitting up of a phenyl ring whereas in the other polymers the ring (phenylene) is bonded via C_{ring}-C_{ester} or C_{ring}-$C_{carbonate}$ bonds within the polymer backbone. The C=O units of the ester groups in PET are much more stable than the C=O unit of the carbonate group in aromatic PC.

Linear PET 1,700 shows in the MALDI spectrum a decrease in molecular weight (M_W), a broadening of the molecular weight distribution and the appearance of 3 new products with their own distributions (Fig. 9).

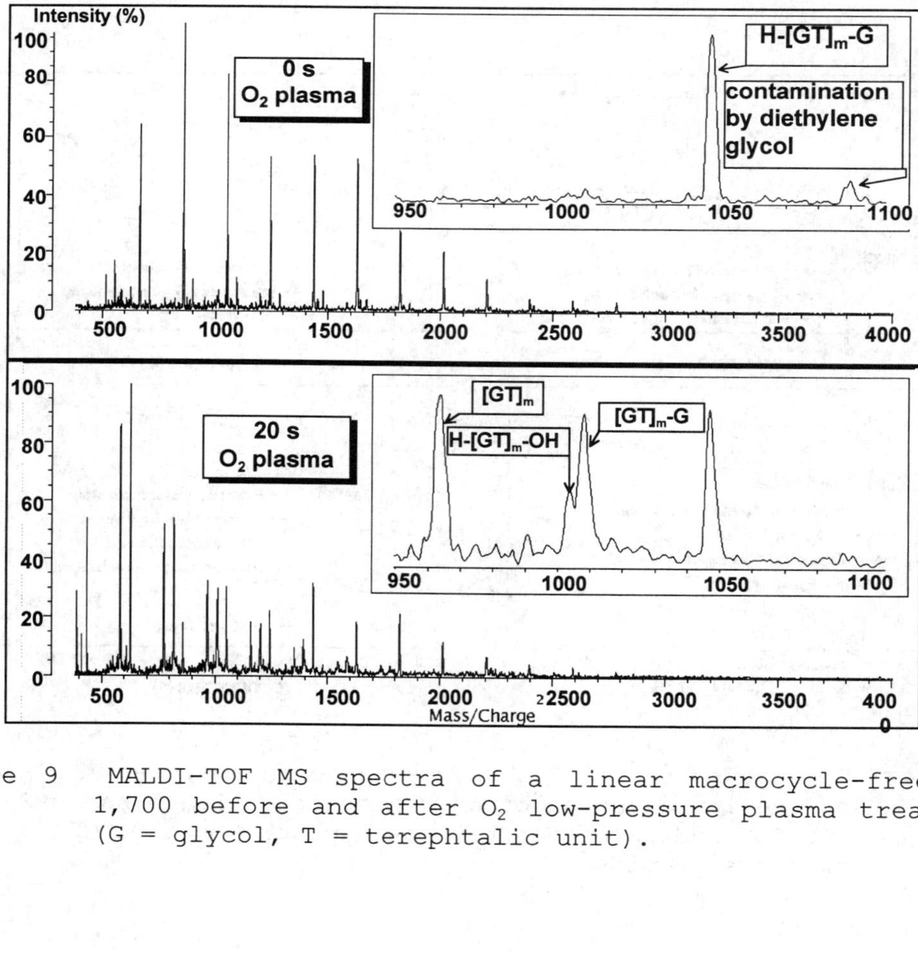

Figure 9 MALDI-TOF MS spectra of a linear macrocycle-free PET 1,700 before and after O_2 low-pressure plasma treatment (G = glycol, T = terephtalic unit).

Among the newly formed products two types of macrocycles were detected. On the other hand, the linear oxidation product with COOH groups at each end of the chain is missing. As example for another degradation behaviour a series of MALDI spectra of PEG are presented in Fig. 10 which differ in the time of exposure to the oxygen low-pressure plasma.

The original molecular weight distribution successively alters. A broad new one at the half of the original molecular weights appears in dependence on treatment time measured by a set of MALDI spectra. PMMA shows a continuous shift in the molecular weight distribution and a successive filling of the valleys between the peaks of neighbouring repeat units by numerously new peaks.

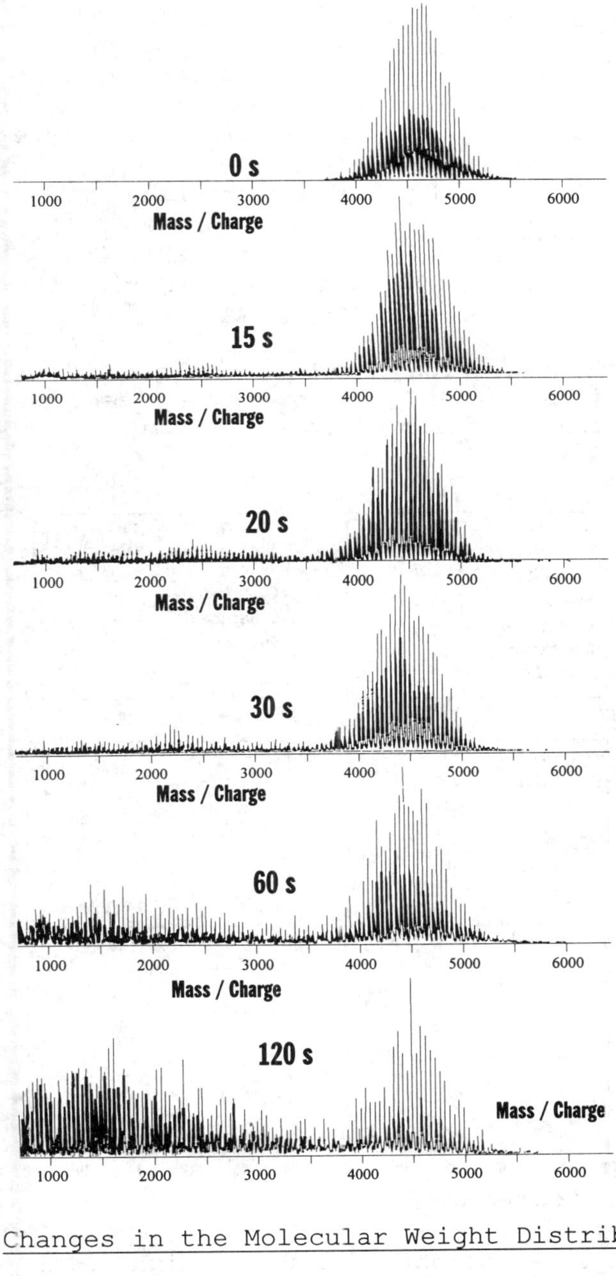

Figure 10
MALDI-TOF MS spectra of O_2 plasma treated PEG.

Influence of Plasma Gas. Whereas the oxygen plasma chemically acts (oxidizes) at polymer surfaces, the He plasma is limited to physical etching (sputtering). By adding some percent helium to the O_2 plasma an enhanced yield of introduced O functional groups at PP surfaces is observed (Fig. 11). This effect is caused by superelastic collisions of metastable excited He species ($He^{*metastable}$ =19.8 eV[56,57]) and oxygen species which allow an energy transfer by ionisation of oxygen (Penning ionisation process).

Changes in the Molecular Weight Distribution in Deeper Layers

SEC Analysis. The low-pressure O_2 plasma and the corona treatment in air cause degradation of PET, PS, PC, PMMA and PEG as shown by

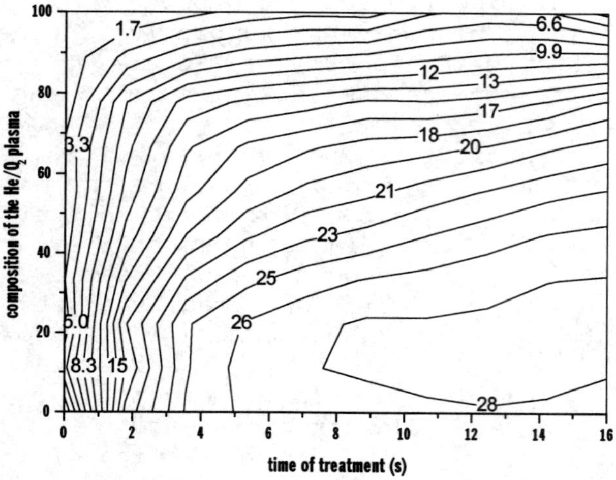

SEC analysis of O_2 plasma exposed polymer layers from 100 nm thickness. This is demonstrated in detail for PET as well as for PS and PC (Fig. 12). SEC and MALDI data allow to distinguish between three types of polymer degradation behaviours on exposure to O_2 plasma: a) random cracking (PEG, PPG), b) photo oxidation (PET, PC), and c) depolymerization (PS, PMMA).

Figure 11 2d plot of O functionalization yield in % oxygen at the surface of PP (XPS) in an O_2 low-pressure plasma in dependence on the concentration of added helium.

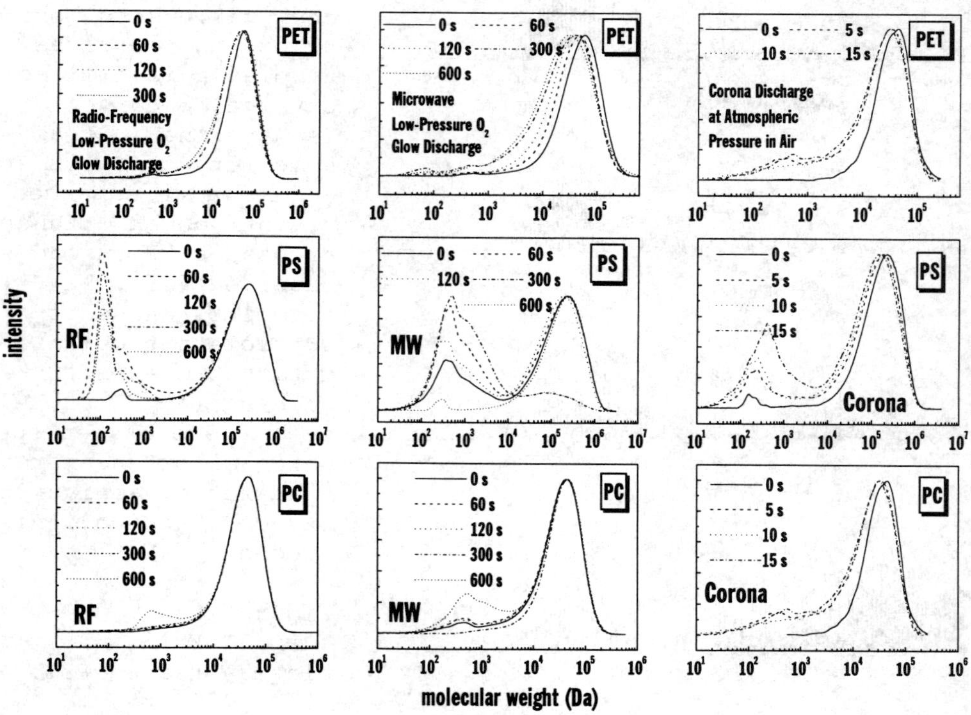

Figure 12 Changes in molecular weight of 100 nm thick films of PET, PC, and PS by MW, RF, and corona treatments

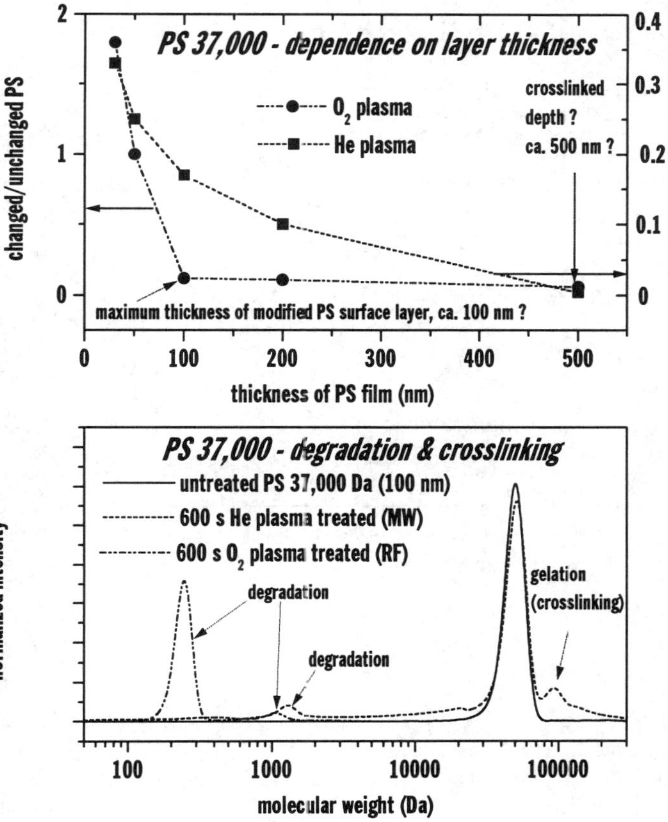

By varying the thickness of PS which was O_2 or He plasma treated it could be shown that the penetration depth of the oxygen plasma was calculated to 20, maximum to 100 nm. The helium plasma modifies the complete polymer bulk with exponentially decreasing crosslinking density in dependence on the distance from the surface into the bulk (see Fig. 13). For a longer treatment time (600 s) the differences between He and O_2 plasma exposures of PS are clearly visible (Fig. 13).

Figure 13 Dependence of the ratio between degraded or crosslinked ("changed") and unmodified ("unchanged") PS on the layer thickness and the type of plasma (top). SE chromatograms of PS exposed to a He or an O_2 plasma (bottom).

Stability of Functionalized Surface Layer. In the MW O_2 low-pressure plasma the O functionalized top layer on PP was stable as shown by extraction with ether (Table I). However, at treatment times of ten second a remarkable decrease of functional groups was found. Furthermore, after O_2 plasma exposure no significant influence of post-plasma oxidations could be measured as shown by varying the post-plasma air contact between 1 hour and two days (see Table I).

Table I. Increase in O Concentration (% O per 100 C-XPS) at the Surface of PP with Exposure to a Low-pressure O_2 Plasma (MW).

Time of exposure *(s)*	*10*	*20*	*30*	*60*	*120*	*300*
in situ	13.8	16.1	18.5	20.6	25.0	29.3
After 1 h	13.9	16.1	18.6	20.8	25.2	29.5
After 2 days	15.8	16.6	19.2	23.5	23.5	28.0
After 2 days and 15 min ether extraction	6.7	11.6	14.2	19.5	20.0	26.9

Crosslinking. The molecular weight measurement by SEC analysis is limited to only ca. 10^6 Da. Higher molecular weights and weakly

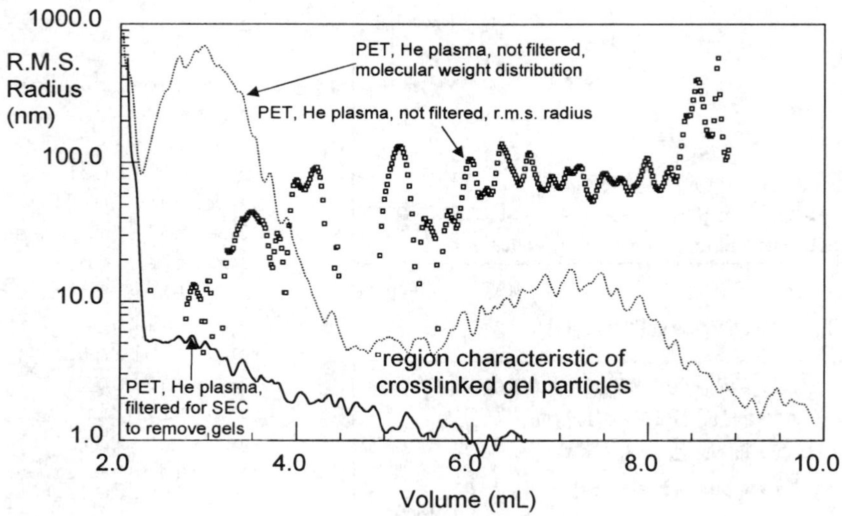

Figure 14 Gel particle formation in PET-Mylar foils after exposure to a MW low-pressure He plasma detected in the unfiltered ThFFF eluate (eluate volume ~ molecular weight) by measuring the r.m.s. radii (dots) with MALLS.

crosslinked gel particles were analyzed by ThFFF. Especially with long exposure times of PET to a low-pressure He plasma significant crosslinking was detected accompanied by some degradation (Fig. 14).

CONCLUSIONS

NEXAFS analysis of low-pressure O_2 plasma exposed polymer surfaces (PET, PC, PMMA, PPE (poly(phenylene ether), PP etc.) as well as NEXAFS and SEIRA measurements on C_{18} SAM and LB monolayers revealed that anisotropic molecular orientations were stable within 2 s, although more than 20% O (per 100 C atoms) was introduced onto the surface. Scanning Force Microscopy (SFM or AFM) pictures of PP show slight first changes in surface topography after 1 s O_2 plasma treatment due to changes in the supermolecular structure (see Fig. 2). The etching rates amount to only some 0.1 nm/s, therefore, etching effects can be nearly neglected; however, rearrangements, relaxations and modifications are more probable.

With the application of XPS, NEXAFS or SEIRA and C_{18} monolayer models the relative stability of the orientation in the first period up to 2 s (see Figs. 1 and 5) is caused by introduction of polar groups at each molecule, which can strongly interact or intraact at this high density. Moreover, the formation of double bonds introduces some stiffness in the C_{18} molecules and, last but not least, some crosslinking can also occur on exposure to oxygen plasma in the case of alkanes and fatty acids as shown by MALDI results with HTC (see Fig. 5).

The topmost layer of the polymer surface is completely oxidized when a steady state between the newly formed groups and their destruction by plasma particle bombardment or radiation is reached. From Fig. 3 it can be deduced that in the low-pressure O_2 plasma ca. 2 s are needed and in the atmospheric corona discharge in air ca. 0.1 s. This was confirmed with a number of polymers as well as with the monolayer models.

The exposure to an O_2 plasma results in degradation as shown by SEC and MALDI-TOF MS analyses for PS, PC, PET, PEG, PPG and PMMA. With HTC (Fig. 5) traces of dimers and trimers were detected by the MALDI technique. This was interpreted to be a possible indication for crosslinking. On exposure of $C_{36}H_{74}$ as model to O_2 plasma 1 or 2 double bonds per molecule are also formed. Furthermore, with PE and PP some indications exist for the formation of C=C conjugated system (Fig. 5).

On the other hand the He plasma produces crosslinking and degradation simultaneously which was demonstrated by SEC and ThFFF measurements (Figs. 13,14). He addition to O_2 plasma gas (10:90 mixture) results in an enhanced yield of O functional groups at PP surfaces due to a higher oxygen ion formation rate by the Penning ionisation.

The types of functional groups detected at the surface of oxygen plasma treated polymers were similar to the spectrum of products after an autoxidation[8]. Therefore, we did not find remarkable changes in chemical composition at the surface of PP after transfer and storage in air. It seems that the plasma oxidation involves all autoxidative reaction pathways.

PP shows the preferred splitting up of CH_3 groups[27] as well as the formation of double bonds. PC has an extremely fragile carbonate group but the aromatic rings are also destroyed by O_2 plasma exposure. Furthermore, in analogy to photooxidation, a Photo-Fries rearrangement is possible with formation of phenolic groups. Plasma oxidative reaction products of PET are in the majority chemically well-defined and correlate with those known from photooxidation (compare with refs. 8,11,14). Therefore, it can be deduced that similar mechanisms exist on oxygen plasma exposure with the exception of energetic hard particle bombardment and the extreme short-wavelength plasma UV irradiation. The bombardment should be limited to a few atomic layers and the UV may penetrate deeper; however, it was shown that both processes were limited at PS to ca. 20 nm calculated on the base of a two phases model. Some plasma-specific reaction products, originating from high-energetic impacts or radiation ($h\nu > 10$ eV), are ring destruction products and peroxyacids found in an earlier investigation[8]. Here, we additionally detect the very important role of macrocycle formation. The macrocycles could be identified by MALDI and using a specially synthesized linear PET oligomer. Two types of macrocycles were identified: one formed by esterification of OH and COOH terminal species and, the other, a ring which has lost one glycol unit. Table II summarises all products identified at the PET surface or in the bulk after exposure to the O_2 plasma.

Table II. Reactions during Oxygen Plasma Exposure of PET Surfaces and Their Proposed Nature.

No.	*Reaction*	*Type*
1	Cracking of aromatic rings	Plasma-specific
2	Formation of ketones	Photooxidative
3	Formation of peroxyacids	Plasma-specific
4	Formation of aliphatic hydroperoxides	Photooxidative
5	Decarbonylation	Photooxidative
6	Decarboxylation	Photooxidative
7	Norrish-I and II chain scissions	Photooxidative
8	Formation of C=C, $C=C_{conj.}$	Plasma-specific, also photooxidative
9	Formation of macrocycles	Plasma-specific
10	Loss in molecular weight (M_W)	Plasma-specific
11	Formation of phenols	Photooxidative
12	Crosslinking	Not found by SEC

Three general reaction ways for the polymers exposed to an O_2 plasma were found: a) random cracking (PPG, PEG), b) photooxidation (PET, PC), and c) depolymerization (PS, PMMA).

The atmospheric pressure plasma introduces oxygen functionalities at least 10 times faster than the low-pressure O_2 plasma, which can be explained by the 10^3 times higher partial oxygen pressure. We have also found that the oxygen introduced by the low-pressure O_2 plasma improves most significantly the adhesion properties of PP (lap shear strength of polyurethane-PP composites), that introduced by the plasma jet acts more moderately and that introduced by the corona treatment has no measurable influence[21,53]. To significantly improve the adhesion properties of PP at least 4 % and maximum 20 to 25 % oxygen should be introduced.

ACKNOWLEDGEMENTS

Thanks are due to G. Feyder (DuPont, Luxembourg), Ch. Wöll, K. Weiss (Ruhr-University, Bochum) and W. Braun (BESSY GmbH, Berlin) for collaboration as well as Deutsche Forschungsgemeinschaft for financial support.

REFERENCES

1 M. Strobel, C. S. Lyons and K. L. Mittal, editors, "Plasma Surface Modification of Polymers: Relevance to Adhesion," VSP, Utrecht, The Netherlands, 1994.

2 J. F. Friedrich, unpublished results.

3 K. Rossmann, J. Polym. Sci. 19, 141 (1956).

4 J. R. Hollahan, B. B. Stafford, R. D. Falb, and S. T. Payne, J. Appl. Polym. Sci. 13, 807 (1969).

5 D. T. Clark, and A. Dilks, J. Polym. Sci., Polym. Chem. Ed. 17, 957 (1979).

6 D. Briggs, D. G. Rance, C. R. Kendall, and A. R. Blythe, Polymer 21, 895 (1980)

7 V. André, F. Tchoubineh, F. Arefi, J. Amouroux, in "Plasma-Surface Interactions and Processing of Materials," O. Auciello, editor, Kluwer, Dordrecht, 1990

8 J. F. Friedrich, I. Loeschcke, H. Frommelt, H.-D. Reiner, H. Zimmermann, and P. Lutgen, J. Polym. Degrad. Stabil. 31, 97 (1991).

9 J. F. Friedrich, G. Kühn, and J. Gähde, Acta Polymerica 30, 470 (1979).

10 S. Wu, "Polymer Interface and Adhesion," Marcel Dekker, New York, 1982.

11 A. Dilks, in "Degradation and Stabilisation of Polymers", H. H. G. Jellinek (Ed.), Elsevier, Amsterdam, 1983.

12 W. Stiller, and J. F. Friedrich, Zeitschrift für Chemie 31, 91 (1981).

13 J. F. Friedrich, Wissenschaft und Fortschritt, 36, 12 (1986).

14 J. F. Rabek, "Polymer Photodegradation," Chapman & Hall, London, 1995.

15 R. H. Hansen, and H. Schonhorn, J. Polym. Sci. B4, 203 (1966).

16 M. Hudis, J. Appl. Polym. Sci. 16, 2397 (1972).

17 J. F. Friedrich, W. E. S. Unger, A. Lippitz, St. Weidner, Sh. Geng, L. Wigant, Ch. Wöll, J. Erdmann, and H.-V. Gorsler, Proc. ECASIA 95, Montreux, Schweiz, 1995, Wiley & Sons, Chichester, 1996, p. 803.

18 J. F. Friedrich, Sh. Geng, W. E. S. Unger, A. Lippitz, J. Erdmann, H.-V. Gorsler, Ch. Wöll, A. Schertel, and K. Bierbaum, Surface Coating Technol. 74-75, 664 (1995).

19 J. F. Friedrich, W. E. S. Unger, A. Lippitz, P. Rohrer, W. Saur, J. Erdmann, and H.-V. Gorsler, Surface Coating Technol. 59, 371 (1993).

20 J. F. Friedrich, Proc. of "Plasmatec 93," Print Service Mühlheim, Mühlheim, 1993, p. 92.

21 J. F. Friedrich, W. E. S. Unger, A. Lippitz, Th. Gross, P. Rohrer, W. Saur, J. Erdmann, and H.-V. Gorsler, J. Adhesion Sci. Technol. 9, 575 (1995).

22 St. Weidner, G. Kühn, W. E. S. Unger, A. Lippitz, and J. F. Friedrich, Proc. ECASIA 97, Gotenburg, Sweden, p. 188.

23 W. E. S. Unger, A. Lippitz, Ch. Wöll, J. F. Friedrich, and L. Nick, Suppl. Revue "Le Vide, les Couche Minces," 272, 95

(1994).
24 J. F. Friedrich, and L. Wigant, Wissenschaftliche Zeitschrift der TFH Wildau 1, 36 (1995).
25 J. F. Friedrich, in "Kleben-Adhesion," Proceedings of 7th Swissbonding, Ed. Schindel-Bidinelli, editor, Rapperswil, Print Service Mühlheim, 1993, p. 92.
26 A. Lippitz, J. F. Friedrich, W. E. S. Unger, A. Schertel, and Ch. Wöll, Polymer 37, 3151 (1996).
27 Th. Gross, A. Lippitz, W. E. S. Unger, J. F. Friedrich, and Ch. Wöll, Polymer 35, 5591 (1994).
28 A. Lippitz, I. Koprinarov, J. F. Friedrich, W. E. S. Unger, K. Weiss, and Ch. Wöll, Polymer 37, 3157 (1996).
29 I. Koprinarov, A. Lippitz, J. F. Friedrich, W. E. S. Unger, Ch. Wöll, Polymer 38, 2005 (1997).
30 W. Unger, A. Lippitz, and J. Friedrich, Materialprüfung 37, 6 (1995).
31 I. Koprinarov, A. Lippitz, J. F. Friedrich, W. E. S. Unger, and Ch. Wöll, Polymer, submitted.
32 St. Weidner, G. Kühn, J. F. Friedrich, W. E. S. Unger, and A. Lippitz, Rapid Comm. Mass Spectrom. 10, 727 (1996)
33 St. Weidner, G. Kühn, J. F. Friedrich, and H. Schröder, Rapid Comm. Mass Spectrom., 10, 40 (1996).
34 St. Weidner, G. Kühn, and G. Pfahler, Rapid Commun. Mass Spectrom., 9, 697 (1995).
35 J. F. Friedrich, W. E. S. Unger, A. Lippitz, I. Koprinarov, G. Kühn, St. Weidner, and L. Vogel, Proc. "Plasmatec 97," Duisburg, Sept. 23rd, Print Service Mühlheim, Mühlheim, 1997, p. 118.
36 M. Karas, and F. Hillenkamp, Anal. Chem. 60, 2299 (1988).
37 F. Hillenkamp, M. Karas, R. C. Beavis, and B. T. Chait, Anal. Chem. 63, 1193 (1991).
38 D. C. Schriemer, and Liang Li, Anal. Chem. 68, 2721 (1996).
39 J. C. Giddings, Science, 260, 1456 (1993).
40 U. Just, B. Werthmann, J. F. Friedrich, R. Hänsel, D. Harzer, and D. Roessner, Proc. HPLC ´97, June 22-27, 97, Birmingham-UK, 1997, p. 343.
41 St. Weidner, G. Kühn, J. F. Friedrich, B. Werthmann, and D. Roessner, Proc. Polymer Analysis, Hamburg, Sept. 22/23, 1997, Shaker-Verlag, Aachen, 1997.
42 J. F. Friedrich, L. Wigant, W. E. S. Unger, A. Lippitz, and H. Wittrich, Surface Coatings Technol., in press.
43 G. Beamson and D. Briggs, "High Resolution XPS of Organic Polymers," Wiley, Chichester, 1992, p. 174.
44 M. P. Seah, and W. A. Dench, Surface Interface Anal., 1, 2 (1979)..
45 M. T. Anthony, and M. P. Seah, Surface Interface Anal. 6, 95 (1984).
46 A. D. Shirley, Phys. Rev. B5, 4709 (1972).
47 J. Stöhr, "NEXAFS Spectroscopy," Springer-Verlag, Heidelberg, 1993.
48 Standard Terminology Relating to Surface Analysis, ASTM E 673-95a, Annual Book of ASTM Standards, Vol. 03.06 (ASTM,

Philadelphia, 1995), p. 789.
49 T. Kamata, A. Kato, J. Umemura, and T. Takenaka, Langmuir, 3, 1150 (1987).
50 M. Osawa, Analytical Sci. 7, 503 (1991).
51 Shiming Geng, PhD Thesis, University of Potsdam, Potsdam, Germany, 1996, p. 22.
52 C. Bubeck, and D. Holtkamp, Adv. Mater. 3, 32 (1991).
53 J. F. Friedrich, W. E. S. Unger, A. Lippitz, Sh. Geng, I. Koprinarov, G. Kühn, and St. Weidner, Surface Coatings Technol., 98, 1132 (1998).
54 J. F. Friedrich, W. E. S. Unger, and A. Lippitz, J. Macromol. Sci. 100, 111 (1995).
55 H. Yasuda, C. E. Lamaze, and K. Sakaoku, J. Appl. Polym. Sci. 17, 137 (1973).
56 J. Friedrich, Contributions to Plasma Physics/Beiträge zur Plasmaphysik 21, 261 (1981).
57 H. Drost, "Plasmachemie," Akademie-Verlag, Berlin, 1978.

Metallized Plastics 5&6: Fundamental and Applied Aspects, pp. 295-317.
K.L. Mittal (Ed.)

STUDY OF HE-NH_3 PLASMA MIXTURES AND THEIR ROLE IN THE STABILITY OF THE PLASMA TREATED POLYPROPYLENE SURFACE

J. KURDI, F. AREFI-KHONSARI, M. TATOULIAN and J. AMOUROUX

Laboratoire de Génie des Procédés Plasmas, Université Paris VI, ENSCP
11, rue Pierre et Marie Curie, 75231 Paris cedex 05, France

The purpose of this work was to minimize the well-known ageing phenomenon of the plasma-treated polymer surfaces, by stabilizing the outermost layer via crosslinking using discharges in mixtures of ammonia (reactive gas) and helium (noble gas). Treatments in He + NH_3 gas mixtures are interesting because they produce a surface nitridation simultaneously with a crosslinking reaction in a one step process. The physicochemical modifications of He+NH_3 treated polypropylene have been studied in terms of the wettability, the acid-base and the mechanical properties. The contact angle measurements reveal that for He-rich mixtures (>96% He), there is an optimum in the stability of the surface wettability with a basic character comparable to the one obtained for pure ammonia plasma. Similarly, stress-strain characteristics show that mixtures with a high percentage of helium (>96%) enhance the cohesive strengthening of the polymer surface by the formation of a thin crosslinked layer that mechanically stabilizes the surface. Such treated surfaces with a predominant basic character and stable surface properties are very useful for applications such as metallization of polymers.

INTRODUCTION

In recent years, we have witnessed a remarkable growth in the use of synthetic organic polymers in many technological applications from adhesive bonding to biocompatibility. Economic reasons as well as the achievable physical bulk properties (high strength to weight ratio, resistance to corrosion) are some of the driving forces for this trend. However, the surfaces of polymers are generally characterized by a low surface energy and hence poor intrinsic adhesion properties. Examples of technological applications where the adhesion properties of polymer surfaces are of concern involve polymer-metal interfaces in microelectronics, decoration, or packaging. The plasma technique is a dry process which can be used to modify the surface of a polymer in order to improve adhesion and wettability, or to provide a diffusion barrier layer[1-4]. A unique feature of plasma modification is that the surface structure of the polymer (only few nanometers) can be selectively modified for a specific application while the bulk properties of the polymer are unaffected.

In this study, polyaliphatics such as isotactic polypropylene (PP) was plasma treated. For the plasma treatment a particular low pressure reactor using asymmetrical geometry of electrodes (high voltage hollow electrode - grounded cylinder) with an excitation frequency of 70 kHz was employed. This special configuration provides the ability to simulate the continuous industrial surface treatment of polymer films, i.e. very short treatment times (a few milliseconds to a few seconds)[5].

In the first part of this paper we shall study the impact of the plasma atmosphere, using two gases namely an oxidizing one such as oxygen and a reducing one like ammonia, on the adhesion properties with respect to a thin aluminium layer (20-30 nm). For this purpose, the plasma treated polymer films were metallized by thermal evaporation in-situ in the same plasma chamber and peel-tests were performed to evaluate the adhesion improvement of the Al-PP systems. An attempt has been made in order to explain the adhesion improvement in terms of acid-base interactions. In fact, the importance of Lewis type acid-base interactions has been demonstrated for polymer - adherend systems where it was observed that basic polymers only chemisorbed on acidic adherends and acidic polymers only chemisorbed on basic adherends[6]. However, one should emphasize that the degradation effects of het plasmas of these two gases (O_2 NH_3) are different. In fact, they affect in a different manner the formation of a weak boundary layer and the mechanical properties of the plasma-treated polypropylene films. The latter would, therefore, affect the adhesion properties of the metal-polymer system[7].

The second part of this paper is devoted to the study of He + NH_3 mixtures in order to obtain stable polymer surface properties. Indeed, reorganization of polymers surfaces upon exposure to a relatively nonpolar medium such as air is well known as the ageing phenomenon[8-10] and this is very important from the viewpoint of adhesion. Many works[11-13] showed that the mobility of polymer chains enabled polymer surfaces to respond to interfacial forces and thus to adapt their surface chemical structure to their environment. By rotational and translational motions of chains and chain segments, the surface composition can change in order to minimize the interfacial free energy between the polymer and its environment. In this way, the surface rendered wettable by the plasma treatment is found to revert to a less wettable state with time. Since the surface properties i.e. wettability and adhesion of such plasma treated surfaces degrade with time, a helium plasma pretreatment has been used in order to stabilize the surface layer via crosslinking[14]. The CASING (Crosslinking via Activated Species of Inert Gases)[1,15] process consists of creating new bonds such as C=C, which improve the bond strength of the surface by forming a very cohesive and dense skin. As a result of this, the ageing effect of an ammonia-plasma treated PP has been shown to slow down by using a helium plasma pretreatment[14]. Helium is the most efficient of the inert gases for the crosslinking of the outermost layer of a polymer[16]. This feature being probably due to the large amount of energy available to transfer to the polymer surface via ion neutralization, Auger de-excitation and Penning ionization in the polymer[17].

In order to carry out the functionalization and crosslinking of the surface layer of the polymer by a one step process, previous works carried out in our laboratory and aiming at biocompatibility applications have treated poly(ethylene-terephthalate (PET) in He + O_2 plasma mixtures[18].

On the other hand, the attachement of amine groups to polymer surfaces is particularly attractive because amines are relatively reactive nucleophiles useful either for the covalent binding of larger chemical compounds such as proteins for different applications, or for the supply of amine reactive sites for adhesive bonding[19]. Wetting of the polymer is, however, only a necessary but not a sufficient condition for good adhesion. This goal is only reached if a stable and strong interaction can occur between the polymer and the coating[14,20]. Treatments in He + NH_3 gas mixtures are interesting because they produce a surface nitridation, simultaneously with a crosslinking reaction in a one step process. These two mechanisms when achieved at the same time allow to obtain a reactive surface which also presents a good stability with ageing time. The competitive action of the two mechanisms will depend, in a critical manner, on the composition of the gaseous mixture.

Taking into account the above-mentioned aspects, the second part of this paper will be devoted to the characterization of the wettability, mechanical and acid-base properties of He + NH_3 treated polypropylene. These properties will be correlated to the chemical modifications occurred at the surface of PP after plasma treatment. In this way, one can hope to point out in a clearer manner the role played by helium, in the plasma mixture, on the stability of the surface properties mentioned above.

EXPERIMENTAL SETUP

Apparatus

The details of the experimental set up used for the plasma treatment of polymer films have been described elsewhere[5]. It consists mainly of a glass reactor with a non-symmetrical configuration of electrodes. Low frequency power (70 kHz), provided by an industrial generator, was capacitively coupled to the blade-type electrode. The polymer film (22 x 22 cm^2) was rolled on the grounded cylinder. The real exposure time of the polymer to the discharge can be calculated from the total duration of plasma discharge and the plasma width on the polymer. For a total plasma duration of 1s, the calculated real time is equal to 0.023s. The gases were introduced through MKS mass flow controllers and the pressure was monitored with an MKS capacitive gauge.

Materials

An isotactic polypropylene (PP, monomer : CH_2=CH-CH_3) film 8 µm thick was used in this study, and X-ray photoelectron spectroscopy (XPS) analysis showed a negligible (<1%) amount of oxygen on the surface of the untreated PP .

Operating Conditions

For He + NH_3 plasma treatments, an effective power of 6.5 W was applied throughout the study. Helium and ammonia, provided by Airliquid at a purity of 99.995% and 99.96%, respectively, were used without further purification, and were introduced uniformly through the HV (high voltage) blade-type electrode. The operating pressure was fixed at 100 Pa. The relative flow rates of the two gases were varied while keeping constant the total flow rate at 100 sccm.

Metallization and Peel Strength Measurements

Metallization of plasma pretreated PP films was carried out in-situ by thermal evaporation in the same chamber[5], which avoided any contact of the pretreated polymer with air prior to aluminium deposition. A thin aluminium coating (20-30 nm) was deposited at 10^{-4} Pa. To evaluate the improvement in adhesion, we used a U-form peel test specially adapted to flexible substrates[5]. After the peel test, the percentage of the metal peeled off was determined by an image processing system on the peeled off films[21].

Stress-Strain Measurements

Stress-strain measurements were performed using an Instron at a crosshead speed of 8 mm/min at room temperature. PP films were cut into specimens of 25 mm width and 70 mm length.

Contact Angle Measurements

The surface properties were determined as a function of ageing time in air, by contact angle measurements with bidistilled water. The apparatus developed in our laboratory has been described in detail elsewhere[21]. It is based on an image processing system which allows the measurements of the contact angle of a drop of liquid on a flat surface. The volume of the drop used was always 2 µl and samples were cut to the size 0.5 x 4 cm. All reported values of advancing contact angles are an average of five measurements taken at different locations on the films and have a maximum error of ± 3°. In order to determine the character of the plasma-treated PP as well as the untreated one, unbuffered aqueous acidic and basic solutions with pH values ranging from 1 to 14 were prepared with double distilled water with NaOH or HCl as the base or acid[22,23]. The surface tension of the solutions was measured by the Wilhelmy technique[24]. It showed that there was no difference between the surface tension of pure water (γ_W = 72.8 mN/m) and those of the acidic and basic solutions.

XPS Analysis

XPS spectra were recorded using a VG Scientific ESCALAB MKI system with Al Kα radiation[25]. The electron take-off angle was 90° with respect to the surface. Charge referencing was determined by setting the main C1s component at 285 eV. Quantification of the XPS data was achieved using peak areas and experimental sensitivity factors in accordance with the following relation :

$$\% A = [(I_A/s_A)/\Sigma (I_n/s_n)] \times 100 \%$$

where I and s are the integrated peak area and the sensitivity, respectively, for a given element. n stands for the nth element considered in the quantitative analysis.

RESULTS AND DISCUSSION

Impact of the plasma atmosphere on the adhesion properties in the case of Al-PP system

Peel strength measurements

The variation of the percentage of the metal peeled off is shown in Figure 1 for both ammonia and oxygen plasma treated PP. For the untreated PP, one notes that all of the metal is peeled off (98%) while in the case of both plasma treatments, two distinct regions are observed:

1- The first one corresponds to where the percentage of the metal peeled off decreases sharply with treatment time. In this region, one can see that for treatment times as short as 23 ms in ammonia plasma only 4-5% of metal was peeled off whereas for oxygen this percentage was about 30% for t = 400 ms.

2- The second region corresponds to where the percentage of the metal peeled off increases (above 1 s for ammonia and 400 ms for oxygen). This is likely to be due to the formation of a weak boundary layer (WBL) in the polymer because of an overtreatment leading to a cohesive failure in the PP[14].

However, a higher adhesion was observed in the case of ammonia plasma treated PP as compared to oxygen plasma treated sample.

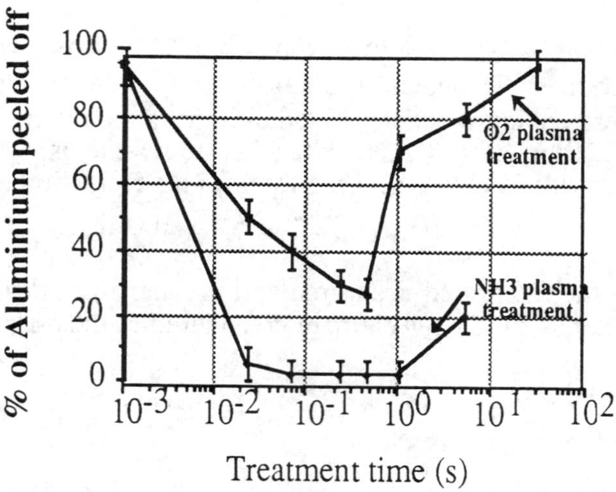

Figure 1 Variation of the amount of aluminium peeled off as a function of the treatment time in the case of ammonia and oxygen plasmas treated PP.
O_2 Plasma : P = 100 Pa, Pw = 5 W, D = 100 sccm, t = 1 s.
NH_3 Plasma : P = 100 Pa, Pw = 9 W, D = 100 sccm, t = 1 s.

Stress-strain characteristics

The overtreatment could affect differently the mechanical properties of the plasma-treated nonmetallized PP in the two different atmospheres. That is why the stress-strain characteristics of PP have been measured as a function of the treatment time. In Figure 2, the role of the plasma gas on the elongation-at-break point of PP films is shown.

First of all, one notes that the plasma treatment does not modify the stress-strain behavior of the polymer, since this is an intrinsic property of the polymer bulk (Young's modulus is the same for the oxygen and ammonia plasma treated PP). Nevertheless, the position of the break point on the stress-strain curve depends on the gas[14]. When one increases the treatment time, the break point is shifted to lower tensile strength and elongation values. This probably results from a weakening of the surface due to an overtreatment. In such a case, the crack starts at the fragile layer and then propagates in the bulk of the polymer. The presence of a weak boundary layer gives rise to a premature break point even though the trends of the stress-strain curves remain unchanged.

The results show that in case of both gases the elongation-at-break point decreases from 160% (untreated PP) to about 140% (ammonia plasma) or 130% (oxygen plasma) for a short treatment time of 1s. However, when the treatment time is increased, in the case of the oxygen plasma treated PP the decrease of the elongation-at-break point is much sharper than in the case of the ammonia plasma treated sample. One notes a drastic decrease of this parameter (30%) for a treatment time of 60s in an oxygen plasma.

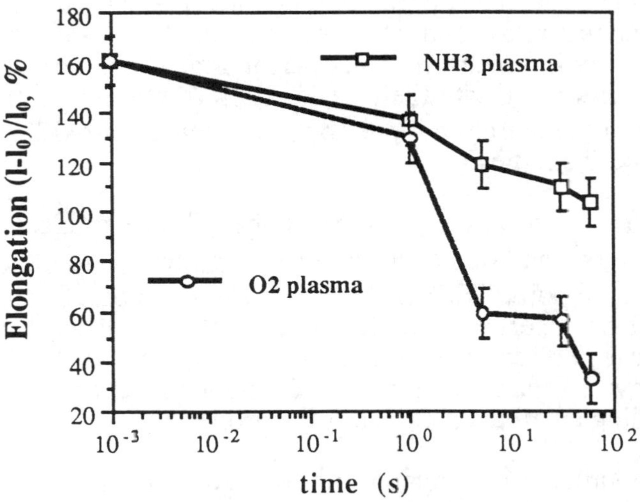

Figure 2 Elongation-at-break $(l-l_0)/l_0$ of PP treated in oxygen and ammonia plasma atmospheres as a function of the treatment time.
l_0 : initial length of the sample, l : length after elongation of the same sample.

Discussion

In the region where the treatment time is between 1 s and 100 s (1 s $<t<$ 100 s), good correlations are obtained between the tensile measurements (Figure 2) and the peel test results (Figure 1). For overtreated samples the unsatisfactory adhesion results obtained in the case of both gases can be related to the formation of a weak boundary layer in the polymer. However, the lower adhesion observed in the case of oxygen plasma-treated PP can be explained by a more intense degradation of the PP layer.

Since the stress-strain behavior in the cases of the two gases is practically the same for short treatment times (<1s), the better adhesion observed in the case of ammonia plasma is explained by favorable acid-base interactions at the metal-polymer interface.

In fact, in our experimental conditions during the metallization (10^{-3} Pa), one can suggest that there exists a high probability that the first monolayers of metal vapour reaching the polymer film were immediately oxidized due to the residual air. Furthermore, in-depth Auger Electron Spectroscopy Analysis (AES) carried out on the metallized ammonia-plasma treated PP showed that the interface was oxidized, and Al-O-C type of bondings were detected[14]. In this way, we should consider the interactions of oxygen or ammonia-plasma treated PP with alumina rather than aluminium.

On the other hand, contact angle titration measurements carried out on a thin metallic aluminium layer, deposited in our metallization reactor and covered by a natural passivating alumina layer (3 nm thick) shows an amphoteric behavior with an acidic tendency[25]. The amphoteric character is shown by an increase of the total work of adhesion for both strong acidic and basic solutions (Figure 3). The contact angle titration measurements performed on the passivation alumina layer are in good agreement with the classification of alumina, a trivalent oxide, as an acidic oxide in the literature[26], in contrast to bivalent oxides considered to have generally basic properties.

However, the acid-base character of the surface oxide depends on its hydration state , purity and defect structures[27]. In fact, these factors are known to affect the isoelectric point (IEP) of the oxide. For example, Parks[27] has shown that the structural or adsorbed anionic impurities as well as dehydration and increased structural perfection decrease the IEP by shifting it to more acidic values. In this way, a broad range of IEP characteristics of the oxides can be established depending on all the parameters mentioned above.

In order to validate the contact angle titration technique as a method to point out the predominant acid-base characteristics, we have used a magnesium oxide surface known to possess pure basic properties[27, 28]. Indeed, one can observe in Figure 3 the predominant basic character of this oxide. Therefore, one can rely on the amphoteric character with an acidic tendency of the passivating layer of alumina observed by this technique (Figure 3) and so to consider the adhesion improvement of Al to PP by ammonia-plasma treatment in terms of favorable Lewis acid-base interactions. In fact, it has been shown in previous works that for treatment less than 1 s, the NH_3-treated PP exhibited a dominant basic character[29]. In this case, good interactions occur between the basic ammonia-treated PP surface and the alumina shown to possess an amphoteric character with acidic tendency. Oxygen treatment performed in the same experimental set-up gave rise to a predominantly acidic property of the surface[29]. The adhesion of such a surface to aluminium was not improved as much as when ammonia treatment was performed under the same experimental conditions (Figure 1), probably because of the unfavorable acid-base interactions.

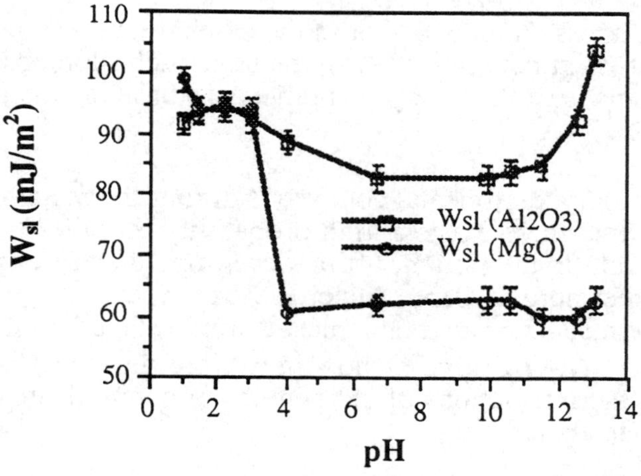

Figure 3 Total work of adhesion calculated from contact angle measurements carried out with pH test liquids on a thin metallic aluminium layer covered by a natural passivating alumina layer and on a magnesium oxide substrate.

As shown above, low molecular weight fragments, resulting from excessve chain scission when the polymer is overtreated, can result in a weak boundary layer leading to poor practical adhesion even if it is highly hydrophilic[30-32]. Crosslinking promotes cohesive strengthening in polymers[1]. It was thus interesting to study the He + NH_3 plasma treated polymers in order to evidence its role in strengthening the topmost layers at the surface and to stabilize the surface properties. The general effect should lead to the pronounced adhesion improvements.

Study of plasmas in He-NH_3 mixtures and their role in the stability of the surface properties

Plasma characterization by Optical Emission Spectroscopy

The He + NH_3 discharges have been studied in the whole range of the variation of the composition of the mixture by Optical Emission Spectroscopy (OES) and Mass Spectrometry (MS). The spectrum of the discharge was recorded in the UV-Visible region between 300 and 510 nm (Figure 4), in order to characterize the different excited species responsible for the simultaneous crosslinking and functionalization of the polymer.

One observes the lines corresponding to the transition between the radiative (3^3p, 3^1p) and metastable states of He (2^3S, 2^1S). It was reported elsewhere[33] that the triplet metastable, situated around 19.8 eV, was the most abundant metastable species in a pure He discharge and played an important role in the crosslinking effect. Placinta et al.[34] have shown that the addition of O_2 to a He discharge reduced the density of the He (2^3S) and increased the quenching rates via the Penning reactions.

The NH line observed at 336 nm, which was shown to catalyze the nitrogen uptake on the surface treated in an ammonia plasma in comparison to a nitrogen one[5], was also detected. In the same way, we note the presence in the spectrum of the characteristic lines of the second positive system of N_2 ($C^3\Pi_u$ ----> $B^3\Pi_g$) as well as the first negative system of N_2^+ line at 391.4 nm.

The emissions of H_β (486.1 nm), H_γ (434 nm) and H_δ (410.2 nm) result mainly from the fragmentation of the NH_3 molecules, but also due to the chain scission of the polymer since hydrogen emission lines have equally been observed in a nitrogen plasma[35]. However, the former is the main source of production of H species in an ammonia plasma.

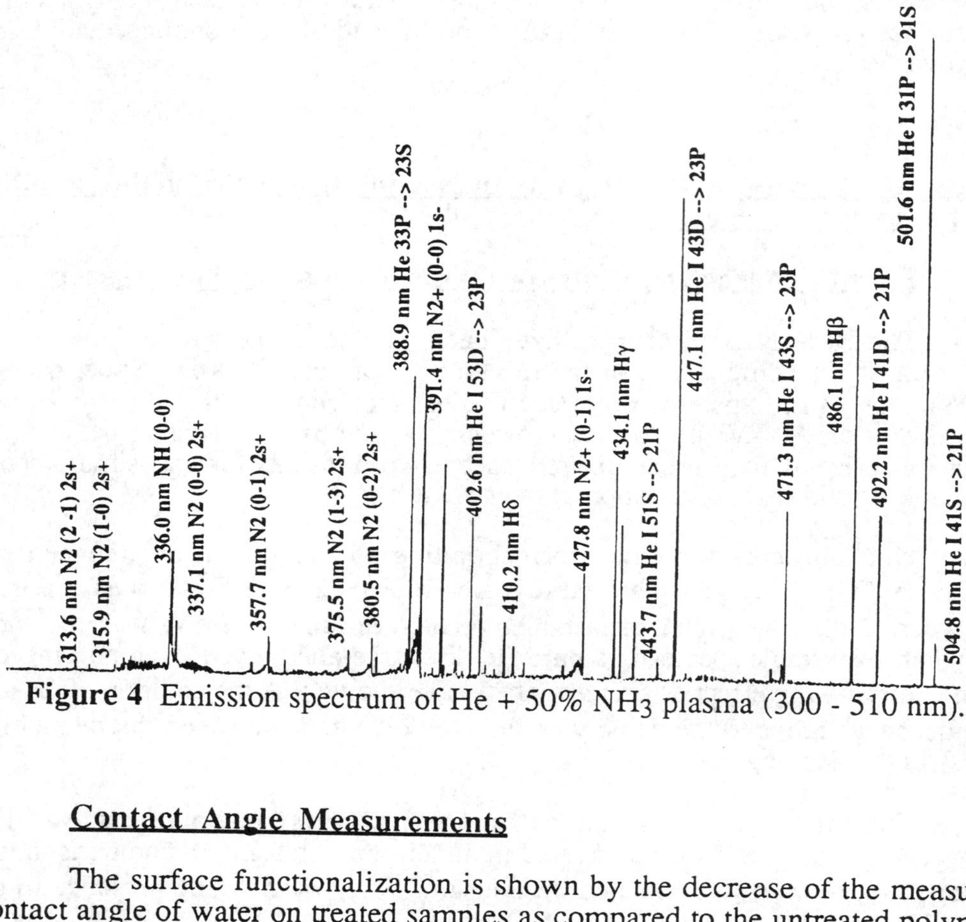

Figure 4 Emission spectrum of He + 50% NH_3 plasma (300 - 510 nm).

Contact Angle Measurements

The surface functionalization is shown by the decrease of the measured contact angle of water on treated samples as compared to the untreated polymer. Θ_{water} shows a steep decrease even for the helium treatment where one notes a decrease of about 40% as compared to untreated PP (Figure 5). However, one notices that for NH_3 in the discharge, Θ_{water} reaches a constant value around 42° throughout the whole composition of the mixture. This value is shown to be comparable to the contact angle measured on ammonia plasma treated PP (38°, cf Figure 5). These results show that the functionalization in the He+NH_3 plasma mixture is efficient as soon as a few percentages of ammonia is introduced in a helium discharge.

between He^+ species and O_2 to give two O* detected by OES, respectively, at λ = 777.1 nm and λ = 844.6 nm :

$$He^+ + O_2 \text{ -----> } He + O^+ + O\ (3^5P) \qquad (1)$$

The Penning reactions are particularly important in plasmas sustained in mixtures, which include gases such as helium which are characterized by long-lifetime metastable states. De Benedictis and Dilecci[37] have shown that N_2 and O_2 are very efficient quenchers of He (2^3S) and that the addition of N_2 to a helium discharge reduces the He (2^3S) density. The process responsible for such a strong quenching is the Penning ionization of N_2 according to:

$$He_M + N_2\ (X^1\Sigma^+{}_g) \text{ -----> } He\ (^1\ 1S) + N_2^+\ (B^2\Sigma^+{}_u) + e^- \qquad (2)$$

With the increase of the percentage of ammonia introduced in the gaseous mixture, the energetic character of the discharge is decreased and thus the density of the excited species of helium (He_M, He^+). As a result of this, the probability of excitation of oxygen molecules decreases in favour of that of ammonia molecules. This occurs either by Penning ionization for helium-rich mixtures (<4% NH_3) or by direct electronic impact. This is observed on the surface by a decrease of the O/C ratio and an increase of the N/C ratio as soon as a few percentages of NH_3 is added to the discharge. XPS results are in agreement with the contact angle measurements as they have shown the efficiency of the surface nitridation for small amount of ammonia in a helium discharge.

As plasma treated polymer surfaces are known to restructure with the ageing time[38], the role of the He + NH_3 plasma on the stability of the surface wettability has been studied by measuring the water contact angle on the plasma treated PP stored in air at room temperature. The measurements were made over 70 days after the treatment and were correlated to the XPS results, i.e. (O+N)/C ratio (Figure 7).

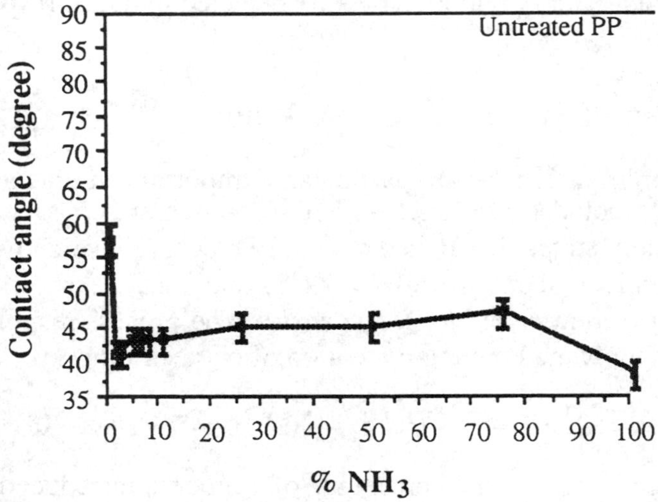

Figure 5 Variation of the water contact angle on treated PP with the composition of the gaseous mixture.

The wettability of the surface is closely related to the number of polar groups grafted by the plasma treatment. This is why the variations of the O/C and N/C ratios with the composition of the mixture have been reported in Figure 6. As can be seen in this figure, the increase of the ammonia percentage in the gas mixture leads to a decrease of the O/C ratio (16.5% for He plasma -----> 5% for NH_3 plasma) while the N/C ratio shows a sharp increase at first (for 1% NH_3) before reaching a plateau around 9%.

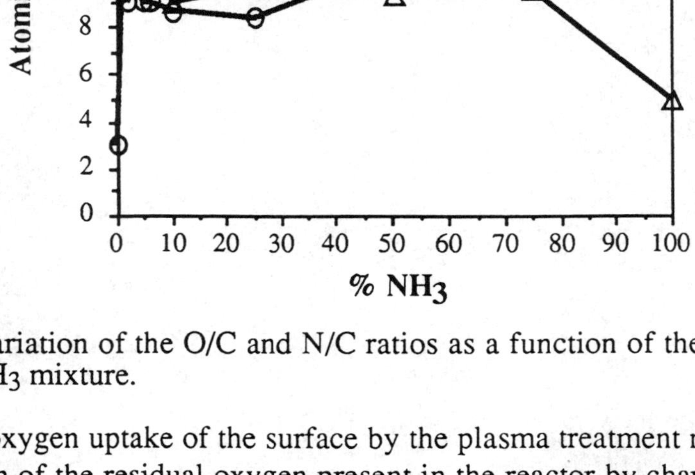

Figure 6 Variation of the O/C and N/C ratios as a function of the composition of the He-NH_3 mixture.

The oxygen uptake of the surface by the plasma treatment may be due to the excitation of the residual oxygen present in the reactor by charge transfer[36]

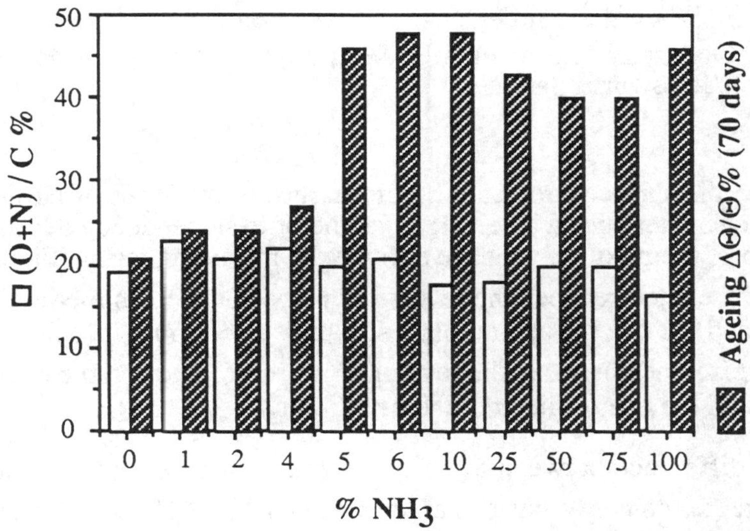

Figure 7 Variation of the water contact angle on treated PP with the ageing time with the composition of the mixture. Correlation to the O+N/C ratio.

A good stability of the wettability is obtained for samples treated in gaseous mixtures which contain less than 5% of ammonia in the discharge. Thus, we can estimate the variation of Θ_{water} for 1% of ammonia in the mixture to be about 22%, while it is around 45% for pure ammonia discharge. Since the (O+N)/C ratio is practically constant over the whole composition of the mixture (≈ 20%), the reorganization of the surface cannot be a function of the number of polar groups grafted by the plasma treatment. In this way, the gain in the stability observed above with only a few percentages of ammonia in the plasma (<5%) can be explained by the lower mobility of the surface polar groups as the crosslink density of the topmost layers of He + NH_3 plasma treated PP is increased. This crosslinked layer will also inhibit the diffusion of low molecular weight (LMW) fragments up to the surface limiting in this way the ageing phenomenon[13,14,39].

The study of the stability of the surface wettability can be considered to be an indirect indication of the surface crosslinking effect. Therefore, in He + NH_3 plasma mixtures, crosslinking and functionalization occur simultaneously but the compromise between these two mechanisms is optimal only when a few percentages of ammonia (<5%) is introduced in the discharge. Placinta et al.[34] have shown in the case of He + O_2 mixtures that a small amount of oxygen was required to carry out both the crosslinking and functionalization in a one-step process.

Acid-base Characteristics

The acid-base properties of He + NH_3 treated polypropylene surfaces were characterized using contact angle titration method. This method employs the sessile drop technique, but instead of pure water, unbuffered aqueous acidic and basic solutions with pH values ranging from 1 to 14 were used. From these measurements, the work of adhesion is defined as a function of pH and the shape of the curve shows the acidic, basic or amphoteric character of the treated surfaces[40]. Generally, the reversible thermodynamic work of adhesion W_{sl} is defined by the following relation :

$$W_{sl} = \gamma_l \, (1 + \cos \Theta)$$

Since the surfaces treated with discharges in mixtures of He + NH_3 show amphoteric characteristics with either an acidic or basic tendency depending on the composition of the mixture, we have defined the following terms (Figure 8) :

ΔW_a = the difference between the reversible work of adhesion measured for acidic test liquids and the one for neutral pH test liquids (pH = 7).

ΔW_b = the difference between the reversible work of adhesion measured for basic test liquids and the one for neutral pH test liquids.

The difference between these two terms will determine the predominant surface character namely a basic character when $\Delta W_a - \Delta W_b > 0$ or an acidic one when $\Delta W_a - \Delta W_b < 0$.

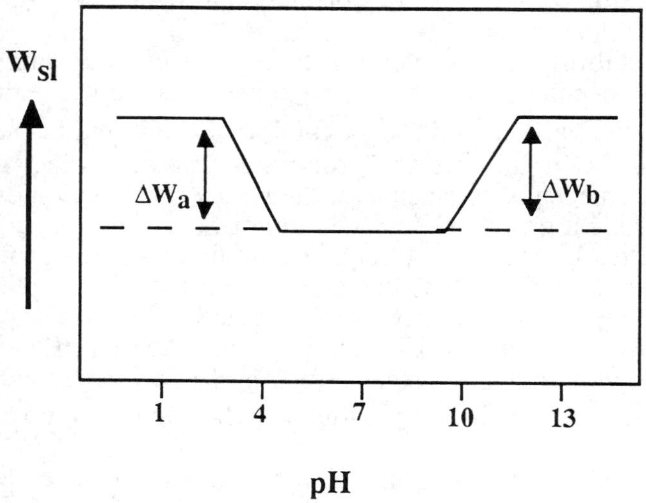

Figure 8 Schematic representation of ΔW_a and ΔW_b.

The contact angles of unbuffered aqueous solutions at the surface were measured immediately after treatments. Figure 9 shows the variation of the total work of adhesion as a function of pH, for He, NH_3 and He + 50% NH_3-plasma treated PP. First of all, one notes that the behaviour of the untreated PP is independent of the pH of the test liquids used. In this case, the substrate being a nonpolar one the acid-base interactions do not exist between the substrate and the aqueous test liquids. For helium treated PP, the surface reveals an amphoteric character compared to the one with a basic tendency obtained in the case of NH_3 and He + 50% NH_3 ($\Delta W_a - \Delta W_b$ = 7 mJ/m^2, $\Delta W_a - \Delta W_b$ = 16 mJ/m^2), respectively.

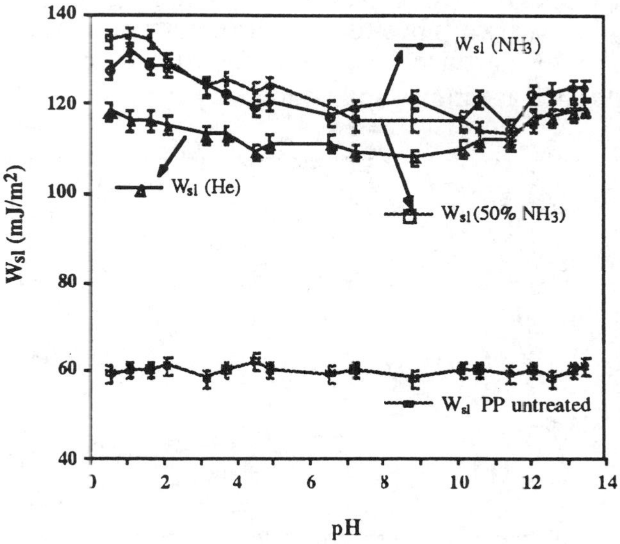

Figure 9 Variation of the total work of adhesion with different pH test liquids for untreated PP, He, NH_3 and He + 50% NH_3 plasma treated PP.

In order to determine the predominant character of the polypropylene surface treated in various He + NH_3 plasma mixtures, Figure 10 shows the variation of the total work of adhesion with the composition of the mixtures for two extreme pH values, an acidic one (pH = 0.5) and a basic one (pH = 13.45). The trend of the curves is similar in both cases. One observes the presence of a maximum in the reversible work of adhesion for 3% of ammonia in the discharge revealing in this way a greater interaction between the treated surfaces and the aqueous solutions. This can be related to the fact that the polar functional groups grafted to the surface by the plasma treatment, carried out in gaseous mixtures with less than 10% of ammonia, are more accessible at the topmost layer. As an explanation for the results given above, we can propose the following interpretations :

1- A good compromise between the functionalization and the crosslinking is reached for less than 10% of ammonia in the gaseous mixture as a result of the CASING process induced by the helium gas.

2- As we have noted before in Figure 6, the O/C ratio decreases with the increase of the amount of ammonia in the discharge. Furthermore, one can suggest that for a few percentages of ammonia introduced in the mixture (<10%), there exists a high probability to form hydrogen bonding between the oxygen containing groups. This effect will contribute to limit the reorganization of the grafted polar groups inside the polymer and thus to keep them more accessible in the the first few monolayers of the surface[41].

However, the variation of the total work of adhesion measured at pH = 0.5 is always higher than the one measured at pH = 13.45 except for a pure helium discharge which shows an amphoteric character with an acidic tendency as mentioned above. In other words, whatever the composition of the mixture, the

He + NH_3 plasma treated PP surfaces show an amphoteric character with a basic tendency i.e. W_a-W_b is always positive for different test liquids. The predominant basic character of the surface has been reported to be due to the nitrogen containing groups such as amines and amides observed by XPS[24].

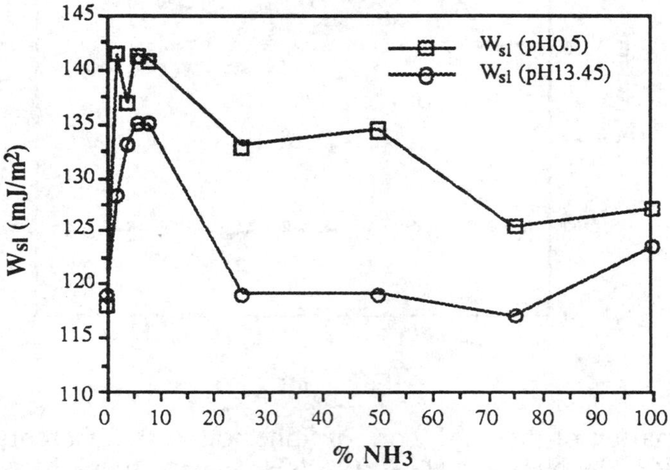

Figure 10 Variation of the total work of adhesion for PP as a function of the composition of the He + NH_3 mixture at pH = 0.5 and pH = 13.45.

In order to point out the functionalization/crosslinking compromise obtained for a few percentages of ammonia in the composition of the gaseous mixture, Figure 11 shows the variation of the global acid or basic character of the surface as well as the ageing of the surface represented by the parameter $\Delta\Theta/\Theta$ $[(\Theta_f-\Theta_i)/\Theta_i]$. $(\Theta_f-\Theta_i)$ corresponding to the difference between the water contact angle measured on the treated plasma PP aged for 70 days and the one taken on the fresh plasma-treated surface.

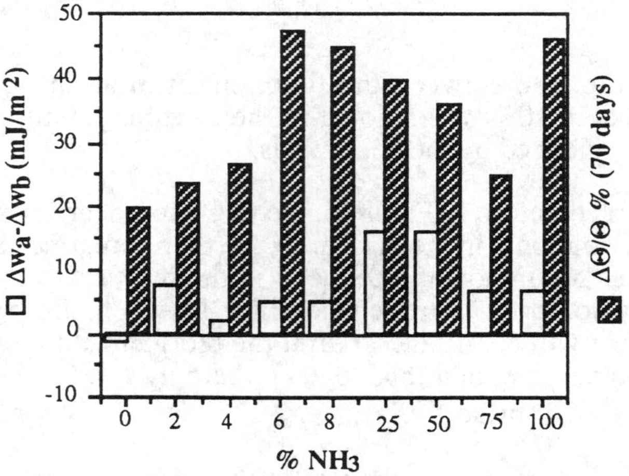

Figure 11 Variation of the acid-base character of the surface (ΔW_a-ΔW_b) as well as the variation of the contact angle with time for different compositions of the He + NH_3 mixture.

One notes in this figure the basic character of the He + NH_3 treated PP surfaces irrespective of the composition of the mixture, as ΔW_a-ΔW_b is positive for percentages as low as 1% of NH_3 in the discharge. One also observes that for less than 3% of ammonia in the mixture, the surface of PP reveals a basic character similar to the one obtained with pure ammonia plasma. However, when PP films are treated in He + NH_3 plasmas for 1 s with less than 3% of ammonia in the gaseous mixture, the stability of the surface wettability is better ($\Delta\Theta/\Theta$ = 22%) than when they are treated in a pure ammonia plasma ($\Delta\Theta/\Theta$ =45%).

The results above show that helium-rich discharges with a small amount of ammonia are suitable for adhesion improvement. Knowing that such surfaces have a predominant basic character, it was interesting to study the stability of their surface properties.The stability of the surface acid-base properties was also studied for ammonia-plasma treated PP.

The stability of the basic character of NH_3 and He + 2% NH_3 plasma-treated PP is reported in Figure 12. In fact, it is for such a mixture that we have noted an excellent compromise between the functionalization and the crosslinking effects as judged by the wettability, the stress-strain and the acid-base characteristics. One notices in this figure, the decrease of the global basic character of the NH_3 and He + 2% NH_3 plasma-treated PP when stored at ambient air. The decrease of the basic character of the surface by ageing can be explained in terms of neutralization of the groups possessing a basic character such as amines, amides and imines, by atmospheric contaminants such as CO_2[24,25] or H_2O[42]. It can also be attributed to a surface rearrangement through contact with air. However, the topmost layers of the He + 2% NH_3 plasma-treated PP films, consisting of the polar groups, are less rapidly reorganized when compared to the pure ammonia treated samples. Such a difference can be explained by the lower mobility of the surface polar groups after the He + 2% NH_3 plasma treatment has crosslinked the polymeric surface, as the mobility of the polymer segments at the surface depends on its crosslink density. Further studies are necessary in order to correlate the loss of the basic character with the adhesion of the Al layers to aged He + 2% NH_3 treated PP.

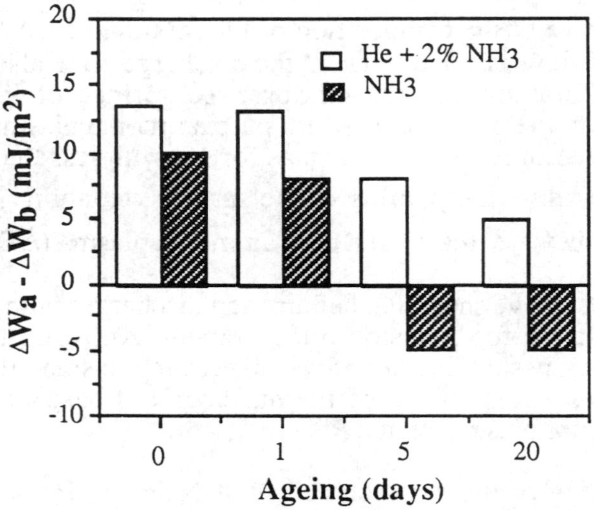

Figure 12 Acid-based character of the NH_3 and He + 2% NH_3 plasma treated PP, measured by the variation of ΔW_a - ΔW_b with the ageing.

In the first part of this paper, we have studied the variation of the contact angles with the composition of the He + NH_3 plasma mixtures as an indirect measurement of the crosslinking of the topmost layers of the surface, the second indirect method used is the stress-strain characteristics explained in the next section. As it was mentioned earlier in this paper, the formation of a weak boundary layer which degrades the adhesion properties of the surface may affect the mechanical properties of the polymers since it can modify the locus at failure and the overall bond strength. It was thus interesting to study the mechanical properties of the He + NH_3 plasma treated polymers in order to evidence its role in strengthening the topmost layers at the surface.

Stress-Strain Characteristics

Figures 13(a, b) show the variation of the elongation-at-break point of He+NH_3 plasma treated PP with the treatment time and the composition of the mixture. First of all, one observes a similar trend in the variation of the elongation curves for the three different treatment times. Also, one notices that by increasing the treatment time without modifiying the plasma composition, the break point is shifted to lower elongation revealing in this way the fragile nature of the sample due to the formation of low molecular weight fragments after an overtreatment. For example, at 6% of NH_3 the elongation-at-break decreases from 160% (untreated PP) to 150% for the first treatment time (t = 1s) and reaches 110% for an overtreated sample (t = 30s).

For the same treatment time, the elongation curves show two distinct regions :

- the first one is situated below 10% of ammonia in the mixture. This region corresponds to a sharp decrease in the elongation-at-break point, i.e. 170% (0%

NH3) to 130% (10% NH3).

- the second region corresponds to ammonia percentages equal or above 10% where one observes a plateau with a constant value for the elongation-at-break.

As the functionalization of the surface remains practically constant over the whole composition of the mixture, therefore the results suggest that when the percentage of ammonia in the gaseous mixture is decreased, the cohesive strengthening of the polymer surface is enhanced. The latter is probably due to the formation of a thin crosslinked layer that mechanically stabilizes the surface.

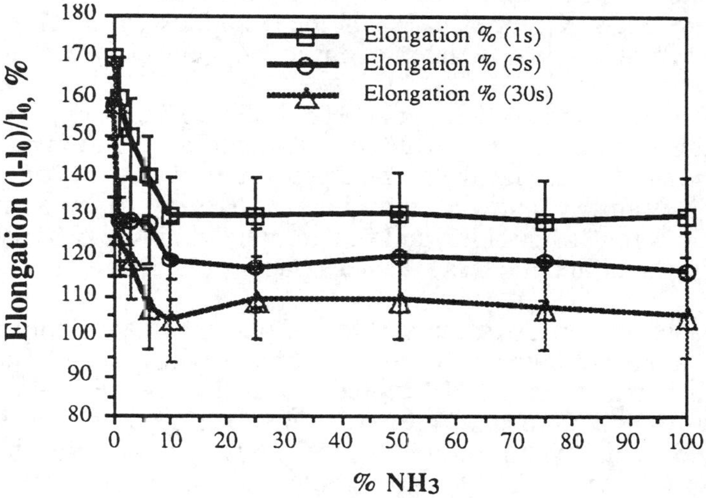

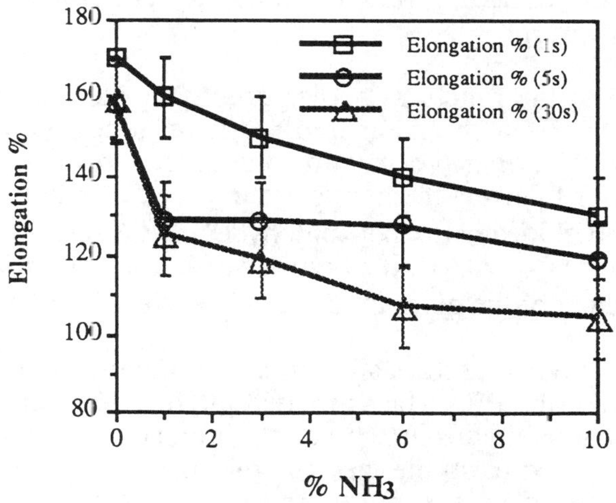

Figure 13 Elongation-at-break of He + NH_3 plasma treated PP as a function of the composition of the mixture and the treatment time (a) for He/NH_3 mixtures.

CONCLUSION

In this paper, plasma process variables such as the treatment time, the nature of the plasma gas (O_2, NH_3) and the composition of the plasma mixtures (He + NH_3) have been studied.

In the first part, the adhesion improvement of Al to PP as well as the elongation-at-break have been studied using two different plasma gases : an oxidizing atmosphere such as oxygen and a reducing one such as ammonia. For long treatment times (>1s) in the case of both gases, a degradation of the adhesion properties of Al to PP has been observed due to the formation of a weak boundary layer. For short treatment times (<1s) the better adhesion of Al to PP has been explained by favorable acid-base interactions at the metal-polymer interface.

In the second part of this paper, He + NH_3 plasma mixtures have been studied in order to evidence their role in strengthening the topmost layers at the treated surface and to stabilize the surface properties. The study of the stability of the surface wettability showed that in He + NH_3 plasma mixtures, the compromise between crosslinking and functionalization was optimal only when a few percentages of ammonia, less than 4%, was introduced in the discharge.

The basic properties of the surface of plasma-treated polymers can give rise to a good adhesion between aluminium and polypropylene. That is why, the acid-base properties of He + NH_3 treated PP surfaces were characterized using contact angle titration method. Therefore, in the case of discharges in mixtures of He + NH_3, and for He-rich mixtures (<3% of NH_3), the results showed a better stability of the surface wettability and a comparable basic character with the ones obtained for pure ammonia plasma. Therefore, it appears that it is more interesting to work with mixtures of helium with a small amount of ammonia in the discharge for adhesion improvement.

The nature of the functional polar groups grafted on the surface should play an important role in the measured work of adhesion with different test liquids. This is why, further works aim to characterize precisely the nature of these polar functions by using the derivatization method (derivatization of amino groups with 4-trifluoromethylbenzaldehyde[43]). Angle dependent x-ray photoelectron spectroscopy (ADXPS) will also be used in order to determine the concentration gradient of the polar groups in the surface[38,44].

The second method used as an indirect measurement of the crosslinking of the topmost layers of the PP surfaces treated in He + NH_3 plasma mixtures was the stress-strain characteristics. Once again, the results show that on decreasing the percentage of ammonia in the gaseous mixture (<4%), one can enhance the cohesive strengthening of the polymer surface by the formation of a thin crosslinked layer that mechanically stabilizes the surface.

REFERENCES

1. E.M. Liston, L. Martinu and R. Wertheimer, in "Plasma Surface Modification of Polymer: Relevance to Adhesion," M. Strobel, C.S. Lyons and K.L. Mittal (Eds.), pp. 3-39, VSP, Utrecht, The Netherlands, 1994.

2. E.M. Liston, in "Polymer-Solid Interfaces," J.J. Pireaux, P. Bertrand and J.L. Brédas (Eds.), pp. 429-442, IOP Publishing, 1992.

3. F.D. Egitto and L.J. Matienzo, Paper presented at the 189th Meeting of the Electrochemical Society, Los Angeles, CA, May 5-10, 1996.

4. F. Arefi-Khonsari, M. Tatoulian, N. Shahidzadeh and J. Amouroux, in "Plasma Processing of Polymer," R. d'Agostino et al. (Eds.), pp. 165-207, Kluwer Academic publishers, 1997.

5. F. Arefi-Khonsari, M. Tatoulian, V. André, J. Amouroux and G. Lorang, in "Metallized Plastics 3: Fundamental and Applied Aspects," K.L. Mittal (Ed.), pp. 243-256, Plenum Press, New York,1992.

6. F.M. Fowkes, in "Surface and Colloid Science in Computer Technology," K.L. Mittal (Ed.), pp. 3-25, Plenum Press, New York, 1987.

7. J.J. Bikerman, "The Science of Adhesive Joints," 2nd Ed., Academic Press, New York, 1968.

8. E. Morra, E. Occhiello and F. Garbassi, in "Polymer-Solid Interfaces," J.J. Pireaux, P. Bertrand and J.L. Brédas (Eds.), pp. 407-428, IOP Publishing, 1992.

9. Ronald C. Chatelier, Ximing Xie, Thomas R. Gengenbach and Hans J. Griesser, *Langmuir,* **11**, 2585-2591 (1995).

10. Thomas R. Gengenbach, Ronald C. Chatelier and Hans J. Griesser, *Surface Interface Anal.*, **24**, 271-281 (1996).

11. M. Morra, E. Occhiello and F. Garbassi, in "Plasma Surface Modification of Polymers: Relevance to Adhesion," M. Strobel, C.S. Lyons and K.L. Mittal (Eds.), pp. 183-195, VSP, Utrecht, The Netherlands, 1994.

12. M. Strobel, C.S. Lyons, J.M. Strobel and R.S. Kapaun, *J. Adhesion Sci., Technol.*, **6**, 429-443 (1992).

13. J.E. Klemberg-Sapieha, L. Martinu, O.M. Kûttel and M.R. Wertheimer, in "Metallized Plastics 2: Fundamental and Applied Aspects," K.L. Mittal (Ed.), pp. 315-329, Plenum Press, New York,1991.

14. M. Tatoulian, F. Arefi-Khonsari, I. Mabille-Rouger, J. Amouroux, M. Gheorgiu and D. Bouchier, *J. Adhesion Sci. Technol.*, **9**, 923-934 (1995).

15. H. Schonhorn and R.H. Hansen, *J. Appl. Polym. Sci.*, **11**, 1461-1474 (1967).

16. D.T. Clark and A. Dilks, *J. Polym. Sci., Polym. Chem. Ed.*, **16**, 911 (1978).

17. D.T. Clark and A. Dilks, *J. Polym. Sci., Polym. Chem. Ed.*, **17**, 957 (1978).

18. G. Placinta, F. Arefi-Khonsari, J. Amouroux and G. Popa, Proceedings of ISPC **13**, 1269-1274 (1997).

19. M. Alami, M. Charbonnier and M. Romand, *Plasmas and Polymers*, **1**, 113-126 (1996).

20. J.J. Bikerman, *J. Appl. Chem.*, **11**, 81-85 (1961).

21. P. Montazer-Rahmati, F. Arefi, R. Borrin, A. Delacroix and J. Amouroux, *Bull. Soc. Chim. de France*, **5**, 811-816 (1988).

22. Stephen Randall Holmes-Farley, Robert H. Reamey, Thomas J. McCarthy, John Deutch and George M. Whitesides, *Langmuir*, **1**, 725-740 (1985).

23. M.D. Vrbanac and J.C. Berg, in "Acid-Base Interactions: Relevance to Adhesion Science and Technology," K.L. Mittal and H. Anderson, Jr. (Eds.), pp. 67-78, VSP, Utrecht, 1991.

24. N. Shahidzadeh, P. & M. Curie Thesis, 1996.

25. F. Arefi-Khonsari, M. Tatoulian, N. Shahidzadeh, M. Chehimi, J. Amouroux, D. Leonard and P. Bertrand, Mittal Festschrift, VSP, in press.

26. J. Bernard, "Oxidation of Metals, I. Fundamental Process," Gauthier-Villars & C., Paris, 1962.

27. George A. Parks, *Chem. Rev.*, **63**, 177-198 (1965).

28. H. Noller, J.A. Lercher and H. Vinek, *Mater. Chem. Phys.*, **18**, 577-593 (1988).

29. N. Shahidzadeh-Ahmadi, F. Arefi-Khonsari, M.M. Chehimi and J. Amouroux, *Surface Sc.*, 352-354 (1996).

30. I. Sutherland, R.P. Popat and D.M. Brewis, *J. Adhesion*, **46**, 79-88 (1994).

31. R. Foerch, G. Kill and M.J. Walzak, in "Plasma Surface Modification of Polymers: Relevance to Adhesion," M. Strobel, C.S. Lyons and K.L. Mittal (Eds.), pp. 99-111, VSP, Utrecht, The Netherlands, 1994.

32. M.R. Wertheimer and R. Bartnikas in "Plasma Processing of Polymers," R. d'Agostino et al. (Eds.), pp. 435-450, Kluwer Academic Publishers, 1997.

33. R. Deloche, R. Monchicourt, M. Cheret and F. Lambert, *Phys. Rev.* A. **13**, 1140-1176 (1976).

34. G. Placinta, F. Arefi-Khonsari, M. Gheorghiu, J. Amouroux and G. Popa, *J. Appl. Polym. Sci*, **66**, 1367-1375 (1997).

35. V. Andre, F. Tchoubineh, F. Arefi and J. Amouroux, in "Plasma Surface Interactions and Processing of Materials," O. Auciello (Ed.), Kluwer Academic Publishers, 1990.

36. R. Johnsen, J.A. MacDonald and M.A. Biondi, *J. Chem. Phys.*, **66**, 4718-4719 (1977).

37. S. De Benedictis and G. Dilecce, *Plasma Sources Sci. Technol.*, **4**, 212-223 (1995).

38. Ximing Xie, Thomas R. Gengenbach and Hans J. Griesser, *J. Adhesion Sci. Technol.*, **6**, 1411-1431 (1992).

39. M.H. Bernier, J.E. Klemberg-Sapieha, L. Martinu and M.R. Wertheimer, in "Metallization of Polymers," E. Sacher, J.J. Pireaux and S. Kowalczyk (Eds.), pp. 147-160, ACS Symp. Ser. No. 440, American Chemical Society, Washington, DC, 1990.

40. K.J. Hüttinger, S. Höhmann-Wien and G. Krekel, *J. Ahesion Sci. Technol.*, **6**, 317-331 (1992).

41. M. Morra, E. Ochiello and F. Garbassi, in "Polymer-Solid Interfaces," J.J. Pireaux, P. Bertrand and J.L. Brédas (Eds.), pp. 407-428, IOP Publishing, 1992.

42. L.J. Gerenser, in "Plasma Surface Modification of Polymers: Relevance to Adhesion," M. Strobel, C.S. Lyons and K.L. Mittal (Eds.), pp. 43-64, VSP, Utrecht, The Netherlands, 1994.

43. J.G.A. Terlingen, L.M. Brenneisen, H.T.J. Super, A.P. Pijpers, A.S. Hoffman and J. Feijen, *J. Biomater. Sci, Polym. Ed.*, **4**, 165 (1993).

44. Thomas R. Gengenbach, Ximing Xie, Ronald C. Chatelier and Hans J. Griesser, *J. Adhesion Sci. Technol.*, **8**, 305-328 (1994).

Metallized Plastics 5&6: Fundamental and Applied Aspects, pp. 319-330.
K.L. Mittal (Ed.)

PLASMA PRETREATMENT FOR METALLIZATION OF POLYIMIDES: EFFECT OF PLASMA SYSTEM CONFIGURATION

F.D. Egitto and L.J. Matienzo

IBM Corporation
Microelectronics Division
Endicott, NY 13760

Plasma modification of polymer surfaces is a common means of improving adhesion to subsequently-deposited metal films. Plasma system configuration and processing parameters greatly influence the degree of adhesion enhancement and, if not designed properly, may even degrade adhesion. Depending on the configuration and plasma parameters, modification can proceed primarily through surface reactions with chemically reactive neutral species generated in the plasma, through physical bombardment by energetic particles (e.g., ions and photons), or, most commonly, a combination of these chemical and physical mechanisms (through ion-assisted surface chemistry). Plasma-induced polymer surface sites that interact chemically with deposited films can improve adhesion by orders of magnitude over that obtained for untreated polymer surfaces. Crosslinking, enhanced by bombardment of ions, electrons and photons from the plasma, improves the polymer cohesive strength while excessive chain scission reactions may lead to weak boundary layers and degrade practical adhesion. In general, effective modification induces "adhesion-promoting" functional groups at the surface without degrading the structural integrity of the near-surface region. This paper focuses on the use of plasmas to modify polymer surfaces for improvement of adhesion to metals. Emphasis is placed on plasma treatment of polyimides. Examples are given to illustrate the effect that plasma reactor configuration has on resulting metal/polymer adhesion.

INTRODUCTION

Plasma treatment has been employed to alter the surface wetting and adhesion behavior of many natural and synthetic polymers. Plasma processes[1,2] and plasma treatment of polymer surfaces[3-7] have been reviewed extensively in the literature. Other reviews have been more specific in nature with respect to the type of polymer films investigated. Such reviews have focused on polymers having unique and desirable properties for industrial applications, for example, the class of polymers known as polyimides[8]. Polyimides exhibit good thermal stability, reproducible film-forming characteristics, and electrical properties well suited to microelectronics applications. Plasma treatment is rarely the last processing step for a polymer surface in an industrial application. More commonly, the plasma process is a means to enhance the performance characteristics of the polymer during subsequent processing steps, for example, adhesion of deposited metal films. Matienzo and Unertl have discussed metallization of polymers and measurement of practical adhesion of the metal/polymer interface with emphasis on polyimides[9].

Metal/polymer adhesion in industrial processes is a result of contributions from a number of parameters related to the polymer, the metal, and the polymer surface modification technique, e.g., plasma (Table I). The present paper primarily addresses the plasma-related parameters and, in particular, the design of the plasma reactor and the effect of the reactor design on the chemical and physical processes that occur during surface modification of polymers. As such, this is not a comprehensive review of plasma modification of polymer surfaces, but a rather focused summary of the relevant literature as it pertains to plasma system configuration. Although the emphasis is on treatment of polyimides, reference is made to treatment of other polymers for comparison and to illustrate concepts for which no data are available for polyimides.

Table I. Some of the parameters that affect measurements of practical adhesion at metal/polymer interfaces.

POLYMER	PLASMA	METAL	ADHESION
Surface Functionality	System Configuration	Type	Test Method
Crosslinking Density	Excitation Frequency	Deposition Technique	Material Properties
Crystallinity	Gas	Deposition Conditions	Stress Conditions (e.g., Temp & RH)
Morphology	Pressure	Thermal Annealing	Etc.
Presence of Additives	Power	Presence of Oxides	
Outgassing	Gas Flow Rate	Etc.	
Etc.	Etc.		

Oxygen-plasma treatment is commonly employed to enhance adhesion of vapor-deposited metal films to polyimide[10-14]. Katnani, et al.[10] investigated oxygen-plasma treatment of polyimide film produced by thermally curing a polyamic acid having the precursors pyromellitic dianhydride (PMDA) and oxydianiline (ODA). Using x-ray photoelectron spectroscopy (XPS), they observed that the surface concentration of carbonyl groups (-C=O) increased with treatment time, and demonstrated that as the carbonyl concentration increased, the polyimide surface became more hydrophilic, and the adhesion of subsequently-deposited chromium films, measured using 90° peel test, was improved. Several years later, Egitto, et al.[15] reported on the effect that oxygen plasma treatment of Kapton[R] polyimide had on the adhesion to subsequently-deposited chromium films. Kapton material is a PMDA-ODA polyimide, like that studied by Katnani, et al.[10], except that it is cured via a chemical imidization reaction and some thermal treatment to form a free-standing dry film. In one set of experiments, films were not positioned directly in the plasma, but were treated in a remote location, downstream from the plasma. In this configuration, the material is exposed to long-lived reactive neutral species generated in the plasma region. As in the study by Katnani, et al.[10], treated films exhibited increased levels of carbonyl functionality and became increasingly hydrophilic. However, practical adhesion of chromium deposited onto the treated Kapton polyimide, as determined by 90° peel test, was reduced as a result of the pretreatment. Obviously, the practical adhesion, or bondability, between polymer surfaces and other materials deposited onto them cannot always be correlated with wetting behavior or surface functionality.

PLASMA SYSTEMS

Plasmas are collections of charged particles, most commonly occurring in the gaseous state. The types of gaseous plasmas used for surface modification of polymers are collections of electrons, ions, neutral particles in ground or excited states, and photons. Processing gases are dissociated into fragments by virtue of collisions with energetic electrons. The fragments from this dissociation can be highly reactive. Each of these plasma-generated species can interact with polymers to alter their surface properties without affecting the characteristics of the bulk (Figure 1). The degree of interaction depends upon selection of processing gas, reactor configuration, and a variety of other processing parameters.

The major structural components of plasma processing systems are a vacuum chamber, a vacuum pumping system, a gas introduction and distribution system, and an excitation source. Materials of chamber construction should not

---

[R]Kapton is a registered trademark of E.I. du Pont de Nemours & Co., Wilmington, DE.

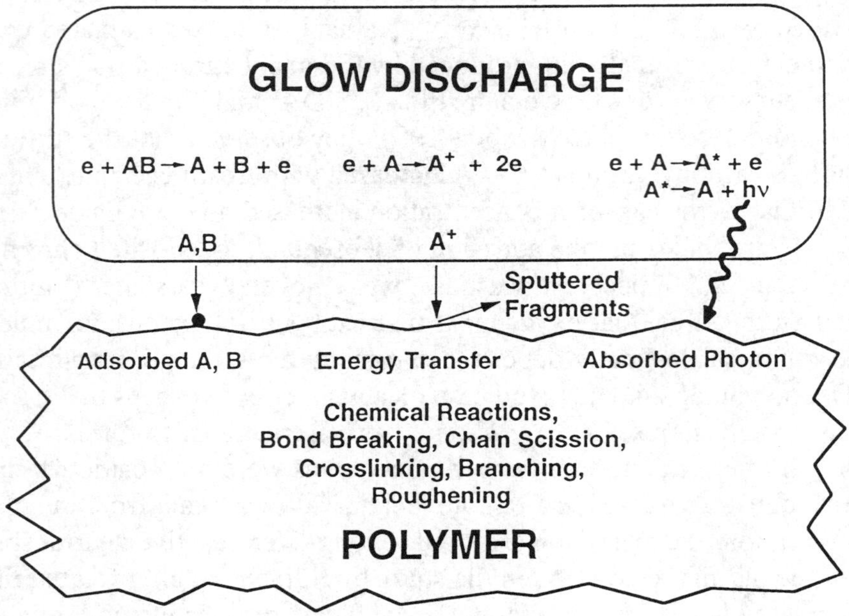

Figure 1. Plasma constituents and their roles in polymer surface modification.

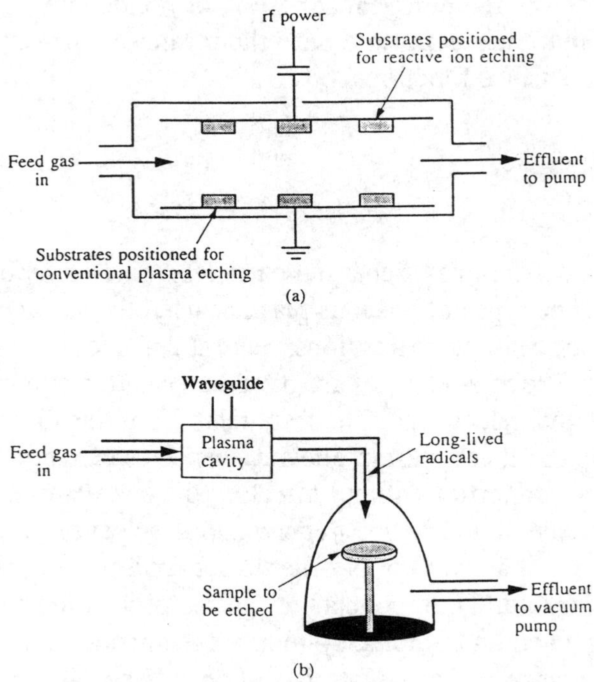

Figure 2. (a) RF-planar diode plasma system configuration with substrates positioned on the rf-driven (top) or grounded (bottom) electrodes. (b) Microwave plasma system configuration with a substrate positioned downstream from the plasma and not exposed to ion bombardment.

interact with the processing gases and their dissociative products. Such interaction could result in contamination of the surfaces being processed. Typically, a constant flow of gas is fed into the vacuum chamber while continuously pumping to maintain a fixed gas pressure.

Free electrons in the plasma interact with atoms and molecules in the gas phase, leading to fragmentation (dissociation) and ionization. These fragments can be chemically reactive. The cross section for dissociation via electron-molecule collisions does not increase linearly (and can even decrease) with electron energy[16]. However, it is generally true in most processing plasmas that higher electron energies will result in greater dissociation efficiencies and, consequently, higher processing speeds. Each of the plasma processing parameters listed in Table I affects the electron energy distribution in the plasma.

Processing of materials can be performed "in the plasma" or "remote from the plasma." In the former configuration, substrates reside directly in the chamber housing the plasma (Figure 2a). Plasma excitation sources in this configuration are commonly either direct current (DC) or radio-frequency (RF, typically 13.56 MHz). The plasma exists at a greater electrical potential than any surface with which it is in contact, and positive ions from the plasma are accelerated in a direction normal to such surfaces. These ions can attain energies from several electron volts (eV) to greater that 1.0 keV and, by virtue of their kinetic energy, can enhance the rates of chemical reactions at polymer surfaces and/or physically remove (sputter) material from the surface. In the "remote" configurations, samples reside in a location outside the influence of electrical potentials and electric fields and, therefore, these samples are not subjected to energetic ion bombardment (Figure 2b). Microwave (MW) frequencies are common (but not exclusive) excitation sources in these systems.

Parameters that affect the concentrations of reactive species and the densities and energies of ions that bombard surfaces in the plasma include feed gas composition, gas flow rates and chamber pressure, frequency and power of the excitation source, and configuration of the reactor. Particularly important attributes of reactor configuration are the size and shape of the electrodes and the position of the sample with respect to the electrodes.

POLYMER SURFACE MODIFICATION

For polymer surface modification to proceed by virtue of interaction with reactive neutral species in the plasma, the reactive dissociation products of the parent gas must serve to weaken polymer chemical bonds and change surface chemical composition. By far, the gas most commonly used for plasma modification of polymers is oxygen, but other commonly used gases include ammonia, nitrogen, noble gases and water vapor. As a result of surface chemical reactions, surface functionality is altered during plasma treatment. The electronic configurations of surface functional groups on the treated surface and of the subsequently-deposited material determine the nature and strength of chemical bonding at the interface. Acid-base interactions have been shown to play an important role in adhesion[17-20]. In a Lewis sense, polymers can be classified as acidic (electron acceptors), basic (electron donors), or neutral. Incorporation of carboxylic acid groups (-COOH) into a polymer imparts an acidic character to the surface whereas incorporation of amino groups (-NHR_2 or -NH_2R) or carbonyls (-C=O) imparts a basic character. Reactive metals, such as chromium, deposited in industrial systems that typically operate at 10^{-6} Torr or above are likely to oxidize in the gas phase[21,22]. Chromium oxides are more acidic than chromium metal[23]. Therefore, electron donors such as carbonyl groups can interact more strongly with the oxide.

The contribution of ion bombardment to polymer surface modification is a function of dose and energy of those ions incident onto the surface. The energy of ions that bombard surfaces in plasma systems typically does not exceed a few hundred eV. The ion penetration depth at this energy is on the order of a few angstroms. Ion interactions with surfaces can induce physical removal of atoms or groups (sputtering), or enhance chemical reactions in the presence of free radicals. The surface properties of polymers subjected to interactions with energetic ions are altered with respect to both chemical composition (e.g., functionality) and structure (e.g., crosslinking).

Ultraviolet and vacuum ultraviolet radiation is abundant in many plasmas and most polymers absorb well in this region. As a result, photon irradiation at such wavelengths can also be used to modify polymer surfaces. Modification by high energy photons in the plasma can occur by virtue of crosslinking, desaturation (e.g., for polyolefins like PTFE and PE), removal of surface atoms or groups, or photoablative roughening[24,25].

Impurities at the polymer surface can affect wetting behavior and surface chemical analyses. These contaminants can affect observed experimental trends in modification-induced surface properties. For example, hydrocarbon contaminants found on fluoropolymer films[25-28] can be modified at conditions that would not alter the uncontaminated fluoropolymer. Hydrocarbon contaminants on commercial films of PTFE have been found using hydrogen forward recoil spectroscopy[25]. Enhancement in the hydrophilic nature of a contaminated surface could lead to erroneous assumptions regarding the treatment conditions required to modify the uncontaminated polymer to achieve optimum adhesion behavior.

PLASMA-INDUCED SURFACE MODIFICATION: EFFECT ON ADHESION

Oxygen plasma treatment of polyimide surfaces often results in improved adhesion to subsequently-deposited metal films. The presence of oxygen-containing groups at the polymer surface can lead to formation of metal-oxides. Burkstrand has shown that the formation of metal-oxygen-polymer complexes formed on oxygen plasma-treated polymer surfaces correlated with adhesion of the metal film[29]. Metals were observed to interact with hydroxyl (-OH), carbonyl (-C=O) and ester (-COOR) groups (typically more basic in nature than the untreated polymer, with respect to the metal) on the plasma-modified surface. However, as noted above, there are instances for which measurements of practical adhesion to plasma-treated polymers are actually lower than obtained on the untreated polymer films[15]. Although oxygen plasmas typically result in an oxygenated polymer surface with increased hydrophilic character, plasmas impart properties to the polymer surface that are not revealed by measurements of contact angle or atomic composition, or by analysis of the bonding environments of these atoms on the surface (for example, as determined by XPS). In particular, plasma treatments in systems having different configurations can yield similar improvements in the hydrophilic nature of the polymer surface, but vastly different behavior with respect to practical adhesion to metal films (Figure 3).

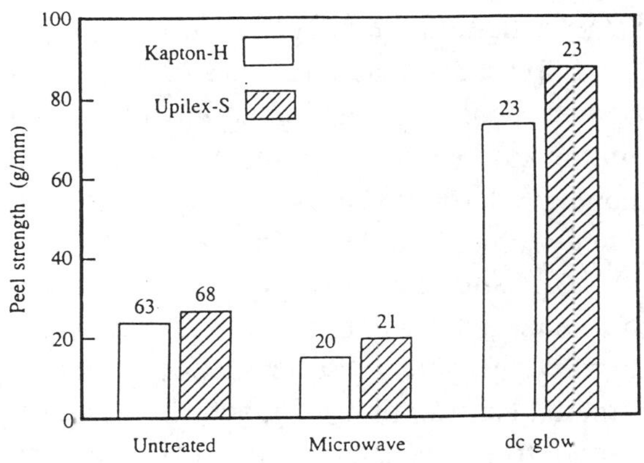

Figure 3. Peel strength measurements used to compare the practical adhesion of chromium lines on Kapton-H and Upilex-S[R] polyimide films. Shown are results for the untreated polymers and for films treated downstream from a microwave plasma and treated directly in a dc-glow discharge. Advancing DI water contact angles are shown above the bars. Smaller contact angles indicate a more hydrophilic polymer surface (From reference 15).

[R]Upilex is a registered trademark of UBE Industries, LTD., Tokyo, Japan.

Some insight into the role that system configuration plays in oxygen plasma modification of polymer surfaces is provided by the investigations of Tead, et al.[30]. This group reported on the relative occurrence of chain scission reactions and crosslinking occurring at the surface of polystyrene (olefinic backbone with aromatic side groups) as a function of oxygen plasma treatment using various plasma system configurations. They studied three plasma systems in which samples were exposed to bombardment by ions having kinetic energies ranging from a few eV to 1.0 keV, and a fourth system in which samples were placed in a location remote from the plasma region and not exposed to energetic ion bombardment. Following treatment, the rate of thermal diffusion (at 178 °C) of a deuterated polystyrene layer into the treated polystyrene samples was tracked using Rutherford Backscattering Spectroscopy. Lower diffusion rates were attributed to a greater degree of crosslinking. It was concluded that configurations conducive to higher degrees of ion bombardment were favorable toward crosslinking events, while systems for which ion bombardment was light were more prone to chain scission events.

Crosslinking promotes improved cohesive strength in polymers and increased resistance to solvents and moisture diffusion. On the other hand, low molecular weight fragments resulting from excessive chain scission can result in a weak boundary layer leading to poor practical adhesion. Certainly, the presence of a weak boundary layer, even if highly hydrophilic, will lead to a low value of practical adhesion.

Egitto, et al.[15] investigated the effect of plasma system configuration on adhesion of chromium to plasma-treated surfaces of Kapton polyimide (Figure 3). It was suggested that treatment of polyimide downstream from an oxygen microwave plasma, in the absence of ion bombardment, resulted in excessive chain scission reactions and yielded lower values of peel strength for sputter deposited chromium than obtained with untreated polyimide. In the same study, treatment performed directly in the plasma (using an oxygen DC glow discharge), where samples are exposed to energetic ion bombardment and, as a result, may be more prone to crosslinking of the polymer, resulted in peel strengths about three times that obtained for untreated Kapton films. In fact, the value of peel strength in the latter instance was shown to increase as the kinetic energy and flux of ions to the surface was increased by variation in plasma parameters (power and pressure). Both configurations resulted in greatly enhanced hydrophilic character of the treated surfaces and increased levels of oxygenated groups on the surface. Configuration-dependent differences in plasma-induced surface functionality were measured. XPS analysis suggested that the major contribution to the microwave (remote) plasma modification was due to the incorporation of carbonyl groups; for modification in the DC glow discharge, peaks attributed to carbonyl groups showed little change from the spectrum for untreated Kapton films, while the contribution due to carboxylate groups increased significantly.

Several other investigators had reported observations related to formation of weak boundary layers on polymer surfaces exposed to oxygen plasmas. Burger and Gerenser[31] proposed that chain scission resulting from overtreatment of polymers in oxygen plasmas resulted in a surface that was rich in low molecular weight material. This produced a a weak boundary layer that was detrimental to metal/polymer adhesion. In addition, Chin and Wightman[32] proposed that although creating a more hydrophilic, polar surface, O_2 RF-plasma treatment of LARC-TPI[R] polyimide caused chain scission leading to formation of a weak boundary layer that was deleterious to adhesion, measured by 180° peel test with pressure sensitive adhesive. In this study, samples were treated on an electrically insulated substrate[33] in a plasma operating at high pressures and relatively low power density. This configuration is conducive to relatively low levels of ion bombardment, that is, low energy and number densities of ions striking the polymer surface. The presence of a weak boundary layer was consistent with the observation that washing of plasma-treated samples with methanol resulted in peel values greater than those observed for both untreated films and as-treated films. For comparison, high density polyethylene, a polyolefin, exhibited improved adhesion with plasma treatment. This was related to the propensity of polyolefin materials to undergo surface crosslinking upon exposure to the high energy radiation present in plasmas. Vacuum ultraviolet radiation present in certain plasmas has been shown to increase the crosslinking density at the surface of some fluoropolymer films like poly(tetrafluoroethylene) (PTFE)[24], a polymer with inherently low cohesive strength. However, no attempt was made in the course of that investigation to correlate the increased crosslinking density with adhesion to metal films.

SUMMARY AND CONCLUSIONS

Plasma-induced surface species that interact chemically with subsequently-deposited metal films can improve adhesion by orders of magnitude. However, excessive chain scission reactions leading to weak boundary layers must be avoided. Crosslinking, enhanced by bombardment of ions, electrons and photons from the plasma can improve the polymer cohesive strength. As a word of caution, excessive crosslinking can embrittle polymers and induce cohesive failure within the polymer. Characteristics of surfaces well prepared for deposition of metals include an abundance of favorable (in terms of acid-base type interactions with deposited-metal atoms) functional groups on the surface, with no treatment-induced degradation of the structural/mechanical integrity (e.g., cohesive strength) of the polymer surface.

It is often difficult to attribute changes in practical adhesion to any one feature of the plasma-modified surface, e.g., incorporation of functional groups,

---

[R]LARC-TPI is a registered trademark of NASA-LaRC, Hampton, VA.

changes in crosslinking density, variation in surface morphology, removal of a weak boundary layer, or any combination of these effects. Indeed, for samples immersed in a plasma environment, any or all of these changes can be induced and both chemical and mechanical changes might contribute to practical adhesion. In addition, although some control of the concentration of surface functional groups can be maintained by variation in plasma parameters, the relative abundance and spatial orientations of these plasma-generated groups on the surface cannot, in general, be controlled.

One prospect for fundamental investigation of the interaction of metals with specific chemical groups in a polymer, in a controlled environment, is the use of self-assembled monolayers (SAMs)[34,35]. SAMs are molecules that adsorb on a surface in a specific conformation. By this means, the orientation and surface density of desirable functional groups that can interact with selected metal atoms is uniform and can be controlled reproducibly. Such a surface might be used to model the relationship between polymer surface functionality and adhesion to metals by investigating interactions of metal atoms with a selected type of functional group. One possible technique for measuring this interaction might be Chemical Force Microscopy (CFM), a variation of the more common Atomic Force Microscopy (AFM) technique that is used to map surface topography. CFM measures the force of interaction between a functionalized group on a probe tip and a group on a surface[36]. It is possible to visualize that a similar approach may be useful for measuring interactions of metal atoms (on the probe) with a particular group on a polymer surface. In this manner, the contribution of a specific chemical group, and the surface density of that group, to adhesion might be determined and separated from the other structural effects that result from plasma surface modification.

REFERENCES

1. B. Chapman, "Glow Discharge Processes," John Wiley & Sons, New York, 1980.
2. D. M. Manos and D.L. Flamm, "Plasma Etching: An Introduction," Academic Press, Boston, 1989.
3. F.D. Egitto, V. Vukanovic, and G.N. Taylor, in "Plasma Deposition, Treatment, and Etching of Polymers," R. d'Agostino (ed.), Ch. 5, Academic Press, Boston, 1990.
4. F.D. Egitto and L.J. Matienzo, Proc. Society of Vacuum Coaters 36th Ann. Tech. Conf., Dallas, TX, 1993, p. 10.
5. F.D. Egitto and L.J. Matienzo, IBM J. Res. Devel., 38, 423 (1994).
6. M. Strobel, C.S. Lyons, and K.L. Mittal, eds., "Plasma Surface Modification of Polymers: Relevance to Adhesion," VSP, Utrecht, The Netherlands, 1994.

7. E.M. Liston, L. Martinu, and M.R. Wertheimer, J. Adhesion Sci. Technol., 7, 1091 (1993).
8. F.D. Egitto and L.J. Matienzo, in "Polyimides: Fundamentals and Applications," M.K. Ghosh and K.L. Mittal (eds.), Ch. 15, Marcel Dekker, New York, 1996.
9. L.J. Matienzo and W.N. Unertl, in "Polyimides: Fundamentals and Applications," M.K. Ghosh and K.L. Mittal (eds.), Ch. 21, Marcel Dekker, New York, 1996.
10. A.D. Katnani, A.R. Knoll, and M.A. Mycek, J. Adhesion Sci. Technol., 3,441 (1989).
11. K.J. Blackwell, F.D. Egitto, and A.R. Knoll, Proc 35th Annual Tech. Conf. Society of Vacuum Coaters, Baltimore, MD, 1992, p. 279.
12. S. Tanigawa, K. Nakamae, and T. Matsumoto, Kobunshi Ronbunshu, 47, 41 (1990).
13. A. Callegari, B. Furman, T. Graham, H. Clearfield, W. Price, and S. Purushothaman, Microelectron. Eng., 19, 575 (1992).
14. K.W. Paik and A.L. Ruoff, J. Adhesion Sci. Technol., 4, 465 (1990).
15. F.D. Egitto, L.J. Matienzo, K.J. Blackwell, and A.R. Knoll. J. Adhesion Sci. Technol., 8, 411 (1994).
16. W. E. Mlynko, S.R. Cain, F.D. Egitto, and F. Emmi, in "Principles of Electronic Packaging," D.P. Seraphim, R. Lasky, and C.-Y. Li (eds.), p. 430, McGraw-Hill, New York, 1989.
17. F.M. Fowkes, J. Adhesion Sci. Technol., 1, 7 (1987).
18. K.L. Mittal and H.R. Anderson, Jr. (Eds.) "Acid-Base Interactions: Relevance to Adhesion Science and Technology," VSP, Zeist, The Netherlands, 1991.
19. M. Tatoulian, F. Arefi-Khonsari, N. Shahidzadeh-Ahmadi, and J. Amouroux, Intl. J. Adhesion Adhesives, 15, 177 (1995).
20. N. Shahidzadeh-Ahmadi, M.M. Chehimi, F. Arefi-Khonsari, N. Foulon-Belkacemi, J. Amouroux, and M. Delamar, Colloids Surfaces A: Physicochem. Eng. Aspects, 105, 277 (1995).
21. J.S. Arlow, D.F. Mitchell, and M.J. Graham, J. Vac. Sci. Technol. A5, 572 (1987).
22. J.A. Thornton, in "Deposition Technologies For Films and Coatings", (R.F. Bunshah, Ed.), p. 173, Noyes Publications, Park Ridge, NJ, 1982.
23. S.R. Cain and L.J. Matienzo, J. Adhesion Sci. Technol., 2, 395 (1988).
24. L.J. Matienzo, J.A. Zimmerman, and F.D. Egitto, J. Vac. Sci. Technol. A 12, 2662 (1994).
25. F.D. Egitto and L.J. Matienzo, Polym. Degrad. Stabil. 30,293 (1990).
26. M.A. Golub, T. Wydeven, and R.D. Cormia, Proc. Technology 2001: 2nd Natl. Technology Transfer Conf. & Expo., NASA Conference Publication 3136, Vol. 2, San Jose, 1991, p. 495.
27. M.A. Golub, T. Wydeven, and R.D. Cormia, Langmuir, 7, 1026 (1991).
28. F. Quaranta, A. Valentini, P. Favia, R. Lamendola, and R. d'Agostino, Appl. Phys. Lett., 63, 10 (1993).

29. J.M. Burkstrand, Phys. Rev. B, 20, 4853 (1979).
30. S.F. Tead, W.E. Vanderlinde, G. Marra, A.L. Ruoff, E.J. Kramer, and F.D. Egitto, J. Appl. Phys., 68, 2972 (1990).
31. R.W. Burger and L.J. Gerenser, in "Metallized Plastics 3: Fundamental and Applied Aspects," K.L. Mittal, (Ed.), p. 179, Plenum Press,New York, 1992.
32. J.W. Chin and J.P. Wightman, SAMPE Q., 23, 2 (1992).
33. J.W. Chin, private communication, 1993.
34. G.C. Herdt, D.R. Jung, and A.W. Czanderna, Prog. Surface Sci. 50, 103 (1995).
35. G.C. Herdt and A.W. Czanderna, J. Vac. Sci. Technol. A, 13, 1275 (1995).
36. C.D. Frisbie, A. Rozsnyia, A. Noy, M.S. Wrighton, and C.M. Lieber, Science, 265, 2071 (1994).

Metallized Plastics 5&6: Fundamental and Applied Aspects, pp. 331-351.
K.L. Mittal (Ed.)

SURFACE MODIFICATION OF POLYMERS USED IN THE LOW EARTH ORBIT SPACE ENVIRONMENT

Jacob I. Kleiman

Integrity Testing Laboratory Inc.;
University of Toronto Institute for Aerospace Studies
4925 Dufferin Street, North York, Ontario, Canada, M3H 5T6

A review of current activities at Integrity Testing Laboratory Inc. (ITL) and University of Toronto Institute for Aerospace Studies (UTIAS) in the field of surface modification of polymers is being presented. A number of new surface modification technologies were developed recently at ITL and UTIAS for protection of polymer materials from the space environment. Using a new approach to silicon functionalization of high-performance polymers, a new chemical surface modification process Photosil™ was developed that allows to incorporate high concentrations of silicon into surface regions of polymer materials. In another process, named Implantox™, a modified ion implantation process was used that allowed to form oxide-based highly protective surface layers on high-performance polymers. A novel, proprietary surface modification process was also demonstrated recently for fluorine-containing polymers. A variety of space bound polymer materials were treated by all three processes and all these demonstrated, through extensive testing, excellent resistance to fast atomic oxygen (FAO) erosion.

In addition, a texturing surface modification process was developed for Teflon FEP used for thermal protection of electronic equipment in space environment. The developed modification process of Teflon surfaces reduced substantially the specularity of Ag-Inconel/Teflon thermal control films. To protect the textured surfaces of Teflon from possible erosion by atomic oxygen and VUV in low Earth Orbit (LEO) environment, the novel, proprietary surface modification process was explored to substitute the SiO_2 coating by a modified surface layer having similar properties.

INTRODUCTION

Atomic oxygen (AO) is the predominant species in the low Earth orbit (LEO) environment at altitudes between 200 and 700 km[1,2]. The number density of AO at about 250 km altitude is 10^9 atoms·cm^{-3}, which corresponds to the density of residual gas in a vacuum of 10^{-7} Torr. However, due to the high orbital velocity of orbiting vehicles (approximately 8 km/sec at Space Shuttle altitude), the flux is high, being of the order of 10^{14} atoms·cm^{-2}·sec^{-1}. Furthermore, this high orbital velocity corresponds to collisions with oxygen atoms having a kinetic energy of about 5 eV. Atomic oxygen having kinetic energy of about 1-5 eV is commonly referred to as fast (FAO) or hyperthermal (HAO) atomic oxygen. When exposed to FAO, polymeric materials, graphite and carbon-fiber reinforced plastic (CFRP) composites have been shown to undergo significant accelerated surface erosion and mass loss[2-6].

The materials eroded in space exhibit strongly altered surface morphology on a micrometer scale (Figs. 1 and 2), having a roughened "carpet-like" texture[3,4] and irreversible degradation of various material properties.

Figure 1. SEM image of a Kapton sample eroded by atomic oxygen beam in a six hour accelerated test (acceleration factor of approximately 100). Magnification 1000x.

Higher energy erosion processes, such as physical sputtering, are well understood for polymers. Chemical reactions of polymer oxidation at thermal energies, including those by thermal atomic oxygen, are also well understood, yet only a few attempts have been made to study the basic aspects of the accelerated mechanism of polymer erosion

Figure 2. SEM cross-sectional image of an epoxy-carbon composite tube eroded during the LDEF mission (approximately 5.5 years in LEO environment). Secondary electron image. a) Magnification 200x. The direction of the incoming atomic oxygen flux, during the flight, is shown with wavy arrows; b) Magnification 750x. Notice a particle (marked with an arrow) that acted as a mask, preventing the erosion of underlying material.

by FAO. This situation is due mostly to the absence of theoretical approaches for this very special energy range. The way the kinetic energy of fast reactive particles is transferred to chemical reactions is still not clear on a micro-level. Correlations were found between fast atomic oxygen erosion yield of hydrocarbon polymers and their chemical content and structure[6]. A very important feature in the relationship found was the observation that carbon films and graphite were more resistant to FAO than hydrocarbon polymers which implied that one could not expect a substantially increased erosion stability in hydrocarbon polymers and composite materials. Most polymers have FAO erosion yields of about $(1\text{-}4)\times10^{-24}$ g/atom of atomic oxygen. Perfluorinated polymers are an exception. Because of the fluorine in their bonding structure, their erosion yields are much lower[7]. Although it was once thought that perfluorinated polymers might be the answer to the problems of polymers in LEO, there are other factors like synergistic effects between atomic oxygen and VUV radiation or the influence of soft-X-ray that increase the erosion yields to unacceptable levels. Materials having erosion yields of the order of 10^{-24} g/atom are unsuitable for long term use in the LEO environment, and for space in general.

In general, the cost of developing new polymer materials for use in space is very high, because of constraints on bulk optical, thermal and mechanical properties. The advantages of using existing organic polymers have forced the development of a wide variety of protection schemes, ranging from simple blankets of glass cloth to sophisticated thin film coatings[4]. The most successful coatings use oxides of silicon as their basis. Silica (SiO_2) and silica-like (SiO_x) coatings deposited using chemical vapour, plasma deposition, electron beam evaporation and other vacuum or spray processes offer considerable improvements in resistance to FAO. Erosion yield is typically reduced from greater than 10^{-24} g/atom to approximately 10^{-26} g/atom. The successful protection demonstrated in short-term missions has led to their incorporation into most space structures, such as the International Space Station. Silicon is known as an excellent material for use as an oxidation barrier. Once oxidized, it is chemically and thermally stable, relatively tough and has good optical properties. A good example of a commercial silicon type coating in current use is Sheldahl's AOR SiO_x coating applied in an RF sputtering process[12]. Thin protective layers based on silicon deposited from volatile compounds via microwave glow discharge have also been tried. The compounds used to form the coatings include: pure silane or silane and inert gas, silane and nitrous oxide, silane and mixture of ammonia and nitrous oxide, organo-silicones such as hexamethyldisiloxane[13]. Coatings of clear RTV silicone have also been tested and used in a number of applications[14]. McGhan NuSil CV-1144-0 RTV silicone was used to protect the original Hubble Space Telescope solar array[15]. Various fluoropolymer filled silica mixtures have also been attempted[16]. However, all coatings have certain limitations[1]. In the most general sense, a coating is a distinctly different material from the substrate it is protecting, and sometimes, its adhesion, especially in thermal or load fatigue conditions, can be a problem. Since the coating is a separate, applied layer, it tends to crack, break and spall, exposing the polymer beneath. To prevent the erosion of polymer materials in space a number of ways were explored in addition to the traditional protective coating's approaches. Some specialized polymeric materials with incorporated siloxane units have been developed that have acceptable erosion resistance for use in space flights. Among the developed materials one can mention polysiloxane-polyimide solid solution[8], polyimides containing pendent siloxane groups[9], epoxy-silane resins[10], and polysiloxane-polyimide copolymers[11]. None of these polymers, however, demonstrated better erosion resistance protection to FAO than coatings.

Surface treatment techniques which can modify the material at the molecular level in a manner which leaves it highly resistant to FAO erosion can alleviate this problem. In accordance with this philosophy, a program was started a number of years ago in our group in which a number of surface modification techniques were developed and applied to polymers and other carbon-based materials to protect them from erosion by FAO[17-20]. An approach for modification of the chemical content and structure of polymer space materials in a thin region of the surface, while retaining most, if not all, of the original properties was developed. Within this region, the degree of modification varies with depth. With a graded modification it is more difficult for cracking or spalling to occur, since the degree of modification decreases gradually with depth. This reduces or eliminates all of the interface problems, such as stress concentration or interlaminar diffusion of destructive elements and compounds. In the sections below, a brief description of three surface modification techniques will be given. It should be noted that while the discussion of the surface modification techniques is centred around space applications, they might find numerous other applications like in metallization of plastics, painting of polymer surfaces, bonding of polymer materials, etc..

EXPERIMENTAL

The surface modification processes were applied to a number of polymer materials that are listed in Table I. In the Photosil™ process, the photo-activation experiments were conducted in a UV/Ozone system that generates UV emission in the 185 and 254 nm range. Liquid-phase silylation was carried out in a beaker kept on a hot plate at 40-60°C by immersing the polymer films in a solution containing organic solvent (for example p-xylene), diffusion promoter, and silylating agent. After the prescribed time, the samples were rinsed with the solvent and blown dry with air. Stabilization of the silylated surface is required for converting organosilicon layer to a more stable silicon-containing structure. The same set-up as for photo-activation was used for stabilization of the silylated surfaces.

In the Implantox™ process, the ion implantation was performed using a high-current Linnelot 3A, monoenergetic ion implanter at Implant Science Corporation, Wakefield, Massachusetts. The energy range required to produce a 50-100 nm thick protective layer was determined to be 20-50 keV through a series of computer simulations using the software package TRIM-90 for the selected species, and the required dose range was established to be 10^{16}-10^{17} ions/cm^2.

The texturing experiments in the third surface modification process were conducted only on Teflon FEP samples. For the texturing experiments, a 58 cm diameter rotating drum fixture was used in the ion-beam system to hold the samples, providing a large surface area of 150 × 50 cm^2 for the ion beam exposure of the samples. A set of beam defining slits was used between the ion source and the sample allowing for formation of two 6 mm wide beams at the sample position. The rotation speed of the drum could be varied from 1-2 and up to 20 rpm. A typical drum speed of 2 rpm was used in the experiments. High purity argon and krypton gases were used in the study. The total beam current density measured at the sample was about 0.4 mA/cm^2, the beam power density was estimated at 0.15-0.20 W/cm^2, and the chamber pressure of $3 \cdot 10^{-2}$ Pa was maintained during the experiments at the above conditions. Typical etch rates of 0.015 μm/min were achieved in rotational experiments under the described conditions.

The details for the rest of the experimental set-up used in this work and the conditions used in analytical testing like x-ray photoelectron spectroscopy (XPS), secondary ion mass-spectroscopy (SIMS), scanning electron microscopy/energy dispersive x-ray microanalysis (SEM/EDS), advanced contact angle measurements, atomic oxygen beam irradiation tests, thermal optical properties, thermal cycling, etc. can be found in[17-20] and references therein, and would not be repeated here.

RESULTS AND DISCUSSION

The Photosil™ Process

The Photosil™ process is based on a silylation reaction and allows Si to be incorporated into the sub-surface region of the original polymer material. The surface becomes a new material and attains new properties. In essence, the process involves three major steps: UV/air pretreatment (activation) of the polymer, actual silylation, and UV/air post-treatment (stabilization) of the modified polymer.

Silylation has been performed on many materials[21] which contain active functional groups, such as hydroxyl (OH), in their original structure. Most polymers, especially those used in space applications, do not contain active functional groups, and a silylating agent will not react with them. Activation of the surface, i.e. production of groups with active hydrogen atoms, can be accomplished in a number of ways. Processes like UV induced oxidation, oxidizing plasmas, flame treatment, ion bombardment, or wet chemical treatment that oxidize the polymer surface can be used to produce functional groups[22]. For the modification processes to be effective in space applications, the active OH groups have to be created in the treated polymer material up to depth of 1 μm or more. It was established[17] that activation of most polymers using

UV radiation was preferable. The activation process is believed to be a result of simultaneous excitement of polymer molecules and attack by molecular oxygen, as well as ozone, atomic oxygen and singlet oxygen generated from molecular oxygen by the UV radiation. The physical and chemical changes occurring under photochemical oxidation involve an increase in concentration of a variety of oxygen groups, cross-linking and chain scission.

Table I: Summary of Data on Materials Used for Modification in the Present Work.

Material	Generally Used Abbreviation	Thickness (μm)	Comments
Kapton 500HN	(PMDA-ODA)	127	
Aluminized Kapton (Al-Kapton)	(PMDA-ODA)	76.2	The back is aluminized (0.1 μm)
Poly(ether ether ketone)	PEEK	50	
Mylar Poly(ethylene terephthalate)	PET	25.4	
Linear-low-density polyethylene	LLDPE	127	
Poly(vinyl chloride)]	PVC	100	
Carbon fibre/PEEK matrix composite sheets	AS4/APC-2, 4-ply,		Prepreg material supplied by Eiberite
Carbon fibre/epoxy matrix composite sheets	AS4/3501-6, 4-ply,		Prepreg material supplied by Hercules;
AS4 PAN based carbon fibre	Magnamite		
	Teflon FEP	127	Supplied by Sheldahl[1]

1. The Teflon FEP material was purchased from Sheldahl Inc., 1150 Sheldahl Road, North Field, Minnesota, 55057, USA.

The competition of these reactions as well as the depth of UV penetration and oxidation all depend on the chemical structure of the polymer, the power and wavelength of the excitation source, time and the sample temperature during irradiation. All these parameters were analysed, and appropriate conditions for activation of each treated polymer material were found. The goal of the silylation step is to replace the reactive hydrogen atom in the formed functional groups with silicon groups from the silylating agent, creating an organo-silicon molecule derived from the original polymer. Of the many silylating agents available commercially, the poly-functional agents are of most interest. Use of a poly-functional agent will increase the efficiency of the process since one molecule will be able to silylate more than one site. In addition, some cross-linking will result. Cross-linking, on one hand, will increase the structural strength of the layer, giving better resistance to thermal or load stress damage, and, on the other hand, should decrease diffusion. After silylation, the newly created polymer structure contains hydrocarbon fragments. These fragments may interact with the atomic oxygen in LEO environment thus increasing further the erosion resistance of the polymer by forming silicon-oxygen enriched compounds. The products of such oxidation processes in real space environment can, however, create additional contamination problems[6]. It was found[17] that by introducing a third, stabilization stage, consisting of exposure of the silylated polymer to UV radiation in the presence of oxygen, it was possible to create new silicon-oxygen-carbon structures that became exceptionally stable under FAO attack. Although the mechanism of this transformation is not clear yet, it is believed to be of fundamental importance. Similar effects were observed for polysiloxane-polyimide (AOR Kapton) film, which became less susceptible to AO attack if the samples were first exposed to VUV radiation[11].

As a result of the Photosil™ process, a newly developed organic polymer structure containing silicon, carbon, and oxygen atoms bonded in a complex configuration is formed that stops the destruction of the polymer in different oxidizing environments. The chemical contents of the organic polymer materials formed in the sub-surface region of the treated polymers include atoms coming partially from the silylating agent and partially from the pristine polymer structure. SIMS analysis was used to estimate the depth of silylation and stabilization (Fig. 3). Considerable amounts of silicon were found, penetrating to a depth of at least 0.5 μm as shown by the ^{28}Si signal. The depth of silylation is indicated by the change in slope of the ^{28} Si and ^{12}C signals at 0.2 μm. The silylating agent appears to have penetrated to this depth, and altered the structure therein. The stabilization zone is indicated by the ^{18}O signal. Within the first 0.05 μm of the surface, the intensity of carbon signal increases, and then the ratio of Si/C signals is constant. The ^{18}O signal indicates that the stabilization process has produced a silicon-oxygen rich protective region within only the first 0.1 μm of the surface. Signal intensity should not be taken as a direct measure of concentration, because the bonding environment will strongly affect the intensities measured. The XPS results (Table II) indicate that considerable changes have occurred in the treated region.

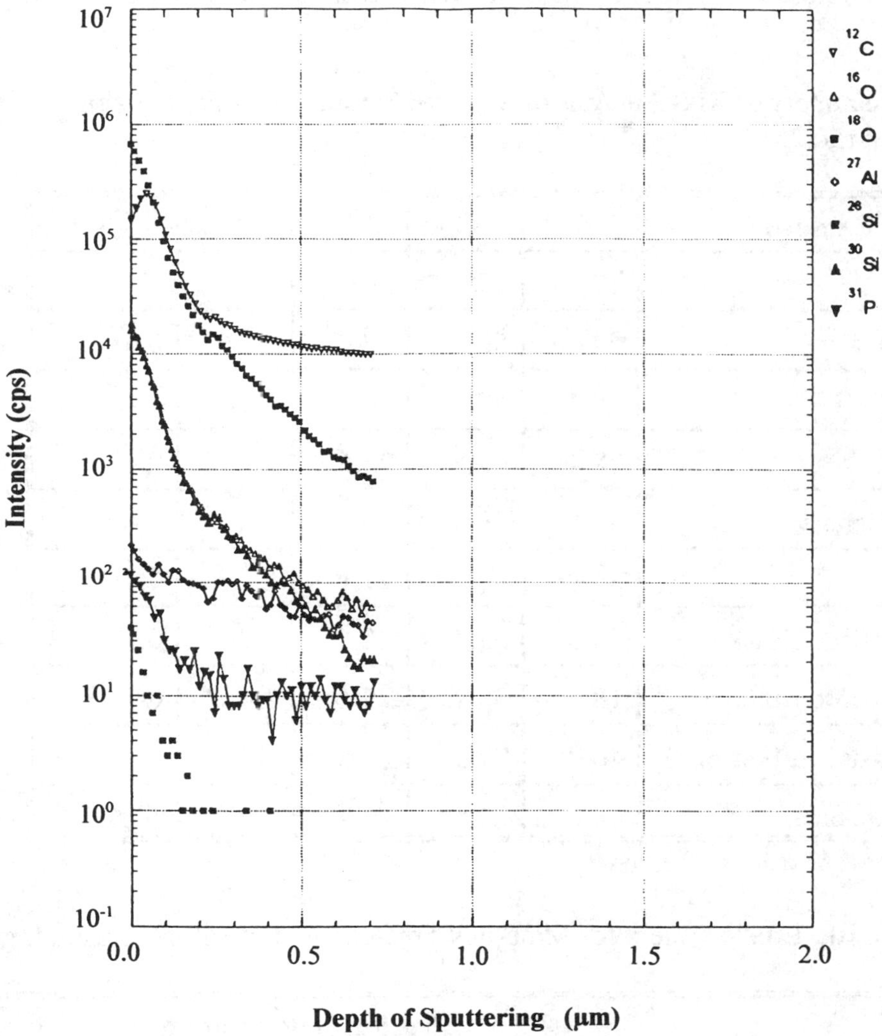

Figure 3. SIMS depth profile results for a Kapton sample after Photosil™ treatment.

Carbon content, for example for Kapton sample, has dropped to 17% and nitrogen content has become negligible. Oxygen content has increased to 49% and silicon makes up 34% of the surface content. The high resolution spectrum of the C 1s region contained a new peak at 283.2 eV which may be indicative of a Si-C bonding in a carbide structure[23]. The ratios between Si, C, and O (Table II) are characteristic for each material and depend on the chemical structure of the original polymer. No mass loss or morphology changes were evident for the Photosil™ treated polymer samples after FAO testing. Moreover, the XPS surface content of samples before and after the FAO testing was practically the same for all analysed samples. The FAO test results for the other Photosil™ treated samples showed that the erosion yields were about two orders

Table II: Summary of XPS Analysis of Silylated Samples after Final Stabilization Stage, At% [from ref. 23).

Sample	C	O	Si	N	Other
Kapton	17	49	34	-	-
Al-Kapton	24	52	22	1.6	0.4
PEEK	26	48	26	-	-
Mylar	28.4	47.5	23.6	-	0.5
Polyethylene	41.1	39.3	19.6	-	-
PVC	30.8	47	21.1	-	1.1
Carbon* fibre/PEEK	20.8	52.7	25.6	0.7	0.2
Carbon* fibre/Epoxy	18	55.1	25.4	1.4	0.1
PAN* based carbon fibre	46.9	27.5	25.6	-	-
Polyamide tape	44	23	19.4	-	13.6 (F)

* samples were wet pre-treated

Table III: Erosion Yield for Materials Treated by Photosil ™ Technology.

Material	Erosion yield (g/atom)	
	Pristine	After the Photosil ™ process
Kapton 500HN	$4.3x10^{-24}$	$<5.5x10^{-26}$
Al-Kapton	$4.3x10^{-24}$	$<5.5x10^{-26}$
PEEK	$2.0x10^{-24}$	$<7.9x10^{-26}$
Polyethylene	$3.6x10^{-24}$	$<5.9x10^{-26}$
Carbon Fibre/PEEK	$2.2x10^{-24}$	$<7.5x10^{-26}$
PAN Fibres	$2.2x10^{-24}$	$<7.9x10^{-26}$

of magnitude lower than the erosion yield of untreated polymer samples (Table III).

The Implantox™ Process

Ion implantation is of growing interest in polymer science and engineering because of its demonstrated capability to modify the molecular structure, morphology, and the physical-chemical properties of polymers[24]. Implantox™ is a new approach to the problem of protection of polymer materials and graphite in space. According to this process[18], by implanting appropriate elements to a depth of several tens of nanometers and then exposing the implanted surface to a reactive oxidizing environment, stable, graded, self-healing surface layers are produced.

The polymer samples were implanted with selected ion species using single, binary or triple combinations of ions, then exposed to a reactive oxidizing atmosphere or tested in the atomic beam facility. The oxidative post-treatment and testing were done to determine the ability of the treated samples for uniform, continued surface conversion, and to evaluate their overall resistance to highly oxidizing environments, including LEO. Based on their oxides' excellent protective properties and the ability for chemical interaction under implantation conditions, Si and Al were chosen for implantation first. In addition to single element implantation, the binary and triple approaches have been considered as well. In an attempt to create borosilicate types of structures, boron (B) was implanted into a number of samples at 10 keV in addition to Si, and at 100 keV in addition to Si and/or Al. All implanted samples demonstrated the development of an atomic oxygen resistant surface, which was confirmed by both mass loss measurements and SEM studies. XPS analysis confirmed surface enrichment in carbon and reconstruction of carbon chemical states.

The mass changes, observed in FAO exposure tests, in the first 1.5-3 hours, for different samples (Fig. 4) can be associated with volatile products formation and mass gain due to oxidation through oxygen uptake. The following stabilization in the mass change curve indicates formation of stable, protective surface structures.

The final surface content of a number of polymer materials after FAO testing is presented in Table IV. The mass loss of the Kapton witness sample indicated that the average FAO fluence during the test was 1.7×10^{20} atoms/cm^2.

After implantation, all samples exhibited a dark brownish surface color. This surface carbonization effect is a well-known phenomenon in high energy implantation processes in polymers. The samples underwent some degree of recovery of the original colour and optical transmission of the material as the FAO dose increased, with some materials showing total recovery by the end of the test. The observed changes in optical transmission is a further indicator of the conversion of at least the top surface of the implanted layer to a colourless oxidized layer in the presence of atomic oxygen. XPS data confirmed the presence of 29% Si in the surface layer of the sample K-9, and 30% Si in the sample K-8 (Table IV). The deconvolution of high resolution Si 2p XPS

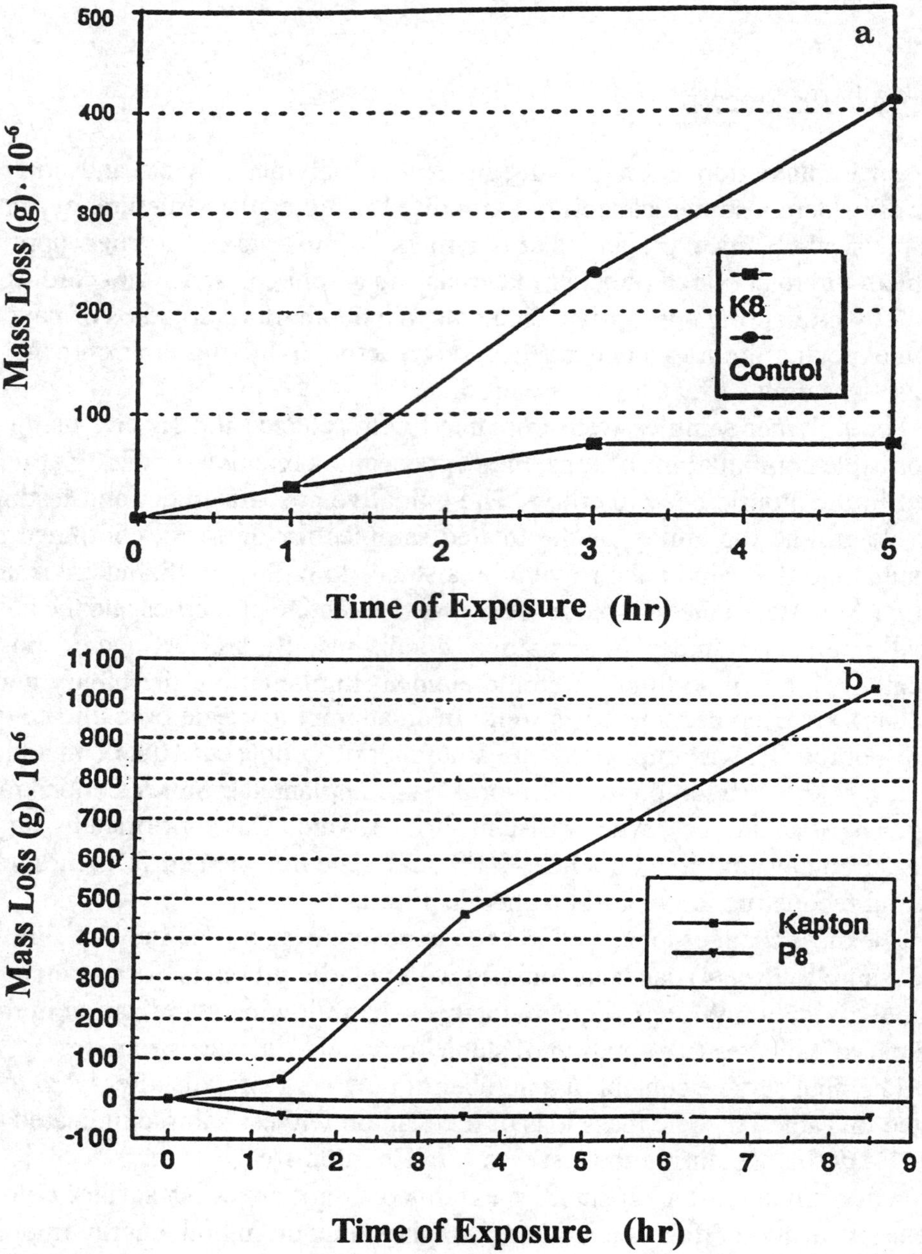

Figure 4. The mass change for binary implanted polymers under FAO exposure.
a) Implanted Kapton sample (K-8) with Si (E=25 keV, $\varphi=5\times10^{16}$ cm^{-2}) and B (E=10 keV, $\varphi=1.7\times10^{16}$ cm^{-2}).
b) PEEK (sample P8) implanted with Al (E=35 keV, $\varphi=5\times10^{16}$ cm^{-2}) and Si (E=35 keV, $\varphi=2\times10^{16}$ cm^{-2})

Table IV: XPS Data for Surface Composition of Implanted Polymers After FAO Testing,

Implanted ion	Element Detected (at. %)				
	C	O	Si	Al	Other
Kapton					
Si, (sample K-9)	17	54	29	-	-
Si-B, (sample K-8)	24	46	30	-	-
Si-Al	30.5	45.6	5.4	17.5	0.9
Si-Al-B	30.7	44.6	6.4	17.1	1.4
PEEK					
Si	41.2	39.7	19.1	-	-
Si-Al	37.4	41	5.6	13.1	2.8
Mylar					
Si-Al	43.3	37.8	5.1	12.2	1.6
Si-Al-B	36.3	42.4	5.3	14.1	1.9
Graphite					
Si	35.6	45	16.7	-	2.7

spectra indicated that Si was present in the surface layer in a number of silicon-oxygen-carbon structures, which can be associated with the peaks at 103.9 eV (56.2%) and 102.8 (43.8%)[23].

Kapton implanted with Si and low energy B and exposed to FAO (sample K-8) also showed a similar surface content (Table IV). While boron was not found in the top surface layer of the sample by XPS, the SIMS depth profile analysis (Fig. 5) revealed that the boron was distributed in the sample together with a silicon tail up to 1 μm deep[23]. The B/Si ratio in the "tail" distribution in Fig.5 was found to be close to the implanted fluence ratio ~ 1/3, as recommended from computer simulations, and chosen to imitate the content of these elements in a borosilicate glass. The long-range penetration of the implanted elements is very important because of the possibility for self-healing of the modified layer in case of damage to the surface. The thin subsurface layer enriched with ^{18}O found to penetrate from the surface to about 0.1÷0.2 μm can be

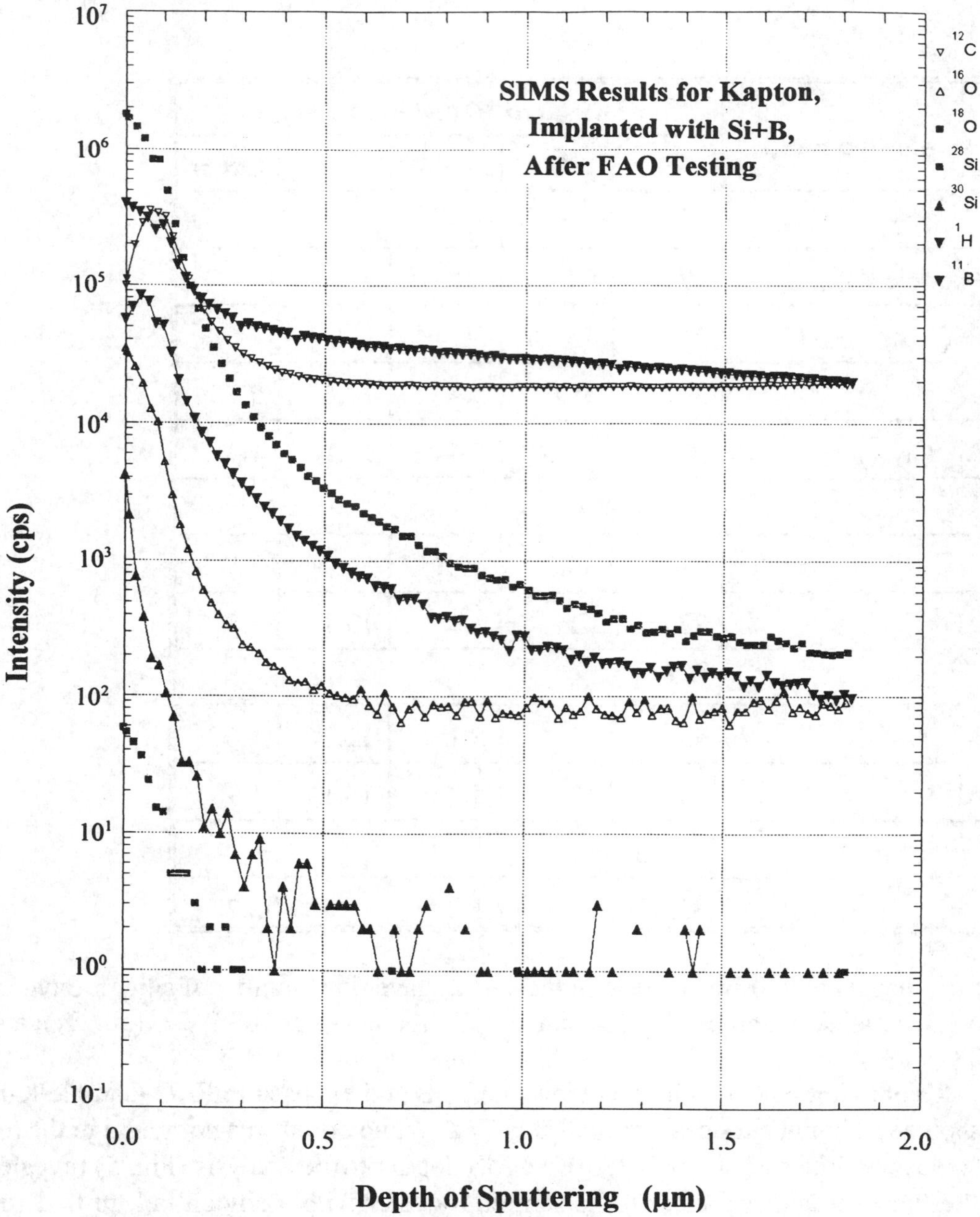

Figure 5. SIMS depth profile results for a Kapton sample after Implantox™ treatment.

accepted as an estimate of the depth of oxygen-enriched modified layer. The reason for an increase in the intensity of the ^{12}C secondary ions signal from an extended subsurface layer (approximately 0.3 μm thick) is unclear, and is in disagreement with the data from XPS analysis that suggest a highly reduced amount of carbon at the surface. It should be noted that the maximum in the concentration profile of the carbon peak occurs in the oxygen-enriched layer. This very special behaviour of ^{12}C can be attributed to the change in ionization yield of carbon sputtered from the modified layer during SIMS analysis, due to formation of new surface structures.

Surface Texturing Modification Process for Teflon FEP

A surface texturing process was developed for Teflon FEP to be used for thermal protection of the electronic equipment on the outside of the Canadian Mobile Servicing Station (MSS) for the International Space Station (SS) ALPHA[20]. The process that is based on ion-beam texturing reduced substantially the specularity of Ag-Inconel/Teflon thermal control films by changing the morphological appearance of their surfaces in a controlled manner from a metallic-like and shiny to complete milky, white appearance without significantly affecting the thermal optical properties.

Ion beam texturing is a well-known phenomenon[25]. Most ion sputtered fluoropolymer surfaces develop cone or spire-like features. The applications of ion beam textured polymers for adhesion improvement, biomedical applications, electrical properties changes, wettability properties changes, etc. was studied extensively and is well described (see, for instance, ref. 25 and references therein).

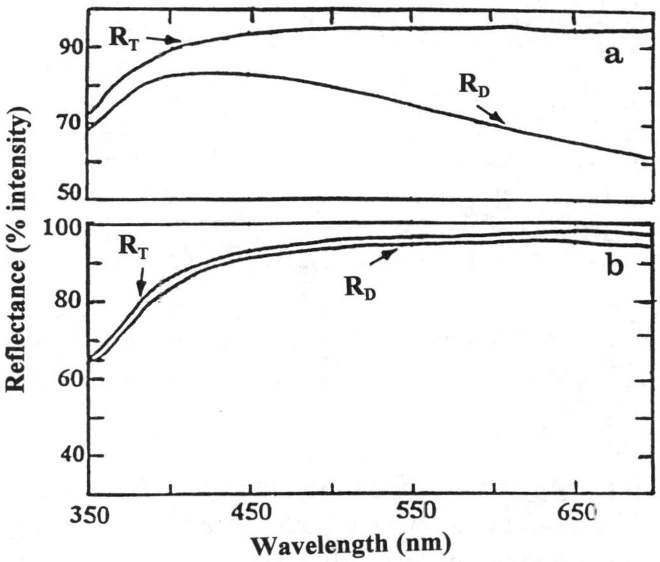

Figure 6. Optical reflectance of 127-μm silver/Inconel coated Teflon FEP film. a) Total (R_T) and diffuse (R_D) reflectance after argon ion beam texturing; b) Total (R_T) and diffuse (R_D) reflectance after krypton ion beam texturing.

Most of the work described in the literature[25], however, was conducted with small area samples that remained stationary during the texturing process. To scale up the ion texturing process for manufacturing of samples with large surface areas, rotation of the sample was introduced to expose large areas of Teflon FEP films to the ion beam[20]. In addition, to protect further the textured Teflon FEP surfaces from the LEO environment, a novel, proprietary surface modification technique was used[19] that allowed to introduce substantial amounts of silicon into the subsurface region of the textured Teflon FEP. It was found that ion beam texturing using krypton ions produced the best results[20] (Fig. 6).

To obtain a texturing effect in rotational experiments, similar to that achieved in stationary experiments, the exposure times to the ion-beam had to be increased considerably. Figure 7 shows, for instance, the dependence of the diffuse reflectance, as measured at a fixed wavelength of 550 nm versus the exposure time to the ion beam found experimentally in runs with krypton gas.

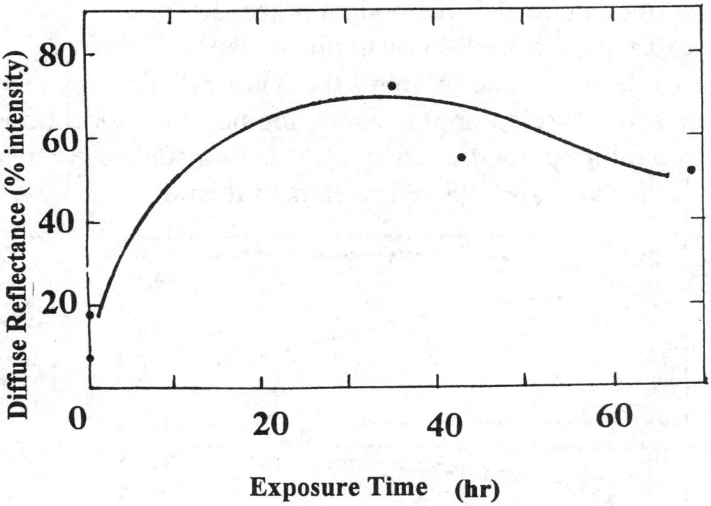

Figure 7. Graph of diffuse reflectance (at 550 nm) versus time of krypton ion-beam texturing in a rotational configuration. The line is drawn for visualization of the trend only.

Because the sputtering yield depends on the angle of incidence of bombarding ion species[26] (with the maximum sputtering yield occurring at an angle of incidence close to 60° from the normal to the surface), slits were introduced in front of the beam, eliminating to a large extent the tangential component of the ion beam, and thus allowing for the development and preservation of extensive surface texturing. X-ray photoelectron spectroscopy and energy dispersive x-ray microanalysis confirmed that the ratio of fluorine/ carbon atoms on the surface changed substantially after ion beam texturing[19].

The textured samples were coated with a layer of SiO_x using a conventional vacuum coating process[20]. During the thermal cycling testing it was noticed that the surfaces of cycled samples attained a yellowish discolouration and that the ratio of the solar absorptance to total emittance coefficients (α/ϵ), used in space industry as a measure of thermal optical properties of materials, increased from 0.10 to 0.21[19]. A proprietary process for surface modification of fluorine-based materials, recently developed in our group, was applied to the textured Teflon FEP surfaces to evaluate and to compare it to the conventional SiO_x - based coating deposited in a vacuum process.

A number of steps are involved in the new proprietary process that usually starts with a plasma activation treatment of the surface followed by steps which include a proprietary treatment (referred in the text as process A), a silylation-based treatment (referred in the text as process B), and a final stabilization stage. Water contact angles with the treated surfaces were used as a sensitive tool for indicating changes at the surface. The results of water contact angle measurements after different stages in the processing of the textured and the control Teflon FEP samples for two different sets of plasma treatment conditions (shown in Table V) suggest changes in the surface energies of the treated samples. For textured and pristine control samples the water contact angles, prior to plasma treatment, were 160° and 108°, respectively. A detailed description of the novel surface modification technique will be given elsewhere.

XPS analysis conducted for all samples shown in Table V confirmed that the Teflon FEP surfaces were chemically modified with formation of a stable silicon-based layer containing up to 21% of Si bonded to oxygen, carbon, and fluorine of the starting material structure[19]. XPS depth profiling analysis was performed on a number of samples described above, before and after FAO testing. The silicon concentration on the surface of the textured sample was estimated to be 21 at%.

An XPS depth profile analysis of the textured and surface modified sample (Fig. 8) suggested that up to 15% of silicon was present at a depth of approximately 17 nm from the surface, and that approximately 10 at% of silicon was still present at 60 -70 nm. The textured surfaces of the Teflon FEP films, without any reflective back coating, appeared grey mat. After a silver or silver/Inconel reflective coating was deposited on the back, the textured FEP film appeared milky mat.

The influence of mechanical abrasion on the stability of the textured surfaces and their optical performance was evaluated by applying a rubbing motion with a cotton swab using different applied forces, and analysing the spectral reflectance of the surfaces afterwards. It was found that even after a very strong abrasive treatment the diffuse reflectance of the samples remained high[20].

It was found that after the thermal cycling tests, the textured samples modified by the proprietary process did not change their colour, while the SiO_x vacuum-coated samples attained a yellow discolouration[19]. The thermal optical properties of both Teflon FEP films textured by the ion bombardment process and those coated with SiO_x and modified with the proprietary process were evaluated before and after a thermal

Table V: Summary Data on Water Contact Angles After Different Processing Steps in the Proprietary Surface Modification Process.

Sample	Plasma Treatment Conditions	Water Contact Angle (WCA) in degrees After		
		Plasma Treatment	Process A +B	Process A+B+ stabilization
Textured, JKR-12-1	Set #1 (H_2/MeOHgas; 30 min; 30-100°C)	123	143	62
Control, JK-12-1		66	86	47
Textured, JKR-12-3	Set #2 (H_2/H_2O gas; 30 min; 30-100°C)	121	137	59
Control, JK-12-3		69	105	48

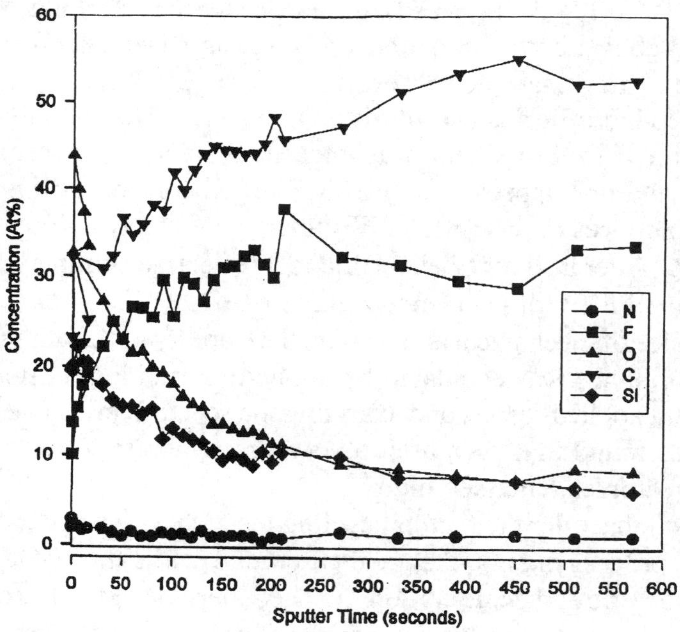

Figure 8. XPS depth profile analysis of a textured Teflon FEP sample after treatment with the proprietary surface modification technique.

cycling test. An increase in the α/ϵ ratio (from 0.1 to 0.2), due to an increase in solar absorptance, was observed for both types of samples which may lead to somewhat higher surface temperatures[27].

In order to evaluate the resistance of the Teflon FEP surfaces to LEO after texturing and surface modification, the samples were tested in the atomic oxygen beam facility at UTIAS. In all conducted tests, the morphology of the surfaces and the mass of the samples remained practically unchanged.

CONCLUSIONS

Using a new approach for protection of polymers and carbon composite materials from the LEO space environment, a number of surface modification techniques were developed and successfully applied to a variety of materials.

Photosil™ is a chemical surface modification process that uses a new approach to the silicon functionalization of high-performance polymers. It allows to incorporate up to 36 at.% of Si into the surface region of many polymers resulting in new properties and the ability to withstand atomic oxygen erosion in accelerated environmental tests.

Implantox™ is a modified ion implantation process that forms a highly protective surface layer in polymers by high-dose implantation of selected metal or semi-metal elements, or their combinations with other selected species and a following oxidative post-treatment. The erosion protection properties of the treated materials depend on the chemical content and super-molecular organization of the sub-surface layers.

Large areas of Teflon FEP films were successfully textured in a continuous ion beam process using krypton as the ion source thus solving the problem of excessive glare of the FEP Teflon upon direct illumination. Highly developed surfaces having milky mat appearance were produced, substantially increasing the diffuse reflectance of the Teflon FEP films without changing significantly their thermal optical properties. Mechanical abrasion treatments did not change the high diffuse reflectance properties of the surfaces allowing for their handling and processing.

A new proprietary process for surface modification of fluorine-based polymers was successfully applied to the textured Teflon FEP surfaces. Silicon incorporation into the subsurface region of the textured Teflon FEP to a depth of 20-60 nm was confirmed. The proprietary Si-based modification of the textured surface allowed to eliminate the yellow discolouration observed after thermal cycling of similar samples of Teflon FEP coated in vacuum with a thin layer of SiO_x. With further development, such textured surfaces may find terrestrial applications for different cooling or heating systems or as solar diffusers.

ACKNOWLEDGMENTS

The author would like to acknowledge his colleagues Dr. Y. Gudimenko, Dr. Z. Iskanderova and Dr. R.C. Tennyson, without whom the work described here would not have been possible. Thanks are also due to Dr. D. Morison for atomic oxygen experiments and participation in the work, Mr. R. Davidson for the SIMS and XPS depth profile analyses, and Dr. R. Sodhi for help in XPS analysis. Part of this work was performed under a SPAR CND Space Station D1 (PO14287) contract to ITL, Inc.

REFERENCES

1. R.C. Tennyson, in "The Behaviour of Systems in the Space Environment," R.N. DeWitt et al., Eds., p.233, Kluwer Acad. Publ., (1993).
2. B.A. Banks, "LDEF Materials Data Analysis Workshop," compiled by B.A. Stein and Ph. R. Young, NASA CP 10046, p. 191 (1990).
3. R.C. Tennyson, Canadian J. Phys., 69, 1190 (1991).
4. R.C. Tennyson, G.E. Mabson, W.D. Morison, and J. Kleiman, "Preliminary Results from the LDEF-69 Months in Space," First Post-Retrieval Symposium, A.S. Levine, editor, NASA CP 3134, Pt.2, p. 1057, June 1991,
5. J.B. Cross, S.L. Koontz, J.C. Gregory, and M.J. Edgell, in "Materials Degradation in Low Earth Orbit (LEO)," V.Srinivasan and B.A. Banks, Eds., p. 1, TMS Minerals, Metals and Materials, Warrendale, PA (1990).
6. Z.A. Iskanderova, Yu. I. Gudimenko, J. Kleiman and R.C. Tennyson, J. Spacecraft Rockets, 32, 878 (1995).
7. S.L.Koontz, L.J. Leger, K. Albin, and J. Cross, J. Spacecraft Rockets, 27, 346 (1990).
8. S.K.Rutledge, J.M. Cooper, and R.M. Olle, The Effect of Atomic Oxygen on Polysiloxane-Polyimide for Spacecraft Applications in Low Earth Orbit, *NASA CP-3103,* Pt.2, 755-762 (1991).
9. J.W.Connell, Polyimides Containing Pendent Siloxane Group, *NASA CP-3136,* Vol.1, 330-338 (1991).
10. S.L. Oldham, "Proceedings of the EOIM-3 BMDO Experiment Workshop," June 22-23, Arcadia, California, (1993).
11. J.A. Dever, E.J. Bruckner, and E. Rodriguez, AIAA paper 92-0794, 1-9 (1992).
12. A. Freeland, J. Fink, G. Meinke, J. Grieser, and H. Takeo, Proc. of International Forum "Protection of Materials and Surface Finishes from the Low Earth Orbit Environment," Toronto, (February 1992).
13. D.G. Zimcik, M.R. Wertheimer, K.B. Balmain, and R.C. Tennyson, J.Spacecraft Rockets, 28, 652 (1991).
14. E.M. Silverman, Space Environmental Effects on Spacecraft: LEO Materials Selection Guide, NASA CR 4661, part 1, 8-10 (1995).
15. D. Petraitis, CSRU/NASA Materials on Space Meeting, Cleveland, Ohio, May 25,

1993.
16. B.A. Banks, M.J. Mirtich, S.K. Rutledge, and H.K. Nahra, AIAA Paper 85-0420, (1985).
17. Y. Gudimenko, J.I. Kleiman, Z. Iskanderova, R.C. Tennyson, and G. Cool, US Patent Application No. 08/493535, June 1995.
18. Z.A. Iskanderova, J.I. Kleiman, Y. Gudimenko, D. Morison, and R.C. Tennyson, Proc. Third International Space Conference , ICPMSE-3, Toronto, 25-26 April, 1996, p.58.; Z.A. Iskanderova, J.I. Kleiman, Y. Gudimenko, R.C. Tennyson, and G. Cool, US Patent #5683757, issued 11 April, 1997.
19. J.I. Kleiman and Y. Gudimenko, Proc. 7th International Symposium on Materials in a Space Environment, 16-20 June, 1997, ENSAE-SUPAERO, Toulouse, France, p.419 (1997).
20. J.I. Kleiman, O. Popov, A. Chang, and D. Molenda, Proc. Third International Space Conference , ICPMSE-3, p.42, Toronto, 25-26 April, 1996.
21. F. Coopmans and B. Roland, SPIE Proc. 631, 34-36 (1986).
22. M. Strobel, C.S. Lyons and K.L. Mittal, editors, "Plasma Surface Modification of Polymers:Relevance to Adhesion," VSP, Utrecht, The Netherlands, (1994).
23. Y. Gudimenko, Z. Iskanderova, J.I. Kleiman, G.C. Cool, D. Morison, and R.C. Tennyson, Proc. 7th International Symposium on Materials in a Space Environment, 16-20 June, 1997, ENSAE-SUPAERO, Toulouse, France, 403 (1997).
24. G.K. Hubler et al., Eds., "Ion Implantation and Ion Beam Processing of Materials," Mater. Res. Soc. Symp. Proc, vol.27, North Holland, (1984).
25. B. A. Banks, in "Handbook of Ion beam Processing Technology," J.J. Cuomo, S.M. Rossnagel, and H.R. Kaufman, eds., ch. 17, p.338, Noyes Publications, (1994).
26. H. Oechsner, Appl. Phys., 8, 185, (1975).
27. M.J. Mirtich and J.S. Sovey , J. Vac. Sci.Technol., 15, 697 (1978).

Metallized Plastics 5&6: Fundamental and Applied Aspects, pp. 353-363.
K.L. Mittal (Ed.).

METAL ADHESION ENHANCEMENT BY PRETREATMENT OF POLY-(BUTYLENE TEREPHTHALATE) SURFACES WITH EXCIMER LASER RADIATION

H. Horn[1], R. Weichenhain[2], D. A. Wesner[1], S. Beil[1] and E. W. Kreutz[2]

[1]*Fraunhofer Institut für Lasertechnik, Steinbachstr. 15, D-52074 Aachen, Germany*
[2]*Lehrstuhl für Lasertechnik der Rheinisch-westfälischen Technischen Hochschule Aachen, Steinbachstr. 15, D-52074 Aachen, Germany*

The pretreatment of polymer surfaces by laser radiation is investigated as an alternative technique to common wet chemical, plasma or other pretreatment methods. Chemical and morphological changes in the polymer surface are investigated with poly(butylene terephthalate) (PBT) as a model substance, containing UV-absorbing (aromatic) chromophoric groups. Chemical changes in the surface are investigated via x-ray photoelectron spectroscopy (XPS) and Fourier-transform infrared spectroscopy (FTIR). Analysis of the morphological changes in the polymer surface is done via optical microscopy and scanning electron microscopy (SEM). The metallization of polymers pretreated by excimer laser irradiation is performed via electroless plating of nickel followed by electrolytically deposited copper. The influence of excimer laser irradiation on the adhesion of the metal overlayers is measured by tape and peel tests.

INTRODUCTION

Metal/polymer interfaces are important for a variety of applications in the high-technology areas of microelectronics and microsystems, as well as for other, more conventional applications such as decorative overlayers, diffusion barriers, and electromagnetic shielding. Depending on the specific metal/polymer system, the interface may exhibit little or no adhesion unless the polymer is pretreated prior to metal deposition [1]. However, stringent demands may be placed on this pretreatment process, especially in the area of microsystem technology. For example, in a sensor module many different materials may be integrated which restrict the possible thermal processing steps and thus do not allow the application of all the conventional methods to enhance the metal/polymer adhesion.

The conventional methods of polymer pretreatment (Table I) rely on plasma, wet-chemical or mechanical methods to produce either chemical or morphological changes in the surface, which strengthen the metal/polymer interfacial bond.

Table I. Common Methods of Polymer Pretreatment.

vacuum techniques	**wet-chemical techniques**	**other techniques**
glow discharge	etching	sand blasting
activation by oxidation with a glow discharge	selective chemical dissolution	mechanical removal by particles
etching with Ar-plasma	chemical surface modification	flaming
ablation by impact processes	reactions of specific functional groups	surface oxidation with an oxidizing gas flame
etching with N_2 or O_2 -plasma	solvent treatment	SO_3 treatment
surface oxidation	stress relaxation	gas phase oxidation

All these methods have the common disadvantage that they are usually not area-selective unless an additional processing step, such as the application of a contact mask, is used. The wet chemical methods may have the disadvantage of causing undesired chemical reactions in areas not intended to be processed. A method such as sand blasting obviously involves mechanical contact to (and stressing of) the surface, and the plasma methods require the use of a vacuum environment, which may limit the productivity.

In comparison to conventional methods, photochemical pretreatment, especially with pulsed UV-laser radiation, offers several distinct advantages. The high photon energy ($h\nu$=5 eV for the KrF-excimer laser) can induce both photolytic (e.g., bond scission) and pyrolytic (thermal degradation) processes. The short pulse duration (about 25 ns) results in a very high instantaneous fluence which is laterally well-confined to the irradiated region because of the beam geometry and is restricted to the near-surface region of the polymer because of the optical penetration depth of typically $\lesssim$10 μm. There may be laser-induced chemical reactions, e.g., oxidation due to reaction with the ambient atmosphere, in these areas. The energy deposited by the laser radiation is sufficient to cause significant material removal by ablation and/or evaporation. The removal rate may reach several μm of material depth per laser pulse, depending on the polymer characteristics (Fig. 1, optical properties e.g. absorption coefficient α, removal threshold ε_t) and irradiation conditions (fluence, pulse number, pulse repetion rate).

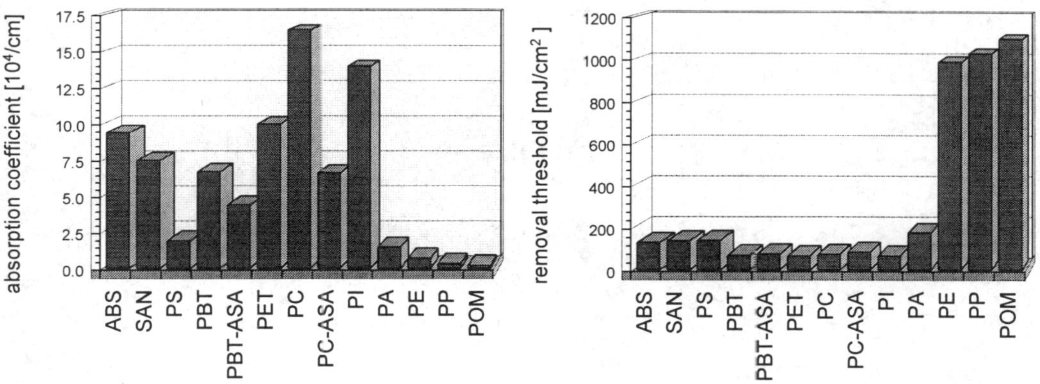

Fig. 1 Absorption coefficient α and removal threshold ε_t of different polymers for excimer laser radiation (laser wavelength 248 nm; pulse duration 25 ns and pulse repetion rate 3 Hz, 50 pulses).

This may also change the surface morphology, e.g., enhancement of the surface roughness in treated areas. All these effects are potentially important for adhesion enhancement of the metal/polymer interface. They can be produced without a vacuum environment or mechanical contact to the polymer surface, and can quite readily be made area-selective by use of a mask projection or direct-writing irradiation scheme. Thus, laser-induced polymer pretreatment seems predestined to become an important method for adhesion enhancement in microsystem technology.

The present work reports on the use of KrF-excimer laser irradiation to enhance the adhesion of copper/nickel overlayers on poly(butylene terephthalate) (PBT), concentrating on the laser-induced chemical and morphological changes and the corresponding adhesion improvement.

EXPERIMENTAL DETAILS

Laser irradiation is performed in ambient air at atmospheric pressure. The KrF excimer laser radiation source emits pulses at a wavelength of λ=248 nm with a 25 ns pulse duration and a maximum pulse energy of 600 mJ (ε = 150 mJ/cm^2). The laser beam, after reduction of its cross section by a telescope, illuminates a mask, which is imaged onto the sample surface, where fluences ε < 400 mJ/cm^2 are achieved. A programmable (x, y) positioning system is used to move the samples relative to the laser beam.

The processed surfaces were analysed using optical microscopy, scanning electron microscopy (SEM), x-ray photoelectron spectroscopy (XPS), Fourier transform infrared spectroscopy (FTIR) and optical profilometry. Processed surfaces were metallized by electroless nickel plating. The deposited nickel films were electrolytically reinforced with a copper overlayer 25 µm in thickness. To experimentally determine the adhesion, many methods are possible[2]. The different methods do not give comparable results because of the different nature of the stress at the polymer/metal interface during testing. In our case the adhesion of the metal overlayers on polymers was investigated by tape and peel tests. The tape test provides a relative, qualitative determination of adhesion, and is applicable mainly to thin metal films less than about 10 μm in thickness. For more quantitative measurements the peel test (as described in the Norm IEC 326) (Fig. 2) is used.

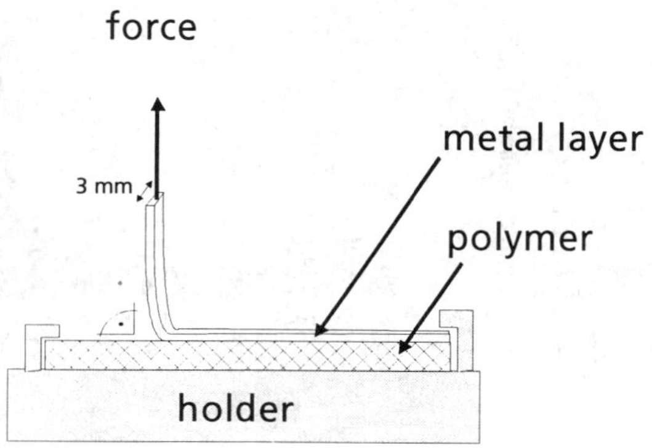

Fig. 2. Schematic representation of peel test (as described in the Norm IEC 326).

PROCESSES IN LASER IRRADIATION

Material removal from polymers due to interaction with pulsed laser radiation may be described in terms of a Beer's law-type model with the parameters ε_t, the threshold fluence for material removal, and α, the absorption coefficient [3] (Fig. 1). In the range of small pulse numbers and small fluence (< 300 pulses, $\varepsilon < 500$ mJ/cm²) the measured removal depths for polymers with small optical penetration depths correlate with Beer´s law. Larger pulse numbers lead through photolytical reactions in the polymers to a change in the absorption coefficient. In the range of large fluence, a developing plume reduces the effective fluence at the polymer surface by absorbing energy. Therefore, the removal depth is smaller than described by Beer`s law. For polymers with large optical penetration depths, non-linear effects in the removal volume cause the material removal to diverge from the Beer`s law prediction. This depends on the chemical structure, especially on the concentration of UV-absorbing chromophores. Polymers with small material removal thresholds ($\varepsilon_t <$ 200 mJ/cm²) are well-suited for laser pretreatment of large areas, because processing speeds are limited by the minimum threshold fluence necessary for material removal. For the polymer used in the present work, PBT, the material removal threshold ε_t is approximately 38±12 mJ/cm² (Fig. 1).

Irradiation at fluences below the material removal threshold leads to chemical and microstructural modification of the polymer. The photochemical reactions may include effects such as bond breaking, chain cross-linking, depolymerisation, induced radical reactions and/or photo-induced oxidation [4]. The possible chemical reactions are related to the chemical structure of the polymers, the processing gas atmosphere, the wavelength of the radiation and the radiation dose. These may affect adhesion of metal overlayers by changing the chemical reactivity of the surface with the metal and/or by changing the surface energy and thus the wetting behaviour at the interface.

Periodic microstructures associated with irradiation at fluences below the material removal threshold have been explained in terms of interference effects between the incident beam and scattered waves along the interface, producing locally enhanced fields[4]. As such, the characteristic lateral dimensions of these structures are related to the laser wavelength, polarisation and the angle of incidence on the surface. Typically the structure depth for irradiation with pulsed UV-lasers [5] is only on the order of several tens of nm, which would probably affect the adhesion properties of the metal/polymer interface only slightly. Chemical changes in PBT surfaces after irradiation with KrF-excimer laser radiation in air start at fluences below the material removal threshold as derived from XPS spectra taken before and after irradiation (Fig. 3). The C1s and O1s line shapes broaden with irradiation, probably due to bond scission in the macromolecule, yielding locally varying bonding environments and fluctuations of the local charge density. This is especially evident for the

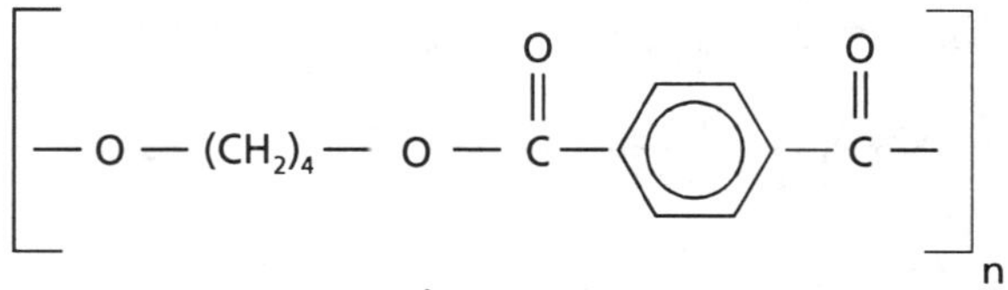

chemical structure of PBT

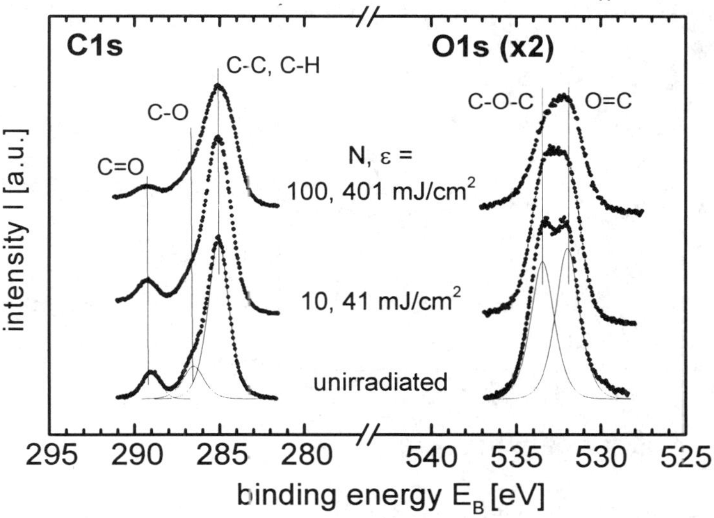

Fig. 3 C1s and O1s core level XPS spectra for unirradiated and irradiated PBT surfaces (10 pulses, $\varepsilon = 41$ mJ/cm^2 and 100 pulses, $\varepsilon = 401$ mJ/cm^2). Spectra are shifted vertically for clarity.

spectra taken after irradiation at a fluence of 401 mJ/cm^2. In addition, specific changes in the peak components due to carbon-oxygen bonds are also evident. For example, in the C1s spectrum the partially resolved shoulder at 286.4 eV corresponding to carbon atoms with ether oxygen (C-O) and the separate peak at 289.0 eV due to the ester (O-C=O) group are both reduced in intensity compared to the main aromatic peak (C-C, C-H) at 285 eV. This change in the O:C stoichiometric ratio is shown in Fig. 4. The ratio of the total integrated O1s and C1s peak intensities, corrected for photo-ionisation cross section, photo-electron mean free path, and analyser transmission, is plotted. The decrease in the O:C ratio starts at fluences below the material removal threshold, reaching a value about 30% below that of the unirradiated surface after irradiation at 401 mJ/cm^2. Laser radiation may selectively split off oxygen from the macromolecule. These results are supported by FTIR measurements (not shown), which indicate that spectral structures due to C-O and C=O excitations are reduced in intensity by irradiation, while, e.g., the C-H bending mode is relatively unaffected [6]. A further change is evident in the relative XPS peak intensities of the ether (O-C) and carbonyl (O=C) oxygen atoms (partially resolved O1s peaks at E_B = 532.1 and 533.3 eV, respectively). While the O1s intensity for the unirradiated surface is evenly distributed between the peaks due to (O-C) and (O=C) bonds (as expected), after irradiation proportionally more carbonyl O is present (Fig. 4), i.e., irradiation especially decreases the amount of ether oxygen.

For irradiation at conditions other than those of Fig. 3 and Fig. 4, other effects might be expected. For much higher fluence, or for smaller fluence but for many more pulses, a stronger warming of the surface (higher mean temperature) might be expected, which could result in a re-oxidation of the surface due to reaction with atmospheric oxygen. For

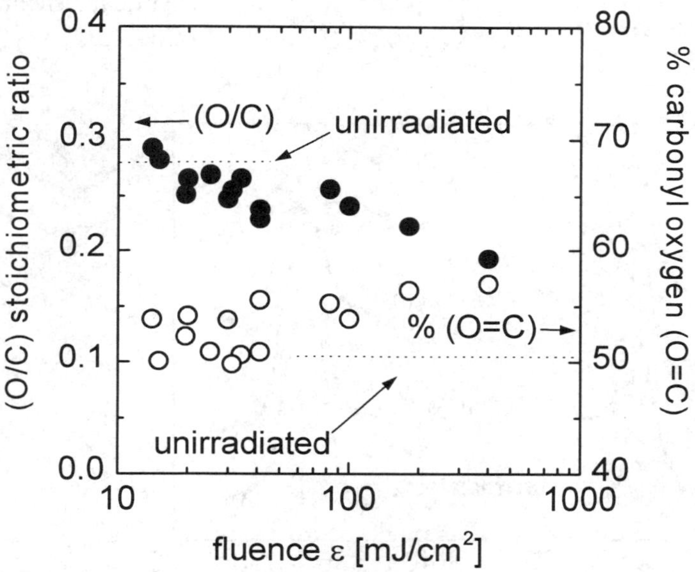

Fig. 4 (O/C) stoichiometric ratio as a function of fluence determined from O1s and C1s core-level spectra and percent contribution of the carbonyl (O=C) O1s peak to the total (carbonyl plus ether) O1s peak intensity (10 pulses).

example, for irradiation of poly(ethylene terephthalate) (PET) with a KrCl* excimer lamp radiation (λ=222 nm) in 0.8 Torr of air, an increase in oxygen content was observed for long irradiation times (20 min.) at a 20 kHz repetition rate [7]. The approximately comparable irradiation conditions for the excimer laser irradiation would be a very small pulse fluence and a large number of pulses.

Irradiation with fluences above the material removal threshold changes the morphology by photochemical and photothermal processes. The extent of modification of the surface morphology is determined by the radiation parameters (fluence, pulse number, repetition rate), the absorption coefficient α of the polymer, the material removal threshold ε_t and by other physical properties of the polymer. Several different types of surface morphologies [8] may result for irradiation of PBT with KrF-excimer laser radiation (Fig. 5).

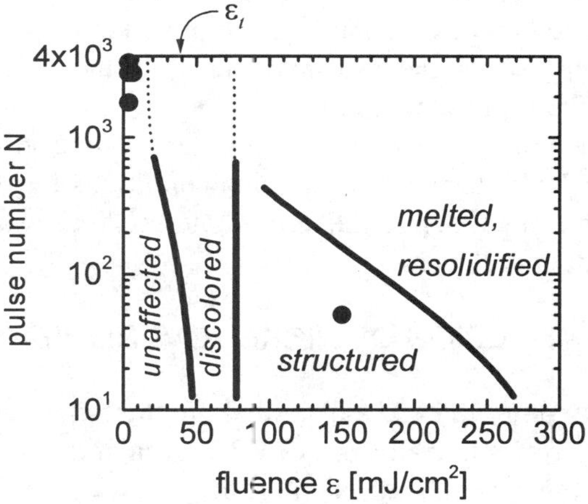

Fig. 5 Phase diagram for the existence regions in irradiation-parameter-space of the various surface morphologies observed after excimer laser irradiation of PBT (pulse repetition rate 3 Hz). The data points indicate exposure conditions for which electroplated nickel overlayers exhibit enhanced adhesion. ε_t indicates the material removal threshold.

For irradiation at fluences slightly above the material removal threshold ε_t, the surface appears smooth but discoloured as seen by optical microscopy. At larger fluences, irradiation produces a structured surface with ripples (Fig. 6).

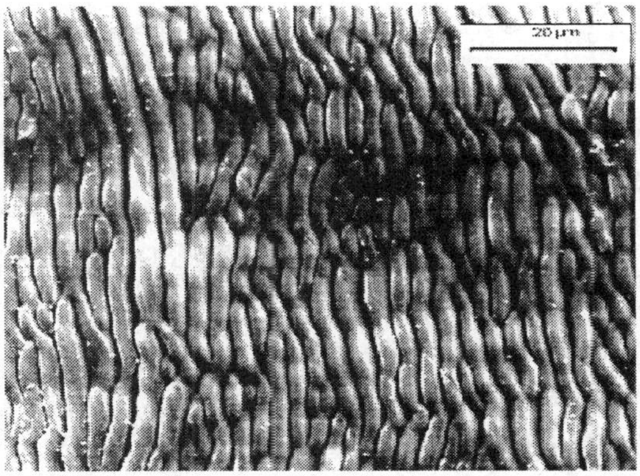

Fig. 6 Scanning electron micrograph of the structured surface (ripples) of PBT, produced by excimer laser irradiation (fluence $\varepsilon = 150$ mJ/cm^2, 10 pulses).

This structure is believed to arise from thermal stressing of the near-surface region coupled with material flow determined by uniaxial mechanical stress within the polymer [4]. The structures, which are much larger than the microstructures originating from interference effects, exhibit lateral periodicity in the range 5-10 µm and structure depths in the range 10-20 µm. For irradiation with a larger number of pulses and/or a larger fluence, the ripple morphology is replaced by a surface having distinct indications of melting and re-solidification. The structure depth, however, does not increase much above the range 10-20 μm. For extreme irradiation conditions (e.g., (N, ε) values from the upper right-hand side of Fig. 5), the surface even becomes relatively smooth, i.e., the surface roughness R_a decreases. In this case the subsurface region of the semitransparent surface still exhibits molten structures by optical microscopy. Thus, the maximum surface roughness achievable by irradiation of a clean polymer surface is limited. For the irradiation conditions which result in an enhanced surface roughness an effect on the metal/polymer adhesion would be expected. In fact, the filled circles in Fig. 5 designate irradiation conditions for which an enhanced metal/polymer adhesion, in comparison to the unirradiated surface, was found in tape and peel tests (see below) of metal overlayers.

SURFACE MORPHOLOGY AND ADHESION OF NICKEL OVERLAYERS ON PBT

The adhesion of metal layers on unirradiated PBT is weak, in the sense that the tape test completely removes the metal overlayers. Irradiation of PBT, even at a fluence below the material removal threshold, does improve the adhesion of thin Ni layers, probably by a chemical modification of the surface. The nickel overlayer is not removed by the tape test,

Table II. Effect of Pretreatment on the Adhesion of Copper/Nickel Layers on Excimer Laser Pretreated PBT.

pretreatment with KrF excimer laser radiation $\lambda = 248$ nm, pulse duration 25 ns, various pulse numbers	adhesion force measured with peel test
none	not measurable
fluence $\varepsilon < \varepsilon_t$	< 0.1 N/mm
fluence $\varepsilon > \varepsilon_t$	<0.2 N/mm
removal with particle mask, $\varepsilon > \varepsilon_t$	0.2-1.6 N/mm

but the adhesion of copper reinforced layers is relatively low (< 0.1 N/mm) (Table II). Irradiation generates changes in the surface morphology and may, depending on the irradiation parameters, improve the adhesion (Table II). The improvement in adhesion after excimer laser irradiation with fluence above the material removal threshold is about two times higher (< 0.2 N/mm) than that achieved for irradiation below the threshold. Still, even this degree of adhesion improvement is insufficient for most technological applications (e.g., in microsystem technology), for which a typical value for "good" adhesion is about 0.8 N/mm, as measured in a peel test.

It is known that adhesion is enhanced by the mechanical anchoring of the metal layer in the rough polymer surface. Irradiation at a fluence above the material removal threshold through a particle mask on the surface of the polymer increases the surface roughness and generates a high density of such mechanical anchoring points with a high aspect ratio (ratio of depth to width) (Fig. 7a and Fig. 7b).

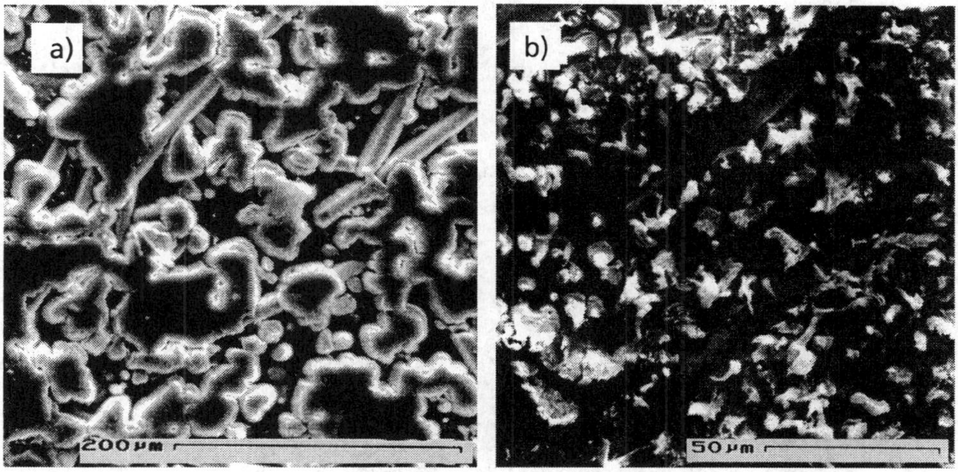

Fig. 7 Scanning electron micrographs of the surface of PBT after excimer laser irradiation with (a) a mask of mica particles ($\varepsilon > \varepsilon_t$, removal depths 15-20 µm, adhesion strength 0.4 N/mm) and (b) a mask of boron particles ($\varepsilon > \varepsilon_t$, removal depths 15-20 µm, adhesion strength 0.8 N/mm).

The mask consists of small particles (1 - 10 µm), deposited on the surface from an alcohol solution, which are not removed by the excimer laser irradiation. Only the polymer surface between the particles is removed. The adhesion with the boron particle mask (Fig. 7b) is better because of the higher aspect ratio achieved compared to the mica particle mask (Fig. 7a).The largest adhesion strength measured (1.6 N/mm) was obtained using a mask of boron particles (Table II).
In the main processing step with the particle mask, a larger removal depth between the

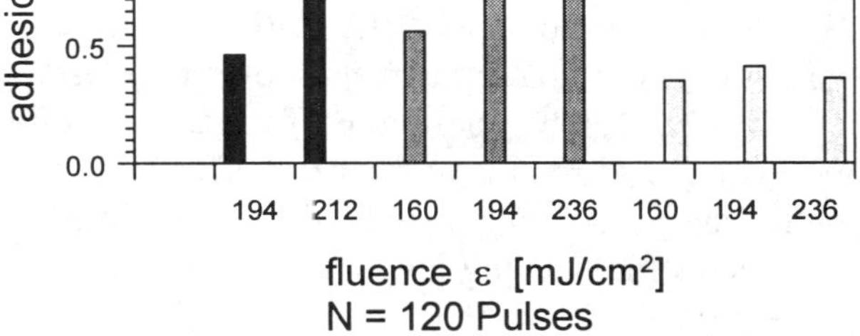

Fig. 8 Adhesion strength (peel test) of metal layers on excimer laser irradiated PBT as a function of fluence and post-processing with an additional radiation dose (removal with a mask of boron particles, $\varepsilon > \varepsilon_t$, 120 pulses).

particles is achieved by excimer laser irradiation with the same number of pulses but at a higher fluence. Because of the larger structure depths, the mechanical anchoring of the metal layer in the polymer surface is improved, and a larger adhesion strength is measured (Fig. 8).
The influence of chemical changes at the surface can be shown by an additional irradiation. This is done subsequent to the main dose with the particle mask, after removal of the particles and at a fluence below ε_t, and improves the adhesion significantly (Fig. 8). For this post-processing apparently an optimal dose is necessary. An additional irradiation higher than the optimal dose reduces adhesion.

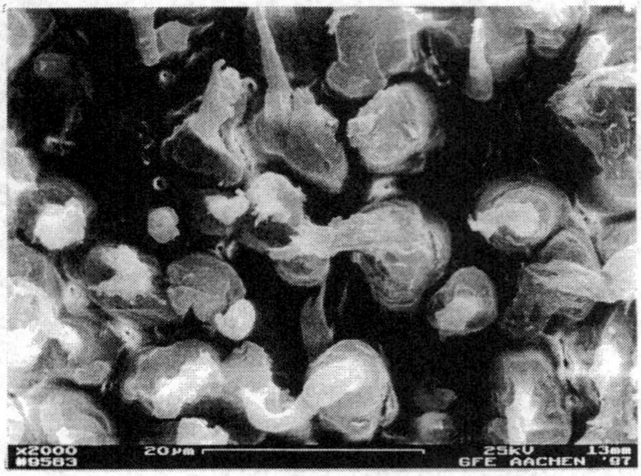

Fig. 9 Scanning electron micrograph of the surface of PBT (excimer laser irradiated with a mask of boron particles, $\varepsilon > \varepsilon_t$) after peel test.

The maximum achievable adhesion strength seems to be limited by the mechanical properties of the polymer. Fig. 9 shows the polymer surface after the peel test. The cones of the rough polymer surface structure have been completely removed by shear forces. Thus, the strength of the metal/polymer interface is, in principle, mainly determined by the bulk shear strength of the polymer.

CONCLUSIONS

With excimer laser radiation pretreatment of PBT using particle masks, it is possible to achieve adhesion improvements sufficient for common applications of metallized polymers. The main cause for the strong adhesion in the nickel/PBT interface is of structural nature, involving mechanical anchoring in the metal/polymer interface. Chemical interaction between the metal and the chemically modified PBT surface does improve the adhesion, but the effect is small in comparison to the mechanical effects. By using a particle mask at fluences above the material removal threshold, it is possible to modify the surface morphology of polymers such as PBT having low material removal thresholds, independently of the chemical structure.

REFERENCES

1. See, e.g., L. J. Gerenser, in "Metallization of Polymers," E. Sacher, J.-J. Pireaux, and S. P. Kowalczyk, Eds., ACS Symp. Ser. No. 440, p. 433, American Chemical Society, Washington, DC, 1990

2 K.L. Mittal, in "Adhesion Measurement of Films and Coatings," K.L. Mittal, editor, pp. 1-13, VSP, Utrecht, The Netherlands, 1995.

3. E. Sutcliffe and R. Srinivasan, J. Appl. Phys. 60, 3315 (1986); J. T. C. Yeh, J. Vac. Sci. Technol. A4, 653 (1986); R. Srinivasan, in "Interfaces Under Laser Irradiation," L. D. Laude, D. Bäuerle, and M. Wautelet, Eds., p. 359, Martinus Nijhoff Publishers, Dordrecht, 1987; G. Pettit and R. Sauerbrey, Appl. Phys. A56, 51 (1993).

4. D. Bäuerle, "Laser Processing and Chemistry," 2nd edition,. Springer-Verlag, Berlin, 1996.

5. C.-M. Chan, T.-M. Ko, and H. Hiraoka, Surface Sci. Rep. 24, 1 (1996).

6. R. Weichenhain, Diploma thesis in physics, RWTH Aachen, Germany (1996).

7. J.-Y. Zhang, I. W. Boyd, and H. Esrom, Surface Interface. Anal. 24, 718 (1996).

8. D. A. Wesner, H. Horn, R. Weichenhain, W. Pfleging and E. W. Kreutz, J. Adhesion. Sci. Technol., 11, 1229 (1997).

Metallized Plastics 5&6: Fundamental and Applied Aspects, pp. 365-387.
K.L. Mittal (Ed.)

ION BEAM INDUCED ADHESION AT METAL-POLYMER INTERFACES

J.E.E. Baglin

IBM Almaden Research Center
650 Harry Road
San Jose, California 95120

Ion beam processing can provide a powerful tool for promoting adhesion at metal-polymer interfaces. In this review paper, the techniques are outlined, and some recent advances are discussed. In particular, the importance of tailored interface chemistry is stressed, together with various means of inducing interface bonding: pre-treatment of the substrate surface with low energy ions or plasma; ion beam mixing at an existing interface; implantation of reactive species at the interface; and ion beam assisted deposition (IBAD) of vapor deposited metal coatings, in order to activate the substrate surface and reduce stress in the metal film. Recent studies of ion beam induced adhesion are reviewed, and analytical studies of the underlying phenomena are discussed. We review evidence of the important role of electronic (ionization) energy delivered by the ions, and of the negative consequences of collisional energy transfer.

INTRODUCTION

Thin coatings of metals on polymer surfaces are required for a variety of technological applications. In most cases, the metal-polymer adhesion must be strong and stable, a condition best reached by the creation of direct chemical bonding at the interface. However, very few metals are sufficiently reactive to spontaneously disrupt the termination bonding at a polymer surface to make new interface complexes, even at elevated temperatures. Thus, the adhesion of a metal film such as gold, vapor deposited directly on a polymer such as Teflon, is weak or non-existent, and depends only on van der Waals forces or topographical interlocking.

Ion beam techniques offer several ways to promote the formation of direct metal-polymer chemical bonding, and also to enhance the effective adhesion of thin film coatings by stress reduction and contaminant removal. Strong, direct adhesion has been achieved for a variety of metal film coatings (e.g., Cu, Al, Ni, Au, Cr, Ag) on polymer surfaces (e.g. Teflon, Mylar, polyimides). The technique and the intrinsic mechanisms of ion enhanced adhesion have been reviewed in depth in other publications[1-5], to which the reader is referred. It is the purpose of the present review (a) to summarize the ion beam processes; and (b) to review recent studies that have made clear the dominant intrinsic mechanisms whereby beam interactions can lead to interface adhesion. In the author's opinion, the technique and its optimization are now quite well understood, and we are at the stage where a reliable, high-throughput, commercially attractive procedure can probably be developed to produce a useful, and possibly high, degree of adhesion for many metal-polymer couples.

ION BEAM PROCESSES

Substrate Surface Treatment

A low energy beam of ions (e.g. 50 eV - 10 keV Ar^+) may be used to "activate" the surface of a polymer substrate, immediately prior to the vapor deposition of a thin metal coating, as illustrated in Fig. 1. Such ions may come from a directed Kaufman ion source whose beam is spread or rastered over large areas as needed, or from a non-directional plasma-immersion ion source, which can readily treat large, complex, non-planar substrate shapes[6]. Either source can be adapted for continuous-feed processing. These processes must be applied in vacuum conditions, and are compatible with standard vacuum metallization tools.

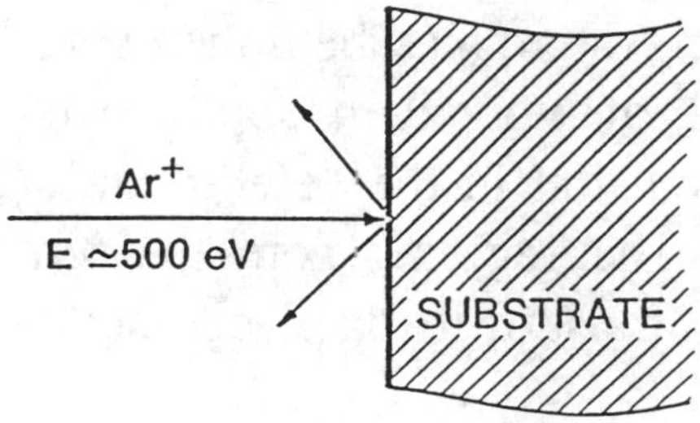

Figure 1. Substrate surface pre-sputtering.

The low energy ions impinging on the polymer surface cause some disruption of the chemical bonding and composition at the surface, as a result of ionization and recoil displacement of surface atoms and some loss of surface species by sputtering. It appears that in many cases (depending on the particular polymer) chemically active sites will be exposed at the surface, where metal atoms arriving may bond to form complexes, thus linking the subsequently deposited metal with the substrate itself. The sputtering loss may also roughen the polymer surface, thereby assisting the attachment of subsequently deposited metal layers.

Most native substrate surfaces support one or two monolayers of physisorbed contaminant species which can well constitute a barrier to bonding of an added metal coating. Ion beam sputter-cleaning of a compound substrate will generally remove such surface contaminants, and this benefit can be expected to accompany ion pre-treatment, in addition to the promotion of chemically active sites at the substrate surface.

Interface Mixing

A penetrating beam of energetic ions ($\geq$ 100 keV) may be used to modify a metal-polymer interface that has already been prepared, (Fig. 2). In this case, each ion serves to create a "collision cascade" within the sample - a volume surrounding the primary ion path, in which dense ionization has occurred, and in which a heavy ion like Ar^+ may have ballistically displaced some host atoms from their original locations in the solid. Quantitative Monte Carlo modelling of this process for a specific system may be done using TRIM software[7], from which may be derived, at any given depth in the solid, the total energy lost by the individual ion in the "electronic" (ionization) or "nuclear" (collisional) form. The region where the ion is slow, near the end of its range, sustains most of its displacement damage.

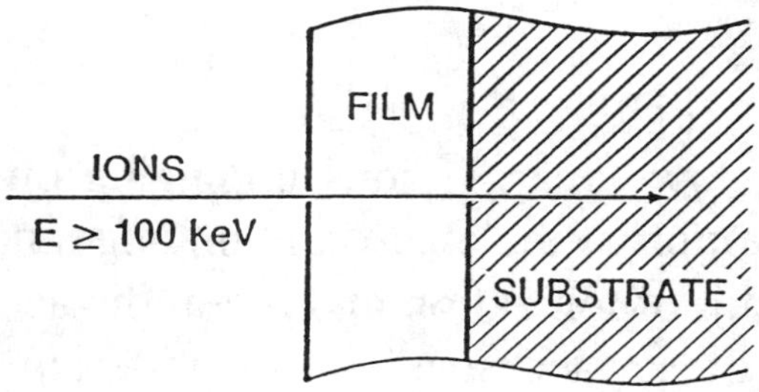

Figure 2. Ion beam mixing at film-substrate interface.

The effect of the cascade at an internal "interface" may be likened to a localized thermal effect: atoms are briefly mobilized and will attempt to relax in a low-energy configuration. Stably bound clusters of neighboring interface atoms may well be formed, depending on the species and their equilibrium chemistry. These new complexes can produce interface adhesion.

Further, if extensive ballistic mixing of the interface has occurred, adhesion performance could be enhanced because of the dispersion of interface stress and the greater fracture toughness of the irregular interface.

Contaminant layers at a pre-formed interface may be dispersed by ion mixing, thus facilitating direct adhesion. The effect has been demonstrated by Bøttiger et al.[8] for the case of bonding Cu to an oxidized Cr surface, and it should be beneficial at metal-polymer interfaces too.

Reactive Ion Implantation

This approach combines the kinetic effects of ion mixing with the added chemical inventory provided by ion implantation of a reactive species through a deposited film, with an energy chosen to match its end of range to the depth of the interface, as illustrated in Fig. 3. Another similar approach is to introduce an extra reactive species at the interface during low energy surface treatment (generally adding O_2 in the Ar plasma).

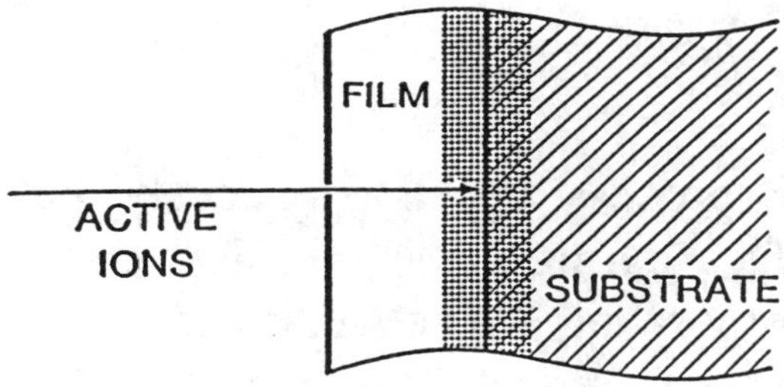

Figure 3. Ion implantation (plus ion beam mixing) in the interface region.

Galuska[9] has demonstrated the possibilities for reactive ion mixing. Attempts to produce adhesion of a Ni film on a Mylar substrate failed when using only an inert species (Kr^+) for interface mixing. However, by implanting Si at the interface, strong adhesion was obtained. Subsequent XPS analysis indicated that simple mixing led to the growth of a Ni_3C layer, which failed to bond with the substrate, while the addition of Si led to the formation of a strongly bonded interface layer containing Ni-Si-O-C complexes. Similar results were obtained for Ni on polyimide substrates.

While this example illustrates an interesting option for ion beam engineering, implantation at large doses and high energy in this way is probably not an attractive economic option for most applications. Doping of the interface during pre-treatment is a more practical and realistic option.

Ion Beam Assisted Deposition (IBAD)

The adhesion performance of a thin film coating of a metal on a polymer will be governed not only by the strength of the interface bonding, but also by the stress within the film. For example, a vapor deposited Cr film a few µm thick can develop sufficient tensile stress to delaminate spontaneously from a weakly attached substrate.

Figure 4. Ion Beam Assisted Deposition (IBAD)

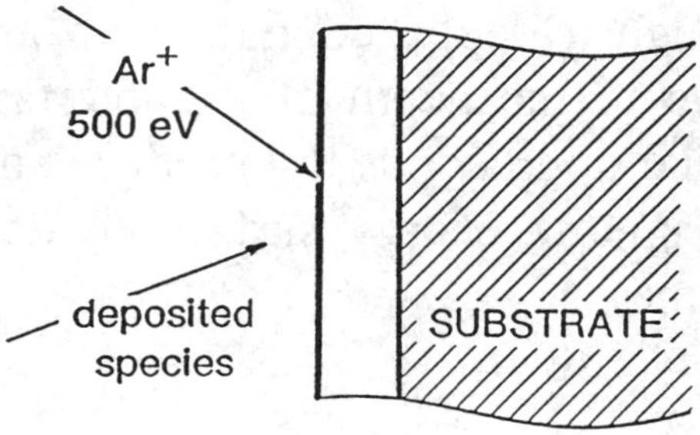

The process of simultaneously ion-beam bombarding the receiving surface during metal film deposition (IBAD) (see Fig. 4) can assist in incorporating a reactive ambient species into the interface. It may also be used to reduce film stress, thereby improving adhesion performance. For example, Barth et al.[10] reported IBAD deposition of Cr (1 µm) on a steel substrate, using 6 keV Ar^+ bombardment. By optimizing the ratio of arriving Ar^+ ions to arriving Cr at the sample, the stress in the Cr was reduced from highly tensile to zero, with the result that measured adhesion increased by a factor of 5.

EXPERIMENTAL PROGRESS

Since the discovery[11,12,13] that ion beam irradiation could improve the interface adhesion between materials having no bulk chemical affinity, a great deal of experimental and modelling work has been presented. Experimentally, it was demonstrated that the irradiation process using MeV ion beams need not cause extensive interface mixing in order to be effective; in fact, the strongest adhesion is the result of new chemical bonding at the atomic layers of the interface, induced by the interaction of the passing ions.

The respective roles of electronic energy deposition and collisional (nuclear) energy loss from the ions were debated, especially since the intuitively visualized "mixing" would result from nuclear collisions, even though a body of experiments had shown adhesion enhancement to be achievable from electron beam processing alone[14].

Yet another series of experiments demonstrated the success (for many systems) of using a low energy ion beam to pre-activate the polymer surface before the coating is deposited. The approach worked well for metals on most compound substrates, whose stoichiometry and surface termination could be disrupted in order to leave active sites or functional groups exposed at the surface, ready to bond chemically with the deposited metal. Not only did this approach prove to yield stronger adhesion for a given system than did ion mixing; it also offered easy compatibility with commercial processing technologies based on vacuum plasma treatment of surfaces. Depending on the specific system, this method may sometimes require that the activated surface be coated immediately in situ, and not be exposed to air before deposition of the metal.

Pre-treatment of polymer surfaces with low energy plasmas has been reviewed in detail by Egitto and Matienzo[15].

The generality first attributed to the ion beam effect (i.e. bonding everything to everything) has been modified by the growing realization that the chemical bonding species and structure available will eventually determine the ability of any irradiation to promote interface chemical bonding. In some cases, the formation of stable chemical complexes could be assisted by the intentional addition of a new, active species, such as oxygen, at the interface. However, another very effective and simpler technique has proved to be low energy ion irradiation or ion-assisted deposition in the presence of an oxygen-loaded ambient.

Other factors that have been seen to affect the adhesion performance of thin coatings include roughening of the substrate surface to promote mechanical "interlocking" with the coating layer. While chemical etching to roughen polymers prior to metal coating has long been standard practice, its benefits now appear to be generally smaller than those offered by interface chemical bonding induced by ion beam processing.

In recent years, quite specific experimental work on ion beam enhanced adhesion of metals on polymer substrates has clarified the mechanisms that are involved in such systems, and led us to a much more complete understanding of the options and the limitations for ion beam processing.

Current Understanding

The accumulated evidence today indicates the following basic features of optimized ion beam enhanced adhesion of metal coatings on polymers:

(i) Intrinsically, the ion beam treatment must (a) activate interface chemical bonding between the metal and functional species exposed at the polymer surface as a result of irradiation, and (b) not weaken the polymer near or below the surface by radiation damage.

(ii) The activation may occur during energetic ion mixing of an interface after film deposition, or by exposing the polymer to an ion beam or plasma before and/or during metal deposition.

(iii) Reactive species may be added to assist the formation of interface complexes most easily by adding them (e.g. O_2) to the pre-deposition ion beam ambient.

(iv) Chemical bonding alone, at an abrupt interface, can be a source of the strongest bonds.

(v) Electronic (ionization) energy deposition by the ion beam is the primary beneficial agent for polymer-metal adhesion enhancement. Collisional (nuclear) energy deposition by heavier, slower ions produces displacement damage that typically weakens the polymer beyond the interface.

(vi) Mechanical interlocking will occur if the polymer surface is roughened before coating. However, an ion-beam roughened interface such as that found for Teflon[16, 17] is by itself relatively weak. There is little evidence that intentional pre-sputtering to create such texture leads to a strong or durable bond on polymers, unless interface chemical bonding has also been created; then such texture can be very beneficial.

(vii) Interface grading (e.g. by high energy ion beam mixing) may improve adhesion, in principle. The extent to which it is achievable without displacement damage weakening the polymer itself appears generally to be small, and its contribution is minor.

(viii) Ion beam assisted deposition of a metal coating can lead to low intrinsic stress in the metal, with consequently less tendency to delaminate. This is probably not generally a major effect in comparison with the strength of chemical bonding, except for thick (several µm) films of a few specific metals.

One study that exemplifies the interplay of several of these processes is that of Paik and Ruoff[18] . They examined the adhesion of Cu to polyimide resulting from pre-deposition treatment by an oxygen plasma. Some of their observed peel test results are displayed in Fig. 5, as a

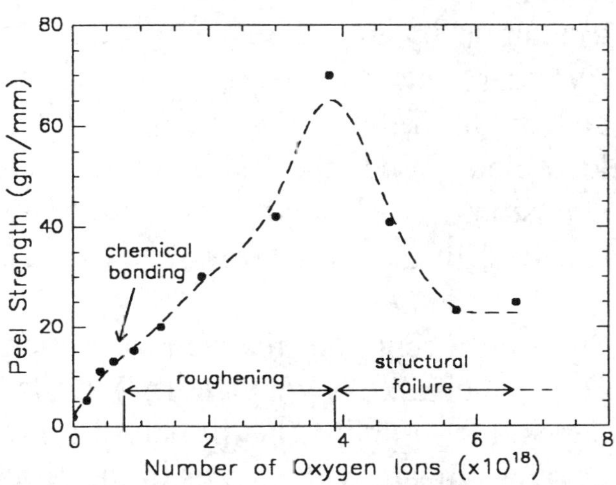

Figure 5. Adhesion of Cu on polyimide (PI), after pre-deposition treatment of the PI by an oxygen ion beam. Chemical bonding enabled by low doses is supported by interface roughening at somewhat higher doses; at high doses, structural failure of PI is evident. *(Based on the results of Paik and Ruoff[18]).*

function of ion dose. Three dose regimes identify the initial production of interface bonding at oxygen carbonyl sites, increased practical adhesion due to roughening at up to 3.6 x 10^{18} ions/cm^2, and declining adhesion at higher doses, as the roughened polymer near the interface becomes physically weak.

REVIEW OF RECENT PUBLICATIONS

Some publications providing important evidence generally consistent with these conclusions will be discussed below. Table I summarizes the contents of these and other selected reports.

Interface Chemistry

Early evidence for the correlation between stable chemical complexing between the metal and the activated polymer surface has been discussed in previous reviews[1-5].

Wäppling and others at Uppsala[19] have reported a series of experiments using Conversion Electron Mössbauer Spectrometry (CEMS) to identify bonding at the Fe-PVC [poly(vinyl chloride)] and Fe-PTFE [poly(tetrafluoroethylene)] interfaces following post-deposition irradiation by low doses (~10^{11}/cm^2) of energetic heavy ions. In the PVC case, clear evidence for the formation of $FeCl_2.2H_2O$ was found, and in the Teflon case, FeF_2, Fe-F-C_x and an Fe-C compound were identified. When the PTFE substrate surface was irradiated prior to Fe deposition, Fe-C bonding was found to dominate. The measurements specifically address a layer of < 10 nm of ^{57}Fe at the interface.

Shi et al.[20] reported treatment of Teflon PFA (poly[tetrafluoroethylene-co-perfluoroalkoxy vinyl ether]) using N_2 or O_2 plasma prior to Cu deposition in situ. This produced a practically non-removable copper coating. In-situ XPS studies of the N_2-plasma exposed surface (after air exposure) indicated the presence of F_xC-O, C=O and C-O groups, and

Table I: Summary of some experimental studies of radiation enhanced adhesion of metal coatings on polymers.

Ref. no.	Polymer	Metal	Process	Irradiation	Beam Energy	Dose ($/cm^2$)	Adhesion [Peel test]	Comments
19	PTFE	Fe	Ion mix or pre-treat	Sulfur	16 MeV	10^{11}-10^{14}	Increased	CEMS: Fe-C, FeF_2, Fe-F.
	PVC			Br	48 MeV	10^{11}	Increased	CEMS: $FeCl_2$. $2H_2O$
20	Teflon - PFA	Cu	Pre-treat + air expose	O_2 plasma microwave	(few eV)	--	Increased	Cu-O, Cu-N bonding. (Incl. adsorbate O_2) ; No surface roughening.
				N_2 plasma	(few eV)	--	Increased	
21	PMDA-ODA	Cu or Au	Pre-treat	Ar	500 eV	10^{13}-10^{16}	Incr. ×10; Au	XPS: Interface chem. bonding, identical for Ar(500 eV) or (10 eV); shows that only the surface matters.
				Ar (ECR plasma)	10-20 eV	10^{16}-10^{17}	stronger than Cu.	
22	PTFE	Au	Pre-treat	Ar, Ar/O_2 glow discharge	--	--	Incr. ×8	Complex chem. bonding, including adsorbate O and H.
23	BPD-PDA	Cr/Cu /Cr	Pre-treat	Ar	800 cV	--	86 g/mm	XPS: Reduction of carbonyl carbon and imide N at polymer surface; new carbon-oxygen, carbon-nitrogen bonding sites; polar surface.
24	Lexan polycar-bonate	Al	Pre-treat, plus IBAD	Ar, N	300 eV	$2x10^{16}$	× 1.7	XPS: No new complexes after pre-treat; attribute adhesion to cleaning.
	ABS					--	Weak	

Table I. *(Continued)*

Ref. no.	Polymer	Metal	Process	Irradiation	Beam Energy	Dose ($/cm^2$)	Adhesion [Peel test]	Comments
5	Teflon	Cu	Mix	He, Ne	250 keV	$3x10^{15}$	20 g/mm	Weaker for high doses.
16	Teflon	Cu	Pre-treat	Ar	500 eV	$1x10^{16}$	50 g/mm	Roughening not the principal mechanism.
17	PTFE	Cu	Pre-treat	Ar, or Ar in O_2	1 keV	10^{14}-10^{17}	Requires $>10^{15}/cm^2$;	XPS: Interface complexes; SEM: Gross roughening of PTFE surface above $10^{16}/cm^2$; much less roughness with O_2.
		Al					Requires $>10^{16}/cm^2$; Best with O_2.	
25	PES	Cu	Mix, plus reactive implant	He, C, N, O, F, Ne, Mg, Si, S, Ar.	60 keV or 120 keV	10^{14}	80 g/mm	Adhesion depends on ion species; Mg was best; electronic energy transfer good; nuclear (damage) transfer weakens adhesion performance.
	PEI						70 g/mm	
	PMDA-ODA						140 g/mm (max.)	
26	PES	Cu	Mix, IBAD	He, Li, B, C, N, O, Ne, Mg, Si, S, Ar, Ca	60 keV or 120 keV	10^{14}	70 g/mm	As above (ref.25); reactive ion species implanted at interface beneficial; irradiation must be low enough to avoid carbonizing polymer surface.
	PEI		Mix, IBAD	(same)	(same)		60 g/mm	
	glassy C		Mix, IBAD	Mg	120 keV		No adhes.	
	CFK compos.		IBAD	Ar	0.2-1.0 keV		50 g/mm	

Table I. *(Continued)*

Ref. no.	Polymer	Metal	Process	Irradiation	Beam Energy	Dose ($/cm^2$)	Adhesion [Peel test]	Comments
27	Poly-imide	Al	Pre-treat	UV pulsed laser, 248 nm	5 eV	10 pulses at 40 mJ/cm^2	Improved	XPS: Laser depletes oxygen at surface; Al reacts at carbonyl oxygen sites.
	PBT						No change	
28	Parylene -N	Cu	IBAD	Partially ionized Cu is deposition source	3 keV	--	70 g/mm	Deeply mixed interface; separates at 300°C; XPS: No chem. bonding; interlocking assumed.
						--		
29	PET	Cu	Pre-treat	Ar,	1 keV	--	40-80 g/mm	Low doses best; high dose weakens. Plasma web processing feasible at several m/s.
				or Ar (ECR plasma),	9 eV	10^{15}-10^{17}		
				or O(ECR plasma)	7 eV	10^{14}-10^{16}		
	Acrylate						30-60 g/mm	
	OPP poly-propyl-ene						<10g/mm	

Table I. *(Continued)*

Ref. no.	Polymer	Metal	Process	Irradiation	Beam Energy	Dose ($/cm^2$)	Adhesion [Peel test]	Comments
30	Teflon AF1600	Al 10nm	Mix; reactive ambient	Ar	300 eV	--	75 g/mm	XPS: Complex Al bonding; N retained, but not bonded to Al.
			Mix only				No effect	
31	Polycar-bonate	Al	Pre-treat	Ar in O_2	1 keV	10^{14}-$5x10^{16}$	Increased	Hydrophilic group on polymer surface bonds to Al.
15	PMDA-ODA	Cr	Pre-treat	O_2 dc glow discharge	--	--	x 3	Active carboxylate groups on polymer surface.
18	Poly-imide	Cu	Pre-treat	Oxygen	500 eV	$4x10^{18}$	70 g/mm	Bonding increased at surface carbonyl oxygen; assisted by polymer roughening.
32	Poly-imide	Al	Pre-treat	Ar discharge	--	--	Increased	Al-C, Al-O-C bonds.
33	PMDA-ODA	Cr	Pre-treat	Ar rf discharge	--	--	x 100	Chemical bonding; negligible roughening.

after Cu deposition they found evidence for Cu-O and Cu-N bonding, supporting the concept of induced chemical bonding with Cu as the source of adhesion. In both experiments, maximum adhesion was reached after a few seconds of irradiation. Longer exposure led to a small decline in adhesion, possibly due to physical weakening of the sub-surface substrate material caused by the ion damage to the polymer.

A further example of chemical interaction was given by Jeong and White[21], in which PMDA-ODA polyimide surfaces were exposed to Ar^+ ion beams (either an ECR plasma at 10-20 eV energy or a 500 eV ion source) before adding either Cu or Au coatings. XPS was used to identify the formation of some chemical complexes at the interface.

Kinbara et al.[22] used pre-treatment of PTFE with low energy plasmas of Ar, O_2 or ($Ar+O_2$) to produce strong adhesion of Au films. Again, XPS displayed the evolution of active surface complexes. In this case, the adhesion produced by the mixed Ar-O_2 plasma was approximately 30% greater than that obtained with Ar alone, showing the added benefit of introducing oxygen to participate in the interface chemical complexing.

Flitsch and Shih[23], in a similar experiment with (BPDA-ODA)-based polyimide exposed to Ar^+ ions at 800 eV, obtained an 80-fold increase in peel strength of post-deposited Cr/Cu/Cr. Using XPS, they found that the ion irradiation caused major changes in the relative surface concentrations of nitrogen and of carbonyl carbon, resulting in the formation of new carbon-oxygen and carbon-nitrogen species, and giving rise to a polar surface. They also reported reduced peel strengths if the irradiated polymer surface were exposed to room air for one or two days, apparently due to chemisorption of contaminants.

Ratchev et al.[24] irradiated the surfaces of Lexan polycarbonate and ABS (acrylonitrile butadiene styrene) polymers with 300 eV Ar^+ or N^+, prior to and during deposition of 200 nm Al films. Contrary to the reports of Ref. 21, they found no changes in chemical state of the polymer surfaces after ion exposure, using XPS. However, the increase in peel-test adhesion strength that they observed was also small ($\times$ 1.7 or less). These authors concluded that the ion energy of 300 eV must be too low to break up the surface carbonyl groups (thus enabling the deposited Al to form Al-O-C complexes at the new dangling bonds), and they attribute the adhesion improvement to sputter cleaning of the polymer.

Koh et al.[17] deposited Cu and Al films on PTFE, following exposure to 1 keV Ar^+ ions, both with and without an oxygen gas ambient. Scotch tape tests showed improved adhesion in each case, although for Cu the adhesion increased with ion dose, up to $>10^{15}/cm^2$, while for Al, good adhesion required the presence of oxygen during the substrate pre-irradiation, improving up to doses of at least $10^{16}/cm^2$. XPS revealed very complex, and very clear, changes in the bonding of C and O on the treated polymer surfaces. The dominant effect of such irradiation, however, as shown in SEM images, was found to be the coarse surface texturing of the PTFE, in the form of spires and canyons of characteristic dimensions 0.1 μm diameter, 0.5 μm deep. (Similar structure was previously reported by Chang et al.[16].) On the basis of this structure, the authors have assumed that interlocking is the main reason for the improved adhesion in this case. However, it is interesting to note that the surfaces irradiated in the O_2 ambient did not develop such deep topographic features, even at 10 times larger dose. The reason for this topographic difference is not clear. Despite this mystery, it seems apparent that adhesion was the result of a combination of chemical bonding and interlocking, but even the gross roughening was probably not the principal determinant of adhesion performance.

Ion Beam Energy Transfer Process

In a series of very illuminating experiments, Tegen et al.[25,26] have examined ion induced adhesion enhancement for Cu films pre-deposited on PES (polyethersulfone), PEI (polyetherimide) and PMDA-ODA polyimide. They explored the dependence of the peel force on irradiation species and energy, using a variety of ions with $Z \leq 20$, always at a dose of $10^{14}/cm^2$, and with incident energies chosen to locate the mean ion range either deep in the substrate or close to the Cu-polymer interface.

Improved adhesion was obtained in all cases. In general, the data were consistent with a model where adhesion is produced as a result of ionization (electronic) energy deposited by the incident ions at the metal-polymer interface. For a given species of ion, the peel strength was greatly dependent on the species of polymer, which is consistent with a model of chemical complexes (unique for each polymer) being formed at the interface due to radiation-induced bond-breaking of the polymer and atomic rearrangements to form new bonding complexes with the metal. The choice of reactive metal ion species implanted at the interface (e.g. Li^+, B^+, Mg^+ and Ca^+) led to even better adhesion in all cases. However, displacement damage to the polymer caused by heavy ions at high energies reduced the overall benefit of irradiation, an effect attributable to the weakening of a layer below the polymer surface where the ions, near the end of their range, deposit energy mainly in the collisional mode, probably liberating hydrogen and other volatile species and carbonizing the region. In an interesting additional experiment, the same group repeated the procedure with Cu films deposited on pure glassy carbon. No adhesion could be produced (without added species). This emphasizes the importance of avoiding too high a dose of ions that can begin to reduce a polymer layer to pure carbon, which would destroy the chance for chemical bonding, since Cu and C alone have no mutual affinity. The optimal ion dose will fall far short of this extreme, while delivering the maximum possible electronic energy transfer, in order to scission the polymer and expose chemically active sites.

Previous experiments[1-5] have also pointed to electronic energy loss as the constructive agent in this process, and displacement damage (nuclear energy loss) as primarily counter-productive. The newer experiments specifically support this model [25,26].

Thus, it would seem that to optimize the adhesion enhancement by inert or transmitted ions, one should plan the energy and dose of ions to maximize the electronic energy transfer to the interface, and minimize the collisional energy deposited in the polymer (at any depth). Light, energetic ions will work well at high dose; heavier ions will require a smaller dose, but may cause damage that offsets the benefit of induced interface chemistry. It would appear that ballistic displacements are not, in general, beneficial to the important formation of new chemical complexes.

Why, then, would not beams of electrons or energetic photons offer an easier and more effective tool for adhesion enhancement? In fact,[14] success with electron beams was reported many years ago for assisting the adhesion of Au films on SiO_2, and in other metal/ceramic applications. Similar processes would be expected to work for polymer substrates too. One practical problem with such beams is the large dose required, which means either long exposures or much heating of the sample by the beam.

In an interesting and related experiment, Weichenhain et al.[27] used a 248 nm KrF-excimer laser to pre-treat the surface of polyimide (PI) prior to coating it with Al. Successful adhesion of Al on treated PI was achieved for laser fluences near 40 mJ/cm^2 (the threshold for material removal). XPS was used to identify the surface chemical changes. The laser irradiation caused oxygen depletion of the surface by desorption of both C=O and C-O-C species, and a new species of nitrogen bonding appeared (C=N-C), consistent with bond breaking in the imide ring and formation of an isoimide structure. Carbonyl oxygen sites were identified as the locations where deposited aluminum binds as Al^{3+} at the interface. It is proposed that laser irradiation enhances the proportion of Al^{3+} species due to the introduction of a ring oxygen species in the isoimide structure, providing new Al binding sites. The authors point out that the photon energy in this experiment (up to several eV) enables photochemical processes such as ionization or bond-breaking; in addition, the peak (pulsed) power density for the laser needed to be set near the threshold for material removal, perhaps indicating the need for a multi-electron cooperative effect to assist surface atom reconfiguration and bond restructuring.

Lee[34] has pointed out the broad spectrum of consequences for a polymer subjected to ion irradiation, which depend not only upon the proportion of energy transferred by ionization or collisional processes, but also upon the local transient energy density delivered to the polymer. For example, sparse ionization (<1 ion pair per nm) will break isolated bonds, producing unsaturated bonds or chain scission. Such densities would be expected along the paths of individual electrons or photons bombarding the polymer. However, an ion in the energy range from a few eV to 5 keV will affect continuous regions along its path, producing a high local density of ionization (~ 1.5 - 20 ion pairs per nm), in which crosslinking or recombination between intermolecular radicals may readily occur.

On the basis of these concepts, it appears that the nature, species and density of active surface sites on a pre-treated polymer where a deposited metal will chemically bond will depend partly on the instantaneous *density of ionization* provided by the radiation, and will also depend upon the specific polymer structure. Although in many cases, simple generation of unsaturated bonds at the surface will enhance adhesion performance for a deposited metal, one can also imagine cases where a degree of crosslinking may be needed, to maintain the cohesive strength of the irradiated polymer, or where recombination processes may produce a preferred type of bond for metal attachment.

Thus, it seems likely that not only electron energy but also the local transient density of ionization delivered may be a significant criterion for obtaining the interface reconstruction that is needed for adhesion enhancement. Knowledge of this criterion would be helpful for optimizing ion beam induced adhesion, allowing us to make an informed selection of ion species and energies (and perhaps dose rates) to suit a particular polymer-metal couple. These arguments also suggest the value of further studies of pulsed lasers or electron beams as simple agents for the beam enhancement of adhesion.

SUMMARY

Our understanding of the mechanism for ion and electron beam enhanced adhesion of metal films on polymers has advanced substantially. Some significant conclusions are presented below.

(a) The strongest attachment is based on interface chemical bond formation.
(b) The adhesion attainable varies depending on the species of polymer and metal.
(c) Ion beam mixing, substrate activation, and addition of chemically active species may all have positive effects. Pre-activation of a polymer surface in situ before metal coating may be the most effective, and most process-compatible.
(d) Electronic energy losses of ions are sufficient to produce interface bond reconstruction.
(e) Nuclear (recoil) damage collisions from heavy ions are harmful to the adhesion, and can weaken the polymer itself.
(f) Pre-treatment of polymer surfaces with a pulsed 248 nm laser shows possibly equivalent effects. The applicability of pulsed electron beams for adhesion promotion warrants further study.
(g) Ion beam processing may also provide the secondary benefit of interface roughening.

The technique is ready for routine commercial application, especially where broad-beam vacuum plasma based manufacturing tools and integrated web processors are now available.

REFERENCES

1. J.E.E. Baglin, in *Materials and Processes for Surface and Interface Engineering,* Y. Pauleau, (Ed.), pp. 111-149, NATO ASI Series, Kluwer Academic Publishers, Dordrecht (1995).

2. J.E.E. Baglin, IBM J. Res. Develop. **38,** 413 (1994).

3. J.E.E. Baglin, in *Fundamentals of Adhesion,* L.-H. Lee, (Ed.), Chapter 13, Plenum Publishing Co., New York (1990).

4. J.E.E. Baglin, in *Ion Beam Modification of Insulators,* P. Mazzoldi and G.W. Arnold, (Eds.), Chapter 15, Elsevier, Amsterdam (1987).

5. J.E.E. Baglin, in *Handbook of Ion Beam Processing Technology,* J.J. Cuomo, S.M. Rossnagel and H.R. Kaufman, (Eds.), Noyes Publications, Park Ridge, NJ, 1989, Chapter 14; see also ref. 2.

6. I.G. Brown, S. Anders, M.R. Dickinson, I.C. Ivanov, R.A. McGill, X.Y. Yao and K.-M. Yu, Nucl. Instrum. and Methods **B80/81,** 1281 (1993).

7. J.F. Ziegler, J.P. Biersack and U. Littmark, *The Stopping and Range of Ions in Solids,* Vol. 1, Pergamon, New York (1985). (The software package TRIM - The TRansport of Ions in Matter - is available from J.F. Ziegler, IBM Research Center, Yorktown Heights, NY 10598).

8. J. Bøttiger, J.E.E. Baglin, V. Brusic, G.J. Clark and D. Anfiteatro, Mater. Res. Soc. Symp. Proc. **25,** 203 (1984).

9. A. A. Galuska, J. Vac. Sci. Technol. **B8**, 470 (1990). Also A. A. Galuska, Nucl. Instrum. and Methods **B44**, 428 (1990).

10. M. Barth, W. Ensinger, V. Hoffmann and G.K. Wolf, Nucl. Instrum. and Methods **B59/60**, 254 (1992).

11. L.E. Collins, J.G. Perkins and P.T. Stroud, Thin Solid Films **4,** 41 (1969).

12. J.E. Griffith, Y. Qiu and T. Tombrello, Nucl. Instrum. and Methods **198,** 607 (1982); also B.T. Werner, T. Vreeland, Jr., M. H.

Mendenhall, Y. Qiu and T.A. Tombrello, Thin Solid Films **104** 163 (1983).

13. S. Jacobson, B. Jonsson and B. Sundqvist, Thin Solid Films **107,** 89 (1983).

14. C.J. Sofield, C.J. Woods, C. Wild, J.C. Rivière and L.S. Welch, Mater. Res. Soc. Symp. Proc. **25,** 197 (1984).

15. F.D. Egitto and L.J. Matienzo, IBM J. Res. Develop. **38,** 423 (1994).

16. C.-A. Chang, J.E.E. Baglin, A.G. Schrott and K.C. Lin, Appl. Phys. Lett. **51**, 103 (1987).

17. S.K. Koh, S.C. Park, C.K. Choi, S.K. Song, H.J. Jung and K.D. Pae, Mater. Res. Soc. Symp. Proc. **396** 335 (1996); also S.K. Koh, S.C. Park, J.W. Seok, S.C. Choi, W.K. Choi and H.J. Jung, Mater. Res. Soc. Symp. Proc. **438,** 523 (1997).

18. K.W. Paik and A.L. Ruoff, J. Adhesion Sci. Technol. **4,** 465 (1990).

19. R. Wäppling, Nucl. Instrum. and Methods **B76,** 22 (1993); also P.A. Ingemarsson, M.P. Keane and U. Gelius, J. Appl. Phys. **66,** 3548 (1989); P.A. Ingemarsson, T,Eriksson, R. Wäppling and G. Possnert, J. Adhesion Sci. Technol. **3**, 503 (1989); P.A. Ingemarsson T. Eriksson, G. Possnert and G. Wäppling, Hyperfine Interactions **46**, 549 (1989).

20. M.K. Shi, A. Selmani, L. Martinu, E. Sacher, M.R. Wertheimer and A. Yelon, J. Adhesion Sci. Technol. **8**, 1129 (1994).

21. H.-S. Jeong and R.C. White, J. Vac. Sci. Technol. **A11,** 2308 (1993).

22. A. Kinbara, A. Kikuchi, S. Baba and T. Abe, J. Adhesion Sci. Technol. **7**, 457 (1993).

23. R. Flitsch and D.-Y. Shih, J. Adhesion Sci. Technol. **10,** 1241 (1996).

24. B.A. Ratchev, G.S. Was and J. H. Booske, Nucl. Instrum. and Methods **B106**, 68 (1995).
25. N. Tegen, J. Wartusch and K.-H. Merkel, Nucl. Instrum. and Methods **B80/81,** 105 (1993).

26. H. Fladry, N. Tegen and G.K. Wolf, Nucl. Instrum. and Methods **B91,** 575 (1994).

27. R. Weichenhain, D.A. Wesner, W. Pfleging, H, Horn and E.W. Kreutz, Appl. Surface Sci. **109/110,** 264 (1997); also E.W. Kreutz, H. Frerichs, J. Stricker and D.A. Wesner, Nucl. Instrum. and Methods **B105,** 245 (1995).

28. P.K. Wu, S. Dabral, G.-R. Yang, B. Gittleman, C. Li, J.F. McDonald and T.-M. Lu, J. Appl. Phys. **80,** 5759 (1996).

29. F. Milde, K. Goedicke and M. Fahland, Thin Solid Films **279,** 169 (1996).

30. P.K. Wu and T.-M. Lu, Appl. Phys. Lett. **71,** 2710 (1997).

31. S.K. Koh, S.K. Song, W.K. Choi and H.J. Jung, J. Mater. Res. **10,** 2390 (1995).

32. M.J. Vasile and B.J. Bachman, J. Vac. Sci. Technol. **A7,** 2992 (1989).

33. T.S. Oh, S.P. Kowalczyk, D.J. Hunt and J. Kim, J. Adhesion Sci. Technol. **4,** 119 (1990).

34. E. H. Lee, in *Polyimides - Fundamentals and Applications,* M.K. Ghosh and K.L. Mittal, (Eds.), p. 471, Marcel Dekker, New York (1996).

Metallized Plastics 5&6: Fundamental and Applied Aspects, pp. 389-407.
K.L. Mittal (Ed.)

CHANGES IN THE ADHESION STRENGTH BETWEEN COPPER THIN FILMS AND POLYIMIDE SUBSTRATES AFTER HEAT TREATMENT

Satoru IWAMORI*, Takehiro MIYASHITA, Shin FUKUDA, Shouhei NOZAKI, Kazufuyu SUDOH, and Nobuhiro FUKUDA

Central Research Institute, Mitsui Toatsu Chemicals, Inc., 1190 Kasama-cho, Sakae-ku, Yokohama, Japan 247

The adhesion strength of a copper (Cu) thin film deposited onto polyimide (PI) substrate deteriorated by heat treatment at 150 ^{o}C in air. After the heat treatment, (1) Cu_2O microparticles diffused into the PI substrate, (2) Cu lumps were present on the peeled PI surface, (3) the PI substrate and the Cu thin film were oxidized, and (4) the weak boundary layer (WBL) shifted towards the Cu/PI interface in the PI with increasing the heat treatment time. In order to prevent the penetration of Cu_2O particles and control the oxidation, several kinds of interfacial layers were introduced at the interface between the Cu thin film and PI substrate. Cobalt (Co) interfacial layer was found to be an effective barrier material against the diffusion of Cu_2O particles, but vanadium (V) interfacial layer was found not to be effective. Quenching the defects in the interfacial layers was also found to be important in achieving this effect. By annealing Co interfacial layers at 280 ^{o}C for 1 hr in vacuum before sputter deposition of Cu thin film, we were able to maintain the adhesion strength between Cu thin film and PI substrate above 10 N/m even after heat treatment at 150 ^{o}C for 3 days.

INTRODUCTION

It is known that some kind of polymer treatment of plastic substrates is often necessary to improve the adhesion strength between metal thin films and plastic substrates.[1,2] Figure 1 shows the measured adhesion strength between a Cu thin film and polymer substrates as a function of O_2-plasma treatment time. Without this plasma treatment, Cu weakly adhered (adhesion strength less than 1.0 N/m)

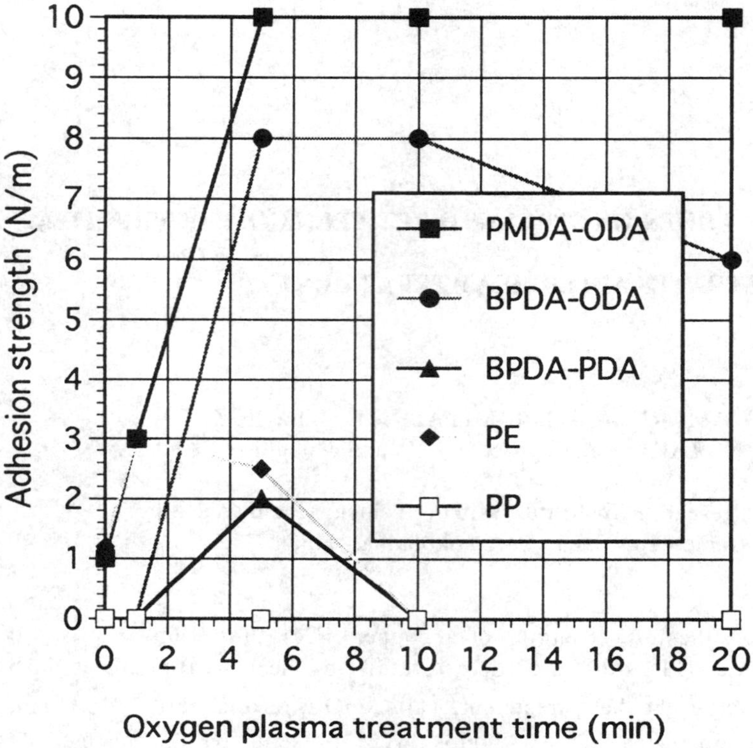

Figure 1. Adhesion strength of Cu/polymer laminates versus oxygen plasma treatment time.

to polyethylene(PE) and PI substrates, signifying that O_2-plasma treatment was effective for improving the adhesion strength. On the other hand, Cu hardly adhered to the polypropylene(PP) substrate even with the O_2-plasma treatment. These results show that oxygen plasma treatment of polymer substrates does not always improve the adhesion strength.

Polyimides have good thermal stability, chemical stability, and excellent mechanical properties such as a high modulus of elasticity.[3] Due to these reasons, Cu-deposited PI films (Cu/PI laminates) are commonly used in printed circuit boards (PCBs).

In this paper, we will focus on the adhesion deterioration phenomenon for the Cu/ PI laminate due to the heat treatment.[4-6] In addition, based on the experimental results, we will report on the effect of the introduction of a metallic interfacial layer as a diffusion barrier and the process to obtain Cu/PI laminates showing only slight thermal deterioration.

EXPERIMENTAL

Kapton-V Film (0.05 mm thick; TORAY-DuPont) with structure shown in Fig.2 was used as the PI substrate. The conditions of oxygen and argon plasma treatments and Cu sputtering are given in Table I. The sputtering conditions of the interfacial layers are also shown in Table II. Cu was electro-plated up to a thickness of 0.03

mm on the sputtered Cu thin film. The patterning of the Cu layer was carried out using 2 mm wide IC tapes as resists for $FeCl_3$ solution.

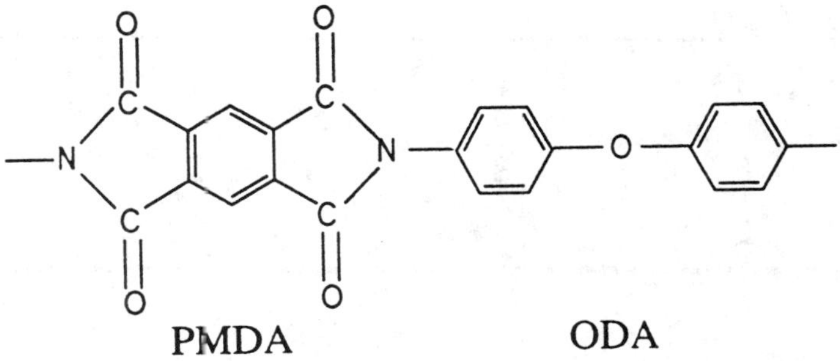

Fig.2 Structure of the PMDA-ODA polyimide

The adhesion strength was measured using a 180° peel test.[7] The peel tests were performed using STROGRAPH M1 Tester (Toyo-seiki, Inc. Tokyo, Japan) at a crosshead speed of 50 mm/min. The heat treatment condition was 150°C in air for 24, 48, and 72 hr. Auger electron spectroscopy (AES) was used for the analysis of the composition of the peeled metal surfaces. AES analysis was performed using MH93-4001 Auger Spectrometer (Perkin-Elmer Inc.) with an ultimate pressure of 2×10^{-8} Pa. To obtain depth profiles, sample surfaces were sputtered with argon ion with an energy of 3 keV. XPS analysis was carried out using PHI 5400MC Spectrometer (Perkin-Elmer Inc.) with a monochromatic AlKα source (1487 eV) with a spot size of 1.1 mm and operated at an electron pass energy of 72 eV. The peeled surface was prepared just before the introduction of the sample into the load-lock vacuum chamber in order to minimize possible contamination originating from the ambient. The depth profiles of the peeled surfaces were obtained using AES by Ar-sputtering at a sputtering pressure of 6.7×10^{-5} Pa. The chemical bonding states of the peeled surfaces were investigated with XPS. In order to study the sputtered films, only the PI substrate was dissolved in an alkaline hydrazine solution (hydrazine : KOH = 4 : 1).

Table I. Experimental Conditions for Plasma Treatment and Copper Deposition.

Gas	*Oxygen*	*Argon*
Flow rate	10 sccm	10 sccm
Pressure	6.7 Pa	0.27 Pa
Power	rf, 400 W	dc, 160 W
Time	20 min	10 min

Table II. Metal Deposition Conditions.

Metal	*Pressure (Pa)*	*Power (W)*	*Time (min)*	*Thickness (nm)*
V	0.27	180	2	20
Co	0.67	200	1	10
Co	0.67	200	30	300

RESULTS AND DISCUSSION

How Does the Adhesion Strength Deteriorate ?

Fig.3 shows the measured adhesion strength of the Cu/PI laminate as a function of the heat treatment time at 150 ^{o}C in air. Up to a heat-treatment time of 24 hours, the adhesion strength decreases sharply, after that it decreases slightly.[8]

The interface of the Cu/PI laminate was analyzed with transmission electron microscopy (TEM). Fig.4 shows a cross-sectional TEM micrograph of the interface after the heat treatment for 24 hr. A high concentration of tiny particles (10-100 nm in size) was found to be dispersed in the PI.[4] These particles were estimated to be Cu_2O from the electron beam diffraction patterns. Fig.5 (A),(B),(C) and (D) show the optical micrographs of the peeled PI surface from the Cu/PI laminate without heat treatment, and after heat treatment at 150 ^{o}C in air for 24, 48, and 72 hr, respectively. The heat-treated samples had many bright spots whose size increased with the heat treatment time and reached a diameter of 2 to 25 μm.[4]

We evaluated the changes in the chemical bonding states of the peeled surfaces with XPS.[8] Fig.6 (A) and (B) show the C(1s) spectra of the Cu/PI laminate (A) without the heat treatment and (B) after the heat treatment at 150 ^{o}C for 24 hr in air. Although only three peaks, due to -C-, -C-N- or -C-O-, and -N-C=O bonds, which represented the PI structure, could be detected at the peeled PI surface of the unheated sample, peaks representing the -O-C-O- or =C=O, -(C=O)-O, and -O-(C=O)-O- bonds appeared in addition to these three peaks at the peeled PI surface of the heat-treated sample. The amount of the oxidized moieties such as the -O-C-O- or =C=O, -(C=O)-O, and -O-(C=O)-O- bonds increased due to the heat treatment.[8] Copper has been shown to catalyze the oxidation of the PI at 350 ^{o}C in air, and also to catalyze the thermo-oxidative degradation of the polyimide film in air.[9,10] In consideration of these reports, the PI substrate at the interface might be oxidized by the copper. Fig.7 (A) and (B) show O(1s) spectra of the peeled PI surface from Cu/PI laminate (A) without the heat tereatment and (B) after the heat treatment at 150 ^{o}C for 24 hr in air. The copper-oxygen bond was found at the peeled PI surface of the Cu/PI laminate only after the heat treatment.[8]

In order to investigate in detail, the peeled Cu surface was analyzed with AES. The distribution of atoms was investigated from AES depth profiles.[4,8] Fig.8 (A) and (B) show the depth profiles for carbon and Cu, respectively, of the peeled Cu

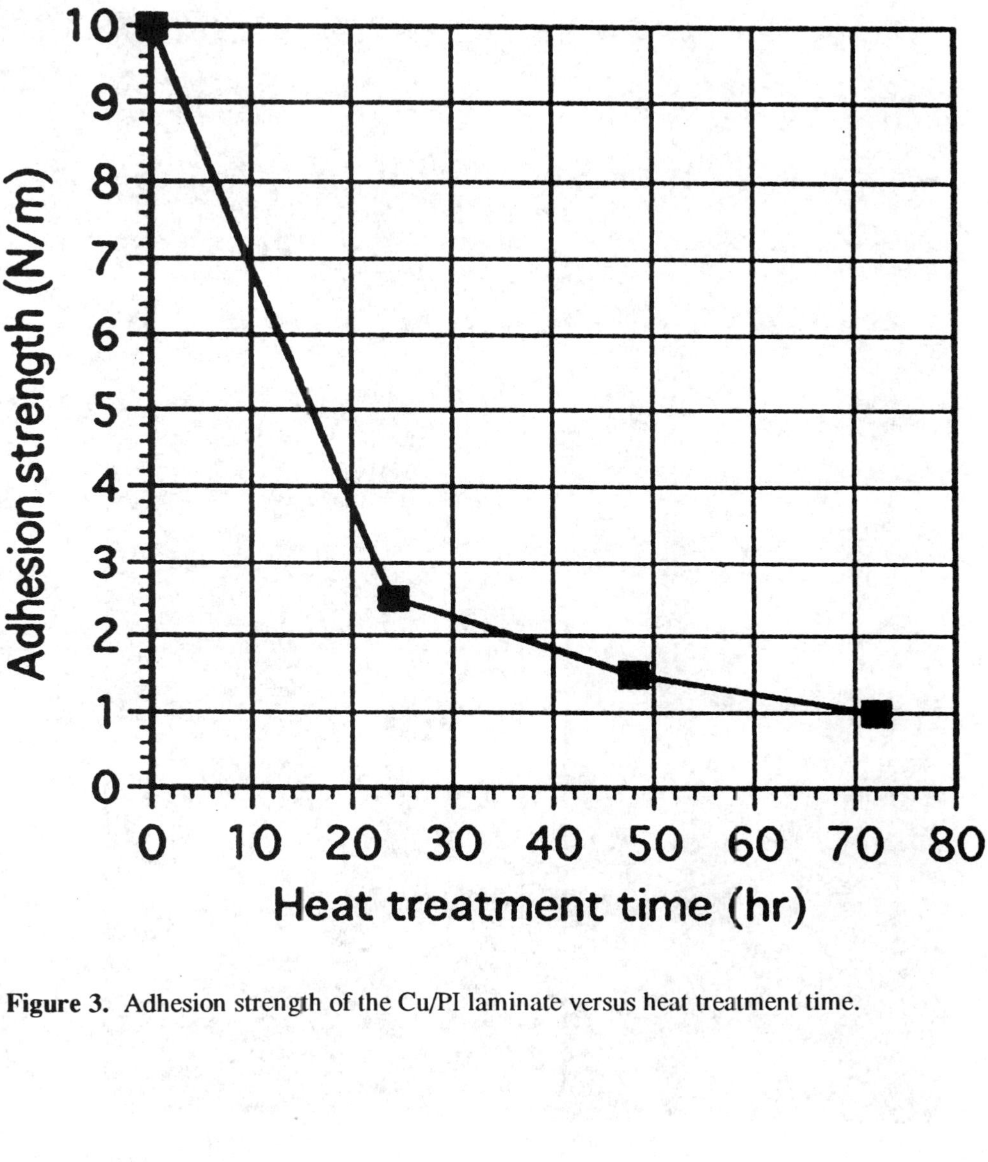

Figure 3. Adhesion strength of the Cu/PI laminate versus heat treatment time.

surface from this laminate. The depth at which the carbon disappeared was shallower with increasing heat treatment time. For the Cu, the depth at which the Cu becomes 100 % was shallower with increasing heat treatment time. Fig.9 shows the relationship between the depth at which the carbon disappeared and the adhesion strength. The weak boundary layer (WBL),[11] where peeling occurs, shifted towards the Cu/PI interface in the PI as the adhesion strength decreased.

These results showed that the position of the WBL was located in the PI layer without the heat treatment, and the WBL shifted toward the Cu/PI interface in the PI after the heat treatment (Fig.10).[8] The amount of this shift was related to the adhesion strength between the Cu and the PI.

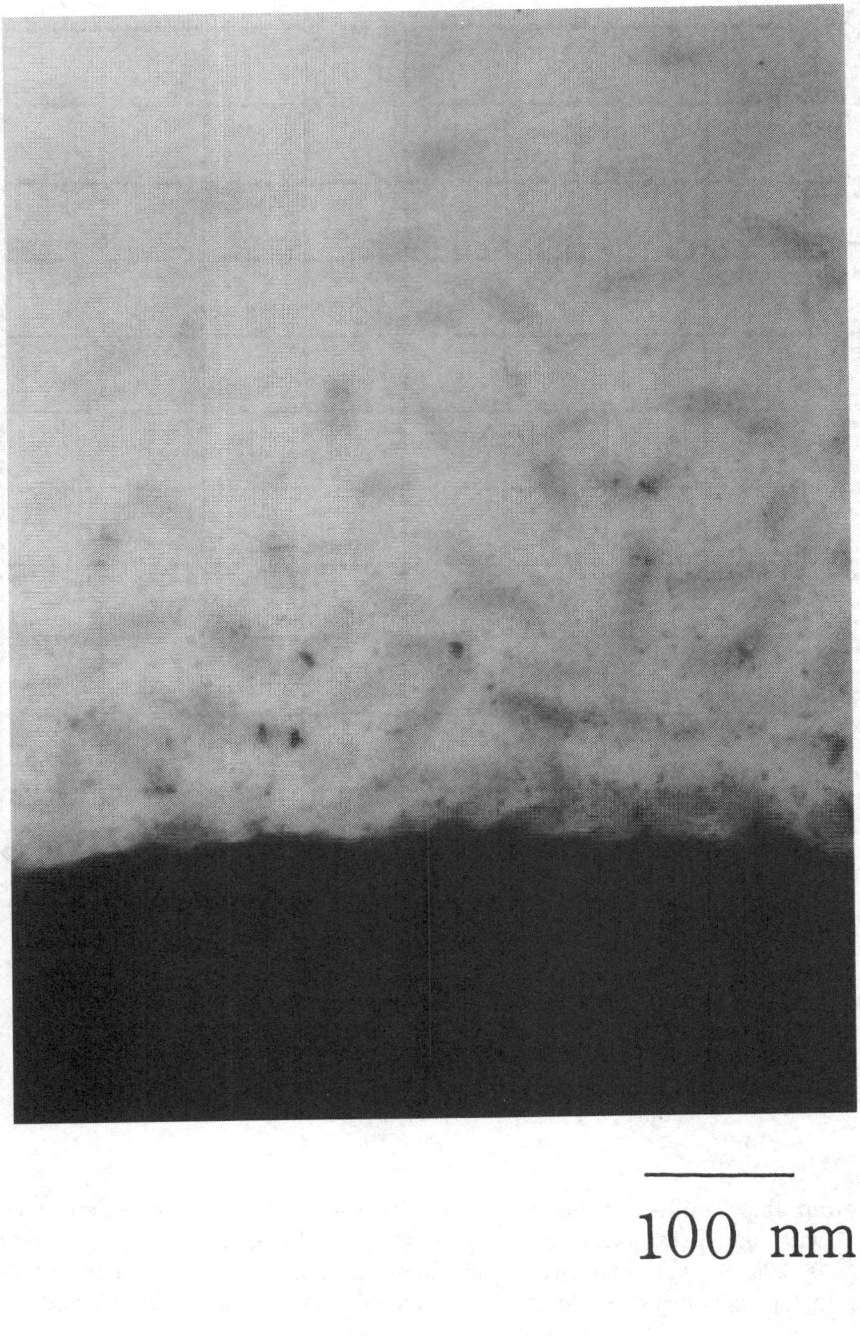

Figure 4. Cross-sectional TEM micrograph of the interface of a Cu/PI laminate after heat treatment time.

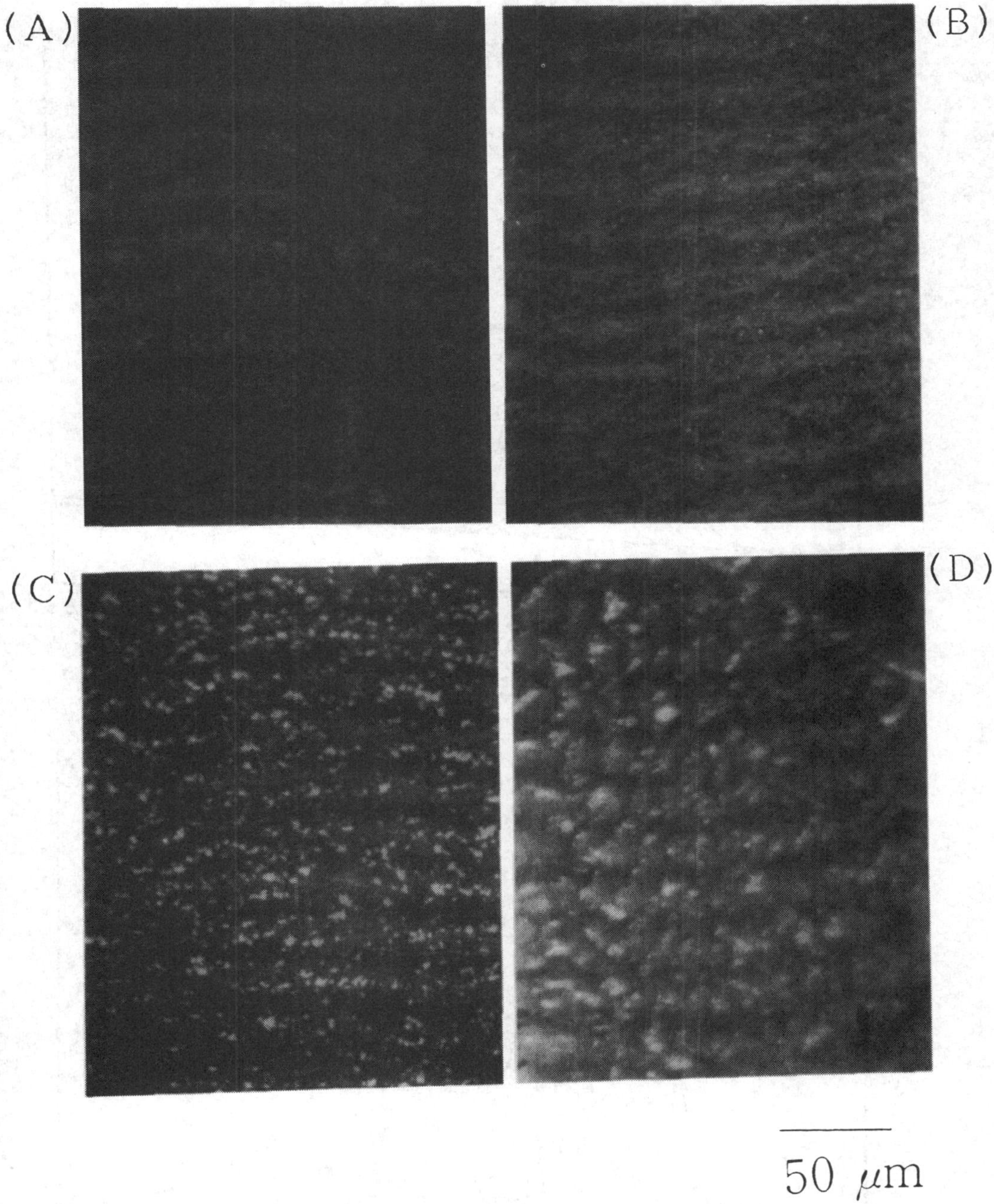

Figure 5. Optical micrographs of the peeled PI surface from a Cu/PI laminate (A) before heat treatment, (B) after heat treatment at150°C in air for 24 hr, (C) for 48 hr, and (D) for 72 hr.

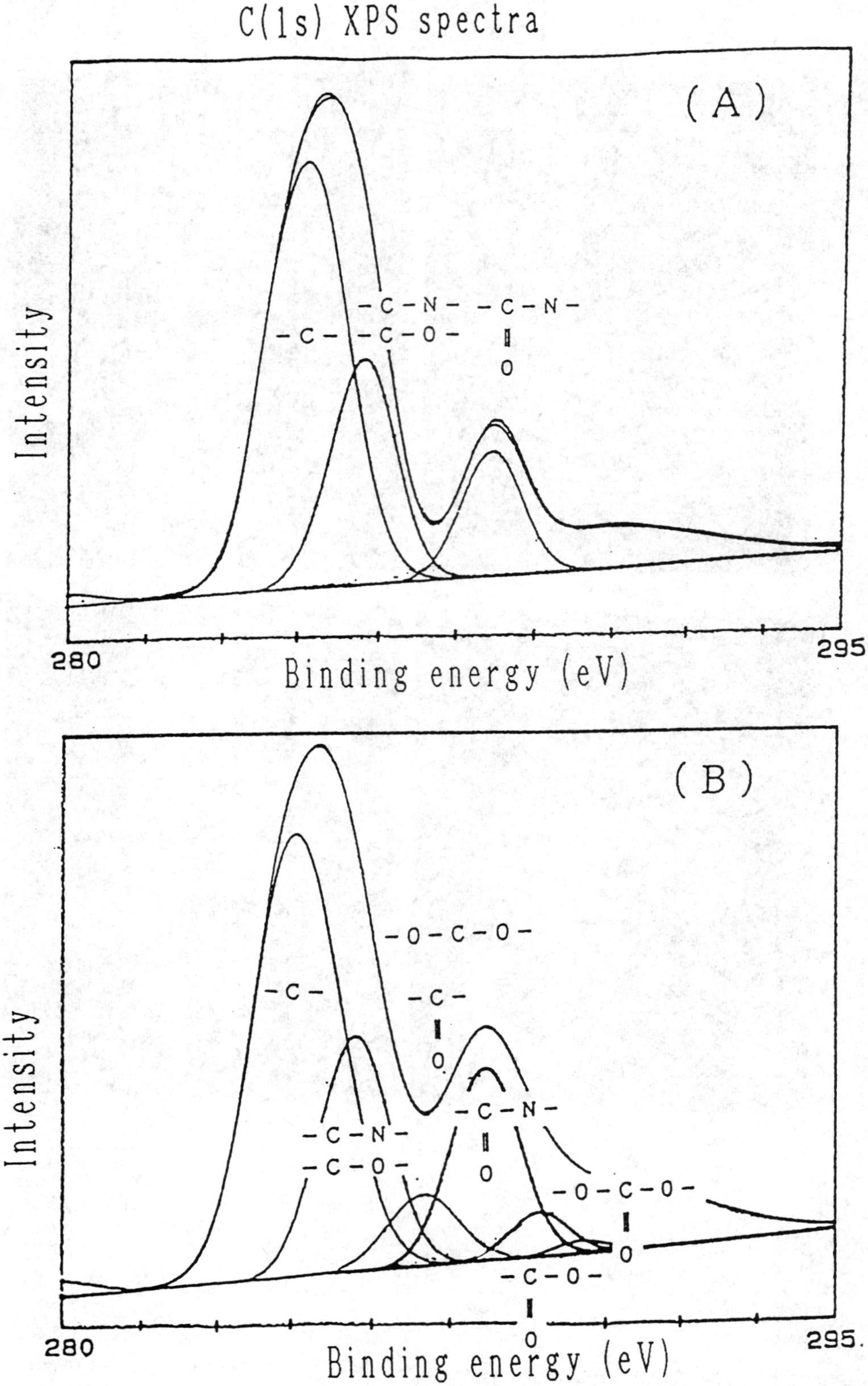

Figure 6. C (1s) spectra of the peeled PI surface from a Cu/PI laminate for (A) unheated sample and (B) sample heat-treated at 150°C for 24 hr in air.

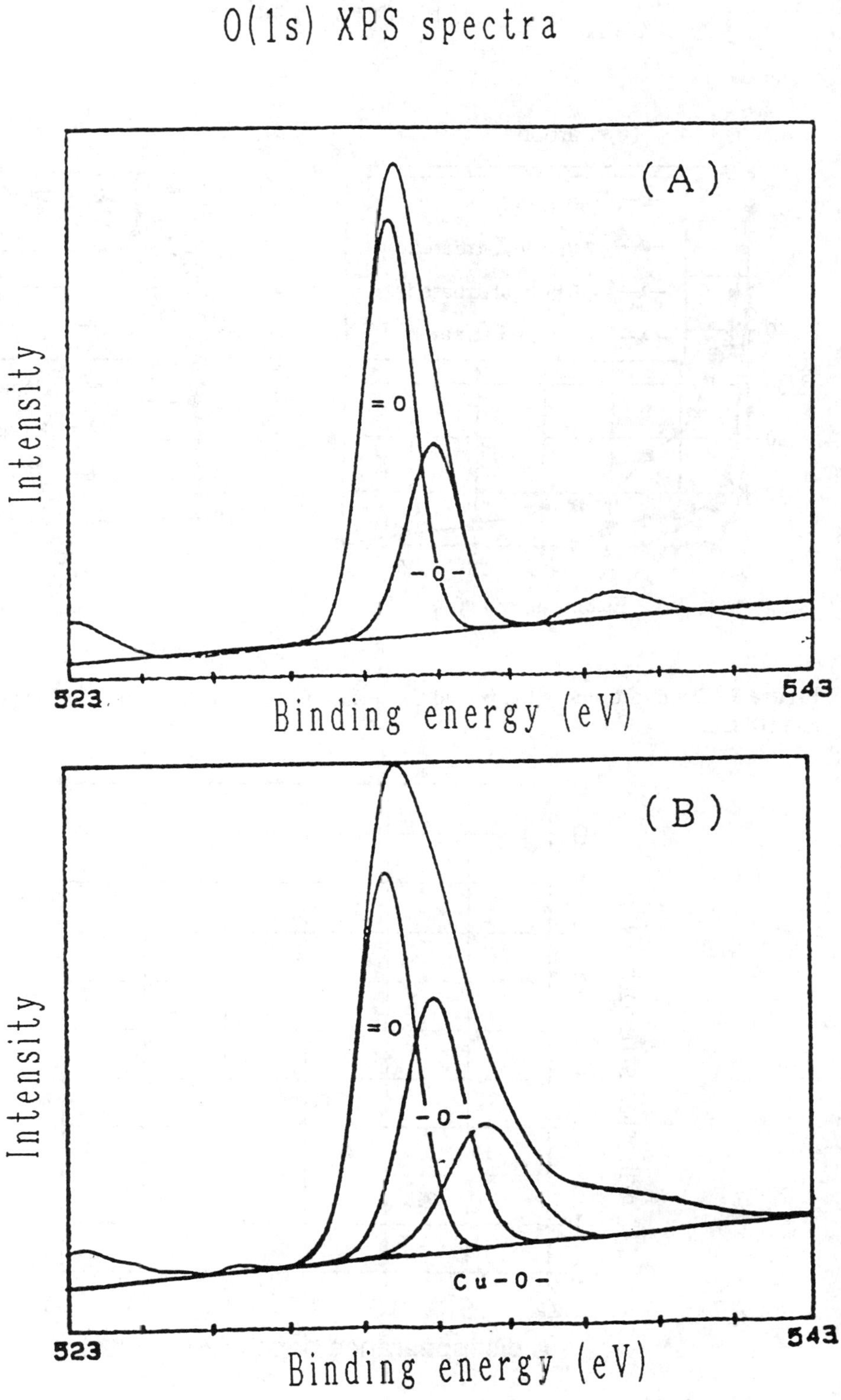

Figure 7. O (1s) spectra of the peeled PI surface from a Cu/PI laminate for (A) unheated sample and (B) sample heat-treated at 150°C for 24 hr in air.

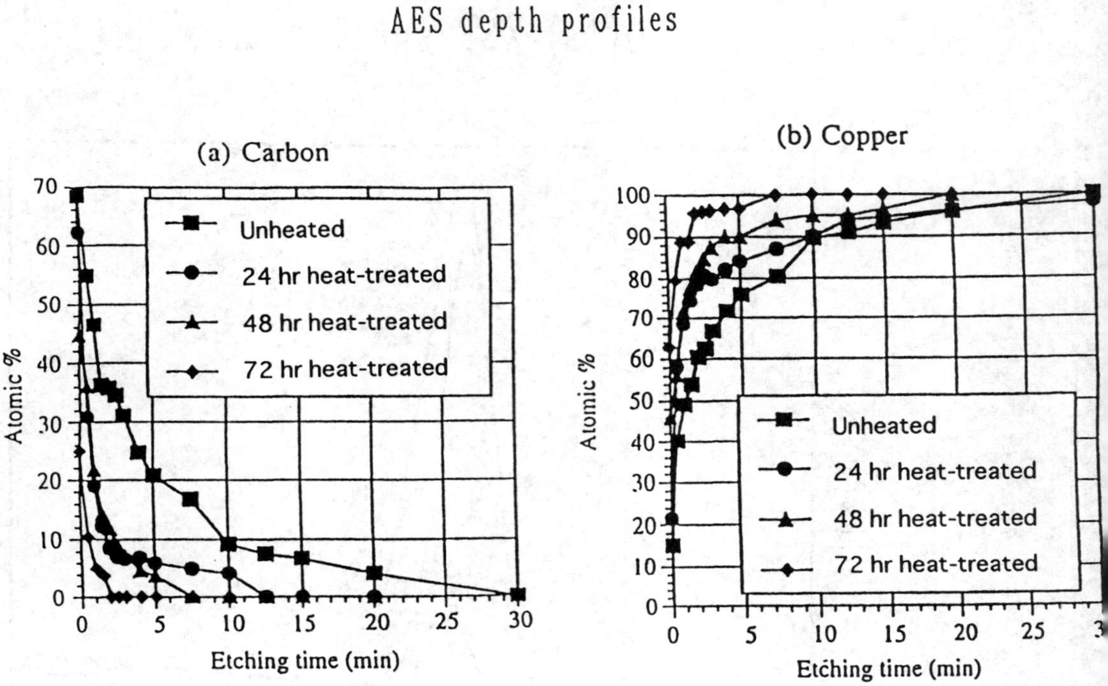

Figure 8. Depth profiles of the peeled Cu surface from a Cu/PI laminate for (A) carbon and (B) Cu.

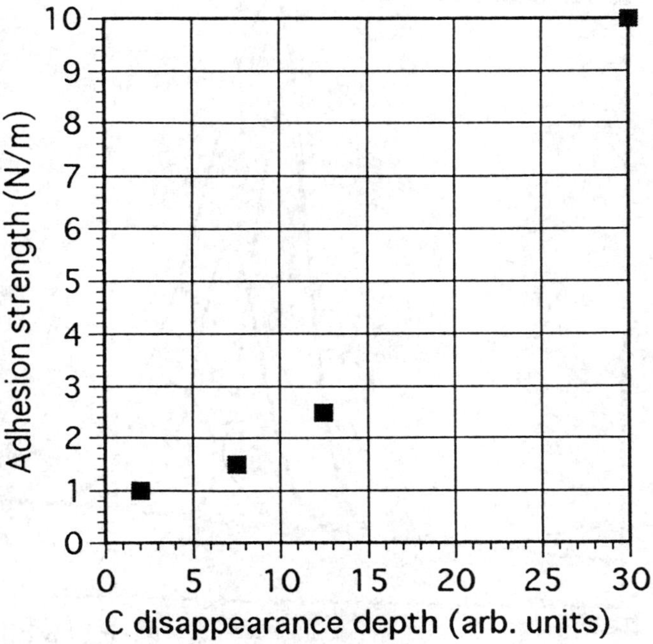

Figure 9. The relationship between the adhesion strength and the depth when the carbon disappeared.

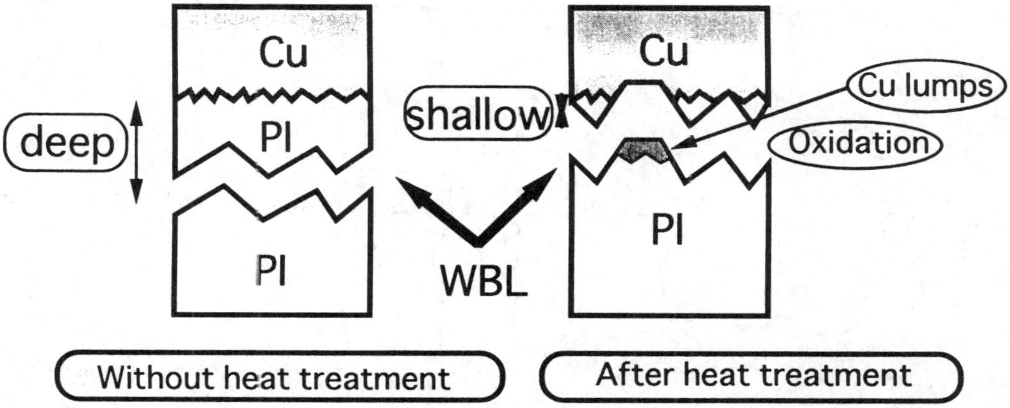

Figure 10. A model of the WBL shift due to heat treatment.

How to Prevent the Adhesion Strength Deterioration?

In order to prevent the Cu diffusion into the PI and the oxidation of both Cu and the PI at the interface, several different metallic interfacial layers were introduced and examined.[6]

First, we introduced the vanadium (V) interfacial layer.[12] The peeled surfaces from the Cu/V/PI laminates without the heat treatment and after the heat treatment for 24 hr were analyzed with the AES (Fig.11).[12] At the peeled metal surface without the heat treatment, vanadium was detected, but copper was not (Fig.11a). The peel strength of the Cu/V/PI laminate without heat treatment was 7.4 N/m. On the contrary, vanadium was not detected, but copper was detected at the peeled metal surface after the heat treatment (Fig11b). The vanadium disappeared due to the heat treatment. The peel strength decreased to 3.2 N/m due to the heat treatment. The depth profiles of the peeled metal surfaces of Cu/V/PI laminate were investigated with AES (Fig.12).[12] Fig.12(B) shows that the copper migrates through the V layer due to the heat treatment. These results meant that the V interfacial layer hardly functioned as a Cu diffusion barrier and could not prevent the adhesion strength deterioration.

Next, the cobalt (Co) interfacial layer was introduced between the Cu and the PI substrate.[12] Fig.13 (A) and (B) show the AES analyses of the Cu/Co/PI laminates without the heat treatment and after the heat treatment for 24 hr, respectively. The copper did not migrate through the Co layer even after the heat treatment for 24 hr. Moreover, the adhesion strength of the Cu/Co/PI laminate without the heat treatment was 6.4 N/m and it slightly decreased due to the heat treatment (4.9 N/m). These results indicated that the Co interfacial layer was effective for the prevention of the copper diffusion and thus the adhesion strength deterioration. However, the adhesion strength after the heat treatment deteriorated slightly. This decrease in the adhesion strength seemed to take place due to the copper diffusion through the cobalt interfacial layer, though copper was below the detection limit of AES.[12]

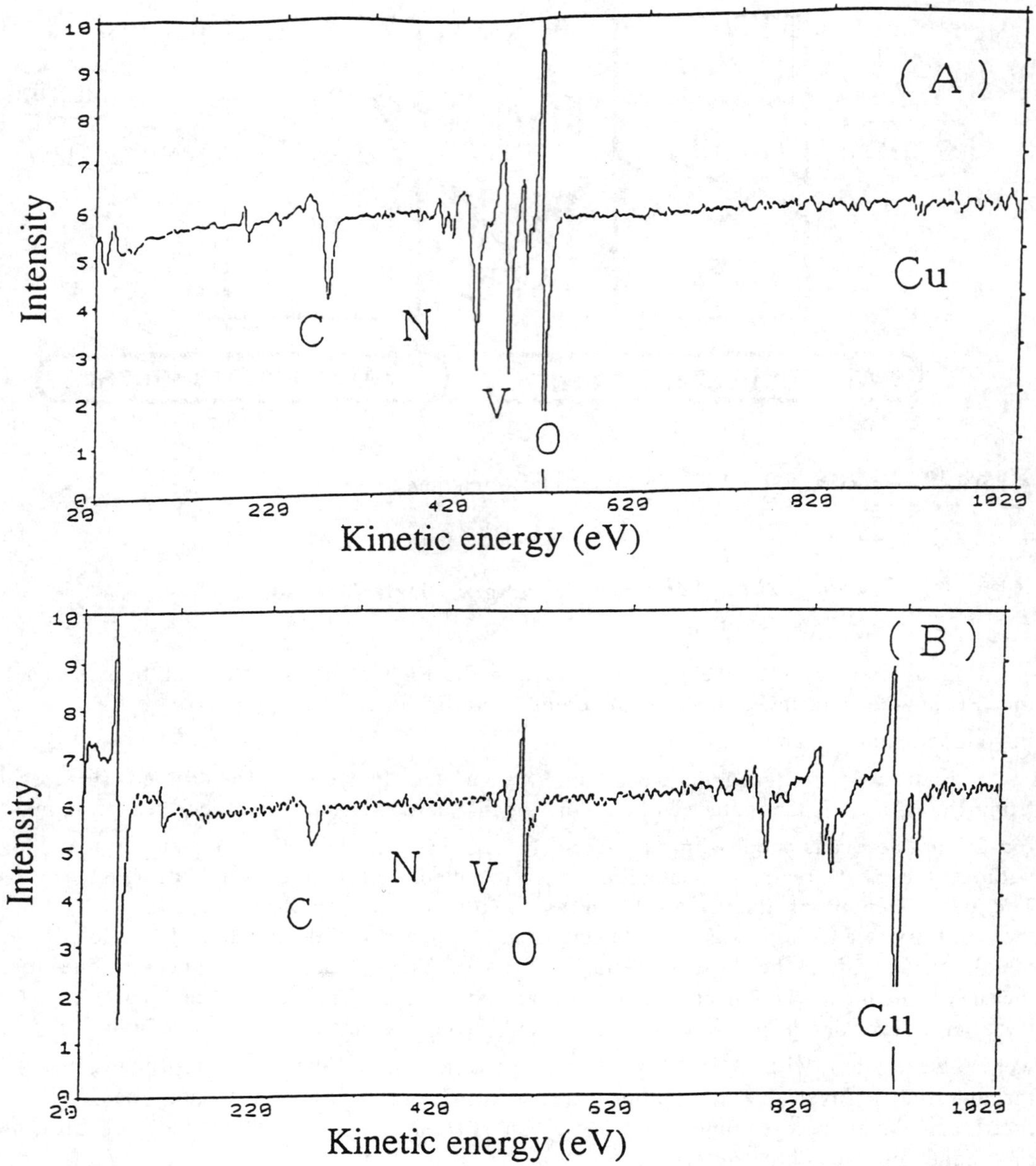

Figure 11. AES analysis of the peeled metal surface from Cu/V/PI laminate (A) without heat treatment, (B) after heat treatment.

In order to prevent the copper diffusion through the cobalt interfacial layer, we considered the generation mechanism of the Cu lumps, which are shown in Fig.5 (B), (C) and (D), on these peeled PI surfaces (Fig.14). Copper diffused into the PI layer due to the heat treatment. This diffusion caused the low density regions which had a large number of voids, and high density regions which had only a few voids in the Cu layer. These low density and high density regions can explain the variations in the uniformity of the Cu layer at the interface. The high density regions generate the Cu lumps on the peeled PI surface (Fig.14-B).

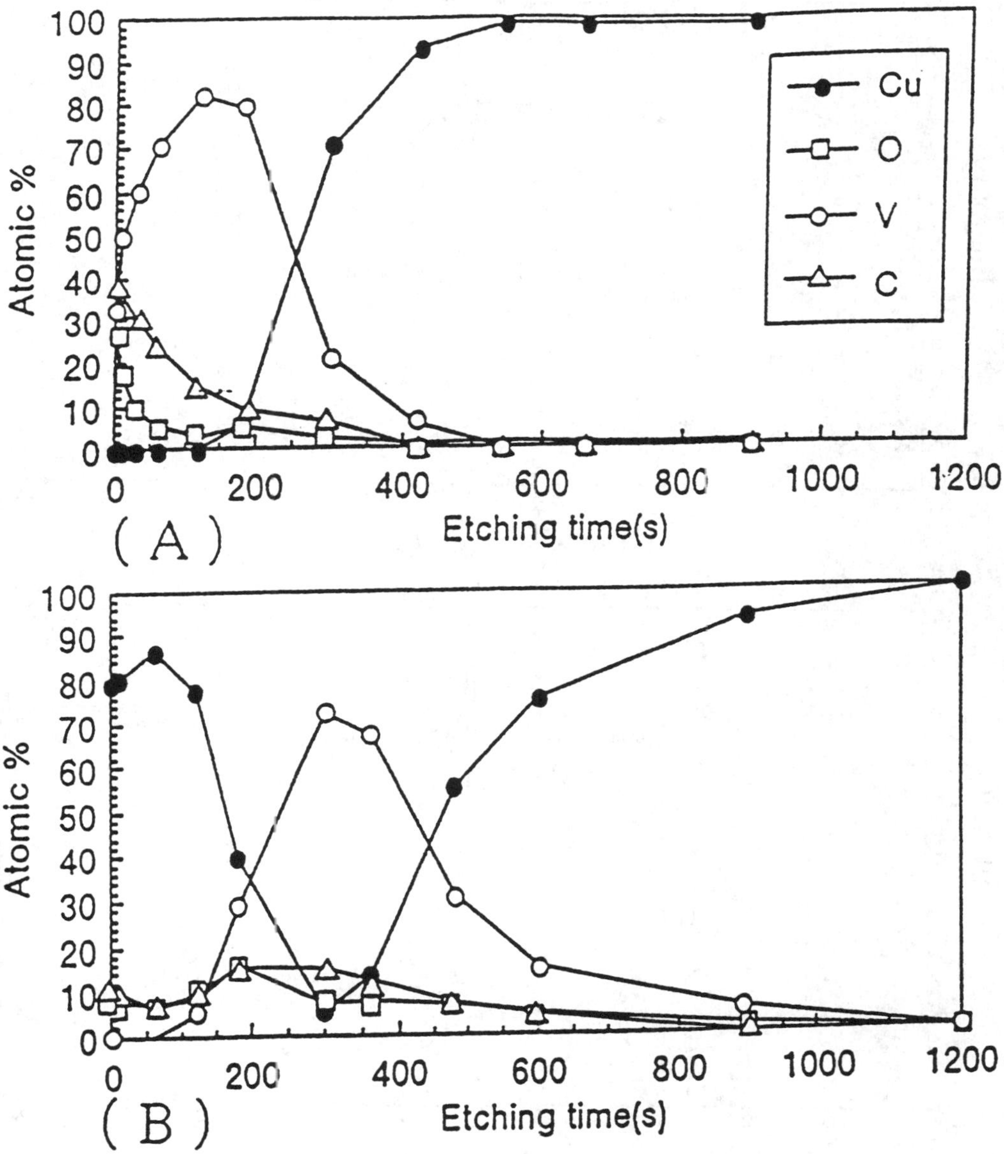

Figure 12. AES depth profiles of the peeled metal surface from Cu/V/PI laminate (A) without heat treatment, (B) after heat treatment.

Copper atoms are considered to migrate through the diffusion channnels such as dislocations and grain boundaries in the Co interfacial layer. It is necessary to enhance the quality of the interfacial layer by annealing these channels in order to modify the property of Co layers as a diffusion barrier (Fig.14-C).

In this experiment, we tried to quench these channels by annealing the Co interfacial layer in vacuum.[12] The Co interfacial layer with a thickness of 300 nm on PI substrate was deposited and annealed at 280 ^{o}C for 1 hr at a pressure of 0.67 Pa. The PI samples with Co deposition with and without annealing were placed in hydrazine solution to remove the PI substrate. Fig.15 (A) and (B) show the Co thin film without and with annealing, respectively. From these results, we

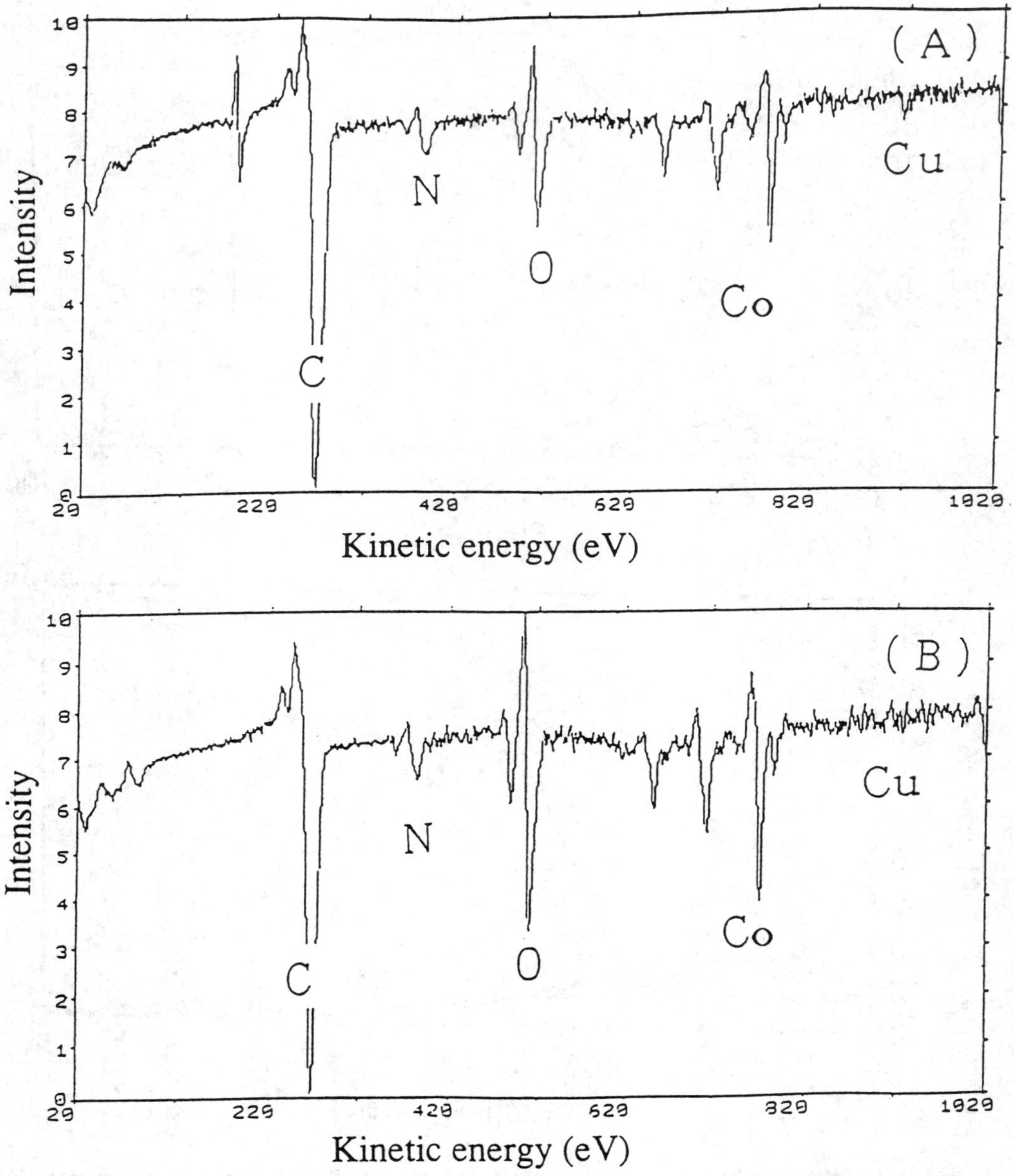

Figure 13. AES analysis of the peeled metal surface from Cu/Co/PI laminate (A) without heat treatment, (B) after heat treatment.

conclude that the Co thin film without annealing was fragile and could not keep its original shape, while the Co layer with annealing could keep its original shape as a film. Fig.15 (A) indicates that the Co thin film without annealing contains many defects and cannot sustain itself without the PI substrate. By annealing, i.e, quenching the defects,[13] the quality of the interfacial layer is improved. As a result, copper diffusion through the Co layer was prevented.

In order to investigate the change in the quality of the interfacial layer before and after annealing, we analyzed the Co interfacial layer crystallographically and the results are shown in Fig.16.[12] Fig.16 shows the x-ray diffraction scans of the

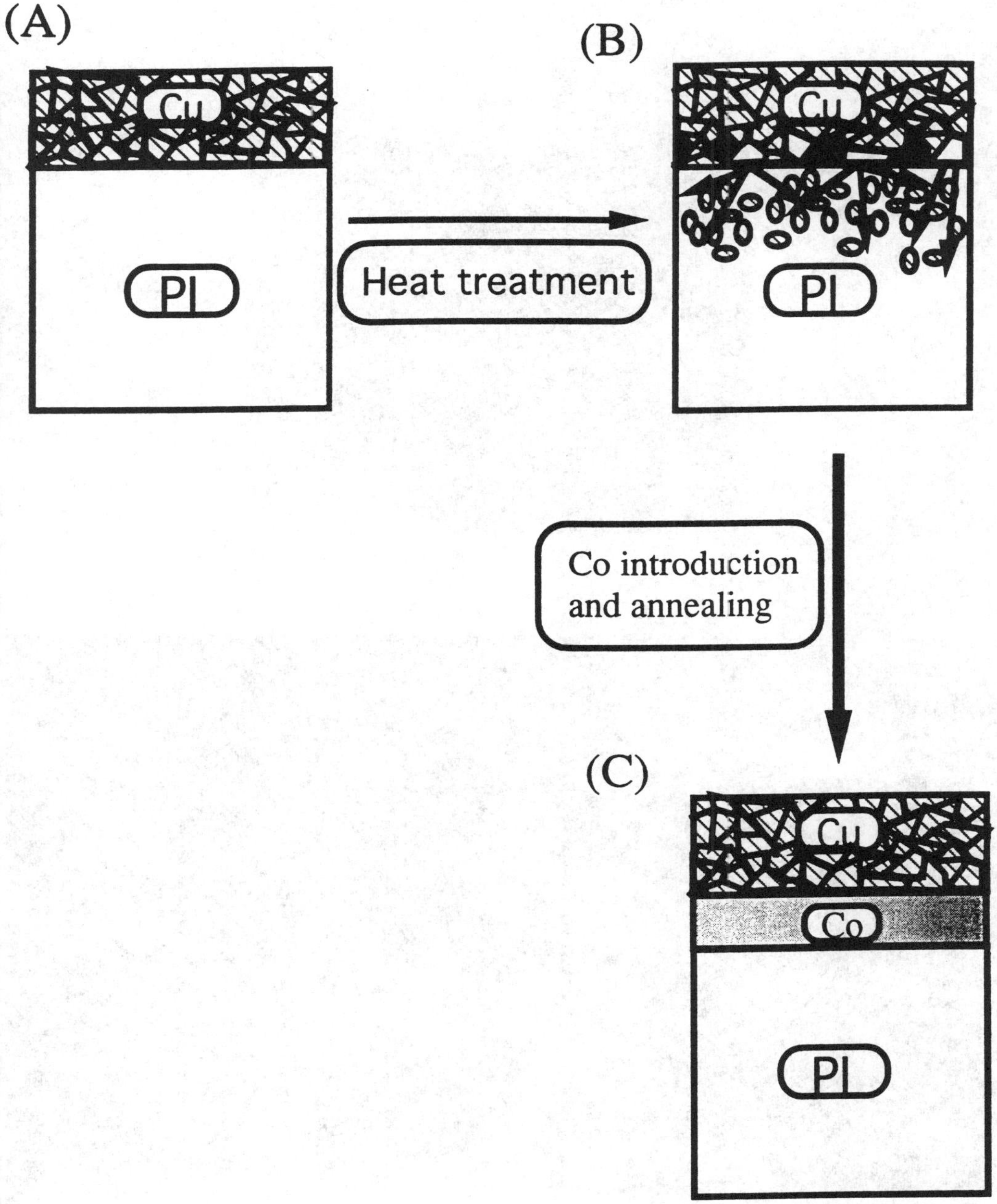

Figure 14. Mechanism of generation of the Cu lumps on the peeled PI surface.

Co films. The peaks Co(002) and Co(101) were observed at 44.0° and 47.4° in the X-ray diffraction scans. It is reported that the peak intensity ratio of Co(002)/Co(101) is 0.6 in JCPDS cards No.5-0727. On the other hand, the peak intensity ratio of the sputtered cobalt film was 1.5 and the full width at half maximum (FWHM) was 0.344. Moreover, this peak intensity ratio increased up to 2.5 and the FWHM decreased to 0.284 by annealing. These results mean that the sputtered Co films had (002) orientation, and that the crystalinity of the Co improved by the annealing.

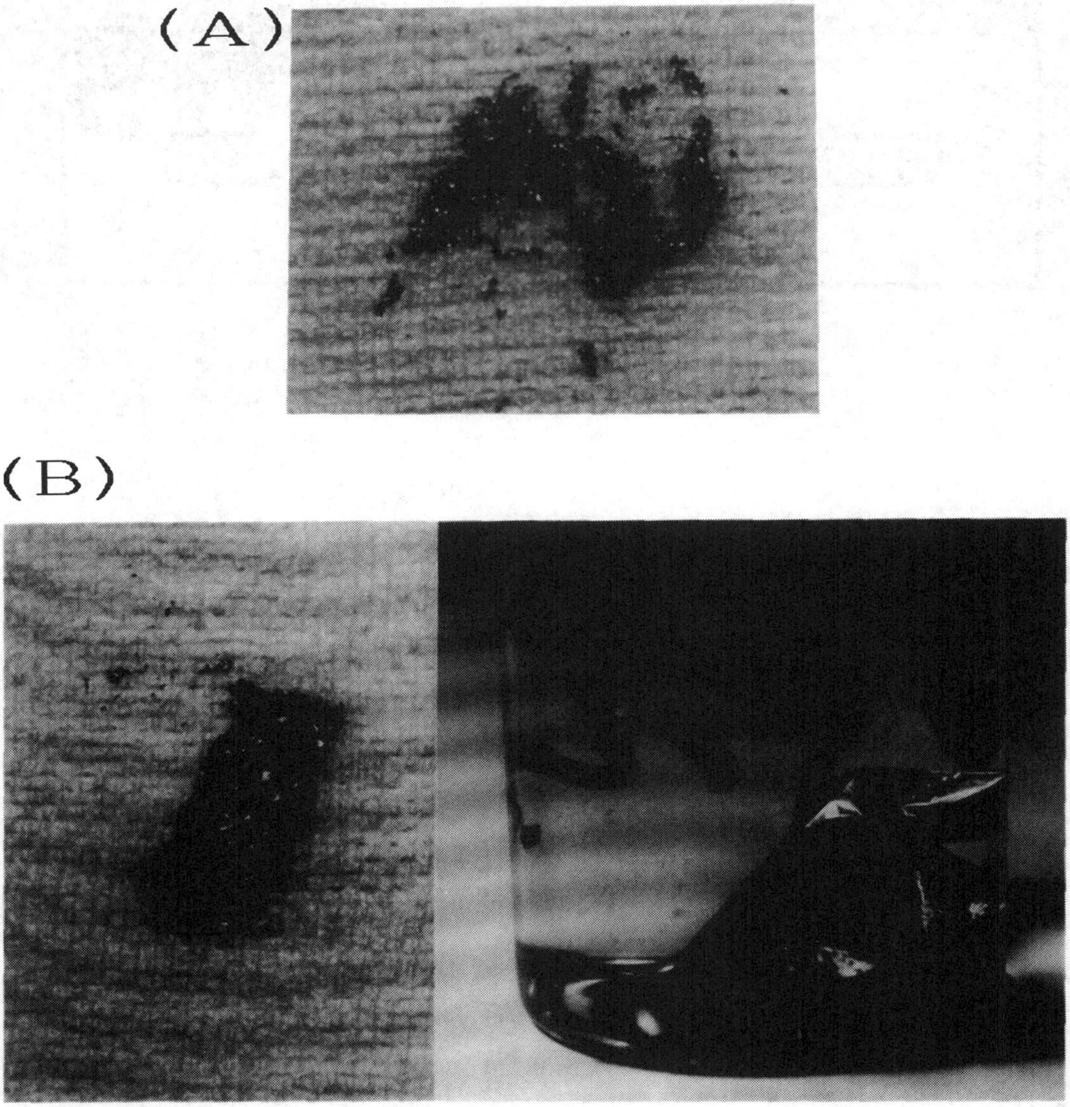

Figure 15. Photographs of Co thin films without polyimide substrate (A) without annealing, (B) with annealing. (B) The Co thin film could keep its original shape even into water (the right photograph).

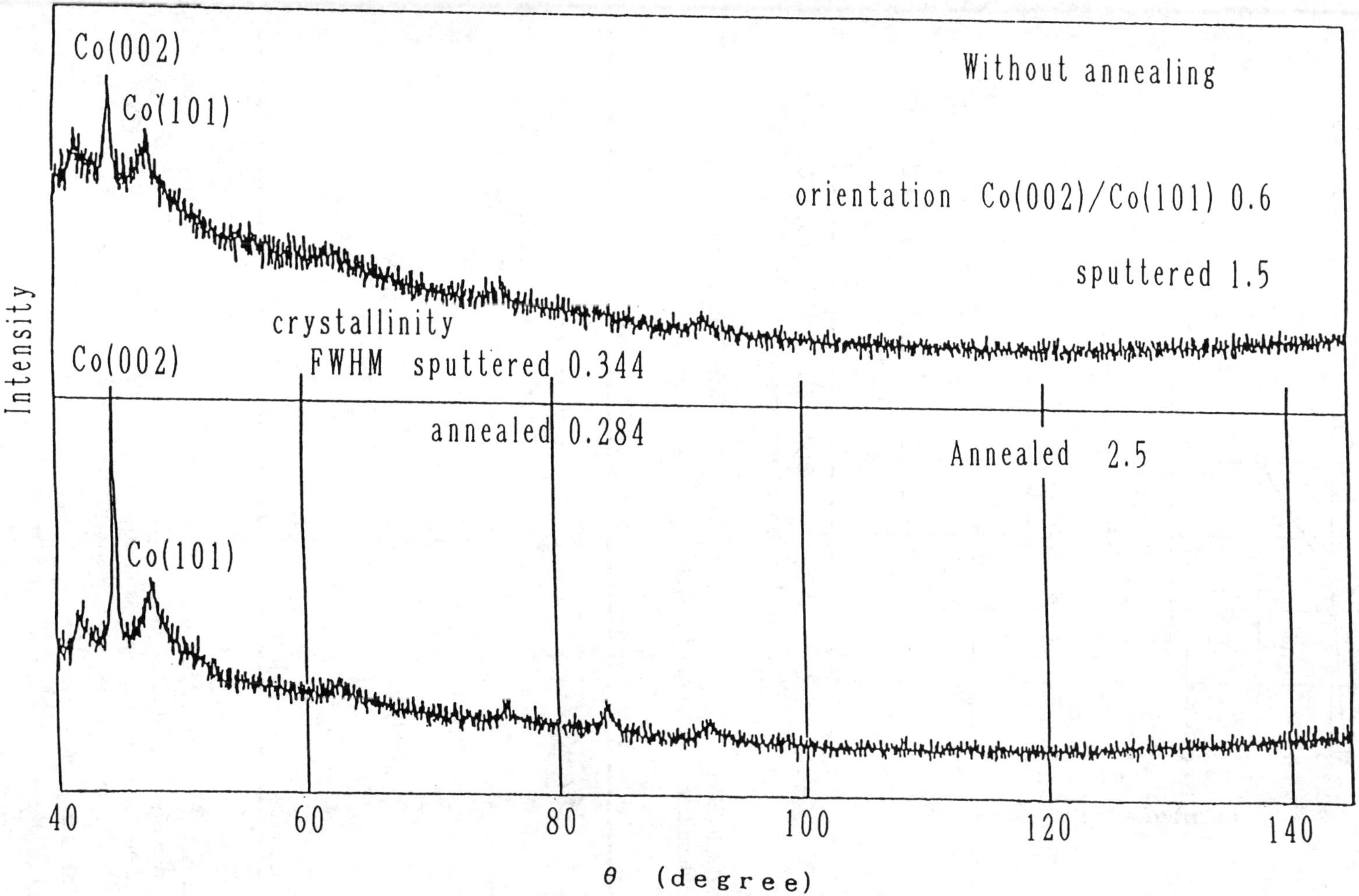

Figure 16. X-ray diffraction (XRD) scans of the cobalt films. Top and bottom show the XRD scans of the Co thin film without annealing and the annealed Co thin film, respectively. The numbers under the crystallinity indicated the full width at half maximum (FWHM) of the Co (200).

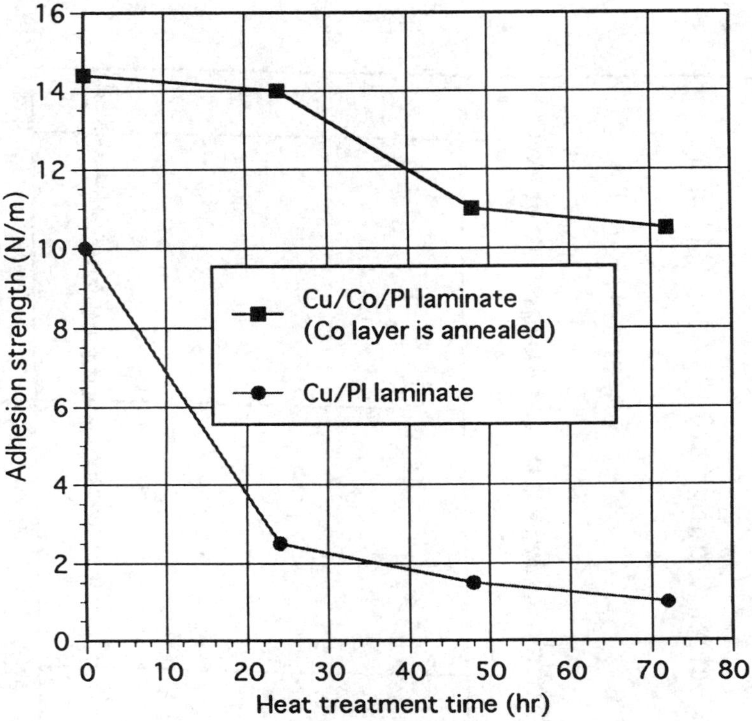

Figure 17. Adhesion strength of Cu/PI and Cu/Co/PI (Co layer is annealed) laminates as a function of heat treatment time at 150°C in air.

Table III. Adhesion Strength Retention Ratio.

Sample	*Adhesion strength retention ratio;* $\frac{\text{After 24 hr heat treatment at 150 °C in air}}{\text{Before heat treatment}}$
Cu/PI	0.25
Cu/V/PI	0.43
Cu/Co/PI	0.77
Cu/Co (Annealed)/PI	0.99

In order to confirm the above assumption, we electroplated Cu onto annealed Co layer deposited on the PI and measured the adhesion strength. The adhesion strength retention ratio increased from 0.77 to 0.99 due to the annealing (Table III). In Fig.17, the horizontal axis corresponds to the heating time and vertical axis to the adhesion strength.[12] It is clearly seen that 10 N/m adhesion strength was maintained even after heating at 150 °C for 72 hr.

CONCLUSIONS

1. A high concentration of Cu_2O particles (10-100 nm) were dispersed in the PI.
2. Cu lumps were found on the peeled PI surface after the heat treatment. The size of these lumps increased with heat-treatment time.
3. The amount of the oxidized moieties such as -O-C-O- or =C=O, -(C=O)-O, and -O-(C=O)-O- bonds increased at the peeled PI surfaces due to the heat treatment.
4. Heat treatment caused the WBL to shift towards the Cu/PI interface in the PI. The amount of this shift was related to the adhesion strength between the Cu and the PI.
5. Co interfacial layer was effective to maintain the adhesion strength.
6. Quenching the defects of the interfacial layers as diffusion barriers was important in preventing the diffusion of copper.
7. We succeeded in developing a Cu/PI laminate, the adhesion strength of which was maintained over 10 N/m even after the heat treatment at 150 ^{o}C for 3 days in air.

REFERENCES

1. N.Inagaki, S.Tasaka and K.Hibi, J.Adhesion, Sci.Technol., 8,395(1994)
2. N.Inagaki, S.Tasaka and K.Hibi, J.Polym. Sci. part A. Polym. Chem., 30,1425(1992)
3. M.K.Ghosh and K.L.Mittal, editors, "Polyimides: Fundamentals and Applications." Marcel Dekker, New York, 1996.
4. S.Iwamori, T.Miyashita, S.Fukuda, N.Fukuda and K.Sudoh, J.Jpn.Vac.Soc.,39(2), 65(1996) (Japanese)
5. S.Iwamori, T.Miyashita, S.Fukuda, N.Fukuda and K.Sudoh, J.Jpn.Vac.Soc.,39(3), 103(1996) (Japanese)
6. S.Iwamori, T.Miyashita, S.Fukuda, N.Fukuda and K.Sudoh, J.Jpn.Vac.Soc.,39(6), 295(1996) (Japanese)
7. Y.H.Kim, G.F.Walker. J.Kim and J.Park, J.Adhesion, Sci Technol.,4,331(1987)
8. S.Iwamori, T.Miyashita, S.Fukuda, K.Sudoh and N.Fukuda. J.Adhesion, Sci.Technol.,11, 783 (1997)
9. D.Y.Shih, N.Klymko, R.Flitsch, J.Paraszczak and S.Nunes. J.Vac.Sci.Technol., A9,2963(1991)
10. M.C.Burrell, P.J.Codella, J.A.Fontana and J.J.Chera, J.Vac.Sci.Technol., A7,1778(1990)
11. C.M.Chan, J.Adhesion, 15,217(1983)
12. S.Iwamori, T.Miyashita, S.Fukuda, N.Fukuda and K.Sudoh, J.Vac.Sci.Technol.,B15(1), 53(1997)
13. R.W.Balluffi, J.Nucl.Mater.,69&70,240(1978)
14. "Powder Diffraction File," Joint Committee on Powder Diffracion Standards (JCPDS) - International Centre for Diffraction Data (ICDD), cards No.5-0727, SWARTHMORE, PA, 1987.

Metallized Plastics 5&6: Fundamental and Applied Aspects, pp. 409-422.
K.L. Mittal (Ed.)

A QUANTITATIVE ADHESICN TEST FOR METAL COATED POLYMER FIBRES

E. K. Chong and M. G. Stevens

Cooperative Research Centre for Cardiac Technology
Department of Materials Science,
University of Technology, Sydney
PO Box 123, Broadway, NSW 2007, Australia

Many adhesion tests have been developed by researchers; however, only a few are applicable to metal coated polymer fibres. The size and shape of the sample restrict the choice of practical adhesion tests to a handful. Two tests were examined in this work: the tape test and the tensile shear test. In the tensile shear test the sample is strained so that the coating cracks, without failure of the substrate. The crack pattern observed is dependent on the amount of shear stress the interface can withstand. A novel tin-free catalyst, used in an electroless platinum deposition on poly(ethylene terephthalate) (PET) fibres, was developed for biomedical applications and tested to compare it with the conventional tin-palladium catalyst. No significant difference in the level of adhesion of the metal film to the substrate using the two different catalysts was observed by both the adhesion tests. A sample aged in saline solution did show a slightly reduced level of adhesion, by the tensile shear test. The two tests were found to compliment each other, with the tape test useful for assessing poor adhesion, and the tensile shear test for assessing good adhesion.

INTRODUCTION

Hundreds of tests have been developed over the years to measure the adhesion of a coating to a substrate[1]. These tests range in complexity from scratching coatings with a fingernail to using an atomic force microscope or electron beams. All of the tests have evolved due to the lack of a single, universally applicable test. This has led investigators to develop adhesion tests specific to their applications. In accordance with the current philosophy on adhesion testing, the best adhesion test is the one which best simulates the service conditions of the sample. As numerous reviews of adhesion testing are available, we will not discuss them here[2-7].

The selection of a suitable adhesion test is dependent on several factors. Some tests may only give qualitative results, others may be very complex or require extremely specialised equipment. Obtaining qualitative results from an adhesion test is relatively straightforward, whereas quantifying adhesion is much more difficult. Most tests can differentiate between good and poor adhesion, and may be able to monitor the effect on adhesion when properties of the system are varied. These are the semi-quantitative tests, where direct comparisons of adhesion strength can only be made using the same system, with the same test. Quantitative tests are those which can yield a number for the amount of stress or energy required to separate the coating from the substrate.

With most adhesion tests, only the practical adhesion can be measured. This measured adhesion is typically a function of many different factors, including the fundamental adhesion. Only when these other many different factors are kept constant, an adhesion test can give an indication of the level of fundamental adhesion. So values from one quantitative test need not be the same as from another test, since these other factors would most likely be different in different tests.

The system under investigation is a thin coating of platinum electrolessly deposited onto fine poly(ethylene terephthalate) (PET) fibres. The thickness of the metal coating is approximately 1 μm, and the diameter of each fibre is approximately 20 μm. The fibres are grouped into yarns containing 48 fibres.

Most adhesion tests were developed for flat samples, and have difficulty when a cylindrical geometry is considered. For large diameter cylinders, an approximation of a flat surface may be made, but this has not been possible when the substrate diameter is very small and, in addition, the coating is thin. The most common adhesion tests, such as the scratch, pull and peel tests, are not able to test these types of samples. This greatly limits the options available when choosing an adhesion test, especially if quantitative results are required.

In this work two tests have been employed to investigate the adhesion of the coatings to the PET fibres. One is the ubiquitous tape test, originated by Strong[8]. It consists of placing a standard adhesive tape over the coating and then removing the tape. The amount of material removed by the tape is indicative of the adhesion of the coating. The test is easily performed on any flat surface and has also been utilised on metallised fibres. The main limitation of the test is that the maximum adhesion measurable is limited by the strength of the adhesive on the tape. In addition, the test is basically qualitative, unless some type of quantification of the amount of material removed is made, which is difficult for small samples.

The second test examined is the tensile shear test, which has been used for thin brittle coatings on ductile substrates[9-15]. From theoretical calculations, the value for the maximum interfacial shear stress for the coating can be taken as a measure of interfacial adhesion. This method has been used by researchers for examining nickel coated PET films[9], as well as for testing of ceramic coatings on metals or other ceramics[10-15]. However, no work has been reported on the testing of metallised polymer fibres by this method.

The overall aim of this work was to establish a useful, practical adhesion test for metal coated polymer fibres. The study of two of these tests indicates that the tensile shear test yields semi-quantitative results for adhesion, and with a few assumptions allows calculation of the maximum interfacial shear stress for the coating-substrate system.

Two different samples were prepared using different catalysts to initiate the plating process. The reference material used a typical tin-palladium catalyst; however, the presence of tin was deemed unsuitable due to the biomedical nature of the application. A new tin-free catalyst was developed and the adhesion of the plated coating using this catalyst was investigated.

The second part of the investigation was the examination of the effect on adhesion of prolonged exposure to saline solution. This solution simulated possible degradative effects of a biomedical environment. Exposure was performed at high temperatures in order to accelerate any detrimental phenomenon.

THEORY OF TENSILE SHEAR TEST

The theory of tensile shear used here is similar to that proposed by Wheeler and Osaki[9], except that the theory has been modified for the cylindrical system under investigation. For tensile shear, the coating-substrate system is placed under uniaxial strain and is extended to an amount above the fracture strain of the coating, but below the fracture strain of the substrate. As the strain increases, the stresses within both the coating and the fibre increase. Once the strain exceeds the fracture strain of the coating it will crack. As the coating is attached to the substrate, once part of the coating is cracked, the only way for stresses to be transferred from the fibre to a section of coating on the fibre is through the interface. It is this interfacial shear stress which governs the tensile stresses within the coating.

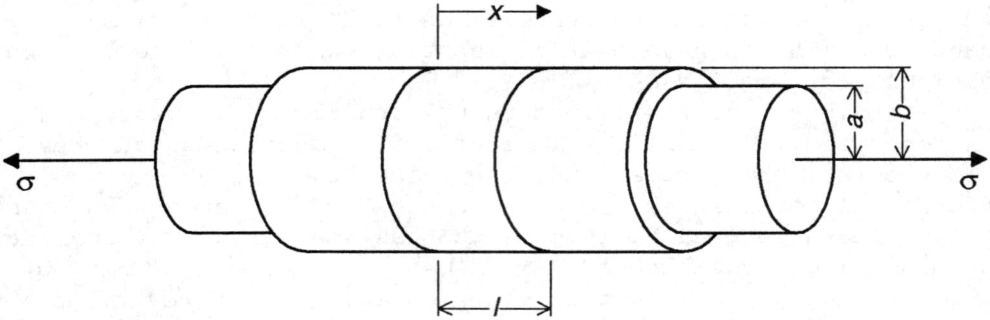

Figure 1. Schematic of coated fibre system.

The amount of shear stress at the interface is governed by the length of the coating in contact with the substrate. In this case it is the length of the uncracked segment, *l*. As the substrate is stretched the shear stress in this section will increase, resulting in tensile stresses within the coating. When the maximum tensile stress of the coating is reached, the coating will crack. The resultant crack pattern reaches an equilibrium of similar sized segments, in which the maximum tensile stress in the coating has been reached. Thus a balance of forces must exist where the tensile stresses in the coating equal the interfacial shear stresses. This is shown by equation (1);

$$\left(\pi b^2 - \pi a^2\right)\sigma(x) = 2\pi a \int_0^x \tau(\xi)d\xi \tag{1}$$

where a and b are the radii of the uncoated and the coated fibres, respectively (Figure 1). $\sigma(x)$ is the tensile stress within the coating, assuming the coating is thin and thus the tensile stress is constant. The function $\tau(\xi)$ is the interfacial shear stress distribution along the length of the fibre. It has been assumed by Wheeler and Osaki[9] that this distribution decreases linearly from a maximum shear stress value, τ_m, at the edge of the uncracked segment, to zero shear stress at a characteristic distance, l_c, from the edge (Figure 2).

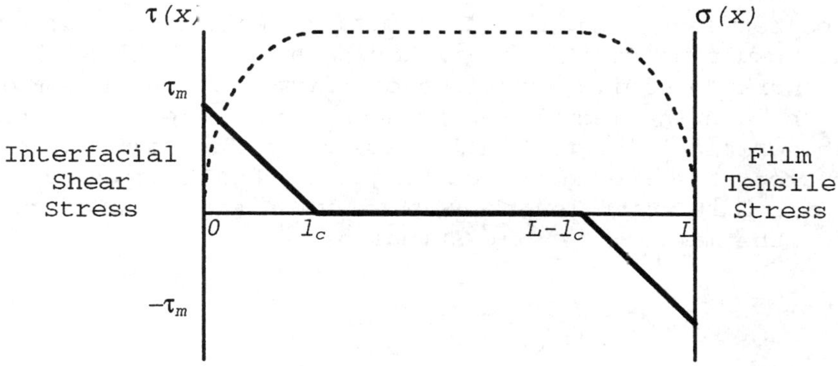

Figure 2. Shear stress distribution along a segment of length, L.

The equation for this distribution is given for each separate region as;

$0 < x < l_c \qquad \tau(x) = -\frac{\tau_m}{l_c}x + \tau_m$ (2)

$l_c < x < L - l_c \quad \tau(x) = 0$ (3)

$L - l_c < x < L \quad \tau(x) = -\frac{\tau_m}{l_c}\left[x - \left(L - l_c\right)\right]$ (4)

Combining these equations with equation (1) gives

For $0 < x < l_c \quad \sigma(x) = \frac{2a}{\left(b^2 - a^2\right)}\left[-\frac{\tau_m}{2l_c}x^2 + \tau_m x\right]$ (5)

For $l_c < x < L - l_c$

$$\sigma(x) = \frac{2a}{\left(b^2 - a^2\right)}\left[\frac{\tau_m l_c}{2}\right] \tag{6}$$

For $L - l_c < x < L$

$$\sigma(x) = \frac{2a}{\left(b^2 - a^2\right)}\left[-\frac{\tau_m}{2l_c}\left[x^2 - 2\left(L - l_c\right)x + L\left(L - l_c\right)\right]\right] \tag{7}$$

The wider the segment, i.e. the larger the l, the higher the tensile stresses in the coating. It is apparent that the maximum tensile forces will occur in the centre of the segment, between l_c and (L - l_c). When the coating is no longer able to be cracked by the transfer of shear stress, the coating begins to debond from the substrate, i.e. interfacial failure begins, rather than coating failure. At this final stage, the maximum interfacial shear stress, τ_m, has reached the maximum value for shear stress the interface can withstand, τ_o, and the widest segments will debond. Theoretically, the length of the final uncracked segments should be between l_c and $2l_c$. In practice, this is seldom the case, with the resulting crack pattern having a distribution of lengths. From the work on similar problems of fragmentation of composite fibres, the average length for fragmentation is equal to 1.337 times the critical length[16]. Thus, assuming the same type of distribution, the average crack length l_d should be equal to 1.337 l_c, the critical crack length. Substitution for l_c into equation (6) gives the maximum tensile stress as;

$$\sigma_o = \frac{a\tau_o l_d}{1.337(b^2 - a^2)} \qquad (8)$$

or

$$\tau_o = \frac{1.337(b^2 - a^2)\sigma_o}{a l_d} \qquad (9)$$

where τ_o is the maximum interfacial shear stress at the interface; σ_o is the maximum tensile strength of the coating; a is the radius of the uncoated fibre; b is the radius of the coated fibre; and l_d is the average uncracked segment length.

Thus knowing the dimensions of the system, the maximum tensile stress in the coating, and the length of the uncracked segments, a value for the maximum interfacial stress at the interface can be determined.

EXPERIMENTAL

Two samples of PET yarn were electrolessly coated with platinum. One sample used a conventional tin-palladium catalyst to initiate coating, whereas the other sample used the newly developed tin-free catalyst. Both samples had continuous coatings approximately 1 μm thick.

The tape test was used as the first test method to determine the relative adhesion strength of the metallised yarn. The yarn was secured at both ends with a self-adhesive tape on a flat sheet of glass. Pressure-sensitive tape (Permacel), specified for the tape tests in ASTM Standards B571 and D3359, was lightly placed on the test specimen. A 1 kg glass bottle, with a smooth and even surface, was slowly rolled forwards and backwards over the tape. The tape was then slowly and steadily peeled at 90° from the sample by hand. Both the tape and the sample were then examined visually and by optical microscopy (Olympus, BH-2) for any removed material. A simple rating system was used for adhesion, indicated in Table I below. The test was performed three times over different areas of the sample.

Table I. Classification Scheme for Tape Test.

Classification	Description
0	No coating removed
1	Speckles of the coating removed
2	Less than half of the coating removed (<50%)
3	More than half of the coating removed (>50%)
4	All of the coating removed (100%)

The tensile shear test was utilised for adhesion measurements of coatings with superior adhesion, as determined by the tape test. Individual fibres were carefully separated from the yarn and placed in a single fibre tensile tester (Micropull Science) located at the Department of Materials Science, UTS. These were secured using soft aluminium grips, set at a gauge length of 20 mm. The strain rate was relatively rapid at 16.7 % per second to assist in maintaining the linearity of extension. The maximum strain of the fibres was approximately 12 - 15 %, which is above the fracture strain of the coating and below the fracture strain of the substrate. The cracked samples were examined by optical microscopy (Olympus, BH-2) and the lengths of the uncracked segments were measured using an image analyser (Analytical Measuring Systems VIDS III) at a magnification of 200×. Ten replicates were used for each sample with 40 ± 3 measurements taken on each fibre.

To examine the stability of the coating using a tin-free catalyst, samples were aged in a 5 % saline (NaCl) solution. Coated yarn, approximately 20 cm in length, was placed in a sealed glass vessel containing the saline solution and heated in an oven at 80 °C for 140 hours. After ageing, the sample was allowed to cool to room temperature before thorough rinsing with deionised water. The samples were dried at room temperature and immediately afterwards adhesion tests were performed.

Selected samples, previously tested for adhesion, were analysed by scanning electron microscopy, SEM (JEOL 6300FEG-SEM) at 5 keV. Imaging allowed determination of the quality and continuity of the coatings, as well as coating thickness. Samples were vacuum coated with carbon to ensure the surface was conductive and to avoid damaging the polymer substrate.

RESULTS

Data obtained from the tape test are summarised in Table II.

Table II. Results from Tape Test.

Sample	**Average Result***
Sn-Pd catalyst	0
Sn-free catalyst	0
Sn-free catalyst (after ageing)	0

*each result is the average of three specimens

For the tensile shear tests, histograms of the uncracked segment lengths are shown in Figures 3, 4, and 5. It is clear from these graphs that the segment lengths do not follow a normal distribution. Thus, the lengths were assumed to follow a two-parameter Weibull distribution. The parameters of the distribution were estimated using the method by Cohen[17]. The average uncracked segment lengths and their standard deviations were calculated from standard equations for Weibull distributions and are summarised in Table III. Note that the standard deviation for the distribution is shown as the error in measurement in Table III.

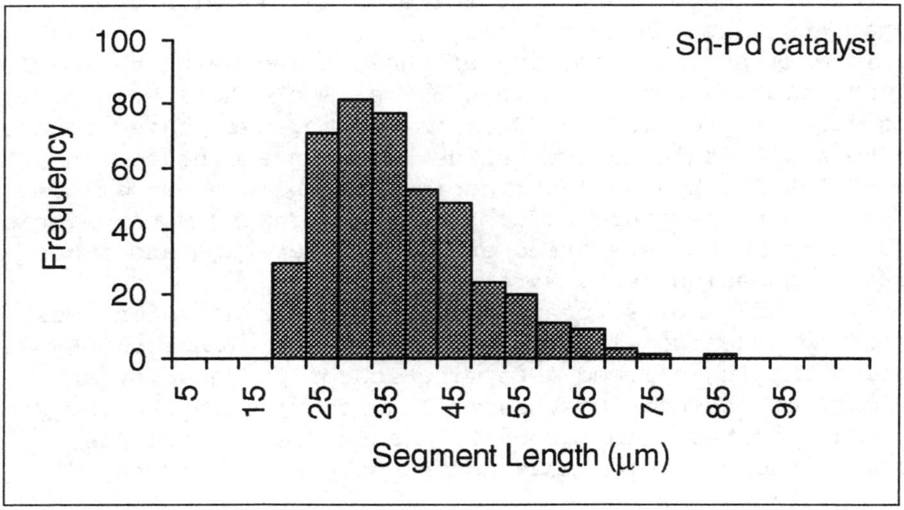

Figure 3. Tensile shear histogram for conventional tin-palladium catalysed coating.

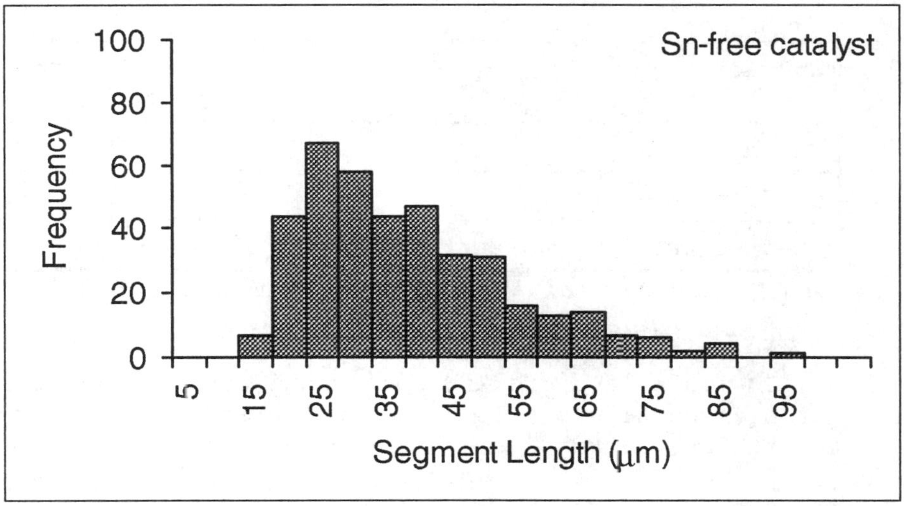

Figure 4. Tensile shear histogram for novel tin-free catalysed coating.

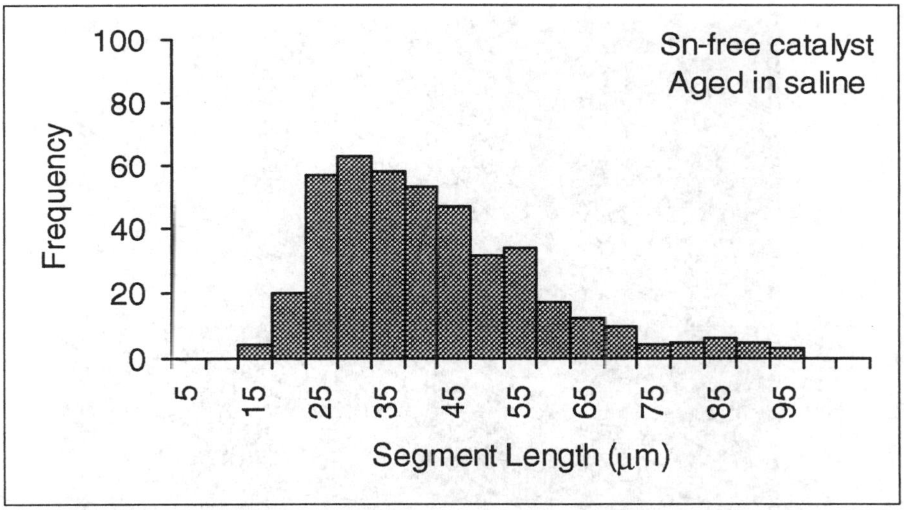

Figure 5. Tensile shear histogram for novel tin-free catalysed coating, after ageing in saline solution.

Table III. Average uncracked segment length, l_d, for coated fibres.

Sample	**Segment Length (μm)***
Sn-Pd catalyst	34 ± 12
Sn-free catalyst	34 ± 15
Sn-free catalyst (after ageing)	40 ± 16

*each result is an average of ~400 measurements

An SEM image of a cracked specimen is shown in Figure 6. This is representative of all the sample crack patterns observed. Estimates of the thickness of the cracked coatings and the sample radius were made. The films were approximately 1 μm thick, and the radius of the fibres was approximately 10 μm.

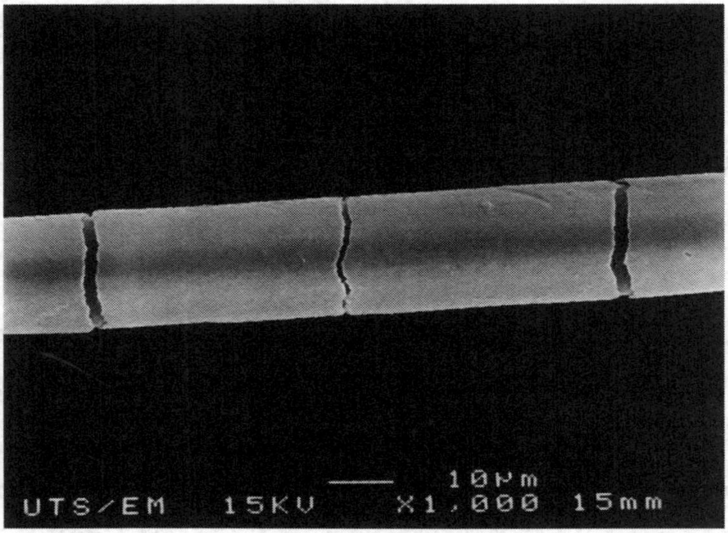

Figure 6. SEM micrograph of metal coated fibre after tensile shear test (conventional tin-palladium catalysed).

The maximum interfacial shear stress, τ_o, was calculated from equation (9), assuming that the maximum tensile strength of the platinum film was 220 MPa[18]. The values for τ_o are given in Table IV.

Table IV. Maximum Interfacial Shear Stress of Coated Fibres.

Sample	**Maximum interfacial shear stress (MPa)**
Sn-Pd catalyst	18 ± 6
Sn-free catalyst	18 ± 8
Sn-free catalyst (after ageing)	16 ± 6

DISCUSSION

The tape test results (Table II) indicate no difference in adhesion between platinum films and PET when using either a tin-palladium or a tin-free catalyst. In both cases the adhesion exceeded the limit of the test, with no material being removed. Rather than assessing the adhesion level, the result demonstrates that the test cannot be used for highly adherent coatings. If the metal to polymer adhesion is greater than the adhesion of the tape to the metal coating, there will be no metal coating removed. Thus above this limit there is no indication of the level of adhesion. For testing of relatively poor adhesion, the tape test is particularly useful as it can effectively separate coatings with poor adhesion from those with good adhesion.

The tensile shear test ensures fracture of the coating into segments, the lengths of which can be measured. The test allows generation of histograms for uncracked segment lengths as shown in Figures 3, 4 and 5. The averages obtained from these data are given in Table III, and are used in the calculation of the maximum interfacial shear stress. Since the average length of the uncracked segment (l_d) is directly proportional to the maximum interfacial shear stress (τ_o), either value may be used to indicate the level of adhesion.

A comparison of the two tests used shows that they compliment each other well. The tape test, although unable to test relatively adherent coatings, is a very quick and effective method for separating poorly adherent coatings from adherent ones. The tensile shear test, which can test good adhesion effectively, cannot test coatings with very poor adhesion. This arises from the assumption of a continuous coating for stress transfer. Poorly adherent coatings are seldom continuous, with constant tensile strength. In addition, the tensile shear test is much more involved and time-consuming than the tape test. Thus for the most effective adhesion testing, both tests should be used. Firstly, the tape test is used as a screening test, to identify samples with poor adhesion and then the tensile shear test is applied to better quantify the adhesion strength of a coating.

There are several assumptions made in calculating the maximum interfacial shear stress. The lengths of the uncracked segments were assumed to follow a distribution similar to that found in fragmentation tests with single fibre composites[16]. The coatings were assumed to be continuous and of a consistent thickness. This was later confirmed from examination by scanning electron microscopy. Shear stress distribution was assumed to be linear, which greatly simplifies the calculation. The tensile strength of the coating was assumed to be the same as for bulk platinum. Although the effect of thickness on the tensile strength of coatings is still not well understood, it can be generally stated that any thickness effect occurs only for coatings below 0.3 μm thickness[19].

The uncracked segment lengths were assumed to follow a two-parameter Weibull distribution. This type of distribution is common in the analysis of fragmentation tests[20] and was confirmed by a Weibull plot, which yielded a relatively linear trend. The parameters were estimated by an iterative process, from which the average segment length and its standard deviation were calculated.

The only apparent source of error, aside from the assumptions, is in the measurement of segment lengths. The precision error using the image analyser is less than 1 %. This is very minor compared with the variation in the segment length.

There was little difference in the coating adhesion obtained for conventional or tin-free catalysts. Both samples exceeded the limit of the tape test, and thus were tested by tensile shear. The shapes of the resultant histograms were quite similar, and the average segment lengths were very close. Statistical testing, assuming a normal distribution rather than a Weibull distribution, which is valid for such a large sample population ($n \approx 400$), showed no significant difference in the results at the 99 % confidence level. Obviously, when the maximum interfacial shear stresses were assessed, the values were virtually identical. By tape testing alone, one cannot reach such a conclusion.

After exposure to saline solution, the samples treated with the tin-free catalyst showed a small decrease in adhesion strength. This is not unusual since saline is such a corrosive environment, and the treatment conditions were severe (80 °C, 140 hours). Such testing is common in biomedical applications, although lower concentrations of salt are used. In this work, the tape test could not detect the loss in adhesion, as the adhesion level of the coating was still greater than that of the tape. Using the tensile shear test, a significant difference, again assuming a normal distribution, was found in the lengths of the uncracked segments. This difference also exists in the resultant maximum interfacial shear stress.

CONCLUSION

Adhesion testing of metal films on polymer fibres is particularly difficult. In this work it was found that to effectively test the adhesion of metal coatings on polymer fibres, both the tape and tensile shear tests should be applied in conjunction. These tests were utilised to demonstrate that a newly developed tin-free catalyst was as effective as plating with a conventional tin-palladium catalyst. The maximum interfacial shear stress for both samples was estimated to be 18 MPa. As a further demonstration of the effectiveness of the tensile shear test, a 15 % loss in adhesion of the platinum coating could be detected after ageing of the coated fibres in saline solution at a high temperature. The maximum interfacial shear stress decreased to 16 MPa, a result inaccessible by tape testing.

ACKNOWLEDGEMENTS

The authors gratefully acknowledge the Cooperative Research Centre for Cardiac Technology for support of this research.

REFERENCES

1. K.L. Mittal, in "Adhesion Measurement of Films and Coatings," K.L. Mittal (Ed.),pp. 1-13, VSP, Zeist, The Netherlands, 1995.
2. K.L. Mittal, Electrocomponent Sci. Technol., 3, 21-42 (1976).
3. P.R. Chalker, S.J. Bull and D.S. Rickerby, Mater. Sci. Eng., A140, 583-592 (1991).
4. A. Kinbara and I. Kondo, J. Adhesion Sci. Technol., 7, 767-782 (1993).
5. J.E. Pawel and C.J. McHargue, J. Adhesion Sci. Technol., 2, 369-383 (1988).
6. J. Valli, J. Vac. Sci. Technol., A4, 3007-3014 (1986).
7. J. Valli, U. Maleka and A. Matthews, Surface Eng., 2, 49-53 (1986).
8. J. Strong, Review Sci. Instrum., 6, 98 (1935).
9. D.R. Wheeler and H. Osaki, in "Metallization of Polymers," E. Sacher, J.-J. Pireaux and S.P. Kowalczyk (Eds.),pp. 500-512, American Chemical Society, Washington, D.C., 1990.
10. M.D. Thouless, E. Olsson and A. Gupta, Acta Metall. Mater., 40, 1287-1292 (1992).
11. R.A. Lowden and D.P. Stinton, Ceram. Eng. Sci. Proc., 9, 705-722 (1988).
12. B. Ollivier, S.J. Dowey, S.J. Young and A. Matthews, J. Adhesion Sci. Technol., 9, 769-784 (1995).

13. Y. Leterrier, Y. Wyser and J.-A.E. Manson, J. Adhesion, 44, 213-227 (1994).
14. M.S. Hu and A.G. Evans, Acta Metall., 37, 917-925 (1989).
15. D.C. Agrawal and R. Raj, Acta Metall., 37, 1265-1270 (1989).
16. A.C. Kimber and J.G. Keer, J. Mater. Sci. Lett., 1, 353-354 (1982).
17. A.C. Cohen, Technometrics, 7, 579-588 (1965).
18. E.D. Zysk, in "Metals Handbook. Volume 2 : Properties and Selection of Non-ferrous Alloys and Special-Purpose Materials," 10th ed.,pp. 707-709, ASM International, 1990.
19. D.A. Hardwick, Thin Solid Films, 154, 109-124 (1987).
20. L.T. Drzal, M.J. Rich, J.D. Camping and W.J. Park, Proc. 35th Annual Technical Conference of Reinforced Plastics/Composites Institute, Section 20-C, pp. 1-7, 1980.

Metallized Plastics 5&6: Fundamental and Applied Aspects, pp. 423-438.
K.L. Mittal (Ed.)

NON-DESTRUCTIVE TECHNIQUES FOR MONITORING METALLIZED PLASTICS ADHESION

Dario Narducci

Istituto Nazionale per la Fisica della Materia
and University of Milan, Department of
Materials Science, via Emanueli 15,
I-20126 Milano, Italy

The aim of this paper is to review and to discuss recent experimental and theoretical work where electrical measurements have been proposed, considered or used as non-destructive techniques to evaluate adhesion. Typically, the monitoring of electrical quantities does not require modifications of the system to be tested. Also, the experiments are fast and can be easily performed routinely for on-line testing. However, the system response often needs complex theoretical analysis to extract relevant adhesion parameters from raw experimental data. Examples showing details of such procedures will be presented. They will be discussed in view of their actual or prospective applications in Adhesion Science.

INTRODUCTION

The use of Metallized Plastics has been expanding over the last years because of their tremendous importance in many technological areas, from printed circuits to microelectronic packaging. Correspondingly, a variety of methods have been used to deposit metals onto polymers, ranging from highly controlled and sophisticated (such as Physical Vapor Deposition) to very simple and economic, such as painting or spraying. In all cases, however, the optimization of the deposition procedure requires a quantitative evaluation of the film adhesion. It is well known that metallized plastics adhesion can be described from two quite different points of view: "practical adhesion" - that is the energy required to detach (e.g.

peel off) the metal in a properly defined geometry; and the so-called "fundamental adhesion"[1] - namely that part of the detachment energy actually spent in breaking the microscopic bonds (mechanical or chemical) responsible for adhesion. While fundamental adhesion is hardly measurable under standard experimental conditions, detachment energies can be evaluated with a huge number of practical methods[1], mostly sharing their characteristics of providing relative estimates of the adhesion strength. Many of these techniques are mechanical, involving the destruction of the adhering system. Thus, they are necessarily off-line methods: adhesion is estimated by measuring it on a statistical sample extracted from the population of all systems under investigation. However, in the last 10-15 years increasing attention has been focussed on the development of non-destructive methods for the investigation of both fundamental and practical adhesion. Several physical techniques have been tentatively applied to specific problems, often yielding interesting results. Among them, electrical methods surely deserve special attention, since they are non-destructive coupled with a remarkable simplicity.

The aim of this paper is to shortly review recent experimental and theoretical work where electrical measurements have been proposed, considered or used as non-destructive test techniques to evaluate adhesion. Emphasis will also be placed on the critical interplay between "practical" and "fundamental" adhesion in view of the information that Electrically-based Non-Destructive Testing (ENDT) can provide to unravel this issue.

EXPERIMENTAL SETUPS FOR ELECTRICALLY-BASED NON-DESTRUCTIVE TESTING

Several experimental setups can be used for ENDT. All of them share the possibility of monitoring electrical quantities without modifying the system, i.e. not requiring special sample preparation for the measurement to be carried out. Actually, electrical contacts are made by using the same metal stripes or dots the adhesion of which on plastics has to be monitored. Since the current flowing through the system is negligible (typically of the order 1 mV/1MΩ = 1 nA) permanent modifications of the interface due to electrochemical processes can be safely ruled out. Also, experiments can be performed under actual operative conditions (i.e. in the environment the metallized plastic structure is supposed to operate in) and can be easily engineered for use as an on-line system for Quality Assurance and/or Failure Analysis.

Fig. 1 shows schematic of an experimental setup for ENDT measurements. Typically, potentiostatic conditions are chosen, which allow simpler mathematical analysis of the

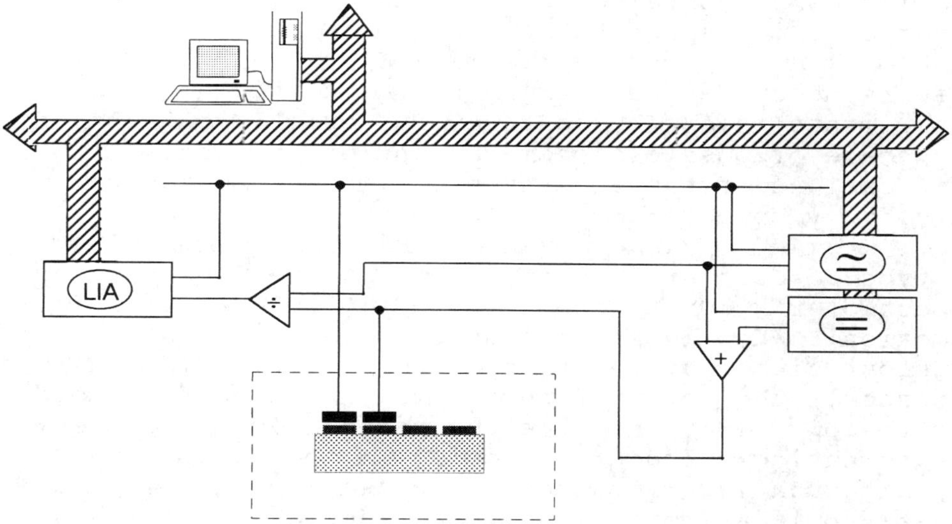

Figure 1. General setup for ENDT measurements.

results. Actually, since the applied voltage controls the thermodynamics of the electrochemical/electronic processes occurring at the electrodes (and around them), constant-voltage conditions force the system to undergo a limited, thermodynamically well-defined range of electrochemical reactions. Thus, the setup implies the use of a constant voltage source (potentiostat) and, almost invariably, the use of time- or frequency-dependent signals. Eventually, a d.c. voltage bias is superimposed to polarize the interface. Under these conditions, either locked-in or multichannel detectors can read the system response to the applied perturbation, separating in-phase and $\pi/2$ out-of-phase components. This separates the dissipative and the conservative parts of the response. Thus, such a measurement allows to reconstruct the detailed dynamics of the processes. Also, since the measurements are carried out as a function of frequency or of time, they also allow to separate out phenomena characterized by different time scales (response or relaxation times). Actually, since we are commonly interested in the dynamics of slow moving charge carriers (ions in the bulk or at the interface between metal and insulator; or electrons (holes) at internal surfaces) further information on the carrier dynamics can be obtained by analyzing the response of the system to time-dependent stimuli.

Since the entire set of measurements (e.g. a scan over five frequency decades, implying 50 individual impedance measurements) has to be carried out quickly compared to the time scale of possible chemical and/or physical modifications of the interface(s), measurements are unavoidably computer-controlled.

The specific experimental setup obviously depends on the type of carrier to be monitored, namely on the factors suspected or known to control adhesion degradation (or promotion). When ionic processes are assumed to control adhesion [either by reactive or non-reactive (diffusional) processes], the setup of choice is that used for impedance spectroscopy (IS). There, a potentiostatic source generates an alternating voltage (typically in the range 1 mHz - 1 MHz, although lower or higher frequencies can be of use for special applications). The signal (possibly after addition of a d.c. bias) is split and sent to both the sample and to a non-capacitive resistor of known impedance. The frequency response of the two devices is compared and the unknown sample impedance is thereby determined. As an alternative, for low frequencies (< 1 Hz) the signal can be digitally generated by the potentiostat, and the response is analyzed in the time domain by a lock-in detector coupled with fast Fourier transforms. Information about ionic motion and/or reactions can also be obtained by using relaxation techniques such as chronoamperometry or chronopotentiometry. As an example, Karas and Foust[2] studied the role played by chemical and mechanical components of adhesion by cathodic charging in acidic solutions, also showing the effect of hydrogen cathodic evolution on metallized plastic interfaces. However, advanced electrochemical methods such as cyclic voltammetry and polarography, although very popular in many areas of applied electrochemistry, have found little application in ENDT.

When the adhesion is related to some extent to electron dynamics at the metal-polymer interface, methods such as Current-Voltage (C-V) characteristics, admittance spectroscopy and Deep-Level Transient Spectroscopy (DLTS) can be invoked to analyze changes in electronic density as well as electron generation-recombination processes occurring at the interface. Admittance spectroscopy is the electronic counterpart of impedance spectroscopy, differing from the latter by its typical frequency range (for admittance spectroscopy, typically ranging from 0.1 MHz to 10 GHz). C-V characteristics are a simpler, more primitive version of admittance measurements, where only the $\pi/2$ out-of-phase (capacitive) component of the impedance response is read. DLTS is a highly sophisticated technique, operating in the time domain using multichannel detection to determine the density of trapped electrons at a metal-dielectric interface based on its typical relaxation time at a certain temperature. In a DLTS experiment temperature is raised (or cycled) from 70 K or less up to room temperature, and the recombination current due to de-trapped (generated) carriers at the interface is read. DLTS provides thereby a map of trap centers labeled by their characteristic generation times. Although extremely powerful and sensitive, DLTS failed to spread out as a general interface diagnostic tool because of the difficulty in properly engineering, controlling and making reproducible contacts on non-standard dielectrics.

APPLICATIONS OF ENDT TO ADHESION SCIENCE

Ionic Systems

In ionic systems, the analysis of frequency spectra requires three sets of processes to be considered:

1. diffusion of chemical species at interfaces (both internal and external surfaces);
2. diffusion throughout the phase, generically due to electrochemical potential ($\tilde{\mu}$) gradients;
3. chemical reactions, usually strongly localized at interfaces.

Furthermore, contributions to the dielectric response also arise from standard polarization currents, associated with dipole orientation under the application of an external (time-dependent) field. Due to the often strong coupling of steps (1) to (3), it is neither possible nor convenient to put forward a general expression including all contributions to the electrical response.

In a previous publication[3] modeling of the impedance in a test system where adhesion failure was mediated by corrosion was presented. Despite the details of the system considered in that work, the theoretical method used for the analysis of experimental IS spectra is worth to be briefly reported and commented here.

It was assumed there as a hypothesis to be verified (or rejected) that adhesion failure at a metal-polymer interface was related to water diffusion under the metal layer. Metal oxidation rate (at the metal-polymer interface) was connected with water diffusion rate underneath the metallic layer. Two questions were, therefore, to be answered: (1) Does water diffusion rate rule the metal oxidation rate? and (2) Does metal oxidation matter to the loss of adhesion with time in humid environments? IS analysis was obviously expected to answer just the first question, while question # 2 (<u>the</u> ENDT question) could be addressed only later by comparing IS and adhesion parameters.

In order to properly state the problem, a mechanism has to be conjectured, splitting the overall process in the following (typical) elemental steps:

1. Physical transport of the reactive species to the interface; more exactly:
   a. Adsorption at the interface between the dielectric system and the environment
   b. Diffusion of the reagent to the metal-polymer interface
   c. Formation of ionic species by chemical equilibration (in our case, by establishment of the water dissociation equilibrium)
2. Electrochemical reaction at the metal-polymer interface
3. Outdiffusion of the reaction products

The model can be made quantitative by recalling that electrochemical kinetics are all of the first order, with their kinetic constant k properly described by Tafel's equation:

$$k \propto \exp\left(-\frac{\alpha n F \eta}{RT}\right) \tag{1}$$

where R is the gas constant, T is the absolute temperature, n is the number of exchanged electrons, α is a parameter reflecting the detailed structure of the potential barrier at the electrode and η is the difference between the applied voltage E and the equilibrium corrosion potential[4].

Water concentration profile $c(x,t)$ can be obtained by resolving the appropriate Fick's equations, bearing in mind the geometry of the specimen. The ionic current can, therefore, be written as a sum of ionic flow components, since $I = dq/dt = ze\, dc/dt$ (where ze is the charge carried by the ion). Finally, an expression for the impedance Z can be written, using

$$Z = \frac{dE}{dI} = \frac{dE}{dc} \times \frac{dc}{dI} = \frac{dE}{dc} \Big/ \frac{dI}{dc} \tag{2}$$

The expression obtained is implicitly a function of the frequency through E, since the perturbing voltage applied to the system periodically depends on t; i.e.

$$E = E_0 + E_1 e^{-i\omega t} \tag{3}$$

Once $Z(\omega)$ has been written, it has to be compared with the experiment in order to check the model agreement with the data and, in the positive case, to evaluate all relevant chemical parameters. Due to the form of $Z(\omega)$, it is often preferred to cast this expression in the form of an equivalent electrical circuit made of standard electrical elements plus some conventional elements such as the Warburg impedance, used when diffusion plays a role in the system. In all cases, no matter how the fitting is carried out, it returns physical and chemical parameters, such as the reagent diffusion coefficient and the actual corrosion rate.

Upon availability of these quantities, answer can be given to question # 2, tentatively relating them to adhesion quantities. Figs. 2 and 3 show examples taken from author's work[3]. Note that the trend is not linear: adhesion strength seems to reach a maximum value when metal oxidation has proceeded to a certain extent, degrading afterward. This is in reasonable agreement with independent findings[5] showing how adhesion is promoted by oxygen bonding between the metal and carbonyl functional groups in the polymer chain.

The procedure shown so far is of absolutely general use and can be adapted to other systems and to the most diverse device structures. However, a thorough analytical treatment of the problem has the disadvantage of being more

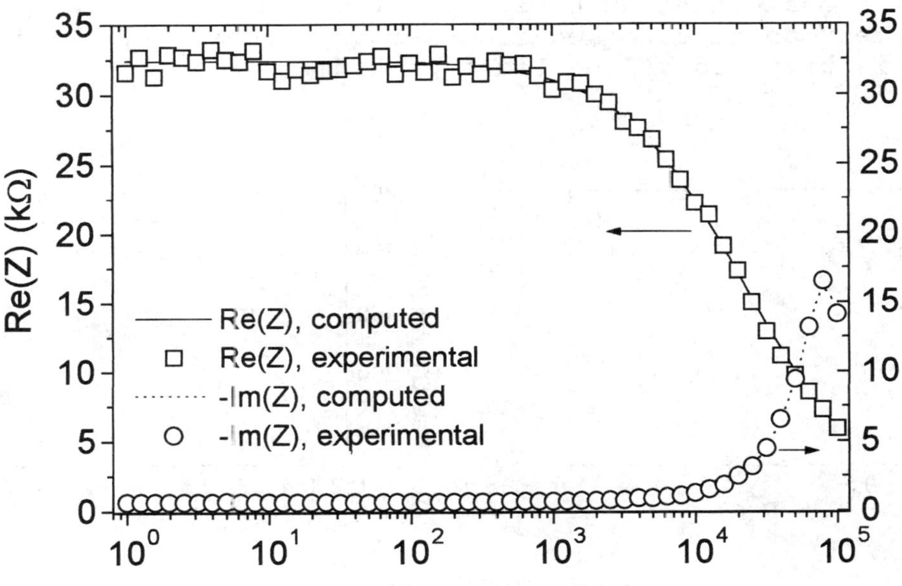

Figure 2. Agreement between computed and experimental impedance spectra.

and more difficult to be applied when the structure complexity increases. In fact, the problem can also be addressed by using numerical methods[6]. Although the level of details achieved is lower, the procedure can be of interest for preliminary analyses of experimental data as well as for outlining limits and capabilities of the technique.

For most of the systems of interest in Adhesion Science, chemical reactions are very localized in space, mostly occurring at internal or external surfaces. The limiting step in the overall process is the diffusion of reactants. Diffusion of charged species under the effect of either transient[7] or alternate electromagnetic fields[8] has been the subject of many studies, mostly devoted to the simple geometries usually encountered in electrochemical cells. In general, one considers that diffusion obeys Fick's second law:

$$\frac{\partial c(\vec{r},t)}{\partial t} = \nabla^2\left[D(\vec{r}) \cdot c(\vec{r},t)\right] \tag{4}$$

where c is the concentration of a chemical species, evolving with time (t) and position ($\vec{r}$), and $D(\vec{r})$ is its (position-dependent) diffusivity.

As seen, the computation of Z requires a knowledge of two functional relationships, namely $E = E(c)$ and $c = c(I)$. The former can be obtained from Nernst's equation, while the latter is provided by the reaction kinetics:

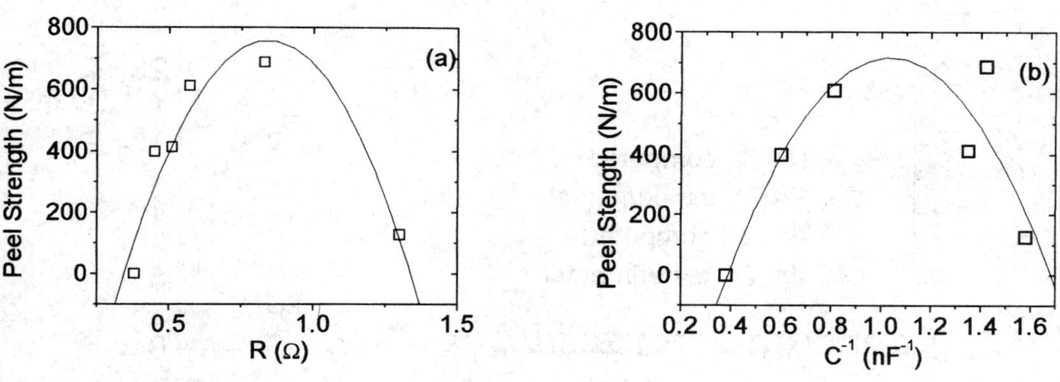

Figure 3. Adhesion strength *vs.* IS fitting parameters. Curves (a) and (b) show the relationship between resistance and capacitance of the equivalent circuit representing the system electrical response.

$$I = nF(k_{ox}c_{ox} - k_{red}c_{red}) \quad (5)$$

where, F is the Faraday constant, and k_{red} and k_{ox} are the kinetic constants for the reduction and the oxidation processes, respectively. Both kinetic constants depend exponentially on the voltage E according to Tafel's equation (1).

The relationship between the concentration and the current can be obtained by solving Fick's equation under appropriate boundary conditions. Time-dependent diffusion equations can be easily solved numerically, for instance, by finite-element methods. However, such techniques normally require medium or large computers to obtain accurate results with reasonable run times. This is particularly true when dealing with structured devices. A different approach was then proposed. Analytical methods[3,9] show that the time dependence of concentration profiles is mostly exponential in character. Solutions of Fick's equation of the form

$$c(\vec{r}, t) = c(\vec{r})e^{i\omega t}\left(1 - e^{-t/\tau}\right) \quad (6)$$

can be searched for, where the term in parentheses accounts for the physical relaxation process (characterized by a time constant τ) and the complex exponential describes the effect of the (fast) electrical perturbation of the system (of frequency ω). Therefore, the quasi-stationary ($t \to \infty$) solution can be written as

$$c_{qs}(\vec{r}, t, \omega) = c(\vec{r})e^{i\omega t} \quad (7)$$

Under the assumptions made above, the numerical resolution of eq. (4) reduces to the solution of the equation:

$$\nabla^2\left[D(\vec{r}) \cdot c(\vec{r})\right] = i\omega + \frac{1/\tau}{e^{t/\tau} - 1} \quad (8)$$

which can be easily integrated numerically. For large τ's, the real part in the right-hand term of eq. (8) can be neglected. Also, since we are interested in the real part of $c(\vec{r})$, we may write

$$\nabla^2\left[D(\vec{r}) \cdot c(\vec{r})\right] = 0 \tag{9}$$

which is simply the stationary form of eq. (4).

Once $c(\vec{r})$ has been numerically computed, it can be placed back into eqs. (6) - (7) in order to obtain an evaluation of the complete (time and space) dependence of the concentration.

Since at a reactive surface the concentration profile can be taken as linear[4], the limiting current equation yields, after differentiation:

$$\left.\frac{\partial c(\vec{r})}{\partial y}\right|_{y=0} = \left.\frac{I}{nFD(\vec{r})}\right|_{y=0} \tag{10}$$

where y is the coordinate normal to the interface. Thus, differentiating Nernst's equation one finally obtains

$$Z(\vec{r},t) = \frac{RTc(\vec{r})_{y=0}}{n^2F^2D(\vec{r})c(\vec{r})}\left[\frac{\partial c(\vec{r})}{\partial y}\right]_{y=0}^{-1} \times \left(1 + e^{-t/\tau}\right) \tag{11}$$

The correctness of the analysis can be checked by setting a constant concentration profile [i.e. $c(\vec{r}) \equiv c_0$] in eq. (11). In that case it reduces to the well-known Warburg's equation[10],

$$Z(\omega) = \frac{RT}{i\omega n^2F^2Dc_0} \tag{12}$$

as one would expect for a diffusion controlled system.

The results of this analysis can be applied to the study of real systems in two ways. As mentioned, it can be used as a tool to predict whether impedance spectroscopy is sensitive enough to monitor a specific phenomenon (e.g. diffusion) in a complex system. An example of contour maps showing the position dependence of Z is displayed in Fig. 4. It clearly shows how the technique reaches its maximum

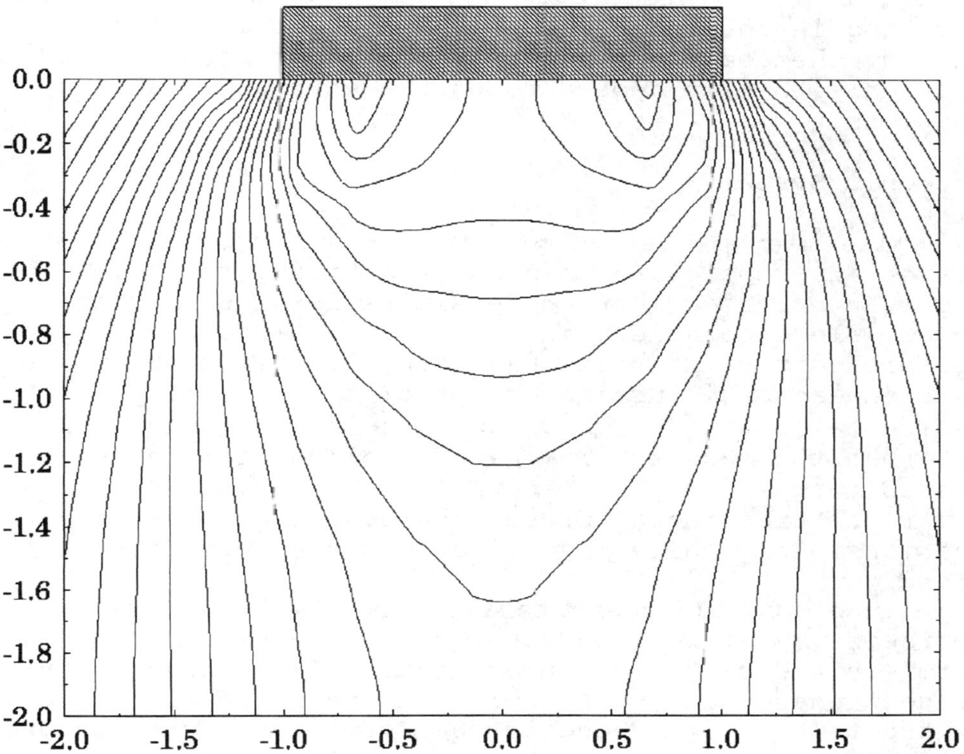

Figure 4. Contour map of impedance as computed by numerical techniques. The metal pad is the dark rectangular area on top while the large light rectangle is the polymer region. Note how force lines reach their maximum density underneath the metallic layer but not in the region immediately underneath it.

sensitivity for modifications occurring underneath the metallic layer but not immediately under it. Also, one can use these results to check the consistency of experimental data with hypotheses concerning the physico-chemical modification of systems on a trial-and-error basis, that is by estimating the initial values of the physical parameters involved in the model, writing and numerically solving the relevant equations to give the concentration profiles and $Z(\omega)$, which is compared with the experimental data. The estimates of the physical parameters are eventually refined, and the previous steps are repeated up to convergence.

The results obtained showed how impedance spectroscopy (operating in the audiofrequency range) could successfully point out chemical factors controlling adhesion and its evolution with time depending on the environmental conditions.

Electronic Systems

While many processes yielding enhancement or loss of adhesion are related to chemical reactions, the role of changes in surface electron densities is often marginal. However, electrostatic (image charge) interactions at a metal-dielectric interface has been recently shown to be of possible diagnostic use in the prediction of incipient adhesion failure[11]. It has to be clearly stated at this point that the research in this field lags far behind as compared to what has been shown for ionic systems. It seems, however, interesting to present here some outlines of the directions along which investigations are currently being carried out.

Simply stated, characterization of electronic interfaces (metal-semiconductor or metal-insulator contacts) determines the distribution of impurity levels/bands in the forbidden energy gap by sweeping the Fermi level inside the gap itself (Fig. 5). Whenever the Fermi level crosses the states inside the gap, electrons empty/fill such states: thus, the macroscopic system responds to the electrical signal as a discharging/charging capacitor. Depending on the ratio between the response time τ_e of the states and the period ($1/\nu$) of the electrical signal, the interface states may or may not equilibrate with the external field. Characteristic delay effects are useful to estimate τ_e and, in turn, allow to chemically identify impurities or defects responsible for the levels. For adhesion application, the possibility of monitoring the chemistry and the structure of the interface can be of theoretical interest but is not expected to immediately provide tools for ENDT. Nevertheless, for especially simple systems some correlations between the gap state density and adhesion has been envisaged.

Actually, wherever no chemical interactions occur, the metal-polymer interface can be seen as a geometrical boundary between non-interdiffused phases. In this case, adhesion takes place by quantum interactions related to the formation of Metal-Induced Gap States (MIGS) in the insulator. As seen, MIGS density and energetics can be easily evaluated experimentally with many techniques. However, theoretical analyses of experimental data often rely on oversimplified semi-classical models which cannot provide a sufficiently realistic description of the interface, yielding a loss of accuracy in the determination of physical quantities of relevance to adhesion studies.

In 1991 Bordier and Noguera[11] presented a quantum-mechanical model of the electronic structure of metal-insulator interfaces aimed at the study of adhesion modifications. Although their work was mainly targeted at inorganic solids, their treatment is of sufficient generality to be useful also in metal-polymer systems.

They analyze the electron density at the bare insulator surface using a one-electron tight-binding Hamiltonian of the form

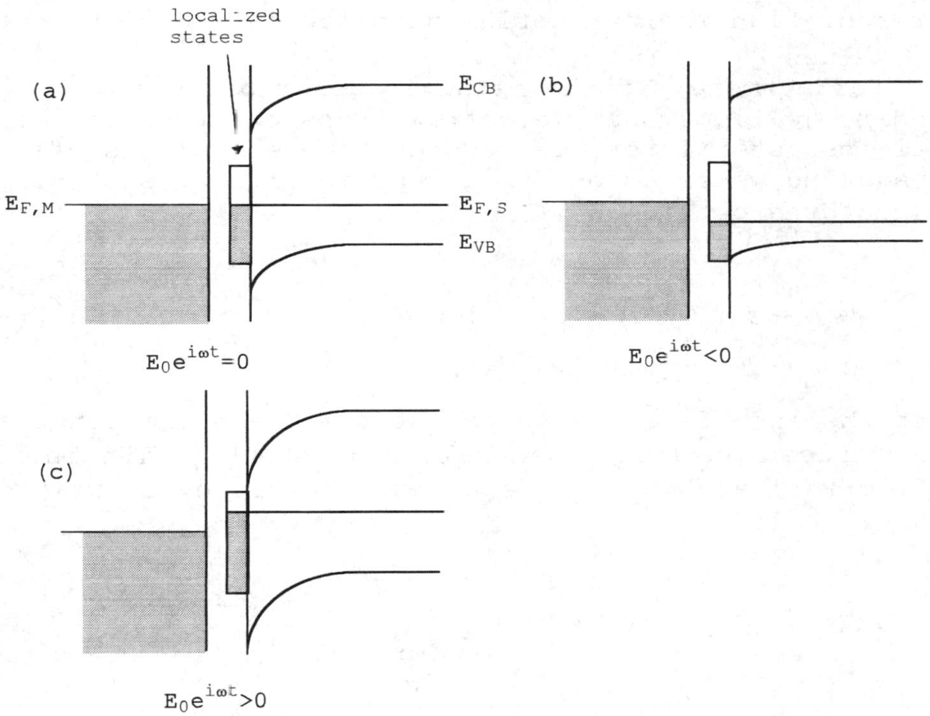

Figure 5. Modification of the electronic structure at a metal-insulator interface as induced by an alternate electrical field: (a) $E(t)=E_0 exp(i\omega t)=0$; (b) $E(t)<0$ and (c) $E(t)>0$. Observe how the Fermi level sweeps through the localized levels at the interface when forced out of equilibrium.

$$\mathcal{H} = \beta \sum_{(R,Y)} \left(| A_R \rangle\langle C_Y | + | C_Y \rangle\langle A_R | \right) + \varepsilon_A \sum_R | A_R \rangle\langle A_R | + \varepsilon_C \sum_Y | C_Y \rangle\langle C_Y | \quad (13)$$

where ε_A and ε_C are the energies for anions and cations of kets $|A_R\rangle$ and $|C_Y\rangle$ at position R and Y, respectively; while β is the hopping energy between neighboring sites. Green's function analysis shows that the electronic structure of the semiconductor at midgap can be stated in terms of the ratio $\Gamma=(\varepsilon_C-\varepsilon_A)/\beta$. This is of special relevance in view of the description of real interfaces, since it allows a universal representation where material specificity is fully contained in Γ values.

Furthermore, if the metal is described using the standard jellium model, electron transport through the metal-insulator interface reduces to the analysis of propagating waves across an abrupt interface. MIGS density $n(E,z)$ turns out to be

$$n\left((\varepsilon_A + \varepsilon_C) / 2, z\right) \approx n_0 \exp\left(-z / l_p\right) \quad (14)$$

where the interface is located at $z = 0$ and the penetration length l_p can be determined self-consistently. The barrier profile $V(z)$ at the interface can, therefore, be written as

$$eV(z) = (E_{ZCP} - E_F) \times \left[1 - \frac{I_0\left(2l_p \exp(-z / 2l_p) / l_i\right)}{I_0(2l_p / l_i) + (l_m / l_i)\varepsilon_i I_1(2l_p / l_i)}\right] \quad (15)$$

where $E_{ZCP}=(\varepsilon_C+\varepsilon_A)/2$, I_0 and I_1 are the two first modified Bessel functions, l_m , l_i indicate the Thomas-Fermi length in the metal and in the insulator sides, respectively, and ε_i is the relative permittivity of the insulator.

As expected, in the presence of MIGS the Schottky barrier height Φ_B becomes independent of the metal used, and the Fermi level is said to be 'pinned', i.e. its position in the band diagram is not changed even in the presence of gap states:

$$\Phi_B^{screened} = \varepsilon_C - E_{ZCP} \quad (16)$$

This is typically the case of strongly covalently-bonded solids. In the extreme ionic limit, on the contrary, screening is inefficient and the Schottky model is recovered:

$$\Phi_B^{unscreened} = \varepsilon_C - E_F \qquad (17)$$

The results obtained are of interest to Adhesion Science since, at least in ideally flat interfaces, the extent of adhesion strength relates to the penetration length l_p of MIGS. Now, since measurements of electrical characteristics of contacts normally return a map of states lying in the forbidden gap (DLTS and admittance spectroscopy) or, at least, an evaluation of the barrier height[12], Bordier and Noguera's model provides a theoretical trait d'union between $n(E,z)$ and the penetration length, the latter possibly being a suitable quantity to be compared with adhesion data.

CONCLUSIONS

It has been shown how metallized plastics can be successfully qualified by using a variety of electrochemical and electronic techniques. Some examples of applications to real systems have been reviewed. In particular, impedance spectroscopy has been found to have the greatest potential for ENDT due to its capability to monitor ionic processes at interfaces. Since electrochemical reactions are known to often control loss or promotion of adhesion, the technique can be applied in a number of practical situations to predict peel energies, and obviously qualifies as a routine testing. When image charge electrostatic interactions play some role in adhesion, also physical methods such as admittance spectroscopy and C-V methods can be of use, although they surely cover a much smaller range of cases. In all cases, it has been shown that a solid theoretical analysis is required in order to extract information on adhesion from experimental data.

REFERENCES

1. K.L. Mittal, J. Adhesion Sci. Technol., 1, 247 (1987).
2. B.R. Karas and D.F. Foust, in "Metallized Plastics 3: Fundamental and Applied Aspects," K.L. Mittal, Editor, p. 319, Plenum Press, New York, 1992.
3. D. Narducci, J.J. Cuomo, D.L. Pappas and K. Sachdev, in "Metallized Plastics 3: Fundamental and Applied Aspects,"

K.L. Mittal, Editor, p. 365, Plenum Press, New York, 1992.

4. see, e.g., K.J. Vetter, "Electrochemical Kinetics," p. 120, Academic Press, New York, 1967.
5. W.N. Unertl and R.G. Mack, in "Metallized Plastics 3: Fundamental and Applied Aspects," K.L. Mittal, Editor, p. 85, Plenum Press, New York, 1992.
6. D. Narducci, in "Metallized Plastics: Fundamentals and Applications," K.L. Mittal, Editor, p. 213, Marcel Dekker, New York, 1998.
7. D.D. MacDonald, "Transient Techniques in Electrochemistry," Plenum Press, New York, 1977.
8. I.D. Raistrick, in "Impedance Spectroscopy," J.R. Macdonald, Editor, p. 27 ff., J. Wiley & Sons, New York, 1987.
9. J. Cranks, "The Mathematics of Diffusion," Ch. 2 and p. 126, Oxford University Press, New York, 1975.
10. I.D. Raistrick, in "Impedance Spectroscopy," J.R. Macdonald, Editor, p. 62, J. Wiley & Sons, New York, 1987.
11. G. Bordier and C. Noguera, Phys. Rev. B, 44, 6361 (1991).
12. E.H. Nicollian and J.R. Brews, "MOS Physics and Technology," Ch. 5, J. Wiley & Sons, New York, 1982.